T0328854

LONDON MATHEMATICAL SOCIETY LECTURE NOTE SERIES

Managing Editor: Professor J.W.S. Cassels, Department of Pure Mathematics and Mathematical Statistics, University of Cambridge, 16 Mill Lane, Cambridge CB2 1SB, England

The titles below are available from booksellers, or, in case of difficulty, from Cambridge University Press.

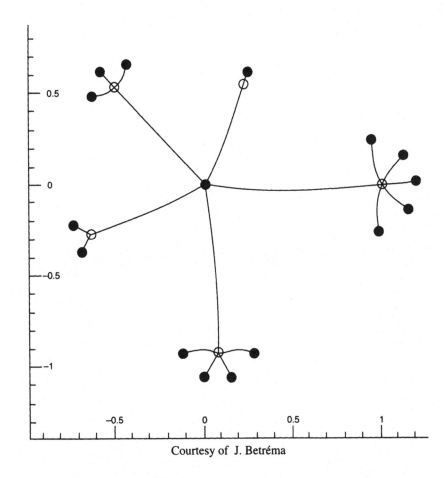

Courtesy of J. Betréma

London Mathematical Society Lecture Note Series. 200

The Grothendieck theory of dessins d'enfants

Edited by

Leila Schneps
CNRS, France

CAMBRIDGE UNIVERSITY PRESS
Cambridge, New York, Melbourne, Madrid, Cape Town, Singapore,
São Paulo, Delhi, Dubai, Tokyo, Mexico City

Cambridge University Press
The Edinburgh Building, Cambridge CB2 8RU, UK

Published in the United States of America by
Cambridge University Press, New York

www.cambridge.org
Information on this title: www.cambridge.org/9780521478212

First published 1994

A catalogue record for this publication is available from the British Library

Library of Congress cataloguing in publication data available

ISBN 978-0-521-47821-2 Paperback

Cambridge University Press has no responsibility for the persistence or
accuracy of URLs for external or third-party internet websites referred to in
this publication, and does not guarantee that any content on such websites is,
or will remain, accurate or appropriate. Information regarding prices, travel
timetables, and other factual information given in this work is correct at
the time of first printing but Cambridge University Press does not guarantee
the accuracy of such information thereafter.

Contents

List of participants at the conference

Yves André
Jean Betréma
Bryan Birch
Bruno Chiarellotto
Henri Cohen
Paula Cohen
Robert Cori
Antoine Coste
Jean-Marc Couveignes
Pierre Dèbes
Pascal Degiovanni
Ralf Dentzer
Michel Emsalem
Mike Fried
Anna Helverson-Pasotto
Yasutaka Ihara
Claude Itzykson
Gareth Jones
Bruno Kahn
Pierre Lochak
Jean Malgoire
Gunter Malle
Georges Maltsiniotis
Alexis Marin
Joseph Oesterlé
Kyoji Saito
Leila Schneps
Jean-Pierre Serre
George Shabat
David Singerman
Gene Smith
Manfred Streit
Zdzislaw Wojtkowiak
Umberto Zannier
Alexandre Zvonkin
Dmitri Zvonkin

DESSINS D'ENFANT

(The theory of cellular maps on Riemann surfaces)

A "dessin d'enfant" is a certain simple type of cellular map on a compact orientable topological surface. It can be visualized as a finite set of points on the surface, connected by a finite set of edges, in such a way that the edges and vertices form a connected set (with a certain property detailed below), which cuts the surfaces into open cells. Typical examples are trees on the sphere; the regular solids also form a well-studied category of dessins on the sphere, considered as such as long ago as 1884 in Felix Klein's book *Das Ikosaeder*. Some visual examples are given in Figure 1.

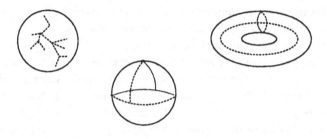

Figure 1

Alexander Grothendieck became interested by the action of the absolute Galois group $\text{Gal}(\overline{\mathbf{Q}}/\mathbf{Q})$ on the set of dessins; this action is highly non-trivial even in the simplest case of genus 0, and depends on a correspondence between dessins and algebraic curves defined over $\overline{\mathbf{Q}}$ which is a direct consequence of Belyi's well-known theorem identifying algebraic curves defined over $\overline{\mathbf{Q}}$ with Riemann surface morphisms $\beta : X \rightarrow \mathbf{P}^1\mathbf{C}$ ramified at most over $\{0, 1, \infty\}$. There are several references on the subject: the necessary basics are contained in the articles by Birch, Schneps, and Bauer-Itzykson in this volume (see their reference lists as well). These facts are at the base of an ambitious program by Grothendieck aiming, among other things, at a complete description of the absolute Galois group. He explains how he was led to this by the problems of research in a provincial university (see the end of the introduction for translations of the French passages):

"Les exigences d'un enseignement universitaire, s'adressant donc à des étudiants (y compris les étudiants dits 'avancés') au bagage mathématique modeste (et souvent moins que modeste), m'ont amené à renouveler de façon draconienne les thèmes de réflexion à proposer à mes élèves, et de fil en aiguille et de plus en plus, à moi-même également. Il m'avait semblé important de partir d'un bagage intuitif commun, indépendant de tout langage technique censé l'exprimer, bien antérieur à tout tel langage – il s'est avéré que l'intuition géométrique et topologique des formes, et plus particulièrement des formes bidimensionnelles, était un tel terrain commun. Il s'agit donc de thèmes qu'on peut grouper sous l'appellation de 'topologie des surfaces' ou 'géométrie des surfaces', étant entendu dans cette dernière appellation que l'accent principal se trouve sur les propriétés topologiques des surfaces, ou sur les aspects combinatoires qui en constituent l'expression technique la plus terre-à-terre, et non sur les aspects différentiels, voire conformes, riemanniens, holomorphes et (de là) l'aspect 'courbes algébriques complexes'. Une fois ce dernier pas franchi cependant, voici soudain la géométrie algébrique (mes anciennes amours!) qui fait irruption à nouveau, et ce par les objets qu'on peut considérer comme les pierres de construction ultimes de toutes les autres variétés algébriques. Alors que dans mes recherches d'avant 1970, mon attention systématiquement était dirigée vers les objets de généralité maximale, afin de dégager un langage d'ensemble adéquat pour le monde de la géométrie algébrique, et que je ne m'attardais sur les courbes algébriques que dans la stricte mesure où cela s'avérait indispensable (notamment en cohomologie étale) pour développer des techniques et énoncés 'passe-partout' valables en toutes dimensions et en tous lieux (j'entends, sur tous schémas de base, voire tous topos annelés de base...), me voici donc ramené, par le truchement d'objets si simples qu'un enfant peut les connaître en jouant, aux débuts et origines de la géométrie algébrique, familiers à Riemann et à ses émules!"[1]

This passage is taken from an unpublished manuscript by Grothendieck called l'*Esquisse d'un Programme*: it dates from 1984 and was written as part of his application for a position in the French Centre National de la Recherche Scientifique for which he applied after many years as a professor at the University of Montpellier, partly out of discouragement with his teaching activities:

"Comme la conjoncture actuelle rend de plus en plus illusoire pour moi les perspectives d'un enseignement de recherche à l'Université, je me suis résolu à demander mon admission au CNRS, pour pouvoir consacrer mon énergie à developper des travaux et perspectives dont il devient clair qu'il ne se trouvera aucun élève (ni même, semble-t-il, aucun congénère mathéticien) pour les développer à ma place."[2]

As indicated by its title, the paper sketches out a vast vision juxtaposing several areas of possible exploration which Grothendieck proposed to study during his tenure in the CNRS, and to write up in an enormous work called *Réflexions Mathématiques*: "il s'agit...de développer des idées et des visions multiples amorcées au cours de ces douze dernières années, en les précisant et les approfondissant, avec tous les rebondissements imprévus qui constamment accompagnent ce genre de travail...et je compte bien laisser apparaître clairement...la démarche de la pensée qui sonde et qui découvre, en tâtonnant dans la pénombre bien souvent, avec des trouées de lumière subite quand quelque tenace image fausse, ou simplement inadéquate, se trouve enfin débusquée et mise à jour, et que les choses qui semblaient de guingois se mettent en place, dans l'harmonie mutuelle qui leur est propre."[3]

One of the most fascinating regions of exploration is contained in sections 2 and 3, where Grothendieck describes an entirely new approach to the problem of describing the structure of the absolute Galois group $\mathrm{Gal}(\overline{\mathbb{Q}}/\mathbb{Q})$. The conference on dessins d'enfant at the CIRM in Luminy, April 19-24, 1993, was organized in an effort to gather together a number of people who were working on or interested in subjects more or less closely related to this part of Grothendieck's *Esquisse d'un Programme*, many of whom were unaware of the fact that others were actively working on the same questions – indeed the *Esquisse d'un Programme* appears to have benefited from a near-universal moratorium among French mathematicians until about a year ago though it was mentioned regularly in articles by Russian and Japanese mathematicians.

The idea of publishing the *Esquisse d'un Programme* in this volume unfortunately had to be abandoned, with regret and reluctance, when it became clear that it was impossible to contact Grothendieck to obtain his written permission to do so. This introduction should serve, in some sense, as a poor substitute, attempting to explain the major underlying ideas of Grothendieck's project to study the absolute Galois group, with a good deal more precision than can be found written in the Esquisse, but certainly much less than could be found in the mind of its author, and with rather more reference to concrete results as presented in various publications including the conference proceedings, but unfortunately a smaller dose of soaring imagination.

* * * * *

In §3 of the Esquisse, entitled *Corps de nombres associé à un dessin d'enfant*, Grothendieck confronts the study of the combinatorical action of $\mathrm{Gal}(\overline{\mathbb{Q}}/\mathbb{Q})$ on the dessins. The main point of departure, as mentioned above, is Belyi's important theorem stating that *every algebraic curve defined over*

$\overline{\mathbb{Q}}$ *can be realized as a covering* $\beta : X \to \mathbf{P}^1\mathbb{C}$ *where the morphism* β *is a Be-lyi morphism, namely ramified at most over the points* 0, 1 *and* ∞ *of* $\mathbf{P}^1\mathbb{C}$. This theorem, and in particular the simplicity of its proof, was astounding to Grothendieck who had hardly dared conjecture it: "une telle supposition avait l'air à tel point dingue que j'étais presque gêné de la soumettre aux compétences en la matière. Deligne consulté trouvait la supposition dingue en effet, mais sans avoir un contrexemple dans ses manches. Moins d'un an après, au Congrès International de Helsinki, le mathématicien soviétique Bielyi annonce justement ce résultat, avec une démonstration d'une sim-plicité déconcertante tenant en deux petites pages d'une lettre de Deligne – jamais sans doute un résultat profond et déroutant ne fut démontré en si peu de lignes!"[4]

To understand the question which was answered by Belyi's theorem, we first set out some background. Let the term "dessin d'enfant" (child's draw-ing), denote a cellular map with a bipartite structure; i.e. a compact topo-logical surface equipped with a finite number of points (vertices) and a finite number of edges connecting them, such that the following properties hold:

i) the set of vertices and edges is connected

ii) this set cuts the surface into open cells

iii) a bipartite structure can be put on the map; it consists of marking the vertices with two different marks, say ● and ⋆, in such a way that the direct neighbors of any vertex are all of the opposite mark.

Two special cases of dessins are frequently considered, in the *Esquisse d'un Programme*, in the introductory article [SV], and in the articles in this volume by Schneps and Couveignes-Granboulan for example. The first, which we may call "pre-clean", are those having ⋆ vertices with valencies (i.e. number of edges coming out of them) less than or equal to 2, and the second, known as clean dessins, are those all of whose ⋆ vertices have valency exactly equal to 2. These dessins can be pictured as cellular maps with all vertices labeled ● and a ⋆ placed either in the middle or at a tail end of each edge in the pre-clean case, only in the middle in the clean case; they have the advantage of simplicity in that ●'s correspond to vertices and ⋆'s to edges. Examples of the three types of dessin are given in Figure 2; they are all in genus zero for simplicity, and drawn on the plane, i.e. on the visible part of $\mathbf{P}^1\mathbb{C}$...

To any Belyi function $\beta : X \to \mathbf{P}^1\mathbb{C}$ one can immediately associate a dessin by considering the inverse image $\beta^{-1}[0,1]$ of the real segment $[0,1]$ on $\mathbf{P}^1\mathbb{C}$, drawn on the topological model of the Riemann surface X; as a convention we consider ●'s to represent the pre-images of 0 and ⋆'s of 1 (also, by convention, we mark with a ○ each pre-image of ∞, so exactly one o falls somewhere into each open cell). Under this association, the pre-clean dessins correspond to Belyi functions whose ramification orders over 1 are all of order less than or equal to 2 and the clean ones to Belyi functions all of whose ramification indices over 1 are exactly equal to 2.

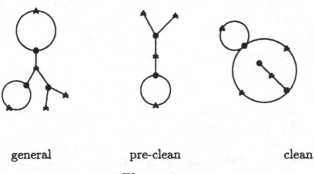

general   pre-clean   clean

Figure 2

Using Belyi's theorem, Grothendieck showed the converse direction, i.e. that to any dessin one can associate a covering $\beta : X \to \mathbf{P}^1\mathbf{C}$ where β is a Belyi morphism and X a Riemann surface. A rigorous combinatorical proof of this result is given in a 1975 article by two of Grothendieck's students, J. Malgoire and C. Voisin [MV]; the result was also proved around the same time in an article by D. Singerman and G. Jones ([JS]). The first such calculations were performed by Atkin, who considered dessins as quotients of the Poincaré upper half plane by subgroups of finite index in $PSL_2(\mathbf{Z})$. In their article [SV] in the Grothendieck Festschrift, in the case of clean dessins, Shabat and Voevodsky give a simple topological argument which we describe here as it makes the association intuitively clear. From a dessin D, one obtains a canonical triangulation of the surface on which D is drawn as follows. In addition to the •'s and the ⋆'s, draw a o somewhere inside each open cell, and add edges to the drawing joining every o to all the neighboring • and ⋆ points. This paves the surface with lozenges of the type shown in Figure 3 (it forms what is known as the *canonical triangulation* of the surface associated to the dessin, which is discussed in the article by Bauer and Itzykson).

Figure 3

Identifying the two sides joining the ● to a ⋆, and also the two sides joining the o to a ⋆ gives a surface homeomorphic to the sphere. One takes β to be the morphism from the topological surface to $\mathbf{P}^1$ sending all these lozenges to $\mathbf{P}^1$ with the ● points going to 0, the ⋆'s to 1 and the o's to ∞. By the Riemann existence theorem, there is a unique complex structure which can be put on the topological surface such that β becomes a rational function for it. This gives the covering $\beta : X \to \mathbf{P}^1\mathbf{C}$ associated to the dessin, and the morphism β is a Belyi morphism under this construction; its ramification indices over 0, 1 and ∞ are given by the number of edges coming out of each ●, ⋆ and o point respectively in the canonical triangulation associated to D (and indeed this triangulation is simply given by $\beta^{-1}(\infty, \infty)$).

Grothendieck had asked the question of whether all algebraic curves defined over $\overline{\mathbf{Q}}$ could be obtained in this way – the affirmative answer is of course a corollary of Belyi's theorem, which gives a natural description – "defined over $\overline{\mathbf{Q}}$" – of the coverings of $\mathbf{P}^1\mathbf{C}$ ramified over just three points. As he describes it: *"il y a une identité profonde entre la combinatoire des cartes finies d'une part, et la géométrie des courbes algébriques définies sur des corps de nombres, de l'autre.* Ce résultat profond, joint à l'interprétation algébrico-géométrique des cartes finies, ouvre la porte sur un monde nouveau, inexploré – et à portée de main de tous, qui passent sans le voir."[5]

With the association of a dessin to any Belyi function described above, we now have a bijection between abstract dessins and isomorphism classes of pairs (X, β) where X is a Riemann surface and β a Belyi morphism; this then gives rise to a bijection between clean dessins and conjugacy classes of the so-called cartographical group C_2^+, which is freely generated by one generator of infinite order and one of order 2. Complete proofs of this, of Belyi's theorem, of the combinatorical proof by Malgoire-Voisin that one can associate a unique isomorphism class of Belyi coverings of $\mathbf{P}^1\mathbf{C}$ to a given dessin and of the topological argument of Shabat and Voevodsky given above; in a word, of the bijection between isomorphism classes of pairs (X, β) and abstract dessins, can be found in the first part of the article by Schneps, and also that of Birch in this volume. This bijection gives a natural way to associate a number field K_D to every dessin D by defining it to be the moduli field of the associated algebraic curve X (for a precise discussion of the terms moduli field and field of definition see the articles by Schneps and Couveignes-Granboulan); this field is also called the moduli field of the dessin. The articles of Birch and Malle are concerned with the study of K_D; Malle gives tables of information for moduli fields for a number of Belyi coverings of degree ≤ 13, whereas Birch considers the question of which primes can be ramified there. The latter article also considers the problem, specifically associated to non-congruence subgroups Γ of $SL_2(\mathbf{Z})$, of the form of the coefficients of a basis of modular forms for $S_{2k}^0(\Gamma)$, the space of cusp forms of weight $2k$, which was studied among others by Atkin, Swinnerton-Dyer and Scholl.

The difficult aspect of the bijection described above is to render it explicit, in the direction dessins→ coverings (the direction coverings→dessins can be handled via Newton's algorithm for example, cf. part III of Schneps). This can be done, in genus zero on small example, using algebraic methods such as the Gröbner basis algorithem to solve a system of equations: variations of this procedure are described in the articles by Birch, Shabat and Schneps (and [C] for some improvements). In his series of introductory lectures, Oesterlé suggested that the use of the Puiseux series as introduced by Ihara could lead to an explicit analytic description of the coverings, giving a method to calculate the Belyi functions associated to dessins of any genus. Following this suggestion, Couveignes and Granboulan describe what information can be obtained from the Puiseux series method, and how this method can be made more effective in the genus zero case where it can be strengthened by the use of approximations, iteration and intuition. They calculate an "optimal" Belyi function associated to a dessin; namely a function defined over the smallest possible number field, shown to be at worst a quadratic extension of the moduli field of the dessin.

Considering more generally coverings of $\mathbf{P}^1\mathbf{C}$ ramified at r points with $r \geq 3$, the article by Michel Emsalem explains the construction of moduli spaces for these objects, known as Hurwitz spaces: the main theorem states that the Hurwitz spaces are algebraic varieties defined over $\mathbf{Q}$.

Paula Cohen's article is a summary of work by Cohen, Itzykson and Wolfart ([CIW]). To any Belyi morphism $\beta : X \rightarrow \mathbf{P}^1\mathbf{C}$, together with the associated triangle group $\Delta(p,q,r)$ where p,q,r are the positive common multiples of the orders of ramification of β over 0, 1 and ∞, they associate the covering group H which is a subgroup of finite index in $\Delta(p,q,r)$, and which determines an arithmetic group Γ which acts on $\mathfrak{H}^t$ for some positive integer t ($\mathfrak{H}$ is the upper half-plane) in such a way that the quotient $V = \Gamma \backslash \mathfrak{H}^t$ is a Shimura variety parametrising isomorphism classes of abelian varieties with certain complex multiplication properties for a subfield K of a cyclotomic field. From the natural injection $H \rightarrow \Gamma$ they define an injection $\mathfrak{H} \rightarrow \mathfrak{H}^t$ which commutes with the respective group actions; by passing to the quotient, this defines a non-trivial morphism $X \rightarrow V$, defined over $\overline{\mathbf{Q}}$, which sends the elements of $\beta^{-1}\{0,1,\infty\}$ onto points of complex multiplication by K.

In the article by Jones and Singerman, certain of these triangle groups, such as the cartographical group C_2^+ mentioned above, are identified with subgroups of $\mathrm{PGL}_2(\mathbf{Z})$: this gives realisations of the associated maps as tessellations of Riemann surfaces or Klein surfaces, and the operations on them of automorphisms of the triangle groups are studied.

Saito's paper gives a real algebraic coordinatization of the Teichmüller space $\mathcal{T}_{g,n}$, the universal covering space of the moduli space $\mathcal{M}_{g,n}$ of Riemann surfaces of genus g with n marked points, whose fundamental group (as an orbifold) is the mapping class group $M(g,n)$. In his article on horizontal divisors, Ihara studies the geometric ramification properties of Belyi coverings. The article by Bauer and Itzykson contains much of the basic information on dessins d'enfants; this is used to give generating functions in the form of matricial integrals which have applications to intersection problems on the moduli spaces $\mathcal{M}_{g,n}$.

The algorithmic methods described in the articles by Birch, Couveignes-Granboulan, Malle, Schneps and Shabat do allow one to calculate examples of the Galois orbits of a good many genus zero dessins, as a first step towards the exploration of the Galois action. In some sense one should not actually lose information about $\mathrm{Gal}(\overline{\mathbf{Q}}/\mathbf{Q})$ by restricting one's attention to the genus zero dessins, since a result due to H. Lenstra (cf. the article by Schneps, part II) shows that $\mathrm{Gal}(\overline{\mathbf{Q}}/\mathbf{Q})$ acts faithfully on the set of genus zero dessins and indeed even on the set of trees which are the simplest example of them. G. Shabat studies the combinatorical properties of trees, and particular the special case of "diameter 4" trees which have several interesting properties, for instance the discriminant of their associated number fields can be studied directly from their combinatorics.

Examples of Galois orbits of genus zero dessins have been much examined with a view to establishing the properties of a given dessin which remain invariant when the Galois group acts on it. Optimistically, one might dream of a "complete" list of such Galois invariants, i.e. a list such that the set of all dessins having the same invariants should constitute exactly one orbit under the action of $\mathrm{Gal}(\overline{\mathbf{Q}}/\mathbf{Q})$. Several Galois invariants have been found, such as the valency lists or the Galois group of a dessin. However the list is known not to be complete since two dessins have also been found which do not belong to the same Galois orbit and as yet continue to defy all attempts to separate them via the known Galois invariants. These two dessins, similar diameter 4 trees (see the frontispiece), are given in Figure 4. For the time being, this curious situation would appear to indicate the limits of our understanding...

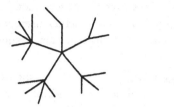 

Figure 4

Grothendieck was led to the question of the action of $\text{Gal}(\overline{\mathbf{Q}}/\mathbf{Q})$ on the dessins via "un effort de compréhension des relations entre groupes de Galois 'arithmétiques' et groupes fondamentaux profinis 'géométriques'. Assez vite il s'oriente vers un travail de formulation calculatoire de l'opération de $\text{Gal}(\overline{\mathbf{Q}}/\mathbf{Q})$ sur $\pi_{0,3}$" [6] (the algebraic fundamental group of $\mathbf{P}^1\mathbf{C} \setminus \{0,1,\infty\}$, which is just a free profinite group on two generators). Because the cartographical group C_2^+ is just a quotient of $\pi_{0,3}$, the action of $\text{Gal}(\overline{\mathbf{Q}}/\mathbf{Q})$ on the fundamental group is reflected in its action on the dessins. We mention briefly here that these ideas led him to what he calls the central theme of anabelian algebraic geometry, namely the idea that certain "anabelian" varieties X defined over a K can be entirely reconstituted from knowledge of their "mixed fundamental group", which is the extension of the fundamental group $\pi_1(X_{\overline{K}})$ by $\text{Gal}(\overline{K}/K)$ ($\overline{K}$ denotes the algebraic closure of K). Much of this part of Grothendieck's program remains mysterious but some results can be found in work by Bogomolov, Pop, Nakamura, Voevodsky and others (cf. for example [B], [N1], [N2], [NT], [P]).

The space $\mathbf{P}^1\mathbf{C} \setminus \{0,1,\infty\}$ is the moduli space of Riemann surfaces of genus 0 with 4 marked points. Grothendieck perceived the study of the action of $\text{Gal}(\overline{\mathbf{Q}}/\mathbf{Q})$ on $\pi_{0,3}$, the algebraic fundamental group of this moduli space, as a special case of the action of $\text{Gal}(\overline{\mathbf{Q}}/\mathbf{Q})$ on all the fundamental groups of the moduli spaces $\mathcal{M}_{g,n}$ of Riemann surfaces of genus g with n marked points, or indeed, on certain "fundamental groupoids" denoted $\hat{T}_{g,n}$, on a certain finite set of base points. These ideas gave rise to §2 of the Esquisse, entitled *Un "jeu de Légo-Teichmüller", et le groupe de Galois de $\overline{\mathbf{Q}}$ sur $\mathbf{Q}$*. Grothendieck considers what he calls the Teichmüller tower of profinite fundamental groupoids $\hat{T}_{g,n}$ of the moduli spaces $\mathcal{M}_{g,n}$ of Riemann surfaces of genus g with n ordered marked points. The groupoids are not explicitly defined in the paper, but many of their properties, including conjectural ones, are enumerated. They should have as "unit group" the usual fundamental groups of these moduli spaces, namely the pure subgroups $K(g,n)$ of the mapping class groups $M(g,n)$, but Grothendieck is adamant on the importance of considering groupoids on several base points:

"...les gens s'obstinent encore, en calculant avec des groupes fondamentaux, à fixer un seul point base, plutôt que d'en choisir astucieusement tout un paquet qui soit invariant par les symétries de la situation, lesquelles sont donc perdues en route."[7]

The set of profinite groupoids $\hat{T}_{g,n}$ should form a *tower* when the different groupoids are linked by certain natural homomorphisms corresponding to degeneration of surfaces. Indeed, surgical operations on the Riemann surfaces which decrease the type g, n, such as erasing points, cutting into pieces along geodesics and so on, induce natural morphisms between the fundamental groups or groupoids of the associated moduli spaces (as do the inverse operations of adding marked points and "gluing" pieces together; however these, in an obvious sense, do not merit the term "degeneration"). In his article [D], which although occasionally quite cryptic for the unwary, sheds tremendous illumination on Grothendieck's ideas, Drinfel'd summarizes this idea, writing $(X, x_1, \ldots, x_n)$ for a Riemann surface X with n ordered marked points: "Since degeneration of the set $(X, x_1, \ldots, x_n)$ results in decreasing g and n, the groupoids $T_{g,n}$ are connected by certain homomorphisms. The collection of all $T_{g,n}$ and all such homomorphisms is called in [2] the Teichmüller tower."

Needless to say, [2] is the *Esquisse d'un Programme*. Although Drinfel'd's goal in this paragraph is not to study the Teichmüller fundamental groupoids but merely to sketch the relation between the rest of the paper and Grothendieck's ideas, he does make some further brief indications about the choice of base points on the moduli spaces for them. He recommends taking tangential base points "at infinity", i.e. base points in a neighborhood of infinity on the moduli space $M_{g,n}$, corresponding to "maximal degeneration" of Riemann surfaces. These base points generalize to all the moduli spaces the well-known tangential base points defined by Deligne in [De] on $\mathbf{P}^1\mathbf{C} - \{0, 1, \infty\}$, which is the first non-trivial moduli space $M_{0,4}$. This construction of a Teichmüller tower with base points near infinity has been carried out ([LS]). It does however seem that this choice of base points corresponds to a simpler version of the Teichmüller groupoids than Grothendieck's. In the Esquisse, Grothendieck suggests using base points on the moduli space corresponding to Riemann surfaces having non-generic automorphism group as well: "ceux-ci s'offrent tout naturellement, comme les courbes algébriques complexes du type (g, n) envisagé, qui ont un groupe d'automorphismes (nécessairement fini) plus grand que dans le cas générique"[8].

Drinfel'd's startling contribution to the development of Grothendieck's ideas comes as a suggested application of his work on quasi-Hopf algebras. In this work, also in the article [D], he defines a group $\widehat{GT}$ (for Grothendieck-

Teichmüller) of deformations of quasi-Hopf algebras; one of the main points of interest of this group, aside from the extreme simplicity of its definition, is that *the absolute Galois group is actually contained in it* if not equal to it. It is this fact which gives the close link with Grothendieck's remark that $\text{Gal}(\overline{\mathbf{Q}}/\mathbf{Q})$ "opère sur toute cette tour de groupoïdes de Teichmüller, en respectant toutes ces structures."[9] In a paper in the Proceedings of the Kyoto ICM in 1990, Ihara very briefly sketched out a geometric proof of this result. The paper by Ihara, Emsalem and Lochak is a complete exposition of Ihara's proof, based on a detailed study of the geometry of the moduli spaces $M_{0,4}$ and $M_{0,5}$.

The elements of $\widehat{GT}$ induce natural automorphisms on such geometric objects as braid groups, the Teichmüller groups and also the Teichmüller groupoids with tangential base points at infinity. Indeed, Drinfel'd suggested that the group $\widehat{GT}$ should be *exactly* the group of automorphisms of the tower consisting of the genus zero profinite Teichmüller groupoids $\hat{T}_{0,n}$, (and possibly the automorphism group even of the full tower in all genera). A full proof of this result is given in [LS]; it is shown there that the automorphism group of $\hat{T}_{0,4}$ (which being the fundamental groupoid of the moduli space $M_{0,4} \sim \mathbf{P}^1\mathbf{C} - \{0,1,\infty\}$, is closely related to the algebraic fundamental group $\pi_{0,3}$ of the sphere minus three points, which is of course free profinite on two generators) is a certain group (which we denote by $\widehat{GT}_0$) containing $\widehat{GT}$, and that these automorphisms extend to automorphisms of $\hat{T}_{0,5}$ respecting the degeneracy homomorphisms between $T_{0,5}$ and $T_{0,4}$ if and only if they actually lie in $\widehat{GT}$ itself, in which case they should extend automatically to automorphisms of the whole genus zero tower. Much of the proof comes from the result proved in the article by Lochak and Schneps in this volume, where instead of considering the groupoids, a tower of profinite completions of the Artin braid groups together with certain connecting homomorphisms is defined (these groups are closely related to the mapping class groups $M(0,n)$, and the connecting homomorphisms to degeneracy homomorphisms) and it is shown that $\widehat{GT}$ is the full automorphism group of this tower; a pro-ℓ analogue is also proved. These assertions, together with the extraordinarily simple, combinatorical definition of the group $\widehat{GT}$, reads surprisingly like a partial realization of the following expression of Grothendieck's vision, where Γ denotes the group $\text{Gal}(\overline{\mathbf{Q}}/\mathbf{Q})$:

"Le groupe de Galois Γ se réalise comme un groupe d'automorphismes d'un groupe profini des plus concrets [i.e. $\hat{\pi}_{0,3}$], respectant d'ailleurs certaines structures essentielles de ce groupe. Il s'ensuit qu'un élément de Γ peut être 'paramétré' (de diverses façons équivalentes d'ailleurs) par un élément convenable de ce groupe profini $\hat{\pi}_{0,3}$ (un groupe profini libre à deux générateurs), ou par un système de tels éléments, ce ou ces éléments étant d'ailleurs soumis à certaines conditions simples, nécessaires (et sans doute non suffisantes) pour que ce ou ces éléments correspondent bien à un élément de Γ. Une des tâches les plus fascinantes ici, est justement d'appréhender des conditions nécessaires et suffisantes sur un automorphisme extérieur de $\hat{\pi}_{0,3}$, i.e. sur le ou les paramètres correspondants, pour qu'il provienne d'un élément de Γ - ce qui fournirait une description 'purement algébrique', en termes de groupes profinis et sans référence à la théorie de Galois des corps de nombres, du groupe de Galois $\Gamma = \mathrm{Gal}(\overline{\mathbb{Q}}/\mathbb{Q})$!"[10]

The fact that the automorphism group of the first two levels $\hat{T}_{0,4}$ and $\hat{T}_{0,5}$ of the genus zero Teichmüller tower is the same as the automorphism group of the full tower is related to Grothendieck's more general conjecture that the entire Teichmüller tower in all genera should be entirely determined by the groups on the two first levels (where the m-th level consists of groupoids of type g, n such that the modular dimension $3g-3+n$ is equal to m), namely the groups with $3g - 3 + n \leq 2$, so (g,n) is one of $(0,4)$, $(0,5)$, $(1,1)$ or $(1,2)$. (It is the idea that all groups at all levels can be reconstructed from these basic pieces, and the analogous idea for the surfaces themselves, which earned the nomenclature "jeu de Légo-Teichmüller" for this construction). A version of this conjecture in terms of group representations was apparently proved by Moore and Seiberg in [MS]; the relations of their work to the present situation are discussed in the summary contained in this volume of work by Pascal Degiovanni.

Many questions remain shrouded in mystery. It is tempting to name the following ones among them: characterizing the Galois orbits of the dessins combinatorially, extending the genus zero Teichmüller tower results to arbitrary genus, characterizing of the image of $\mathrm{Gal}(\overline{\mathbb{Q}}/\mathbb{Q})$ in the group $\widehat{GT}$ (dare one conjecture surjectivity of the inclusion map?), obtaining a new characterization of $\mathrm{Gal}(\overline{\mathbb{Q}}/\mathbb{Q})$ as an automorphism group of the Teichmüller tower, continuing in the direction of the anabelian dream.

Leila Schneps *
January 1994

Translations of the French quotations

[1] The demands of university teaching, addressed to students (including those said to be "advanced") with a modest (and frequently less than modest) mathematical baggage, led me to a Draconian renewal of the themes of re-

---

* I would like to extend my heartfelt thanks to the Bunting Institute of Radcliffe College for their welcome and support during the preparation of this book

flection I proposed to my students, and bit by bit to myself as well. It seemed important to me to start from an intuitive baggage common to everyone, independent of any technical language used to express it, and anterior to any such language – it turned out that the geometric and topological intuition of shapes, particularly two-dimensional shapes, formed such a common ground. This consists of themes which can be grouped under the general name of "topology of surfaces" or "geometry of surfaces", it being understood in this last expression that the main emphasis is on the topological properties of the surfaces, or the combinatorial aspects which form the most down-to-earth technical expression of them, and not on the differential, conformal, Riemannian, holomorphic aspects, and (from there) on to "complex algebraic curves". Once this last step is taken, however, algebraic geometry (my former love!) suddenly bursts forth once again, and this via the objects which we can consider as the basic building blocks for all other algebraic varieties. Whereas in my research before 1970, my attention was systematically directed towards objects of maximal generality, in order to uncover a general language adequate for the world of algebraic geometry, and I never restricted myself to algebraic curves except when strictly necessary (notably in etale cohomology), preferring to develop "pass-key" techniques and statements valid in all dimensions and in every place (I mean, over all base schemes, or even base ringed topoi...), here I was brought back, via objects so simple that a child learns them while playing, to the beginnings and origins of algebraic geometry, familiar to Riemann and his followers!

[2]As the present state of things makes any prospect of teaching research at the University more and more illusory, I have resolved to apply for admission to the CNRS, in order to devote my energy to developing work and perspectives which, it is becoming clear, will not be developed by any student (nor, it seems, any mathematical colleague) in my stead.

[3]It is about...developing the ideas and the multiple visions begun during these last twelve years, to make them more precise and deeper, with all the unexpected rebounds which constantly accompany this kind of work...and I do intend to clearly show the process of thought, which feels and discovers, often blindly in the shadows, with sudden flashes of light when some tenacious false or simply inadequate image is finally shown for what it is, and things which seemed all crooked fall into place, with that mutual harmony which is their own.

[4]...such a supposition seemed so crazy that I was almost embarrassed to submit it to the competent people in the domain. Deligne when I consulted him found it crazy indeed, but didn't have any counterexample up his sleeve.

Less than a year later, at the International Congress in Helsinki, the Soviet
mathematician Belyi announced exactly that result, with a proof of a disconcerting simplicity which fit into two little pages of a lettre of Deligne –
never, without a doubt, was such a deep and disconcerting result proved in
so few lines.

[5] ...there is a profound identity between the combinatorics of finite maps
on the one hand, and the geometry of algebraic curves defined over number
fields on the other. This deep result, together with the algebraic-geometric
interpretation of maps, opens the door onto a new, unexplored world –
within reach of all, who pass without seeing it.

[6] ...an effort to comprehend the relations between "arithmetic" Galois groups
and "geometric" profinite fundamental groups. Quite rapidly it was oriented
towards the job of giving a combinatorial formulation of the operation of
$\mathrm{Gal}(\overline{\mathbf{Q}}/\mathbf{Q})$ on $\pi_{0,3}$.

[7] ...people still obstinately persist, when calculating with fundamental groups,
in fixing a single base point, instead of cleverly choosing a whole packet of
points which is invariant under the symmetries of the situation, which are
thus lost on the way.

[8] ...these offer themselves quite naturally, as the complex algebraic curves of
the type (g, n) under consideration, having automorphism group (necessarily finite) larger than in the generic case.

[9] operates on the whole tower of the Teichmüller groupoids, respecting all
these structures.

[10] The Galois group $\mathbb{\Gamma}$ can be realised as an automorphism group of a per-
fectly concrete profinite group, which respects certain essential structures
of this group. It follows that an element of $\mathbb{\Gamma}$ can be "parametrized" (in
various equivalent ways) by a suitable element of this profinite group $\hat{\pi}_{0,3}$
(a free profinite group on two generators), or by a system of such elements,
these elements being subject to certain simple necessary (but doubtless not
sufficient) for this or these elements to really correspond to an element of
$\mathbb{\Gamma}$. One of the most fascinating tasks here is precisely to discover necessary
and sufficient conditions on an exterior automorphism of $\hat{\pi}_{0,3}$, i.e. on the
corresponding parameter(s), for it to come from an element of $\mathbb{\Gamma}$ – which
would give a "purely algebraic" description, in terms of profinite groups and
with no reference to the Galois theory of number fields, to the Galois group
$\mathbb{\Gamma} = \mathrm{Gal}(\overline{\mathbf{Q}}/\mathbf{Q})$.

References

[B]   F. Bogomolov, in "Algebraic Geometry and Analytic Geometry", ed.
       Fujiki, Kato, Katsura, Kawamata, Miyaoka, Springer Verlag Proc. of
       a conference held in Tokyo, Japan 8/13-17 (1990).

[C]   J-M. Couveignes, Calcul et rationnalité de fonctions de Belyi en genre
       0, *Annales de l'Institut Fourier* **44** (1), 1994.

[CIW] P.B. Cohen, C. Itzykson, J. Wolfart. Fuchsian triangle groups and
       Grothendieck dessins: variations on a theme of Belyi, to appear.

[De]   P. Deligne. Revêtements de la sphere moins trois points, *Galois
       groups over* **Q**, Ihara, Ribet, Serre, eds.

[D]   V. G. Drinfel'd. On quasitriangular quasi-Hopf algebras and a group
       closely connected with $\mathrm{Gal}(\overline{\mathbf{Q}}/\mathbf{Q})$, Leningrad Math. J. **2** (1991), 829-
       860.

[JS]   G. Jones, D. Singerman, Theory of Maps on Orientable surfaces,
       Proc. London Math. Soc. (3) **37** (1978), 273-307.

[LS]   P. Lochak, L. Schneps, On the Teichmüller tower with base points
       near infinity, preprint.

[MS]   G.W. Moore, N. Seiberg, Classical and quantum conformal field the-
       ory, *Comm. Math. Phys.* **123** (1989), 177-254.

[MV]   J. Malgoire, C. Voisin, Cartes Cellulaires, Cahiers Mathématiques de
       Montpellier No. 12, 1977.

[N1]   H. Nakamura, Galois rigidity of the etale fundamental group of punc-
       tured projective lines, *J. Crelle* **411** (1990), 205-216.

[NT]   H. Nakamura, H. Tsunogai, Some finiteness theorems on Galois cen-
       tralizers in pro-ℓ mapping class groups, *J. Crelle* **441** (1993), 115-144.

[P]   F. Pop, On the Galois theory of function fields. *J. Crelle* **406** (1990),
       200-218.

[SV]   G. Shabat and V. Voevodsky, Drawing Curves over Number Fields,
       The Grothendieck Festschrift, Vol. III. Birkhäuser, 1990.

Abstracts of the talks

The fields of definition of coverings: *Bryan Birch.* The author's introduction to coverings and dessins was through the paper of Atkin and Swinnerton-Dyer, on non-congruence subgroups of the modular group; from that point of view, it is natural to consider coverings of $\mathbf{P}^1 \setminus \{0, 1, \infty\}$ with a point marked above ∞, and so forth. For a fixed n, there is a well-known correspondence between subgroups of $\Gamma(2)$, modulo translation; triples (R, ϕ, O) where $\phi : R \to \mathbf{P}^1 \setminus \{0, 1, \infty\}$ is a covering by an n-sheeted Riemann surface and O is a point of R above ∞; and quadruples $(\sigma_\infty, \sigma_0, \sigma_1, \mu)$ where the σs are permutations in S_n satisfying $\sigma_\infty \sigma_0 \sigma_1 = 1$, μ is a marked cycle of σ_∞, and the whole thing is taken modulo simultaneous conjugation in S_n. Such a quadruple may be more engagingly described by a dessin D with 2-coloured vertices, one colour corresponding to each of σ_0, σ_1, and faces corresponding to σ_∞, with one face marked (naturally, the outside one!). The marking has the advantage that the standard theorems about the field of definition may be stated and proved somewhat more easily. Define the *type* $\tau(D)$ as the quadruple $(\pi_\infty, \pi_0, \pi_1, m)$ where each π_i is the partition of n corresponding to the cycles of σ_i and m is the length of μ; and let $G(D)$ be the group generated by σ_0, σ_1. The first theorem, probably first stated though not completely proved by Atkin and Swinnerton-Dyer, states that if there are $M(D)$ dessins D' with $\tau(D') = \tau(D)$ and $G(D') = G(D)$ then (R, ϕ, O) has a minimal field of definition K, which is a number field K of degree at most $M(D)$. (Of course, this is quite different from the theorems of inverse Galois theory, which ask for regular normal extensions of $\mathbf{Q}(t)$.) The second theorem, which in essence goes back to Grothendieck, states that if P is a prime of $\mathbf{Z}_K$ that does not divide the order of G then (R, ϕ, O) has a model with good reduction modulo P; in particular, the only primes that may ramify in K are those dividing the order of $G(D)$. Conversely, if P divides the order of one of the σs then the reduction mod P has to be bad: in the circumstances, it is likely to be even harder to solve the tame inverse Galois Problem, that is, given a group H to construct a *tame* extension L of $\mathbf{Q}$ with $\mathrm{Gal}(L : \mathbf{Q})$ isomorphic to H.

Dessins d'enfant and Shimura varieties: *Paula Beazley Cohen.* The article [Coh] is a summary of joint work by Cohen, Itzykson and Wolfart [CoItzWo]. To any Belyi morphism $\beta : X \to \mathbf{P}^1$, together with the associated triangle group $\Delta(p, q, r)$ where p, q, r are the positive common multiples of the orders of ramification of β over 0, 1 and ∞, they associate the covering group H which is a subgroup of finite index in $\Delta(p, q, r)$, and which determines an arithmetic group Γ which acts on $\mathcal{H}^t$ for some positive integer t ($\mathcal{H}$ is the upper half-plane) in such a way that the quotient $V = \Gamma \backslash \mathcal{H}^t$ is a Shimura variety parametrising isomorphism classes of abelian varieties with certain complex multiplication properties for a subfield K of a cyclotomic field. Moreover, there is a natural injection of H into Γ for which they define an injection of $\mathcal{H}$ into $\mathcal{H}^t$ which commutes with the respective group actions. By passing to the quotient, this defines a non-trivial morphism from X to V which is defined over $\overline{\mathbf{Q}}$ and which sends the elements of $\beta^{-1}\{0, 1, \infty\}$ onto points of complex multiplication by K.

Introduction to the enumeration of planar dessins d'enfant: *Robert Cori.* This talk consists in a presentation of the combinatorical aspect of dessins d'enfant. We consider a *hypermap*: it is a couple of permutations σ, α, which generate a group acting transitively on a finite set B whose elements are called *brins*. A hypermap is a map if α is an involution with no fixed points. We can define a notion of genus for these objects from the inequality

$$z(\sigma) + z(\alpha) + z(\alpha^{-1}\sigma) \leq card(B) + 2$$

where $z(\alpha)$ denotes the number of orbits of the permutation α. We set $2g$ to be the difference between the two sides of this inequality. When $g = 0$, the hypermap is said to be "planar".

Beautiful enumeration formulae are known for pointed planar maps since the work of W.T. Tutte in the 60's. For example, the number of pointed planar maps with $2m$ brins is equal to

$$3^m (2m)!(m-1)!(m+2)!$$

In order to enumerate the non-pointed maps, we give a result on the automorphisms of a planar map which we prove using a combinatorical version of the Riemann-Hurwitz formula. An enumeration formula due to Liskovets for non-pointed planar maps est then presented.

Belyi functions of genus zero and questions of rationality: *Jean-Marc Couveignes.* We give a numerical method for the computation of Belyi functions associated to a given dessin of genus zero. For such a dessin, it is possible to find such a function from $\mathbf{P}^1$ to $\mathbf{P}^1$, defined over at most a quadratic extension of the field of definition. We show how to find such a model explicitly in all cases. In the case where the dessin is a tree, this happens to be particularly simple. When there are no automorphisms, the dessin is canonically associated to a certain isomorphism class of genus zero curves (or quadratic forms); these curves are the only ones on which it is possible to find a rational model for the dessin. In the case where the automorphism group is cyclic of even order there may be an obstruction to

finding a model for the dessin over its own field of definition. We give an
example of this situation.

Rational points on Hurwitz spaces: *Pierre Dèbes.* The geometric form
of the inverse Galois problem – does each finite group G occur as the Galois
group of a regular extension of $\mathbf{Q}(T)$ – amounts to finding a Galois cover of
$\mathbf{P}^1$, of automorphism group G and defined over $\mathbf{Q}$, together with all of its
automorphisms. Using moduli spaces for branched covers $\mathbf{P}^1$, the problem
can be reduced to finding $\mathbf{Q}$-rational points on certain algebraic varieties
$\mathcal{H}$, called Hurwitz spaces. Thanks to results of Conway-Parker and Fried-
Völklein, one can restrict to the case where these varieties are irreducible and
defined over $\mathbf{Q}$. As a consequence, each finite group is a Galois group over
$K(T)$, where K is a Pseudo-Algebraically Closed field. By definition, PAC
fields are such that each irreducible variety defined over K has K-rational
points. A recent result of Pop shows that the field $\mathbf{Q}^{tr}$ of all algebraic totally
real numbers is Pseudo Really Closed, that is, each non-singular irreducible
variety defined over $\mathbf{Q}^{tr}$ has $\mathbf{Q}^{tr}$-rational points provided it has real points.
Consequently, the field $\mathbf{Q}^{tr}(i)$ is PAC. Pop's result can also be used to show
that each finite group is a Galois group over $\mathbf{Q}^{tr}(T)$, a result of Fried and
the author.

Representations of the modular tower and topological theory: *Pascal Degiovanni.* In the *Esquisse d'un Programme*, Grothendieck mentions
the existence of a "principle of reconstruction of the Teichmüller tower from
its two first levels". The goal of this talk was to show that these ideas
are also present in the study of certain topological theories in 3 dimensions. We recalled the definitions of these objects and defined the notion
of a solution to the Moore-Seiberg equations which allow us to construct
three-dimensional projective topological theories. We attempted to indicate
the relations between these objects, and their connexions with the ideas in
§2 of the *Esquisse*.

Deformation of coverings of the projective line: *Michel Emsalem.*
The theory of deformation of coverings of the projective line ramified at r
points, when the ramification points vary, can be traced back to Hurwitz. It
has been the object of much work by M. Fried, who introduced the notion
of Hurwitz spaces, which parametrize the isomorphism classes of coverings
of $_1$ ramified at r points with certain ramification restrictions.

This talk had two goals. First, to rework the topological construction of
the Hurwitz spaces H as coverings of $(\mathbf{P}^1)^r/S_r$ unramified outside of the
discriminant site and in particular, to explain the monodromy action of

the braid group on it. Second, to show that these spaces H, considered as algebraic coverings of $(\mathbf{P}^1)^r/S_r$ thanks to the GAGA principle, are definable over $\mathbf{Q}$. Thus, the solution of the inverse Galois problem is reduced to the question of the existence of rational points on the Hurwitz spaces parametrizing the isomorphism classes of G-coverings. The latter result, whose proof utilises Weil's descent criterion, can also be found in work by Fried and Völklein. In their article they use the existence of a finite moduli space when the coverings considered have no non-trivial automorphisms. They deal with the general case by a procedure of devissage, which to my knowledge has no real geometric interpretation. The approach proposed here is different. We fix an ordinary point a_0 and rigidify the situation by considering families of pointed coverings over a_0. We then let a_0 vary in $\mathbf{P}^1\mathbf{Q}$ and we cover $(\mathbf{P}^1)^r/S_r$ by the open sets $\left(\mathbf{P}^1\setminus\{a_0\}\right)^r/S_r$.

Horizontal divisors and fundamental groups of arithmetic surfaces associated with Belyi uniformizations: *Yasutaka Ihara.* To each Belyi function $t: X_k \to \mathbf{P}^1$ on a curve X_k over a number field k, one can associate a normal arithmetic surface X over $\mathcal{O}$, the ring of integers of k. We shall first show that each of $t^{-1}(0)$, $t^{-1}(1)$ and $t^{-1}(\infty)$ is connected as an $\mathcal{O}$-scheme. In other words, different points of X_k over 0 (etc.) "tend to meet" each other when reduced modulo primes f of $\mathcal{O}$. Applying this, we show that when t is so chosen that $t^{-1}(\infty) = P$ (which is always possible if X_k contains a k-rational point P), then $\pi_1(X) \to \pi_1(\mathrm{Spec}(\mathcal{O}))$. In other words, every algebraic curve X_k over k with a k-rational point has a normal model X over $\mathcal{O}$ which has no non-trivial finite etale connected coverings other than those obtained from unramified extensions of $\mathcal{O}$. (Etale coverings of X correspond to those of X_k having good reduction (as coverings) everywhere).

The absolute Galois group: *Yasutaka Ihara.* As is well-known, $\mathrm{Gal}(\overline{\mathbf{Q}}/\mathbf{Q})$ acts on the profinite completion of the geometric fundamental group of $\mathbf{P}^1\setminus\{0,1,\infty\}$ which is free of rank two. In his well-known paper, Drinfel'd shows that the image of $\mathrm{Gal}(\overline{\mathbf{Q}}/\mathbf{Q})$ in $\mathrm{Aut}(\hat{F}_2)$ is contained in the group $\hat{G}T$. The purpose of this talk is to explain a geometric proof using the moduli space X_5 of the 5-pointed projective line see [Y. Ihara, Proc. ICM 1990]. The main point is how to define tangential base points on X_5.

Triangulations: *Claude Itzykson.* Since G't Hooft (1974), methods of matrix integration have been developed which give generating functions for (possibly decorated) maps. This talk presented three aspects:

   (i) the model with one matrix and the octogonal polynomial method;

   (ii) the enumeration of maps given the valencies of the vertices and the faces, based on the duality between linear and symmetric groups with an application to the calculation of the virtual characteristic of $M_{g,n}$ according to Harer and Zagier and Penner;

(iii) the combinatorics of Witten and Kontsevitch's theorem giving the properties of the generating function for the intersection of the stable classes on $M_{g,n}$: equations of the KdV hierarchy, vector of highest weight of a module of the Virasoro algebra.

A series of references can be found in a compte rendu of the 54-th colloquium of the RCP of Strasbourg in honor of P. Cartier.

Automorphisms and functors on maps and hypermaps: *Gareth Jones.* Various categories of combinatorial structures on surfaces correspond to transitive actions (on coset-spaces $M \backslash \Delta, M \leq \Delta$) of certain triangle groups $\Delta = (p, q, r)$ or extended triangle groups $\Delta = [p, q, r]$. Thus for triangulations we take Δ to be the extended modular group $\Pi = PGL_2(\mathbf{Z}) = [\infty, 2, 3]$, while for triangulations on oriented surfaces without boundary we use its even subgroup, the modular group $\Gamma = PSL_2(\mathbf{Z}) = (\infty, 2, 3)$. Similarly, hypermaps correspond to the principal congruence subgroups $\Pi(2) = [\infty, \infty, \infty]$ and $\Gamma(2) = (\infty, \infty, \infty)$ of level 2 in Π and Γ, while maps correspond to the non-principal congruence subgroups $\Pi_0(2) = [\infty, 2, \infty]$ and $\Gamma_0(2) = (\infty, 2, \infty)$. The actions of these groups Δ on certain universal tessellations of the upper half-plane $\mathcal{U}$ allow us to realise the objects $M \backslash \Delta$ in the corresponding categories as structures on Riemann or Klein surfaces $\mathcal{U}/M$.

The Grothendieck-Teichmüller group: *Pierre Lochak.* In view of introducing the Grothendieck-Teichmüller group $\widehat{GT}$ in the original context in which it was defined by Drinfel'd, the following objects were described: Hopf and quasi-Hopf algebras, braided tensor categories, the Yang-Baxter equation, associativity and commutativity constraints (giving rise to the familiar hexagonal and pentagonal diagrams). The group $\widehat{GT}$ was then defined as a group of transformations obtained by varying the associativity and commutativity constraints of a quasi-Hopf algebra in a "natural" way.

Rigidity: *Gunter Malle.* A short introduction into the basic rigidity criterion for three point ramified Galois extensions of $\overline{\mathbf{Q}}(T)$ was given, recalling the main results on Galois realizations of non-abelian simple groups over abelian number fields. Then the easiest versions of embedding theorems by Matzat were cited.

In the end the known constraints on possible fields of definition of Galois extensions of $\overline{\mathbf{Q}}$ were recalled. A computer experiment on 375 non-trivial fields of definition was described. The result shows that at least in the range considered, the rigidity criterion gives a rather sharp estimate on the degree of fields of definition of three point ramified field extensions of $\overline{\mathbf{Q}}(T)$ of genus zero.

Introduction à la théorie des dessins d'enfants. *Joseph Oesterlé.* In a series of four talks, we gave a list of various aspects from which one can consider the coverings of $\mathbf{P}^1 \setminus \{0, 1, \infty\}$: combinatoric, topological, complex analytic, algebraic over $\mathbf{C}$ or over $\mathbf{Q}$, quotients of the Poincaré upper half-plane. Particular care was taken to define the categories considered in each situation, to describe their objects and morphisms and to explicitly give the equivalences between these categories. From these dictionaries we deduce algorithms to:

a) determine the bicoloured graph associated to a finite extension of $\overline{\mathbf{Q}}(T)$, unramified outside of 0, 1 and ∞:

b) calculate, modulo every open subgroup of finite index of $\widehat{F}_2$ (free profinite group on two generators), the image of an element of $\mathrm{Gal}(\overline{\mathbf{Q}}/\mathbf{Q})$ in $\widehat{F}_2$ by the map associated (by Anderson, Belyi, Deligne and Ihara) to the action of $\mathrm{Gal}(\overline{\mathbf{Q}}/\mathbf{Q})$ on $\pi_1^{\mathrm{alg}}(\mathbf{P}^1\overline{\mathbf{Q}} \setminus \{0, 1, \infty\})$, where the base point is the tangent vector d/dx at 0.

Teichmüller space from a viewpoint of group representations: *Kyoji Saito.* The Teichmüller space $T_{g,n}$ for $g \in \mathbf{Z}_{\geq 0}$ and $n \in \mathbf{Z}_{\geq 0}$ is the "universal covering space" of the moduli space $M_{g,n}$ of the pairs $(C, \underline{p})$ where C is a smooth projective curve defined over $\mathbf{C}$ and $\underline{p}$ is an n-tuple of mutually distinct marked points $\underline{p} = \{p_1, \dots, p_n\}$ on C. In particular we write T_g when $n = 0$.

Following an original idea of Fricke, Helling and Seppälä, we see that $T_g \times (\mathbf{Z}/2\mathbf{Z})^{2g}$ can be naturally identified with an open and closed subset of a real affine algebraic variety $\mathrm{Hom}(R_g, \mathbf{R})$, where R_g is a Noetherian ring given by

$$\mathbf{Z}[s(\gamma); \gamma \in \Pi_g]/\big(s(e) - 2, s(\gamma\delta) + s(\gamma^{-1}\delta) - s(\gamma)s(\delta); \gamma, \delta \in \Pi_g\big)$$

where $\Pi_g = \langle a_1, \dots, a_g, b_1, \dots, b_g; \prod_{i=1}^g [a_i, b_i] = 1\rangle$. (A similar description also holds for $T_{g,n}$ but we omit it.) This description naturally leads to an action of the group $\mathrm{Out}^+(\Pi_g)$ (known as the mapping class group) on T_g and a description of complex and Kähler structures on T_g which are well-known in the classical theory of Teichmüller space developed by Ahlfors and Bers. We do not know a characterization of the ring R_g. Hopefully, it will give a smallest (in a proper sense) real coordinate ring of the space T_g.

$\widehat{GT}$ **and automorphisms of braid towers:** *Leila Schneps.* In the *Esquisse d'un Programme*, Grothendieck suggests studying a tower of fundamental groupoids (Teichmüller groupoids) of moduli spaces of Riemann surfaces of genus g with n marked points and homomorphisms coming from degeneracy of these surfaces, and comparing the automorphism group of this tower with the absolute Galois group $\mathrm{Gal}(\overline{\mathbf{Q}}/\mathbf{Q})$. As a first step towards understanding this idea, we define a simpler tower consisting of Artin braid groups and certain particular homomorphisms between them, and show that the automorphism group of this tower is the Grothendieck-Teichmüller group $\widehat{GT}$ defined by Drinfel'd, which contains $\mathrm{Gal}(\overline{\mathbf{Q}}/\mathbf{Q})$.

The classification of plane trees by their Galois orbit: *George Shabat.* Some problems of effectivisation of the Grothendieck correspondence between the combinatorial-topological and arithmetical objects, such as understanding the Galois group action on these objects and reduction modulo primes, have been pointed out. They turn out to be already interesting in the case of plane trees.

The results of joint work of the author with Nikolai Adrianov were presented. According to their Galois orbits, all the plane trees are assigned to one of four types. The first is very special and consists of "hedgehogs" (trees of diameter 2) and of "chains" (all the valencies no greater than 2); they correspond to guaranteed one-point orbits of the Galois group. The second consists of some special trees of diameter 6; they correspond to cyclotomic fields. The third consists of all the trees of diameter 4; the corresponding arithmetical structures are not yet clear and are very important to understand. All the rest constitute the fourth class; the main finiteness theorem claims that the trees of this type the size of whose Galois orbit is bounded by R have no more than $8R + 1$ edges.

Some suggestions for the development of a Galois-like combinatorial theory were formulated. The relations with pseudo-elliptic integrals, Toda lattices and analytical capacity theory were mentioned.

Combinatorial structures associated with cartographic groups: *D. Singerman.* We investigate permutation representations of the cartographic group and closely related groups, the oriented cartographic group, the hypercartographic group and the oriented hypercartographic group. These give rise to all maps and hypermaps on all surfaces; these surfaces may be non-orientable and may have boundary.

We also describe the universal map. This is a map that covers all maps on all surfaces.

This talk is closely related to the one by Gareth Jones.

Shabat Polynomials for Trees of Diameter 4: *Alexander Zvonkin.*
Any plane tree can be obtained as an inverse image of a segment via a very special kind of a complex polynomial, the so called *Shabat polynomial.* These polynomials have at most two critical values. Their coefficients are algebraic numbers from some number field which we call the *field of definition* of the tree.

We have studied the particular case of trees of diameter 4. For several infinite families of trees a minimal polynomial of some primitive element of the field of definition is computed. Miraculously, in all the cases considered the discriminant of this polynomial is split completely into factors that are linear combinations of vertex valencies, with the coefficients being non-negative integers that are not greater than the number of times the valency enters into the picture.

The above factorization made it possible to construct explicitly many interesting examples of fields of definition. Infinite series of trees with cyclic cubic fields and purely cubic fields as the fields of their definition are constructed. In the same manner, N.Adrianov (Moscow) constructed trees with arbitrary quadratic field of definition, as well as an infinite series of examples of "combinatorial orbits" that are split into more than one Galois group orbit.

The above factorization is verified (with the aid of computers) for more than 20 infinite series of trees and is conjectured in general case. It implies that the primes of "bad reduction" are not greater than the number of edges of the tree.

Noncongruence Subgroups, Covers and Drawings.

Bryan Birch*

The theory of congruence subgroups, more precisely the theory of the action of congruence subgroups of the modular group on the upper half plane, is an area of mathematics in which the mathematical structure is well understood and wonderfully intricate, and beautiful numbers appear as if by magic (see for instance [GZ]). When one turns to noncongruence subgroups, the mathematical structure must be built with fewer bricks, since in particular the Hecke theory is missing, but though the structure is more mysterious it seems to be almost as rich, and the 'ballet of numbers' continues to be just as beautiful. There is an enormous literature on the action of general discontinuous groups on the upper half plane; and there is an enormous literature concerned with the arithmetic theory of the action of *congruence* subgroups of the modular group; in contrast, the arithmetic theory of subgroups which are not congruence subgroups is surprisingly little developed. In a neighbouring area, there is a beautiful corpus of recent work concerned with the arithmetic of Galois coverings of the projective line, ramified in a prescribed way above a finite set of places; a motivation for this has been its application to the inverse Galois problem. As I understood before the Luminy meeting, the theory of dessins was a drawing together of these various strands; as I learnt at Luminy, I had underestimated the number of strands!

This essay is a summary (or a concatenation) of various talks I have given on the subject during the past couple of years. Each time, I became more and more conscious that, because the various fields involved are quite large, there were always members of my audience who knew much more about some part of what I was saying than I did. At the same time it was clear that though various groups of people were working on quite closely related problems, the points of view were very different, communication between groups was rather poor, and so theorems and methods were invented and reinvented independently on several occasions. I thank the audiences (at Essen, Caen, OberWolfach, Tempe and elsewhere) and in particular Tony Scholl and Mike Fried for what they have taught me. In what follows, I have tried to get the history right, but this is well nigh impossible; I apologise in advance to those who think I have got it wrong, and refer to other papers in this volume.

Section 1. Drawings, and equivalent objects.

First, we spend a paragraph fixing some standard notation. We denote by H the upper half plane $\{z = x + iy | y > 0\}$; $\overline{C}$ will be the Riemann sphere $C \cup \{\infty\}$, which has the same points as the complex projective line $P^1(C)$; and as usual $\Gamma(1)$ is the modular group

$$\left\{ z \to \frac{az + b}{cz + d} \middle| a, b, c, d \in \mathbf{Z}, ad - bc = 1 \right\},$$

acting on H and on $\overline{H} = H \cup Q \cup \{\infty\}$. Consider the congruence subgroup $\Gamma(2)$, consisting of the modular transformations with b, c even; it has fundamental region

$$\mathcal{F}_2 \; : \; \{-1 < x \le 1, x^2 + x + y^2 > 0, x^2 - x + y^2 \ge 0\},$$

and is the free group generated by the transformations $z \to z + 2$, $z \to \frac{z}{2z+1}$ that identify the sides of $\mathcal{F}_2$; it has three cusps, equivalent to $0, 1$ and ∞ (the spikes of $\mathcal{F}_2$ at -1 and 1 are of course equivalent to each other). The quotient $H/\Gamma(2)$ is a Riemann surface of genus 0 punctured at three points corresponding to the cusps; its closure $\overline{H}/\Gamma(2)$ is a Riemann sphere, which we may uniformise by a function $J(z)$. In the literature, this particular function is more usually called λ; like everyone else, I denote the standard modular function by $j(z)$, and I tend to use the notation $J(z)$ for the uniformising function of any triangle group other than $\Gamma(1)$. This determines the function $J(z)$ up to a Möbius transformation; we may fix any three distinct points in $P^1(C)$ to be the images of the cusps, so we take $J : (\infty, 0, 1) \to (\infty, 0, k)$ for suitable non-zero k and then $J(z)$ is determined uniquely. It is convenient to refer to the complex projective line, parametrised by J, as the J-line. Note that the constant k could have been simply taken as 1, but it is more convenient to leave it available to be chosen to make things look nice when we work out some equations (in the context of $\Gamma(1)$, $k = 1728$ is traditional).

It is convenient to state a theorem, which says that various families of objects are in 1-1 correspondence, so may be regarded as different starting points for a single theory.

Theorem 1. *For each positive integer n, the following families of objects are in 1-1 correspondence:-*
(i) Triples $(\mathcal{R}, \phi, O)$ where $\mathcal{R}$ is an n-sheeted Riemann surface, $\phi : \mathcal{R} \to \overline{C}$ is a covering map branched at most above $\{\infty, 0, k\}$, and O is a point of $\mathcal{R}$ above 0;
(ii) Quadruples $(\sigma_\infty, \sigma_0, \sigma_1; \nu)$ where $\sigma_\infty, \sigma_0, \sigma_1$ are permutations of S_n such that $\sigma_\infty \sigma_0 \sigma_1 = id$ and such that the group generated by σ_0, σ_1 is transitive on the symbols permuted by S_n, and ν is a marked cycle of σ_∞; all modulo equivalence corresponding to simultaneous conjugation by an element of S_n;

(iii) Subgroups $\Gamma \leq \Gamma(2)$ of index n, modulo conjugacy by translation;
(iv) Drawings with n edges.

Remark. The families considered in this theorem may not quite correspond to those considered by other speakers at the conference; in particular, the data describing a covering includes the marked point O of $\mathcal{R}$ above ∞. My reason for doing it this way is mainly that my original interest was in subgroups of the modular group — and there the cusp at ∞ plays a very special role. There are further advantages, since marking the point O tends to kill unwanted automorphisms; so the statement and proof of Theorem 2 below becomes somewhat simpler, and the same is likely to apply to counting theorems.

I have not yet said what I mean by a *drawing*, so we must clearly leave (iv) to the end. The equivalence of the first three families is well known and goes back to Hurwitz, if not Riemann; we sketch the reasons.

First, let us start with a subgroup of $\Gamma(2)$ of index n; let coset representatives for $\Gamma(2)/\Gamma$ be $\{\gamma_1 = id, ..., \gamma_n\}$; stick together the regions $\gamma_i\mathcal{F}_2$ for $i = 1, ..., n$ by identifying edges by means of Γ and compactify by filling in the cusps; then we get an n-sheeted Riemann surface $\mathcal{R}$, there is a natural projection ϕ of $\mathcal{R}$ onto the J-line $\overline{\mathbf{H}}/\Gamma(2)$, and there is a marked point $O \in \mathcal{R}$ corresponding to the cusp at ∞ of $\overline{\mathbf{H}}/\Gamma$. Also, we obtain permutations σ_0, σ_∞ in S_n by going round little anticlockwise loops on $\overline{\mathbf{H}}/\Gamma$ round the cusps above $0, \infty$ respectively, and considering the effect on the set $\{\gamma_i\mathcal{F}_2\}$; and we take σ_1 so that $\sigma_\infty\sigma_0\sigma_1 = id$. (We chose $\mathcal{F}_2$ so that it is neat to define σ_0, σ_∞ this way; σ_1 is a little less tidy since the corresponding cusp is split as ± 1 in $\mathcal{F}_2$.) The cycles of σ_0 (resp. σ_∞, σ_1) correspond to the points of $\overline{\mathbf{H}}/\Gamma$ above $J = 0, \infty, k$, and there is a marked cycle in σ_∞ corresponding to going round the cusp ∞ of $\overline{\mathbf{H}}/\Gamma$.

To go back again, we may start with a triple of permutations and obtain a surface by taking copies of $\mathcal{F}_2$ labelled from 1 to n and sticking them together round 0 and ∞ as directed by the corresponding permutations; by standard theory, the compactification of this surface is a Riemann surface $\mathcal{R}$, and the points of $\mathcal{R}$ above the branch points clearly correspond to the cycles of the permutations σ. To get from (i) to (iii), we recollect that $\Gamma(2)$ may be identified with $\pi_1(\overline{\mathbf{C}} - \{0, k, \infty\}, z_0)$ for any $z_0 \in \overline{\mathbf{C}}$ — it is a free group on the generators $z \to z + 2$ and $z \to z/(2z + 1)$ — so if we are given $\mathcal{R}, \phi, O$ and a basepoint ε on $\mathcal{R}$ then $\pi_1(\mathcal{R} - \phi^{-1}(\{\infty, 0, k\}), \varepsilon)$ is a subgroup of $\pi_1(\overline{\mathbf{C}} - \{0, k, \infty\}, \phi(\varepsilon))$, leading to a subgroup Γ of $\Gamma(2)$ determined up to conjugacy; by taking a conjugacy, we may ensure that the cusp of Γ at ∞ is the marked one, and then Γ is determined up to conjugating by a translation. (We may also go from $(\mathcal{R}, \phi, O)$ to a triple of permutations by cutting and going round little paths.)

Note that if $\sigma_0^3 = \sigma_1^2 = id$ then a similar procedure may be made to lead to a subgroup of the full modular group $\Gamma(1)$, which is freely generated by $z \to -1/z$ of order 2 and $z \to -1/(z+1)$ of order 3. These are of course the transformations that identify the edges of the standard fundamental region

$$\mathcal{F}_1 \; : \; \left\{ -\frac{1}{2} < Re(z) \leq \frac{1}{2}, |z| \geq 1, Re(z) \geq 0 \text{ if } |z| = 1 \right\}.$$

It is simple to check that with the equivalences as given, the correspondences really are one to one.

After this routine stuff, I must say what I mean by a *drawing*. I will be descriptive rather than formal.

A drawing is a finite connected 1-complex which is the skeleton of an oriented 2-complex. It has two sets of vertices, coloured respectively black and white, and a set of edges. Each edge is incident to a single vertex of each colour, and for each vertex there is a cyclic ordering of the edges incident to it.

Even more informally, one may draw a little picture on a plane piece of paper (or, as Cori did in his lecture, on the blackboard); the white and black vertices are represented by small noughts and crosses, and we join them up by edges as specified by the cyclic orderings; since we are not assuming that this 1-skeleton is a planar graph, it may of course happen that edges cross.

From a drawing, one may immediately read off permutations σ_0 and σ_1; one labels the edges, and then σ_0 is the product of the cycles corresponding to the white (nought) vertices, and σ_1 is the product of the cycles corresponding to the black (cross) vertices. One may then define σ_∞ to be $\sigma_1^{-1}\sigma_0^{-1}$.

The cyclic ordering of the edges incident to each vertex enables us to regain the oriented surface $\mathcal{D}^*$ of which a drawing $\mathcal{D}$ is the 1-skeleton. In detail, in order to obtain a face of $\mathcal{D}^*$, we select a white vertex v_0 of $\mathcal{D}$ and an incident edge, e_0 say; we obtain a closed polygon by leaving v_0 along e_0, turning left at the far end of e_0, and proceeding by going along each edge we come to and turning left at each vertex, and stopping when we return to v_0 and our next left turn would take us out again along e_0; filling in the polygon gives a face of $\mathcal{D}^*$. We repeat the process until every directed edge has been used in some such closed polygon; the surface $\mathcal{D}^*$ obtained by filling in all these closed polygons by faces is clearly a closed oriented surface with $\mathcal{D}$ as 1-skeleton. The permutation σ_∞ may now be described equivalently as a permutation of the edges corresponding to going round the faces of $\mathcal{D}^*$; but we should be a little careful about the details, or else there is a danger of using the edges twice. In detail, we should label the edges by placing the label on the *left* of each edge as we go from the white to the black end. The permutations σ_0 and σ_1 then correspond respectively to going round the white and the black vertices anticlockwise, and the permutation σ_∞ corresponds to going the faces turning left at each corner.

(The reader should draw some pictures. It will be found that one goes round the outside face clockwise, but one goes round inside faces in an anticlockwise direction, at least when they 'lie flat' in the plane).

There is automatically a marked cycle of σ_∞, corresponding to going round the outside face. Accordingly, a drawing $\mathcal{D}$ gives a quadruple $(\sigma_0, \sigma_1, \sigma_\infty; \nu)$ very easily, and the permutation group generated by the σ's is transitive because the drawing is connected. Conversely, from such a triple of permutations with one cycle marked one may clearly regain the drawing. This is enough to prove the theorem (I realise I have not said when two drawings are equivalent: it is probably best to be unscrupulous, and say that drawings are equivalent when they correspond to equivalent permutation data). It is nice to note a nice way of going from a subgroup Γ to a drawing:— Take n copies of $\mathcal{F}_2$, and cover the semicircular parts of the boundaries with red ribbons. Stick the copies together, just as we do when going from Γ to the corresponding Riemann surface $\mathcal{R}$; think of this surface as being flexible, and put it down on a flat surface, being careful to preserve the orientation, particularly near the vertices. Puncture it at the points above ∞, and allow it to collapse onto the red ribbons: we have constructed the drawing!

Despite this pretty construction, it is clear that a drawing as I have defined it contains no more and no less information than a triple of permutations (with a marked cycle), and that this is true for rather trivial reasons. However, even from the very particular point of view from which I myself approached them — a nice way of getting on terms with subgroups of the modular group — there are considerable advantages. In particular, compared with other ways of specifying subgroups, they are friendly and fun; when n is small enough for the human eye to be a useful tool for looking at them, they are easy to tell apart, easy to count, and almost acquire individual characters. They have further virtues — Shabat has spoken much more eloquently than I on such matters — in particular, Zvonkin and his collaborators have shown how drawings give insight into the descriptive geometry of the corresponding Riemann surfaces.

My drawings are of course pretty much the same as what Shabat and Voevodsky [SV] call a dessin: their dessins are a particular case of drawings where σ_1 has to be a product of 2-cycles, with no fixed point. Conversely, every drawing may be made into a dessin by re-colouring all the vertices black, and putting a new white vertex in the middle of each edge. I (obviously) prefer my own version, with a marked point, partly because I am used to it, but mainly because of the equivalences described in Theorem 1. From some points of view, one may make a good case for restricting oneself to drawings (dessins) for which $\sigma_0^3 = \sigma_1^2 = id$, since, as we have remarked, these correspond to subgroups of the full modular group $\Gamma(1)$, rather than subgroups of its subgroup $\Gamma(2)$; moreover, these drawings tend to be the prettiest! Referring to Grothendieck's Esquisse [G2], it seems that drawings are possibly a bit closer to what he meant by dessins d'enfants than are the Shabat-Voevodsky dessins. The idea of representing subgroups of modular groups (or coverings) by little pictures, under various names, has occurred to lots of people: the earliest time I met them was in a manuscript of Tom Storer, dated about 1970; and the drawings for the subgroups $\Gamma_0(N)$ of the modular group occur in the thesis of Dave Tingley [T]. (After the case $\Gamma_0(23)$, he referred to them affectionately as 'cows' stomachs'.)

Having set up our apparatus, we list three Questions, essentially the ones asked by Atkin and Swinnerton-Dyer [ASD].

Question 0. Describe how to list the subgroups of $\Gamma(2)$, in order (say) of increasing index.

Question 1. The Riemann surfaces $\mathcal{R}$ are algebraic curves, and the projections ϕ are algebraic maps. What can we say about the equations of these curves and maps; in particular, what is their field of definition?

Question 2. For any subgroup Γ of finite index in $\Gamma(1)$, and any integer $k \geq 1$, we may define cusp forms of weight $2k$ on $\mathbf{H}/\Gamma$. Such a form $f(z)$ has a Fourier expansion $f(z) = \sum_1^\infty a_r q_\mu^r$, where $q_\mu = \exp(2\pi i z/\mu)$ and $z \to +\mu$ is the least translation in Γ. What can be said about the coefficients a_n? (In the congruence subgroup case, the Hecke algebra action implies that there is a basis of 'eigenforms', whose coefficients are multiplicative.)

In trying to answer such questions, one can work at more than one level. One likes to prove general theorems; but one also wants to establish a decent sized library of particular cases, so that one can see what really happens. Most of all, one would like a good geometric interpretation for the theory.

Question 0 has been answered in effect by (ii) of the Theorem, since it is not too hard to train a computer to list pairs of permutations up to conjugacy; for small n, drawings give a more enjoyable enumeration method; for general n, one should refer to the beautiful enumeration theorems of Itzykson and his collaborators [BIZ]. In Section 2, we will recall what is known in general about the nature of the curves $\mathcal{R}$ and the covering maps ϕ. Almost the only answers to Question 2 are due to Tony Scholl; we recall some of his results in Section 3. Finally, in Section 4, we say a very little about calculation, and give tables of examples.

Section 2. Equations.

In this section, we will consider what the covering $(\mathcal{R}, \phi, O)$ is like as an algebraic curve covering the projective line — we would like to know its field of definition, and the nature of the equations defining it. So we want to know how these equations depend on, for instance, the subgroup Γ or the drawing $\mathcal{D}$ that correspond to the curve according to the recipes of Theorem 1, and in particular we would like to know whether there are intrinsic restrictions on the pair $(\mathcal{R}, \phi)$.

We will shortly state two basic theorems and a disagreeable fact (Theorems 2,3 and 4), but first we pay our respects to the remarkable theorem of Belyi [B]. This tells us that the Riemann surface $\mathcal{R}$ is almost unrestricted: if C is any curve defined over a number field then there is a covering map $\psi : C \to P^1$ defined over the algebraic numbers ramified above just three points; so by Theorem 1 there is a subgroup $\Gamma \subseteq \Gamma(2)$ corresponding to the pair (C, ψ) so that $\overline{\mathsf{H}}/\Gamma$ is a model for the curve C. At first sight, this seems to kill the problem: but it doesn't really. The map ψ constructed by Belyi may be complicated even when C is quite simple, though with care one may ensure that ψ is defined over a field of definition for C. Accordingly, though C by itself is almost unrestricted, the pair C, ψ is fairly special.

We need some notation. If $\mathcal{D}$ is a drawing and $(\sigma_\infty, \sigma_0, \sigma_1; \nu)$ is the associated quadruple of permutations, the *type* $\tau(\mathcal{D})$ is the data $(\pi_\infty, \pi_0, \pi_1; \mu)$, where each π_i is the partition of n corresponding to the cycles of σ_i (that is, to the valences of the faces and of the two sorts of vertices of $\mathcal{D}$) and μ is the length of the marked cycle ν of σ_∞ (that is, the number of sides of the outside face of $\mathcal{D}$, which is the same as the width of the cusp at ∞ of $\overline{\mathsf{H}}/\Gamma$; Since we have several equivalent families of object, the type may of course be described in several different ways!)

We denote by G the group generated by $\{\sigma_\infty, \sigma_0, \sigma_1\}$; we may think of G as just an abstract group, but strictly speaking it is a subgroup of S_n determined up to conjugacy. We denote by E the extension of $\mathbf{Q}$ generated by the characters of the σ's , viewed as elements of G.

With these notations, we state the standard theorem that asserts that there is a minimal field of definition for a model of $(\mathcal{R}, \phi, O)$, and gives an upper bound for its degree.

Theorem 2. *Let $\mathcal{D}$ be a drawing; write $M(\mathcal{D})$ for the number of drawings $\mathcal{D}'$ with $\tau(\mathcal{D}') = \tau(\mathcal{D})$ and $G(\mathcal{D}') = G(\mathcal{D})$. Then there is a minimal field of definition K such that the corresponding covering $(\mathcal{R}, \phi, O)$ has a model* *defined over K, and $[K : \mathbf{Q}] \leq M$.*

This theorem was stated, and proved in the case $g = 0$, by Atkin and Swinnerton-Dyer [ASD]; a correct proof for any genus of a theorem essentially containing it was given by Fried [F]. (M is closely connected to Fried's Hurwitz number.) Since then, similar theorems have been proved and re-proved many times, but unhappily I have been unable to find this particular statement in print; the theorem proved by Couveignes elsewhere in this volume is very similar, and actually implies Theorem 2; the statement is a trifle simpler than that of Couveignes because of the marked point. Theorem 2 as stated is not hard to prove using Weil's theory of descent, cf [W]; we give a brief sketch. Take a model $(\mathcal{C}, \pi, O)$ for $(\mathcal{R}, \phi, O)$ defined over a number field L; we may assume that L is normal over $\overline{\mathbf{Q}}$, and by a translation we may assume that O is a rational point of $\mathcal{C}$. Up to now, $(\mathcal{R}, \phi, O)$ has stood for an equivalence class of models, so in order not to abuse notation too much we had best use a different notation to stand for its models. Let us suppose first that $(\mathcal{R}, \phi, \mathcal{O})$ has no automorphisms (this is the usual case). If $\sigma \in \mathrm{Gal}(\overline{\mathbf{Q}}/\mathbf{Q})$ it takes the model $(\mathcal{C}, \pi, O)$ to $(\mathcal{C}^{\sigma}, \pi^{\sigma}, O)$, corresponding to a drawing of the same type. There are at most M possibilities; so, $\mathrm{Gal}(\overline{\mathbf{Q}}/\mathbf{Q})$ has a subgroup, V say, of index at most M, such that each $\sigma \in V$ maps the equivalence class $(\mathcal{R}, \phi, O)$ to itself; write K for the fixed field of L. If $\sigma \in V$ then it maps $(\mathcal{C}, \pi, O)$ to $(\mathcal{C}^{\sigma}, \pi^{\sigma}, O)$ where there is an isomorphism $f_{\sigma} : \mathcal{C}^{\sigma} \to \mathcal{C}$ fixing O with $\pi^{\sigma} = \pi \circ f_{\sigma}$. We are now essentially in the situation treated in [W]; following Weil's arguments, we obtain a model of $(\mathcal{R}, \phi, O)$ defined over K. It remains to deal with the case where $(\mathcal{R}, \phi, O)$ has an automorphism: but this is easy. It is convenient to picture $(\mathcal{R}, \phi, O)$ as $(\overline{\mathrm{H}}/\Gamma, \infty)$; any possible group of automorphisms is generated by a translation $z \to z + r$, with r dividing μ, the width of the cusp ∞; and now $(\overline{\mathrm{H}}/\Gamma, \infty)$ is a cyclic cover of degree μ/r of $(\overline{\mathrm{H}}/\Gamma', \infty)$, where Γ' is generated by Γ and the translation $z \to z + r$.

The statement of Theorem 2 differs from the more usual statements in the literature in important ways. In the literature, the motivation is usually the application to the inverse Galois problem, so that one wishes to construct a *regular* model for the *Galois closure* of the function field of $\mathcal{R}$; more precisely, one wants a normal extension $\mathcal{L}$ of $L(J)$, for an appropriate number field L, so that $\mathrm{Gal}(\mathcal{L}/L(J)) \cong G$ and $\overline{\mathbf{Q}} \cap \mathcal{L}$ is L; if this can be done then Hilbert irreducibility allows one to specialise J and obtain a Galois extension M/L with $\mathrm{Gal}(M/L) \cong G$, and if it can be done with $L = \mathbf{Q}$ then one has solved the inverse Galois problem for the group G. However, to construct such a regular model for the Galois closure, it is necessary to include the values of the characters in the base field, so all we are allowed to conclude is that there is a Galois model defined over a field L with L an extension of E

(rather than of **Q**) of degree at most M. (For an explicit example of what is happening see for instance [Se2] p86.) In our present context, we are not looking for a normal extension of function fields, only for a model of the covering; accordingly, we do not need to adjoin the values of the characters to the base field.

Theorem 2 is also less general than the versions most usually to be found in the literature, where such results are most usually stated for a covering with three *or more* branch points — and often in even greater generality, for a covering of a curve with genus possibly exceeding 0; the most conveniently accessible version is possibly that of Matzat [Ma], but there is a book in preparation by Matzat and Malle that is probably to be preferred. Note finally that it is important to specify a marked point O, otherwise, as Couveignes shows, Theorem 2 simply isn't true.

The next theorem limits the set of bad primes of the covering, and in particular limits the ramification of the extension $K/\mathbf{Q}$; it was in essence stated by Grothendieck at the end of [G1], and proved by Fulton [Fu]; and it lies at the roots of Grothendieck's theory of the arithmetic fundamental group [G2]. The language used is a bit different, and it is quite hard for a non-geometer to be sure that one is quoting correctly. A clear proof of the theorem we want, with almost the same statement, was given by Sybilla Beckmann in her thesis, and published in [B]; since there are many theorems of Grothendieck, might it be convenient to refer to this one as the Grothendieck-Beckmann theorem?

Theorem 3. *Let K be the minimal field of definition as in Theorem 2, and suppose that $\wp$ is a prime ideal of K that does not divide the order of G. Then $(\mathcal{R}, \phi, O)$ has a model defined over K with good reduction modulo $\wp$; moreover primes that do not divide the order of G are never ramified in K.*

The crux of the matter is in the final phrase (and the crux of the proof is an application of Abhyankar's lemma): given this, the rest of the result follows, essentially by the theorem of purity — but one may use instead the Riemann-Hurwitz formula.

Beckmann actually proves a lot more than Theorem 3; in particular, she considers coverings ramified above more than three points, when of course primes modulo which two of the branch points coincide will be additional primes of bad reduction for the covering; and she obtains quite detailed information about the various specialisations of the field extensions involved. Her theory is stated in terms of Galois coverings, that is to say in terms of regular normal extensions, so the primes that ramify in E are likely to be bad; however, one doesn't need to put this into the statement of Theorem 3, since the primes that ramify in E certainly divide the order of G.

To one who was brought up on [G2] it may be that Theorem 3 is un-surprising; however, if one approaches the coverings by computing a few of them, it seems quite extraordinary. Note that it doesn't just say that the only primes that can ramify in K are those dividing $\#(G)$, one also con-cludes that in case $\mathcal{R}$ has positive genus its conductor involves only primes in $\#(G)$, and that the covering map ϕ never branches or degenerates except above $\{\infty, 0, 1\}$ and modulo primes dividing $\#(G)$. These are very strong conditions, and a model satisfying them comes as a welcome relief after a typically tortuous computation involving rather disgusting numbers!

A particularly beautiful application for Theorem 3 was exhibited by A. Zvonkin in his lecture at the conference. He discussed Shabat trees of diam-eter 4: these are drawings with one white vertex of valency n, $N - n$ white vertices of valency 1, and n black vertices of valencies $l_1, ..., l_n$ say, where $\sum l_i = N$ and of course $N - n, n, l_1, ..., l_n$ are all natural numbers. He ob-serves (by calculation) that the discriminant of the field of definition of such a dessin is a product of linear forms in $l_1, ..., l_n$ with coefficients 0 or 1 ! This actually follows from Theorem 3: the point is that (for generic $l_1, ..., l_n$) the discriminant D (or more precisely a square multiple of the discriminant) is calculated as a polynomial $D(l_1, ..., l_n)$ in $l_1, ..., l_n$, and by Theorem 3 D is never divisible by a prime exceeding N. By a standard theorem essentially due to Chebyshev, if the polynomial $D(X_1, ..., X_n)$ had an irreducible factor of degree exceeding 1, then one could choose natural numbers $l_1, ..., l_n$ so that $D(l)$ had a prime factor exceeding $\sum l_i$. This would contradict The-orem 3, so all the factors of $D(X)$ are linear; and by applying Theorem 3 again we see that the coefficients of these linear forms may only be 0 and 1.

Theorem 3 has an easy partial converse, which I state as a theorem though it doesn't deserve it.

Theorem 4. *If $\wp$ divides the order of one of the permutations σ then the modulo $\wp$ reduction of ϕ is bad.*

The answers given by Theorems 3 and 4 are by no means complete, but it is not easy to guess what more can be said. The examples at the end make it clear that sometimes primes which divide $\#(G)$ but do not divide the order of any σ are good — and sometimes they are bad. Serre ([Se2] p90) asked in particular about the set $S(\mathcal{R})$ of bad primes for the curve $\mathcal{R}$, in cases where it has genus at least 1; he seems mainly interested in cases where the field of definition K is $\mathbf{Q}$. The examples at the end make it clear that $S(\mathcal{R})$ is likely to be hard to predict: there are examples where $S(\mathcal{R})$ contains all the primes dividing $\#(G)$, there are cases where it contains only primes dividing the orders of the branch cycles, and there are even cases where it contains a prime that does not divide the order of any branch cycle but it fails to contain all the primes that do.

I end this section with a somewhat malicious question. By theorem 4, any prime dividing the order of a branch cycle is bad for the covering: so if we specialise J to a value in K, it is clear that such a prime is likely to be wildly ramified in the corresponding field extension. So extensions of $\mathbf{Q}$ with given Galois group constructed by the rigidity method are likely to be wildly ramified. We ask:

Problem. *Given a finite group G, is there a tamely ramified normal extension $F/\mathbf{Q}$ with $\mathrm{Gal}(F/\mathbf{Q}) \cong G$?*

It is easy to construct such an extension when G is abelian, dihedral or a symmetric group; and I have no reason to suppose that Shafarevich's construction of extensions with G a given soluble group cannot be adapted to give a tamely ramified extension; indeed, I have been told that in applying Shafarevich's methods it is important to keep the ramification tame. However, non-abelian simple groups are liable to be difficult. (Serre was able, immediately, to show how to construct tame A_5 extensions by the method of Emmy Noether; but that such a construction is possible is quite a deep fact.)

The examples in [Be3] help to put this question in context.

Section 3. Modular forms.

In this section, Γ will be a subgroup of the modular group $\Gamma(1)$; in the language of Section 1, it corresponds to a drawing with all white vertices having valencies dividing 3, and all black vertices having valences dividing 2; but we must immediately confess that this section is not really about drawings or coverings, except in so far as they specify subgroups. In this section we are concerned with the spaces of modular forms on $\overline{\mathbf{H}}/\Gamma$, and these don't just depend on a covering $(\mathcal{R}, \phi, O)$ of a projective line, they depend on the specific model, namely $\overline{\mathbf{H}}/\Gamma$, that $(\mathcal{R}, \phi, O)$ is a covering of.

We recollect that, for integral k, the space $S_{2k}^0(\Gamma)$ of cusp forms of weight $2k$ consists of forms f(z) holomorphic in $\mathbf{H}$ that satisfy
(i) $\left. \left(f(z)dz^k \right) \right|_{\gamma} = f(z)dz^k$ for each $\gamma \in \Gamma$,
(ii) at each cusp, $f(z)$ has an expansion with no constant term in terms of the relevant local parameter.

In particular, $f(z)$ has a Fourier series

$$f(z) = \sum_{1}^{\infty} a_n q_{\mu}^n$$

where μ is the least positive integer such that Γ contains the translation $z \to z + \mu$, and $q_\mu = exp(2\pi i/\mu)$.

Just as in the classical case, $S^0_{2k}(\Gamma)$ is a finite dimensional vector space whose dimension is easy to calculate using the Riemann-Roch theorem. We would like to know about the coefficients a_n; in the case where Γ is a congruence subgroup of $\Gamma(1)$, Hecke theory tells us that S^0_{2k} has a basis consisting of eigenforms, whose coefficients a_n are multiplicative functions of n. In the non-congruence case no similarly neat and tidy result is true; however, guided by experiment, Atkin and Swinnerton-Dyer [ASD] conjectured that there should be a p-adic analogue, and proved it (with the aid of a computer error!) in the case $k = 1$.

Suppose for a moment, for simplicity, that S^0_{2k} has dimension 1, spanned by $f(z)$ with a_1 normalised to be 1; and suppose that the covering $\mathcal{R}, \phi, O$ is defined over K. One might think that the coefficients a_n should be in K, but that is not usually true; there is a constant κ with $\kappa^\mu \in K$, where μ is the width of the cusp at ∞, such that each $a_n = \kappa^{n-1} b_n$ with, in this case, $b_n \in K$. Suppose for simplicity that K is $\mathbf{Q}$, as is the case in all cases that have been worked out. Their computations led Atkin and Swinnerton-Dyer to the conjecture that for each 'good' prime p (where 'good' means prime to $|G|$) there is an integer $A(p)$ such that *if $p^\alpha|n$ then*

$$a_{pn} - A(p)a_n + p^{2k-1}a_{n/p} \equiv 0 (\mathrm{mod}\ p^{(\alpha+1)(2k-1)}),$$

with the standard convention that $a_{n/p}$ is taken as zero if n/p is not an integer. Here, they do not assert that the a_n's are integers, just that the formula makes sense. (The powers of κ do not actually cause trouble. The big difference from the classical case is that in general the a_n are not integral; indeed, it has been conjectured that the coefficients a_n turn out to be algebraic integers if and only if G is a *congruence* subgroup.) Atkin and Swinnerton-Dyer assert further that the $A(p)$'s have certain properties analogous to properties of the coefficients of Hecke eigenforms in the congruence subgroup case; in particular, $|A(p)| \leq 2p^{k-\frac{1}{2}}$.

The case $k = 1$ of their conjecture turned out to be provable with not too much difficulty; the best proof, given by Cartier [Ca], uses formal groups; and the $A(p)$'s were those associated with the L-function of the curve $\mathcal{R}$. The general case remained open for several years until it was resolved by Scholl using crystalline cohomology.

Scholl [Sch1,Sch2] confirmed the conjectures; the result, when the dimension of S_{2k} is d, is that there are integers $A_1(p), ..., A_{2d}(p)$ so that for every $f \in S^0_{2k}$ with $a_1 = 1$ and every n with $p^\alpha|n$

$$b_{np^d} + A_1(p)b_{np^{d-1}} + ... + A_{2d}(p)b_{n/p^d} \equiv 0 (\mathrm{mod}\ p^{(\alpha+1)(2k-1)}).$$

There is thus an L-function

$$L(\Gamma, 2k, s) = \prod_p \Big(\sum_{r=0}^{2d} \frac{A_r(p)}{p^{rs}} \Big)^{-1}$$

associated with Γ and the weight $2k$. My understanding is that this L-function is indeed the L-function associated with the Kuga variety with base the curve $\mathcal{R}$, fibred with powers of elliptic curves, just like the associated L-function in the congruence case; and Scholl has understanding of the associated representations. However, the situation appears intrinsically less benevolent than in the congruence case, even if one bypasses the 'p-adic' theory and goes straight to the fairly down-to-earth L-function of the Kuga variety; there is no reason to expect the representations to be direct sums of 2-dimensional ones (as they are in the congruence subgroup case), and when d and k are large one expects the representation to be fairly horrible.

Scholl has worked out at least three examples. For the group $\Gamma_{4,3}$, the unique subgroup of $\Gamma(1)$ of type (43,331,2221;4), Scholl shows that $L(\Gamma_{4,3}, 4, s)$ is very nearly the same as a factor of $L(\Gamma_0(28), 4, s)$ (the dimension of the space of cusp forms of weight 5 on $\Gamma_0(28)$ exceeds 1), but the Euler factor at 7 is different. Similarly, if we denote by $\Gamma_{5,2}$ and $\Gamma_{7,1,1}$ the unique subgroups of $\Gamma(1)$ of types respectively (52,331,2221;5) and (711,333,22221;7), it appears that $L(\Gamma_{5,2}, 4, s)$ and $L(\Gamma_{7,1,1}, 4, s)$ resemble factors of $L(\Gamma_0(35), 4, s)$, and $L(\Gamma_0(14), 4, s)$ respectively. Unfortunately, such recognisable L-functions are unlikely to be typical.

(The equations of the coverings corresponding to these examples are in the tables at the end of this paper.)

Though Scholl's results have laid the field open, there is clearly a great deal more to be understood; more examples could help.

Section 4. Some computations and examples.

In this final section, I give tables containing a number of examples: all drawings with up to 5 edges, and a few others which seem useful to have readily available; I could have given many more, but severe restraint seems appropriate. I give separate tables for genus 0 and genus ≥ 1 cases; in each table, I go through the types in what seems a natural dictionary order; I have avoided duplication of information by almost always sorting the partitions so that π_∞ precedes π_0 precedes π_1; and I will *always* take the 'marked cycle' of σ_∞ as (one of) the longest cycle(s) of σ_∞, so it is not specified explicitly in the tables. Many of the computations were performed a very long time ago, and have been performed by many other people, from [ASD] onwards; in particular, Noam Elkies has shown me nearly as much genus 1 information, and Couveignes, Malle and Zvonkin each own large tables in the genus 0 case; PARI is recommended! When the genus is 0, the covering is fully described by an identity like (5.1) below, and that is often all the information I list; in a few cases, following the precedent set by [ASD], I will only give the field of definition F. Note that the tables have been copied from various sources, rather than just machine output, so they cannot be guaranteed against error.

Section 4.1. The genus 0 case.

The standard algebraic method for finding the equations in the genus 0 case was described long ago by Atkin and Swinnerton-Dyer. We describe how to set up the equations, tedious though it is; solving them is pleasanter. Suppose we are looking for the coverings of type $(\pi_\infty, \pi_0, \pi_1; \mu)$, and for each l suppose that π_0, π_1, $\pi_\infty - \{\mu\}$ have respectively a_l, b_l, c_l parts of size l; so $\sum la_l = \sum lb_l = \mu + \sum lc_l = n$, and if the genus is zero then the total number of parts is $\sum a_l + \sum b_l + 1 + \sum c_l = n + 2$. Suppose that a uniformising parameter for $\mathcal{R}$ is x; we may suppose that O corresponds to $x = \infty$. Above $J = 0$, there are for each l just a_l values of x taken with the multiplicity l; and above $J = \infty$ there are for each l just c_l values of x other than ∞ taken with the multiplicity l. Accordingly, J is, up to a constant multiple, of the shape

$$J = \prod p_l^l \Big/ \prod r_l^l ,$$

where p_l is a polynomial in x of degree a_l and r_l is of degree c_l; normalising x by a multiplicative constant we may suppose that all these polynomials are monic. Similarly

$$J - k = \prod q_l^l \Big/ \prod r_l^l ,$$

where q_l has degree b_l, so we have a polynomial identity of the form

$$\prod p_l^l - k \prod r_l^l = \prod q_l^l \tag{5.1}$$

for which we want to find all the solutions in pairwise coprime polynomials p, q, r of the given degrees (it is in the nature of the covering that no p and q have a root in common, and different p's, etc, are coprime by the way we have defined them). *In the genus 0 case, describing the coverings of the given type is thus equivalent to displaying the solutions of the appropriate identity (5.1).*

There is often considerable scope for finding the equations of the covering by use of geometric ingenuity. When this fails, one gives the polynomials indeterminate coefficients, and sets about the lugubrious process of solving a set of polynomial equations between them. A point to remember is that we have the privilege of translating the variable x, so as to make a single coefficient of one of our polynomials vanish; skilful use of this privilege can save much labour. Another point is that it is sensible to regard k as one of our variables, to be chosen (probably near the end of the computation) to make the identity look nice. It is usually best to start with the Atkin-Swinnerton-Dyer 'differentiation trick' which enables us to get rid of roughly half the indeterminate coefficients before starting serious work: one takes the derivative of (5.1), and then eliminates the k term between (5.1) and its derivative. I work out a couple of easy cases at length, to remind one how the trick works.

Example 1. We work out the type $[n , r.n - r , 21^{n-2} ; n]$ with $2r \neq n$, also discussed by [SV]. We are looking for an identity of the shape

$$(t - a)^r(t - b)^{n-r} - k = (t - c)^2 p_{n-2}(t)$$

where p_{n-2} is monic of degree $n - 2$. Differentiating gives

$$(t-a)^{r-1}(t-b)^{n-r-1}(r(t-b)+(n-r)(t-a)) = (t-c)(2p_{n-2}+(t-c)p'_{n-2}) .$$

Now a, b, c must be distinct, so by unique factorisation the only possibility is

$$r(t - b) + (n - r)(t - a) = n(t - c) ,$$

$$n(t - a)^{r-1}(t - b)^{n-r} = 2p_{n-2} + (t - c)p'_{n-2} .$$

We may as well normalise $a = 0$ by translating t and then we get $b = nd, c = rd$ for some constant d. The last identity then determines the coefficients of p_{n-2}, and the value of k comes out in the wash.

Example 2. The type $[31, 31, 22; 3]$. We want to obtain an identity of the form

$$(x - a)^3(x - b) - k(x - c) = q_2(x)^2 ,$$

where $q_2 = x^2 + dx + e$ is a monic quadratic form. Use our privilege to normalise $a = 0$, and differentiate, then we have a pair of identities

$$x^3(x - b) - k(x - c) = q_2(x)^2, \quad x^2(4x - 3b) - k = 2q_2 q'_2 .$$

Eliminating k gives

$$x^2((4x - 3b)(x - c) - x(x - b)) = q_2(2q'_2(x - c) - q_2)$$

and factorisation leads to

$$3x^2 - 2b - 4c + 3bc = 3q_2, \quad 3x^2 = 2q'_2(x - c) - q_2 ,$$

which is easily solved.

Note however that I did the computation this way to illustrate the differentiation trick — once one has normalised $a = 0$ it is actually quicker to notice that the coefficient of x^2 on the left is zero, so $2e + d^2 = 0$. Taking $d = 2$, $e = -2$, we reach the identity

$$x^3(x+4) - 8\left(x - \frac{1}{2}\right) = (x^2 + 2x - 2)^2 .$$

Table for Genus 0.

Type [1 , 1 , 1]
  $J = x$, of course.

Type [2 , 2 , 11]
  2-cover branched at ∞ and 0 , so $J = x^2$. $G \cong C_2$.

Type [3 , 3 , 111]
  3-cover branched at ∞ and 0 , $J = x^3$. $G \cong C_3$.

Type [3 , 21 , 21]
  $J = (x+m)^2(x-2m) = (x-m)^2(x+2m) - 4m^3$. $G \cong S_3$.

Type [4 , 4 , 1^4]
  4-cover branched at ∞ and 0, $J = x^4$. $G \cong C_4$.

Type [4 , 31 , 211]
  $(x+3)^3(x-1) + 27 = x^2(x^2 + 8x + 18)$ as in Ex 1. $G \cong S_4$.

Type [4 , 22 , 211]
  This is a 2-cover of [2,2,11] , $J = x^2$, ramified above $x = \infty$ and $x = 1$;
  we get $J = (y^2 + 1)^2 = y^2(y^2 + 2) + 1$. $G \cong S_4$.

Type [31 , 31 , 31]
  $J = (x+1)^3(x-3)/x$, $(x+1)^3(x-3) + 16x = (x-1)^3(x+3)$. $G \cong A_4$.

Type [31 , 31 , 22]
  Example 2: $J(x - \frac{1}{2}) = x^3(x+4) = (x^2 + 2x - 2)^2 + 8x - 4$. $G \cong A_4$.

Type [22 , 22 , 22]
  A 2-cover of $J = x^2$ ramified at $x = \pm 1$; this leads to $J = x^2$ where
  $x^2 - 1 = y^2$. We can, if we like, parametrise $x^2 - 4 = y^2$ by $x + y = 2t$,
  $x - y = 2/t$ leading to $Jt^2 = (t^2 + 1)^2 = (t^2 - 1)^2 + 4t^2$. $G \cong C_2 \times C_2$.

Type [5 , 5 , 1^5]
  5-cover branched at 0 and ∞, $J = x^5$. $G \cong C_5$.

Type [5 , 41 , 2111]
  $J = (x+4)^4(x-1) = x^2(x^3 + 15x^2 + 80x + 160) - 256$ as in Ex 1.

Type [5 , 32 , 2111]
$$J = (x+3)^3(x-2)^2 = x^2(x^3 + 5x^2 - 5x - 45) + 108 \text{ similarly.}$$

Type [5 , 311 , 311]
There is a unique drawing with obvious symmetry; $J = (x+3)^3(x^2 - 9x + 24) = 1296 + (x-3)^3(x^2 + 9x + 24)$.

Type [5 , 311 , 221]
$$J = x^3(x^2 + 5x + 40) = (x-3)(x^2 + 4x + 24)^2 + 1728.$$

Type [5 , 221 , 221]
$$J = (x-2)(x^2 + x - 1)^2 = (x+2)(x^2 - x - 1)^2 - 4.\ G \cong D_{10}.$$

Type [41 , 41 , 311]
$$(x+3)^4(x-2) + (135x + 162) = x^3(x^2 + 10x + 30) .$$

Type [41 , 41 , 221]
There are two conjugate coverings, defined over $\mathbf{Q}(i)$, given by $x^4(x-1+2i) - (1-2i)x + 1 = (x+1)(x^2 - (1-i)x + 1)^2$. $G \cong S_5$.

Type [41 , 32 , 311]
A pair of coverings defined over $\mathbf{Q}(\sqrt{6})$.

Type [41 , 32 , 221]
$$x^3(x+5)^2 - \tfrac{27}{4}(5x-2) = (x+6)(x^2 + 2x - \tfrac{3}{2})^2.$$

Type [32 , 32 , 311]
$$x^3(x-5)^2 - (5x-1)^2 = (x-1)^3(x^2 - 7x + 1).$$

Type [32 , 32 , 221]
$$x^3(4x-5)^2 + (5x-4)^2 = (x+1)(4x^2 - 7x + 4)^2.$$

Types [6 , 51 , 21^4], [6 , 42 , 21^4], [6 , 33 , 21^4]
$$x^5(5x+6) + 1 = (x+1)^2(5x^4 - 4x^3 + 3x^2 - 2x + 1),$$
$$x^4(x+3)^2 - 16 = (x+2)^2(x^4 + 2x^3 - 3x^2 + 4x - 4),$$
$$(x^2 - 1)^3 + 1 = x^2(x^4 - 3x^2 + 3).$$

Type [6 , 321 , 31^3]   There are two coverings given by $J = x^3(x+4+3i)^2(x+4) =$
$$(x+3+i)^3(x^3 + (3+3i)x^2 - (3+6i)x + 1 + 7i) + 164 - 152i.$$

Type [6 , 321 , 2211]
Three coverings defined over the field $\mathbf{Q}(\sqrt[3]{2})$

Type [51 , 321 ,321]
There are 7 coverings; an easy symmetric one corresponds to the identity $(x+1)^3(x-4)^2(x+5) - (x-1)^3(x+4)^2(x-5) = 432x$; the other six are conjugates defined over a field F, a quadratic extension of a cubic field $\mathbf{Q}(z)$ where $5z^3 + 5z^2 - 5z + 1 = 0$; we need the square root of $3(z+1)(3z+5)$.

Type [42 , 321 , 321]

Four coverings, defined over the quartic field $\mathbf{Q}(\sqrt{10\sqrt{6}+15})$ and its conjugate.

Type [33 , 321 ,321]

Six coverings defined over the fields $\mathbf{Q}(z)$ with $(z+1)^6 + 16z^3 = 0$.

Type [43 , 331 , $2^3 1$]

The covering $\Gamma_{4,3}$ of [ASD], with equations $J = (x+16)(x^2+4x-3)^3/(7t+4)^3$,

$(x+16)(x^2+4x-3)^3 + \frac{27}{4}(7x+4)^3 = x(x^3+14x^2+\frac{35}{2}x+63)^2$.

Type [52 , 331 , $2^3 1$]

$\Gamma_{5,2}$ of [ASD], with equations $j = (x^2+5x-20)^3(x+20)/(7x+15)^2$,

$(x^2+5x-20)^3(x+20) - 1728(7x+15)^2 = (x-7)(x^3+21x^2+84x+280)^2$.

Type [711 , 3^3 , $2^4 1$]

$\Gamma_{7,1,1}$: $(x^4+8x^3+54x^2+160x+329)^2(x+5) = (x^3+7x^2+35x+77)^3 + 1728(2x^2+7x+49)$.

Higher genus.

In the genus 1 case, it is natural to take x as a function on $\mathcal{R}$ with a double pole at O, and y as a function with a triple pole at O, so that $\mathcal{R}$ is provided with a Weierstrass type equation $y^2 = P_3(x)$, where P_3 is a square free cubic polynomial. If we wish, we may normalise P_3 to be monic, but that may not always be beneficial. If now there is only one cusp above ∞, J is a polynomial in x and y, that is to say, a linear form in y with coefficients polynomials in x.

For most curves, I list the 'name' in Cremona's tables [Cr]; in the context, this is really only meaningful up to twists.

Table for Higher Genus.

Type [3 , 3 , 3]

A 3-cover branched at ∞, 0 and k ; $J = x^3 = k + y^3$. $G \cong C_3$.

Type [4 , 4 , 31]

$J = x^2 + 4x + 18 + y$, $y^2 = 4(2x+9)(x^2+2x+9)$, 48A6.

$(x^2+4x+18)^2 - y^2 = x^4$, $(x^2+4x-9)^2 - y^2 = (x+3)^3(x-9)$. $G \cong S_4$.

Type [4 , 4 , 22]

This is a 2-cover of $J = x^2$ branched above $x = \infty, 0, 1$ and -1; we get $J = x^2$, $y^2 = x(x^2-1)$, 32A2. G is dihedral and the only bad prime is 2.

Type [5 , 5 , 311]

There are two cases:

Case 1 :- $y^2 = (x - \frac{5}{9})(x^2 - \frac{15}{4}x + \frac{15}{4})$, 150A3, $J = xy + (\frac{5}{6}x^2 - \frac{5}{2}x + 1)$,
$(x-1)^5 + (\frac{5}{6}x^2 - \frac{5}{2}x + 1)^2 = x^2y^2 = (x - \frac{5}{3})^3(x^2 + \frac{5}{3}) + (\frac{5}{6}x^2 - \frac{5}{2}x + \frac{25}{9})^2$.

Case 2 :- $y^2 = x^3 - \frac{20}{9}x^2 + \frac{5}{3}x - \frac{5}{12}$, 75C1, $J = xy + (\frac{5}{3}x^2 - \frac{5}{2}x + 1)$,
$(x-1)^5 + (\frac{5}{3}x^2 - \frac{5}{2}x + 1)^2 = x^2y^2 = (x - \frac{2}{3})^3(x^2 - 3x + \frac{8}{3}) + (\frac{5}{3}x^2 - \frac{5}{2}x + \frac{8}{9})^2$.

Type [5 , 5 , 221]
$y^2 = x^3 + \frac{5}{4}(x^2 + 18x - 7)$, 50B2, $J = xy - \frac{5}{2}(x^2 + x) + 1$,
$(x-1)^5 + (\frac{5}{2}x^2 + \frac{5}{2}x - 1)^2 = x^2y^2 = (x-9)(x^2 + 2x + 21)^2 + (\frac{5}{2}x^2 + \frac{5}{2}x + 63)^2$.

Type [5 , 41 , 41] There are three dessins; one of them has as double the
double cover of (5,41,2111) branched at ∞ and the 'tips', so we take $J = xy$
with $y^2 = x^3 + 15x^2 + 80x + 160$; 400H1; we note that $x^2y^2 - 256 =
(x+4)^4(x-1)$, so there really are branch points above $J = 16$ and $J = -16$

The other pair are given by $J = xy + i(\frac{5}{2}x^2 + 15x + 22 + 4i)$, $y^2 =
x^3 + \frac{35}{4}x^2 + 25x + 25$, with $x^2y^2 + (\frac{5}{2}x^2 + 15x + 22 + 4i)^2 = (x+3+i)^4(x+3-4i)$.
We note the lovely symmetry, that the complex conjugate of this cover is
actually the same, but with the branch points interchanged.

Type [5 , 41 , 32]
A pair of coverings defined over the field $\mathbf{Q}(\sqrt{6})$.

Type [5 , 32 , 32]
There is a unique dessin whose double is the double cover of [5,32,2111]
branched at infinity and the tips; we are led to $J = xy$ with $y^2 = x^3 + 5x^2 -
5x - 45$, 300A1.

Type [5 , 5 , 5], genus 2.
There is one easy dessin, the 5-fold cyclic cover branched over $\infty, 0, k$; we
get $J * (J - k) = u^5$.
There is also the more interesting cover $x(J - 3x) = y^2 - 8y$, $y(J + 4x) =
x^3 - 8x$; this has 5-fold branching above $J = \pm 8$. Elkies pointed out that
this curve also has the equation $v^3 = (u^2 - 9u + 24)/(u^2 + 9u + 24)$, as a
3-covering of [5,311,311].

Type [6 , 6 , 3111]
First, we have a triple cover of [2,2,11], $J = x^2$, ramified at $x = 0, 1$; so
we take $J = x^2$, $y^3 = x(x - 1)$.
Secondly, we have three coverings defined over the field $\mathbf{Q}(\sqrt[3]{2})$ and its
conjugates.

Type [6 , 6 , 2211]

Again, there is the possibility of a ramified cover; this time, we may take a double cover of the dessin [3 , 3 , 111] , and obtain $J = x^3$, $y^2 = x(x^2 - x + 1)$, 24A4. There are other cases.

Type [6 , 42 , 411]

There is a double cover of [3,21,21], $J = (x - 1)^2(x + 2)$, branched over $x = 1, -1, -2, \infty$; we simply take $y^2 = (x^2 - 1)(x + 2)$. There are also other cases.

Type [6 , 42 , 222]

The only possibility is a double cover of [3,21,21], $J = (x - 1)^2(x + 2)$, branched over $x = 2, 1, -2, \infty$; we take $y^2 = (x - 1)(x^2 - 4)$.

Type [6 , 33 , 222]

The only possibility is a double cover of [3,3,111], ramified at the tips: this is actually the imprimitive drawing obtained by doubling [3,3,3]. We may take it as $J = x^3 = 1 + y^2$.

Type [33 , 51 , 51]

Take a double cover of [6,51,21111] branched at the tips, getting a [66,5511,222222]; this is then the double of a [33,51,51]. The equations work out to $J' = (x + 1)y$ with $y^2 = 5x^4 - 4x^3 + 3x^2 - 2x + 1$, so that $(J' + 1)(J' - 1) = x^5(5x + 6)$.

Type [33 , 42 , 42]

A possibility is a double cover of [3,21,21], $J = (x - 1)^2(x + 2)$, branched over $x = \pm 1, \pm 2$; we take $y^2 = (x^2 - 1)(x^2 - 4)$.

The second possibility is to take a double cover of [6,42,21111], branched at the tips, leading to a [66,4422,222222], and then this is the double of a [33,42,42]. The equations work out to $J' = (x + 2)y$, $-y^2 = x^4 + 2x^3 - 3x^2 + 4x - 4$, so that $(J' + 4)(J' - 4) = -x^4(x + 3)^2$.

Type [33 , 33 , 33]

We take a double cover of [6,33,21111] branched at the tips, and this is the double of a [33,33,33]. The equations are $J' = xy$, $y^2 = x^4 + 3x^2 + 3$; there is branching above $J' = \pm 1$, with $x = \pm i$.

References

[ASD] A.O.L.Atkin and H.P.F.Swinnerton-Dyer, *Modular Forms on Non-congruence Subgroups*, Amer. Math. Soc. Symposia XIX (1971), 1-25.

[Belyi] G.V.Belyi, *On Galois Extensions of the Maximal Cyclotomic Field*, Izvestiya Ak. Nauk. SSSR, ser. mat., **43:2** (1979), 269-276.

[Be1] Sybilla Beckmann, *Ramified Primes in the Field of Moduli of Branch-
       ed Coverings of Curves*, J.Algebra 125 (1989) 236-255.

[Be2] ———, *On extensions of number fields obtained by specialising branch-
       ed coverings*, J. reine angew. Math. 419 (1991) 27-53.

[Be3] ———, *Is every extension of* $\mathbf{Q}$ *obtained by specialising branched cov-
       erings?*, preprint 1993, 21 pp.

[BIZ] D. Bessis, C. Itzykson, J.B. Zuber, Quantum Field Theory Tech-
       niques in Graphical Enumeration, *Adv. in App. Math.* $\mathbf{I}$ (1980),
       109-157.

[Ca]   Pierre Cartier, *Groupes formelles, Fonctions automorphes et Fonc-
       tions zéta des Courbes elliptiques*, Proc. I.C.M. Nice (1970), Tome
       2, 291-299.

[Co]   Jean-Marc Couveignes, these proceedings.

[Cr]   J.E.Cremona, *Algorithms for Elliptic Curves.* Cambridge University
       Press, 1993.

[Fr]   M.D.Fried, *Fields of definition of function fields and Hurwitz families
       — Groups as Galois groups.* Comm. Alg. 5 (1977), 17-82.

[Fu]   W.Fulton, *Hurwitz schemes and irreducibility of algebraic curves*,
       Ann. Math.(2) 90 (1969), 542-575.

[G1]   A.Grothendieck *Géométrie formelle et géométrie algébrique*, Sém.
       Bourbaki exposé 182 (1958-9).

[G2]   A.Grothendieck *Revêtements Etales et Groupe Fondamental*, Sem.
       Géom. Alg. 1960/1, Lecture Notes in Math. 224 (Springer 1971).

[G3]   A.Grothendieck *Esquisse d'un Programme*, typescript circa 1982.

[Ma1] B.H.Matzat, *Konstruktive Galoistheorie*, Lecture Notes in Math.
       1284, Springer 1987.

[Ma2] ———, *Zöpfe und Galoissche Gruppen*, J. reine angew. Math. **420**
       (1991), 99-159.

[MM]   R.Malle and B.H.Matzat, forthcoming book.

[Sch1] A.J.Scholl, *Modular forms and de Rham cohomology; Atkin-Swinner-
       ton-Dyer congruences.* Invent. Math. **79** (1985), 49-77.

[Sch2] ———, *The l-adic representations attached to a certain noncongru-
       ence subgroup.* J. reine angew. Math. **393** (1988), 1-15.

[Se1]   J-P.Serre, *Algebraic Groups and Class Fields,* GTM117 (Springer 1988).

[Se2]   J-P.Serre, *Topics in Galois Theory.* Research Notes in Math. 1, (Jones and Bartlett, Boston 1992.)

[SV]   G.Shabat and V.A.Voevodsky, *Drawing Curves over Number Fields.* Papers in honour of A. Grothendieck, Progress in Math **88** 1990, 199-227.

[T]   D.J.Tingley, *Elliptic Curves and Modular Functions.* Oxford D.Phil thesis 1975.

[Z]   Alexander Zvonkin, *Shabat Trees of Diameter 4.* Preprint 1993.

*Oxford University, Oxford, GB

Dessins d'enfants on the Riemann sphere

Leila Schneps*

Abstract

In part I of this article we define the Grothendieck dessins and recall the description of the Grothendieck correspondence between dessins and Belyi pairs (X, β) where X is a compact connected Riemann surface and $\beta : X \to \mathbf{P}^1\mathbf{C}$ is a Belyi morphism. In part II we discuss the action of $\mathrm{Gal}(\overline{\mathbf{Q}}/\mathbf{Q})$ on dessins and show that it is faithful on genus 0 and genus 1 dessins, and on trees. In part III we consider the genus zero case, i.e. dessins on the Riemann sphere. Given a dessin D on $\mathbf{P}^1\mathbf{C}$, we discuss the explicit association of a rational Belyi function to a genus zero dessin and vice versa. In IV we give a few basic examples.

*

I. The Grothendieck correspondence

The aim of this section is to define the Grothendieck dessins and use Belyi's theorem to give a description of the bijection between the set of isomorphism classes of dessins and the set of isomorphism classes of algebraic curves defined over $\overline{\mathbf{Q}}$. The ideas in this first section originate in Grothendieck's unpublished paper [G]: it was he who suggested the possibility of associating an algebraic curve defined over $\overline{\mathbf{Q}}$ to a cellular map on a topological surface, and remarked that all algebraic curves over $\overline{\mathbf{Q}}$ can be obtained in this way as a consequence of Belyi's theorem, which we state and prove below although it is already well-known. All of the material in this first section is essentially already known, however we provide it here for reference.

I would like express my warmest gratitude to G. Shabat who explained much of the material in section I to me, as well as his own ideas (see section III, §2). Further thanks are due to R. Nauheim and R. Dentzer for introducing me to the Gröbner basis algorithm. Particular thanks are due to H. W. Lenstra, Jr. for help with theorem II.4. Finally, I am grateful

to Jean-Marc Couveignes, Pierre Lochak and Gunter Malle for many interesting discussions and suggestions. Needless to say, the inspiration for this
work is entirely due to A. Grothendieck.

§1. Algebraic curves defined over $\overline{\mathbf{Q}}$

Let X be an algebraic curve defined over $\mathbf{C}$. We recall that via a (non-
trivial) classical theorem, X is defined over $\overline{\mathbf{Q}}$ (i.e. possesses a model defined
over $\overline{\mathbf{Q}}$) if and only if there exists a non-constant holomorphic function
$f : X \to \mathbf{P}^1\mathbf{C}$ all of whose critical values lie in $\overline{\mathbf{Q}}$.

Denote by π_1 the fundamental group of $\mathbf{P}^1\mathbf{C} - \{0,1,\infty\}$, generated by
loops l_0, l_1 and l_∞ around 0, 1 and ∞, with the relation $l_0 l_1 l_\infty = 1$. The
following lemma is entirely classical:

Lemma I.1: *There is a bijection between the conjugacy classes of subgroups
of finite index of π_1 and isomorphism classes of finite coverings X of $\mathbf{P}^1\mathbf{C}$
ramified only over 0, 1 and ∞.*

Let B be a subgroup of finite index of π_1 and let $\tilde{X}$ be the universal
covering of $\mathbf{P}^1\mathbf{C} - \{0,1,\infty\}$. We recall that the correspondence is based on
the identification of the quotient space $B \setminus \tilde{X}$ with an unbranched cover
X' of $\mathbf{P}^1\mathbf{C} - \{0,1,\infty\}$ of degree $[\pi_1 : B]$, where the morphism f of X'
to $\mathbf{P}^1\mathbf{C} - \{0,1,\infty\}$ is given by quotienting by the action of π_1. Note that
considering f as an analytic map on X equips X with a unique analytic
structure. A classical theorem (see Forster [F] Thm. 8.4 for example) shows
that the unbranched cover $f : X' \to \mathbf{P}^1\mathbf{C} - \{0,1,\infty\}$ can be extended to a
branched cover of $\mathbf{P}^1\mathbf{C}$, in a unique way up to biholomorphic fiber-preserving
maps from X' into itself. We note that the ramification indices over 0, 1
and ∞ are given by the lengths of the orbits in π_1/B under the action of
l_0, l_1 and l_∞ respectively. Let us denote by $e_{1,l_i}, \ldots, e_{k_i,l_i}$ the orbit lengths
for $i = 0, 1, \infty$. Then we recall that the degree d of the covering is given by
the index $d = [\pi_1 : B]$ and the genus g of X is given by Hurwitz's formula

$$2g - 2 = -2d + \sum_{i \in \{0,1,\infty\}} \sum_{j=1}^{k_i} (e_{j,l_i} - 1).$$

Let $\pi_1' = \pi_1/\langle l_1^2 \rangle$. Then we have the following as an immediate consequence of lemma I.1:

Corollary: *There is a bijection between the conjugacy classes of subgroups
of finite index of π_1' and the isomorphism classes of coverings of $\mathbf{P}^1\mathbf{C}$ ram-*

ified only over 0, 1 and ∞, such that the ramification over 1 is of degree at most 2.

We now give the theorem which is essential to the description of the Grothendieck correspondence given in §3, Belyi's theorem characterizing algebraic curves defined over $\overline{\mathbf{Q}}$ (see [B]).

Theorem I.2: *Let X be an algebraic curve defined over $\mathbf{C}$. Then X is defined over $\overline{\mathbf{Q}}$ if and only if there exists a holomorphic function $f : X \to \mathbf{P}^1\mathbf{C}$ such that all critical values of f lie in the set $\{0, 1, \infty\}$.*

Proof: Clearly, if there exists such a morphism (called a Belyi morphism) X is defined over $\overline{\mathbf{Q}}$. Suppose now that X is defined over $\overline{\mathbf{Q}}$, and let $g : X \to \mathbf{P}^1\mathbf{C}$ be such that all critical values of g lie in $\overline{\mathbf{Q}}$. We first construct a morphism $h : X \to \mathbf{P}^1\mathbf{C}$ all of whose critical values lie in $\mathbf{Q}$. Let S be the set of all critical values of g and all their conjugates under $\mathrm{Gal}(\overline{\mathbf{Q}}/\mathbf{Q})$. Set $f_0(z_0) = \prod_{s \in S}(z_0 - s) \in \mathbf{Q}[z_0]$, and set

$$f_{j+1}(z_{j+1}) = \mathrm{Res}_{z_j}\left(\frac{df_j}{dz_j}, f_j(z_j) - z_{j+1}\right).$$

By construction, the roots of f_{j+1} are exactly the finite critical values of f_j. All the f_j are defined over $\mathbf{Q}$ and their degrees decrease successively until for some n we have $\deg(f_n) = 0$ (we note that what is happening here is a concentration of the ramification of the original function at ∞ via successive compositions with well-chosen polynomials). Set $h = f_{n-1} \circ f_{n-2} \circ \cdots \circ f_1 \circ f_0 \circ g$. Then the critical values of h are contained in $\mathbf{Q}$, as can be easily seen by induction using the formula $C_{g_1 \circ g_2} = C_{g_1} \cup g_1(C_{g_2})$, where C_g denotes the set of critical values of g. Denote by $S' \subset \mathbf{Q}$ the set of finite critical values of h.

If $|S'| \leq 3$, a linear fractional transformation suffices to take its elements onto a subset of $\{0, 1, \infty\}$. Suppose $|S'| > 3$. Choose three ordered points of S': then we can always find integers m and n such that they go to 0, $m/(m+n)$ and 1 by a linear fractional transformation. Then the transformation

$$z \mapsto \frac{(m+n)^{(m+n)}}{m^m n^n} z^m (1-z)^n$$

transforms both 0 and 1 to 0, and $m/(m+n)$ to 1. In this way we obtain a morphism whose set of critical values has cardinal less than or equal to $|S'| - 1$ (and which contains 0 and 1). Repeating the procedure a finite number of times produces an explicit morphism from X to $\mathbf{P}^1\mathbf{C}$ having set of finite critical values contained in $\{0, 1\}$. ◇

Definition 1: A morphism $\beta : X \to \mathbf{P}^1\mathbf{C}$ all of whose critical values lie in $\{0, 1, \infty\}$ is called a *Belyi morphism*. We call β a *pre-clean* Belyi morphism if all the ramification orders over 1 are less than or equal to 2, and *clean* if they are all exactly equal to 2. We will see that a dessin corresponding to a pre-clean Belyi function has the visually agreeable property that it has exactly one edge corresponding to every pre-image of 1 under β; a dessin corresponding to a clean Belyi function has a vertex at both ends of every edge. We also use the words pre-clean and clean to describe the dessins.

Corollary (to theorem I.2): *An algebraic curve defined over $\mathbf{C}$ is defined over $\overline{\mathbf{Q}}$ if and only if there exists a clean Belyi morphism $\beta : X \to \mathbf{P}^1\mathbf{C}$.*

Proof: If $\alpha : X \to \mathbf{P}^1\mathbf{C}$ is a Belyi morphism, then $\beta = 4\alpha(1 - \alpha)$ is a clean one.   ◇

If X is an algebraic curve defined over $\overline{\mathbf{Q}}$ and β is a Belyi morphism on it, we call the couple (X, β) a *Belyi pair*. Two Belyi pairs (X, β) and (Y, α) are said to be isomorphic if there is an isomorphism $\phi : X \to Y$ such that $\beta = \alpha \circ \phi$. If β is clean we call (X, β) a *clean Belyi pair*.

§2. Dessins d'enfants

Grothendieck [G] gives a sketch of an exploration of the connections between algebraic curves defined over $\overline{\mathbf{Q}}$ and their fields of definition, and what he calls "dessins d'enfants", which might be conveniently described as scribbles on topological surfaces. the precise definition here.

Definition 2: A *Grothendieck dessin* is a triple $X_0 \subset X_1 \subset X_2$ where X_2 is the topological model of a compact connected Riemann surface, X_0 is a finite set of points, $X_1 \setminus X_0$ is a finite disjoint union of segments and $X_2 \setminus X_1$ is a finite disjoint union of open cells, such that a bipartite structure can be put on the set of vertices X_0; namely the vertices can be marked with two distinct marks in such a way that the direct neighbors of any given vertex are all of the opposite mark.

Definition 3: Two dessins $D = X_0 \subset X_1 \subset X_2$ and $D' = X_0' \subset X_1' \subset X_2'$ are *isomorphic* if there exists a homeomorphism from X_2 into X_2' inducing a homeomorphism from X_1 into X_1' and one from X_0 into X_0'. We sometimes use the terminology *abstract dessin* for an isomorphism class of dessins. This indicates that the structure of the dessin is determined, but it is not associated with any particular embedding into a complex topological surface.

Because of the corollary given above, in studying the correspondence of
Belyi pairs to dessins we may restrict ourselves to the clean dessins as does
Grothendieck in [G]; we then have the simpler definition

Definition 2': A *pre-clean* Grothendieck dessin is a triple $X_0 \subset X_1 \subset X_2$
where X_2 is the topological model of a compact connected Riemann surface,
X_0 is a finite set of points, $X_1 \setminus X_0$ is a finite disjoint union of segments and
$X_2 \setminus X_1$ is a finite disjoint union of open cells.

In this definition it is to be understood that all vertices in X_0 are considered to have the same "mark", and somewhere on each edge is a vertex
of the opposite mark (it can be anywhere on the edge, even on the end if
the edge is a tail, as long as it does not coincide with any of the vertices in
X_0); this puts a natural bipartite structure on the dessin.

In order to prepare the ground for the Grothendieck correspondence, we
need to introduce the flag set of a pre-clean dessin and the action of the
cartographical group on it.

Definition 4: A *marking* on a pre-clean dessin is a fixed choice of one point
on each component of $X_1 \setminus X_0$, and one point in each open cell of $X_2 \setminus X_1$.
We will always use the notation • for a point in X_0, ⋆ for a point in $X_1 \setminus X_0$
and o for a point in $X_2 \setminus X_1$.

Definition 5: Let D be a pre-clean dessin with a fixed marking. Then the
flag set $F(D)$ of D is the set of triangles whose three vertices are marked
•, ⋆ and o in such a way that • is in the closure of the segment containing
⋆, and that segment is in the closure of the open cell containing o. The
oriented flag set $F^+(D)$ is the set of flags the order of whose vertices is
o − • − ⋆ when read counterclockwise.

Definition 6: The cartographical group C_2 is given by three generators σ_0,
σ_1 and σ_2 together with the relations $\sigma_0^2 = \sigma_1^2 = \sigma_2^2 = 1$ and $(\sigma_0\sigma_2)^2 = 1$.
The oriented cartographical group C_2^+ is the subgroup of index 2 of C_2 given
by all even words of C_2. A generating set is given by $\rho_0 = \sigma_1\sigma_0$, $\rho_1 = \sigma_0\sigma_2$
and $\rho_2 = \sigma_2\sigma_1$, with the relations $\rho_1^2 = 1$ and $\rho_0\rho_1\rho_2 = 1$.

The generators σ_0, σ_1 and σ_2 of the group C_2 act on $F(D)$ as follows: if
F is a flag given by

then $\sigma_0(F)$, $\sigma_1(F)$ and $\sigma_2(F)$ are given by

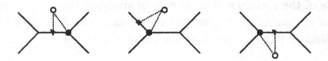

If we restrict our attention to the oriented flags, it suffices to denote them by a vertex • and an edge, since given the • point and the ⋆ edge point of an oriented flag, the position of the o open cell point is determined. Given an oriented flag F, we deduce from the action of σ_0, σ_1 and σ_∞ on it that the flags $\rho_0(F)$, $\rho_1(F)$ and $\rho_2(F)$ are given by

which completely describes the action of C_2^+ on $F^+(D)$ (note that we consider the elements of C_2 as acting on the left, so that in $\rho_0 = \sigma_1\sigma_0$, for instance, σ_0 acts first). We usually consider only the set $\mathcal{F}^+(D)$ of positively oriented flags. In this set there are exactly two flags for each edge of the dessin. It is clear that, viewed as a set together with the action of a certain group C_2^+, the set $F^+(D)$ is independent of the marking on D: in fact it depends only on the abstract dessin which is the isomorphism class of D. Indeed, considering only the oriented flag set makes the marking unnecessary since each flag is exactly equivalent to giving one vertex and one specific edge coming out of it.

Lemma I.3: *Let D be a pre-clean dessin and $F \in F^+(D)$ a fixed flag. Let $B_{F,D}$ be the set of elements of C_2^+ fixing F. Then $B_{F,D}$ is a subgroup of finite index in C_2^+ and the stabilizing subgroup $B_{F',D}$ for any other flag $F' \in F^+(D)$ is conjugate to $B_{F,D}$ in C_2^+. Moreover $B_{F,D}$ depends only on the abstract dessin D.*

Proof: The orbit of F under C_2^+ is necessarily finite since $F(D)$ is finite, thus $B_{F,D}$ is of finite index. Let $F' \in F(D)$ be different from F. Then by applying the different transformations in C_2^+ one can construct an element $\sigma \in C_2$ such that $\sigma(F) = F'$, so it is clear that $B_{F',D} = \sigma^{-1}B_{F,D}\sigma$.   ◇

The following theorem is due to Malgoire-Voisin [MV] (and similar results appear in work of Jones and Singerman, cf. [JS]).

Theorem I.4: *There is a bijection between the isomorphism classes of clean dessins and the conjugacy classes of subgroups of C_2^+ of finite index.*

Proof: By lemma I.3, we can associate to an abstract dessin a conjugacy class of subgroups of finite index of C_2^+. We now let B be a subgroup of C_2^+ of finite index and show how to entirely reconstruct a unique dessin from it. It will be a direct consequence of that argument that the two directions correspond and that changing the subgroup to a conjugate subgroup is the only way to obtain an isomorphic dessin.

Let B be considered as a subgroup of finite index of C_2. Let $H = C_2/B$ denote the coset space. We will construct a dessin D whose flag set $\mathcal{F}(D)$ will be bijective to H (so $\mathcal{F}^+(D)$ will be bijective to C_2^+/B), and such that the action of C_2 on $\mathcal{F}(D)$ is given by the action of C_2 on H by left multiplication. The flag corresponding to the coset B will be fixed by the action of B. The elements σ_0, σ_1 and $\sigma_2 \in C_2$ all act on H, dividing its elements into orbits. From the action of these elements, it is clear that two cosets, i.e. two flags will be in the same σ_0-orbit if their $\circ - \star$ segment is the same. They will be in the same σ_1-orbit if their $\circ - \bullet$ segment is the same, and in the same σ_2-orbit if their $\bullet - \star$ segment is the same. Note that each σ_i-orbit contains at most two elements. We can begin to reconstitute the dessin by noting the number of each of the three types of edge, as the orders of the quotient spaces $\langle \sigma_i \rangle \setminus H$.

Now let us consider the actions of σ_1 and σ_2 on the quotient space $\langle \sigma_0 \rangle \setminus H$. Identifying each element of $\langle \sigma_0 \rangle \setminus H$ with a $\circ - \star$ edge, the σ_0-orbits which are in the same orbit under the action of σ_1 should be those having the same $\circ$ point and those identified under σ_2 have the same $\star$ point. When σ_0 and σ_2 act on $\langle \sigma_1 \rangle \setminus H$, considered as the set of $\circ - \bullet$ edges, they should identify edges having the same $\circ$ and $\bullet$ points respectively, and when σ_0 and σ_1 act on $\langle \sigma_2 \rangle \setminus H$ considered as the set of $\star - \bullet$ edges, they should identify sides having the same $\star$ and $\bullet$ points respectively. We use this information in two steps. The first is to give the orders of the sets of vertices, edges and open cells of D respectively by the orders of the double-orbit sets $\langle \sigma_1, \sigma_2 \rangle \setminus H$, $\langle \sigma_0, \sigma_2 \rangle \setminus H$ and $\langle \sigma_0, \sigma_1 \rangle \setminus H$. The second step is the information on how to glue these components together. Take a point and an edge given by an element x of $\langle \sigma_1, \sigma_2 \rangle \setminus H$ and an element y of $\langle \sigma_0, \sigma_2 \rangle \setminus H$ respectively. The element x can be considered as a σ_1-orbit of σ_2-orbits and the element y as a σ_0-orbit of σ_2 orbits. They can be glued together if and only if there is some σ_2-orbit occurring in both x and y. The same thing works to glue together edges and cells, and vertices and cells. The whole situation is summarized by the similarity of the following two diagrams, where $\mathcal{F}_{\circ\star}$, $\mathcal{F}_{\circ\bullet}$ and $\mathcal{F}_{\bullet\star}$ denote the sets of $\circ - \star$, $\circ - \bullet$ and $\bullet - \star$ edges respectively, while $\mathcal{F}_\bullet$, $\mathcal{F}_\star$

and $\mathcal{F}_0$ denote the sets of vertices, edges and open cells:

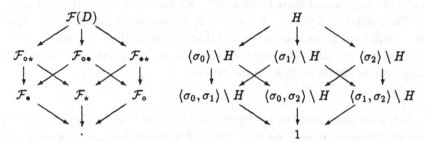

To conclude, we note that changing B to $\sigma^{-1}B\sigma$ for any $\sigma \in C_2^+$ does not change any of the above objects considered as sets together with the action of the elements of C_2, because the action of C_2 on $C_2/\sigma^{-1}B\sigma$ is the same as on C_2/B. So the dessin is independent of the choice of representative of the conjugacy class of B. Moreover, if we construct a dessin as above from a subgroup $B \subset C_2^+$ and then consider the subgroup fixing some given flag of the dessin, we find exactly a subgroup conjugate to B. For the set $H = C_2/B$ is the flag set of the dessin and thus B fixes one flag (the one corresponding to the coset B), and we saw in lemma I.3 that all subgroups fixing the different flags of a dessin are conjugate.   ◇

§3. The Grothendieck correspondence

We now have all the necessary ingredients to prove the main result of section I.

Theorem I.5: *There is a bijection between the set of abstract clean dessins and the set of isomorphism classes of clean Belyi pairs.*

Proof: The groups π_1' and C_2^+ are canonically isomorphic. Let $\phi : C_2^+ \to \pi_1'$ be defined by $\phi(\rho_i) = l_i$, $i = 0, 1$, and $\phi(\rho_2) = l_\infty$. Then the theorem is an immediate consequence of Lemma I.1 and Theorem I.4.   ◇

The Grothendieck correspondence can be described more concretely via the following topological construction.

Given a clean Belyi pair (X, β), we let X_2 be the topological model of X, $X_0 = \beta^{-1}(0)$ and $X_1 = \beta^{-1}([0,1])$, where $[0,1]$ is the segment of the real line on $\mathbf{P}^1\mathbf{C}$. Note that $\beta^{-1}(\infty)$ gives a point in each open cell of the dessin.

Requiring the Belyi function to be clean is equivalent to asking that there be a vertex at each end of every edge. If a rational Belyi function has ramification order 1 or 2 over 1 then one can obtain dessins having edges with no vertex at one end.

We may interpret the association of a curve to a dessin with a marking directly on the topological surface as follows. The flags, considered as triangles with vertices •, ⋆ and o, pave the topological surface X_2 with lozenges made of pairs of adjacent flags, one positively and one negatively oriented, where the common side is of the o − • type. Joining the two ⋆ vertices and the sides of the lozenge gives something homeorphic to the sphere. Identifying all these lozenges with $\mathbf{P}^1\mathbf{C}$ by identifying the ⋆ point with 1, the o with ∞ and the • with 0 gives a morphism $\beta : X_2 \to \mathbf{P}^1\mathbf{C}$, ramified only over 0, 1 and ∞, with the ramification orders corresponding to the dessin. We put a Riemann surface structure on X_2 by requiring β to be a rational function.

It is important to remark that the Belyi function associated to a given abstract dessin is not well-defined. Indeed, since the dessin corresponds to an isomorphism class of Belyi pairs (X, β), β is defined only up to automorphisms of the Riemann surface X. In genus 0, this means that β is defined up to $PSL_2(\mathbf{C})$, in genus 1, up to affine transformations, in genus 2, up to a finite automorphism group generically of order 2 and in genus greater than 1, up to a finite automorphism group which is generically trivial.

§4. Ramified coverings of $\mathbf{P}^1\mathbf{C} - \{0, 1, \infty\}$

Suppose that X is a finite covering of $\mathbf{P}^1\mathbf{C}$ ramified only over 0, 1 and ∞, such that all ramification over 1 is of order at most 2. Let x be a point on $\mathbf{P}^1\mathbf{C}$ different from 0, 1 and ∞, and let $\{x_1, \ldots, x_d\}$ be the fiber over x, where d is the degree of the covering X. Then loops originating from x and going clockwise once around 0, 1, and ∞ respectively induce permutations σ_0, σ_1 and σ_∞ of the points $x_1, \ldots, x_d$, such that $\sigma_0\sigma_1\sigma_\infty = 1$. Note that σ_1 is of order 2, and that σ_0 and σ_1 generate a subgroup of S_d which is transitive if the covering is connected. Indeed, given any σ_0 and $\sigma_1 \in S_d$ such that $\sigma_1^2 = 1$ and the subgroup generated by σ_0 and σ_1 is transitive, there exists a connected covering X of $\mathbf{P}^1\mathbf{C}$ ramified only over 0, 1 and ∞ corresponding to it, such that the ramification orders over 1 are at most 2 (in order for all these orders to be exactly 2, d must be even and σ_1 must be a product of $d/2$ disjoint transpositions).

Given such an X, or equivalently, given an even positive integer d and two permutations σ_0 and σ_1 in S_d such that σ_1 is the product of $d/2$ disjoint transpositions and the subgroup $\langle \sigma_0, \sigma_1 \rangle$ is transitive, we show how to draw the pre-clean dessin associated to X. Set $\sigma_\infty = (\sigma_0\sigma_1)^{-1}$. Recall that the genus g of X can be calculated from the decomposition of the σ_i into disjoint cycles as follows by Hurwitz's formula:

$$2g - 2 = d - n_0 - n_1 - n_\infty,$$

where n_i is the number of disjoint cycles occurring in σ_i.

To draw the dessin, begin by writing σ_∞ as a product of l disjoint cycles $s_1 \cdots s_l$. For $1 \leq j \leq l$ let k_j be the length of s_j and write $s_j = (i_{1,j}, \ldots, i_{k_j,j})$. For each s_j, $1 \leq j \leq l$, draw a k_j-gon. Orient the edges of every k_j-gon by going around it in a counterclockwise direction. Going around the edges of each k_j-gon in order (starting from any edge), label them with transpositions $(i_{1,j}, \sigma_1(i_{1,j})), \ldots, (i_{k_j,j}, \sigma_1(i_{k_j,j}))$. Each such transposition is one which actually occurs in the disjoint cycle decomposition of σ_1.

Glue together the l polygons as follows: identify sides labelled by the same transposition, in the same direction. Clearly every edge is identified with exactly one other, so the result is a compact topological surface Y with no boundary, and a natural dessin drawn on it by the identified edges of the polygons. There is a natural morphism β from this surface to $\mathbf{P}^1\mathbf{C}$ which is the one described at the end of §3, marking the dessin and identifying lozenges with $\mathbf{P}^1\mathbf{C}$. By construction, the covering $\beta : Y \to \mathbf{P}^1\mathbf{C}$ has the same ramification properties as X, and is therefore isomorphic to X.

II. The action of $\mathrm{Gal}(\overline{\mathbf{Q}}/\mathbf{Q})$ on dessins

In the *Esquisse d'un Programme*, Grothendieck notes that the action of $\mathrm{Gal}(\overline{\mathbf{Q}}/\mathbf{Q})$ is faithful on the profinite completion $\hat{\pi}_1$ of the fundamental group of $\mathbf{P}^1\mathbf{C} - \{0, 1, \infty\}$. This means that $\mathrm{Gal}(\overline{\mathbf{Q}}/\mathbf{Q})$ acts with no kernel, i.e. for every element $\sigma \in \mathrm{Gal}(\overline{\mathbf{Q}}/\mathbf{Q})$, there is an element $\gamma \in \hat{\pi}_1$ such that the action of σ on γ is non-trivial. In this section we show that more can be said. In fact, $\mathrm{Gal}(\overline{\mathbf{Q}}/\mathbf{Q})$ acts faithfully on the set of dessins in genus 1, on the set of dessins in genus 0 and even on the set of trees. Given any element $\sigma \in \mathrm{Gal}(\overline{\mathbf{Q}}/\mathbf{Q})$ and a number field on which σ acts non-trivially, one can explicitly construct a tree on which σ acts non-trivially.

The genus 1 case (and thus the faithfulness on $\hat{\pi}_1$) is a well-known result (cf. for example [M, I.6, proof of Satz 2]).

Proposition II.1: *The action of $\mathrm{Gal}(\overline{\mathbf{Q}}/\mathbf{Q})$ on the set of dessins in genus 1 is faithful.*

Proof: A genus 1 dessin corresponds to a $\overline{\mathbf{Q}}$-isomorphism class of genus one curves, and as is well-known these isomorphism classes are classified by the

j-invariant, such a curve being defined over $\overline{\mathbf{Q}}$ if its j-invariant is. Clearly for every $\sigma \in \mathrm{Gal}(\overline{\mathbf{Q}}/\mathbf{Q})$ there exists $j \in \overline{\mathbf{Q}}$ such that σ does not act trivially on j. Let E be a genus 1 curve having j-invariant equal to j. We construct a genus one dessin associated to E simply by using Belyi's procedure to transform some function on E, say x, into a function $\beta : E \to \mathbf{P}^1\mathbf{C}$ ramified only over 0, 1 and ∞, and then letting the dessin be $\beta^{-1}([0,1])$. Then, since β will be defined over a field containing $\mathbf{Q}(j)$, the element σ cannot act trivially on the function β nor on its corresponding dessin. ◇

The genus 0 case is actually more difficult. The elegant proof of theorem II.4 for trees, based on the technique of the proof of Belyi's theorem, is due to H.W. Lenstra, Jr. We first need two technical lemmas.

Lemma II.2: *Let F be a polynomial of degree n and let $d|n$. Suppose there exists a polynomial H such that $H(0) = 0$, H is monic, $\deg(H) = d$ and for some polynomial G, $F = G \circ H$. Then H is unique.*

Proof: Let $\deg(G) = m$ so $n = md$ and write $G = \lambda_m z^m + \cdots + \lambda_0$ and $H = T^d + h_{d-1}T^{d-1} + \cdots + h_1 T$. Then

$$F = \lambda_m H^m + \lambda_{m-1} H^{m-1} + \cdots + \lambda_0.$$

The terms of the right-hand polynomial of degrees $n, \ldots, n - d + 1$ are contributed entirely from the leading term $\lambda_m H^m$. But from these terms one can uniquely solve for the d highest coefficients of H. For the leading term is 1 since H is monic, and for $n - d + 1 \leq i \leq n - 1$, the coefficient of the term of degree i in H^m is a polynomial in $h_{i-n+d}, h_{i-n+d+1}, \ldots, h_{d-1}$ which is linear in h_{i-n+d}. Thus the d highest coefficients $1, h_{d-1}, \ldots, h_1$ of H are determined, and since by assumption the constant term $h_0 = 0$, H is completely determined. ◇

Lemma II.3: *Let G, H, $\tilde{G}$ and $\tilde{H}$ be polynomials such that $G \circ H = \tilde{G} \circ \tilde{H}$ and $\deg(H) = \deg(\tilde{H})$. Then there exist constants c and d such that $\tilde{H} = cH + d$.*

Proof: Let μ be the leading coefficient of H, and ν the constant coefficient of H/μ; let $\tilde{\mu}$ be the leading coefficient of $\tilde{H}$ and $\tilde{\nu}$ the constant coefficient of $\tilde{H}/\tilde{\mu}$. Then there exist polynomials G_1 and G_2 such that $G \circ H = G_1 \circ (H/\mu - \nu) = \tilde{G} \circ \tilde{H} = G_2 \circ (\tilde{H}/\tilde{\mu} - \tilde{\nu})$. But both $H/\mu - \nu$ and $\tilde{H}/\tilde{\mu} - \tilde{\nu}$ are monic, their constant terms are 0 and their degrees are equal, so by lemma II.2 they are equal. Then setting $c = \tilde{\mu}/\mu$ and $d = \tilde{\mu}(\tilde{\nu} - \nu)$ we have $\tilde{H} = cH + d$. ◇

Theorem II.4: *The action of* $\mathrm{Gal}(\overline{\mathbf{Q}}/\mathbf{Q})$ *on the set of trees is faithful.*

Proof: Let $\sigma \in \mathrm{Gal}(\overline{\mathbf{Q}}/\mathbf{Q})$. We will exhibit a tree such that the action of σ on it is non-trivial. Let K be a number field and α a primitive element for K, such that the action of σ on α is non-trivial. In order to show that there is a tree on which σ acts non-trivially, it suffices to show that there is a tree defined over K, i.e. that there exists a Belyi function $\beta(z)$, corresponding to a tree, defined over K and such that $\beta^\sigma(z)$ is not equal to $\beta(\frac{az+b}{cz+d})$ except when $\frac{az+b}{cz+d} = z$. Now, a rational Belyi function β corresponds to a tree when ∞ has exactly one pre-image under β, corresponding to the fact that a tree is a dessin on the sphere possessing a unique open cell. In particular, this will be the case whenever $\beta(z)$ is a Belyi polynomial, in which case the unique point over ∞ will be ∞; β corresponds to a tree whose unique open cell contains ∞. If a polynomial satisfies $\beta^\sigma(z) = \beta(\frac{az+b}{cz+d})$ then we must have $c = 0$ (and $d = 1$, up to replacing a and b by a/d and b/d) since $\beta^\sigma(z)$ is also a polynomial. So we will exhibit a Belyi polynomial $\beta(z)$, defined over K and such that if a and b are such that $\beta^\sigma(z) = \beta(az + b)$, then $a = 1$ and $b = 0$.

We construct such a $\beta(z)$ explicitly as follows. Let $f_\alpha(z) \in K[z]$ be a polynomial whose derivative $f'_\alpha(z)$ is given by

$$f'_\alpha(z) = z^3(z - 1)^2(z - \alpha).$$

By the proof of Belyi's theorem, there exists a polynomial $f(z) \in \mathbf{Q}[z]$ such that $f \circ f_\alpha$ is a Belyi polynomial which we call g_α. Let $\beta = \alpha^\sigma$ (by assumption, $\beta \neq \alpha$). Since f is defined over $\mathbf{Q}$, we obtain another Belyi polynomial $g_\beta = f \circ f_\beta$ where $f_\beta = f_\alpha^\sigma$.

Let T_α be the abstract tree corresponding to the Belyi polynomial g_α, and T_β the tree corresponding to g_β, so $T_\beta = T_\alpha^\sigma$. In order to prove that σ acts non-trivially on T_α, we must show that T_α and T_β are distinct. As mentioned above, this is equivalent to showing that we cannot have $g_\beta(z) = g_\alpha(az + b)$ for any constants a, b.

Suppose we do have such a and b. Then $g_\beta(z) = g_\alpha(az+b)$, i.e. $f(f_\beta(z)) = f(f_\alpha(az + b))$. Now applying lemma II.3 with $G = \tilde{G} = f(z)$ and $H = f_\alpha(az + b)$, $\tilde{H} = f_\beta(z)$, we see that there exist constants c and d such that $f_\alpha(az+b) = cf_\beta(z)+d$. Consider the critical points of both these functions. The right-hand function has the same critical points as f_β, namely the point 0 (of order 3), the point 1 (of order 2) and the point β (of order 1). The left-hand function has three critical points x_i, $i = 1, 2, 3$, where each x_i is of order i and $ax_1 + b = \alpha$, $ax_2 + b = 1$ and $ax_3 + b = 0$, since $az + b$ must take these three critical points to the critical points of f_α, respecting their orders. By equality of the two sides, we must have $x_1 = \beta$, $x_2 = 1$

and $x_3 = 0$. But the two equations $ax_2 + b = 1$ and $ax_3 + b = 0$ then give $a = 1$ and $b = 0$, so the equation $ax_1 + b = \alpha$ gives $\beta = \alpha$, contrary to the assumption that $\beta \neq \alpha$. Therefore, we cannot have $g_\beta(z) = g_\alpha(az + b)$ for any constants a, b other than $a = 1$, $b = 0$, which shows that the trees T_α and $T_\beta = T_\alpha^\sigma$ are distinct. ◇

Corollary: $\mathrm{Gal}(\overline{\mathbf{Q}}/\mathbf{Q})$ *acts faithfully on the set of genus 0 dessins.*

III. The genus zero case

The goal of part III is to make the Grothendieck correspondence completely explicit for dessins of genus 0, i.e. such that X_2 is a sphere, in order to determine the action of $\mathrm{Gal}(\overline{\mathbf{Q}}/\mathbf{Q})$ on them (from now on, we denote $\mathrm{Gal}(\overline{\mathbf{Q}}/\mathbf{Q})$ by $\mathrm{I\!\Gamma}$). There are two directions in the procedure. The first is the theoretically easier direction, i.e. how to calculate the pre-image under a Belyi function β of the segment $[0, 1]$. That is the purpose of §1.

The other direction is, given the dessin, to calculate an associated Belyi morphism. This can be done in various ways. The most complete exposition of the problem, for any genus, is given in the article by Couveignes and Granboulan in this volume. In the genus zero case, when the dessin is sufficiently small for the algorithm to work, it can be done by reducing the problem to that of solving a system of polynomial equations in several variables. The simplest way of doing this, due to Atkin and Swinnerton-Dyer, is described in the article by Birch in this volume, with a simplification for trees described in the article by Shabat. Such calculations were also performed earlier by others such as Matzat or Malle in order to calculate defining equations and Belyi functions for field extensions whose existence was known by rigidity (cf. [M,II.3,III.5,III.6], also [Ma], also Malle's article in this volume.).

In §2, we give a procedure similar to the original one of Atkin and Swinnerton-Dyer, which however replaces the use of the roots of a polynomial as unknowns by its coefficients, giving a slight improvement in efficiency. We then describe in §3 the use of the Gröbner basis method to solve the equations. For a given genus zero dessin D this algorithm yields a finite set of solutions to the equations, each of which gives rise to an explicit Belyi function. We then use the methods of §1 to identify the Belyi function actually associated to the given dessin D. The $\mathrm{I\!\Gamma}$-conjugates of this function then give the $\mathrm{I\!\Gamma}$-conjugates of D. In all, for a genus zero dessin D, these methods yield

(i) the set $\mathcal{O}(D)$ of dessins in the orbit of D under the action of $\mathrm{I\!\Gamma}$

(ii) the number field $K_{D'}$ associated to each dessin $D' \in \mathcal{O}(D)$

(iii) a set of Γ-conjugate Belyi functions corresponding to the dessins in $\mathcal{O}(D)$

(iv) the action of Γ on $\mathcal{O}(D)$.

§1. Reconstruction of a dessin from a Belyi function

In order to explicitly reconstruct the dessin D associated to a given rational clean Belyi function $\beta(z)$ we would like to simply calculate the pre-image of the segment $[0,1]$ under the $\beta(z)$. In order to avoid studying what happens near the ramification points, we use Picard's method to reconstruct the permutations σ_0 and σ_1 associated to the dessin as in I, §4. We proceed explicitly as follows.

Suppose that $\beta : \mathbf{P}^1\mathbf{C} \to \mathbf{P}^1\mathbf{C}$ is a rational clean Belyi function of degree $2d$. Let A_0 be the set of pre-images of 0 under β, A_1 the pre-images of 1 and A_∞ the pre-images of ∞. Choose open sets $\{U_\alpha \mid \alpha \in A_0 \cup A_1 \cup A_\infty\}$, where each U_α is a neighborhood of α. For $i \in \{0,1,\infty\}$, set $V_i = \cup_{\alpha \in A_i}\beta(U_\alpha)$. Then V_0, V_1 and V_∞ are open neighborhoods of 0, 1 and ∞ respectively: we may choose all the open sets concerned small enough so that $d(V_i, V_j) > 0$ for $i,j \in \{0,1,\infty\}$, $i \neq j$, where d denotes the usual distance in $\mathbf{P}^1\mathbf{C}$.

Let
$$X = \mathbf{P}^1\mathbf{C} \setminus (\cup_{\alpha \in A_0 \cup A_1 \cup A_\infty} U_\alpha)$$

and $Y = \mathbf{P}^1\mathbf{C} \setminus (V_0 \cup V_1 \cup V_\infty)$ (note that since V_∞ is a neighborhood of ∞, Y is a compact set). Choose a base point x_0 in Y and let $x_1, \ldots, x_{2d}$ be the pre-images of x_0 under β. Choose a loop γ_0 starting from x_0 and going once clockwise around 0 and a loop γ_1 starting from x_0 and going once around 1, where both γ_0 and γ_1 lie entirely in Y.

The pre-image of the path γ_0 is given by $2d$ non-intersecting paths $g_1, \ldots, g_{2d} \subset X$, where each γ_i starts at the point x_i and ends at a point x_j also in the fiber over x_0. No two of the g_i can end at the same point (since the g_i must be non-intersecting), so these paths induce a permutation in S_{2d} sending each $i \in \{1, \ldots, 2d\}$ to the $j \in \{1, \ldots, 2d\}$ such that the path starting at x_i ends at x_j. This is the permutation σ_0. The same procedure applied to the path γ_1 gives the permutation $\sigma_1 \in S_{2d}$ (which, because $\beta(z)$ is clean, consists of the product of d disjoint transpositions). Thus, if we explicitly determine the complete pre-image of the paths γ_0 and γ_1 we immediately obtain the permutations σ_0 and σ_1.

We use the fact that the second derivative β'' of β is bounded on Y, say $|\beta''(z)| < C$ for $z \in Y$. We need the following proposition (Picard's method):

Proposition III.1: *Let w_0 be a point of Y and $z_0 \in X$ be such that $\beta(z_0) = w_0$. Let $r < |\beta'(z_0)|/2C$ and $r' < r|\beta'(z_0)|/2$. Let U and V be the open balls $B(z_0, r) \subset \mathbf{P}^1\mathbf{C}$ and $B(w_0, r') \subset \mathbf{P}^1\mathbf{C}$ respectively. Let $w \in V$ and let $\phi_w(z)$ be defined on X by*

$$\phi_w(z) = z + \frac{1}{\beta'(z_0)}(w - \beta(z)).$$

Then $\phi_w(U) \subseteq U$ and for all $z \in U$, $|\phi_w(z) - \phi_w(z_0)| \leq \frac{1}{2}|z - z_0|$.

Proof: We first show that $|\phi_w(z) - \phi_w(z_0)| \leq \frac{1}{2}|z - z_0|$ for all $z \in U$. By the Mean Value Theorem, we know that

$$|\phi_w(z) - \phi_w(z_0)| \leq \sup|\phi_w'(z)||z - z_0|$$

where the sup is over $z \in U$. Now,

$$\phi_w'(z) = \frac{1}{\beta'(z_0)}(\beta'(z_0) - \beta'(z))$$

so we again apply the Mean Value Theorem to obtain

$$|\beta'(z) - \beta'(z_0)| \leq \sup|\beta''(z)||z - z_0| \leq C|z - z_0|.$$

So

$$\sup|\phi_w'(z)| \leq \frac{C|z - z_0|}{|\beta'(z_0)|}.$$

Now, since $z \in U$, $|z - z_0| \leq r$ so

$$|\phi_w(z) - \phi_w(z_0)| \leq \frac{Cr}{|\beta'(z_0)|}|z - z_0| \leq \frac{1}{2}|z - z_0|$$

as desired.

We now show that $\phi_w(U) \subseteq U$. It suffices to show that for $z \in U$, $|\phi_w(z) - z_0| \leq r$. Now,

$$|\phi_w(z) - z_0| \leq |\phi_w(z) - \phi_w(z_0)| + |\phi_w(z_0) - z_0| \leq \frac{1}{2}|z - z_0| + \frac{|w - \beta(z_0)|}{|\beta'(z_0)|}$$

$$\leq \frac{1}{2}r + \frac{|w - w_0|}{|\beta'(z_0)|} \leq \frac{1}{2}r + \frac{r'}{|\beta'(z_0)|} \leq r$$

by definition of r'. ◇

Note that the proof of the proposition shows that although the ball $B(z_0, r)$ (resp. $B(w_0, r')$) may not lie completely in X (resp. Y), it cannot

contain any of the critical points (resp. critical values) of β. The point of
the proposition is to show that β is injective on $B(z_0, r)$ and surjective onto
$B(w_0, r')$.

We return to the problem of calculating the permutations σ_0 and σ_1.
Since we have excluded from X neighborhoods of all points z such that
$\beta'(z) = 0$, we must have a lower bound for $|\beta'(z)|$ on X, say $K < |\beta'(z)|$ for
$z \in X$. Choose $r < K/2C$ and $r' < rK/2$. These numbers depend only on
β and on the original choice of open sets $\{U_\alpha\}$.

Let $x_0 \in Y$ be the base point for the curves γ_0 and γ_1 as before. For any
point w_0 on γ_0, let z_0 be a fixed pre-image of w_0 under β. Set $U = B(z_0, r)$
and $V = B(w_0, r')$. Choose any $w \in V$ and let

$$\phi_w(z) = z + \frac{1}{\beta'(z_0)}(w - \beta(z)).$$

Then by proposition III.1, for any $z \in U$, the sequence $\{\phi_w^n(z)\}$ must lie
entirely in $U = B(z_0, r)$. Therefore it must converge to the unique fixed
point of ϕ_w in U, namely the unique element $z \in U$ such that $\beta(z) = w$.
This means that in order to calculate each of the $2d$ paths in the pre-image of
γ_0 it suffices to cut γ_0 into pieces of length $< r'/2$. Suppose there are m such
pieces. Let x_0 be a base point for γ_0 as before. Choose m distinct points
$w_0 = x_0, w_1, \ldots, w_{m-1}$ on γ_0, one in each piece, so $|w_{i+1} - w_i| < r'$, i.e.
$w_{i+1} \in B(w_i, r')$. Set $w_m = x_0$. Now by proposition III.1, for $1 \leq i \leq 2d$,
we can apply iteration of the function ϕ_{w_1} to $z_0 = x_i$ to obtain the unique
point z_1 on the path g_i lying over w_1, then iteration of the function ϕ_{w_2}
to the point z_1 to obtain the unique point z_2 on g_i lying over w_2, and so
on, until we have entirely reconstructed the path g_i. In particular if x_i is
the starting point of the path g_i, the endpoint $x_{\sigma_0(i)}$ of g_i will be given by
iterating the function ϕ_{w_m} starting from the point $z_{m-1} \in g_i$ lying over
$w_{m-1} \in \gamma_0$. The same procedure obviously works for the path γ_1 to give
the permutation σ_1; note that r and r' do not need to be changed. Once the
permutations σ_0 and σ_1 have been calculated, the dessin is reconstructed
following the procedure in I, §4.

§2. The Belyi function associated to a genus zero dessin

We now start the second part of the procedure outlined at the beginning
of part III, that of reducing the construction of the Belyi function of a genus
zero dessin to a set of polynomial equations in several variables having only
a finite number of solutions. Let us recall again that this type of exact
algorithm will only give an explicit result in the cases where the dessin is
of genus zero and reasonably small – improved methods in the other cases

are discussed at length in the article by Couveignes and Granboulan in this volume. We separate the genus 0 dessins into two categories: trees, i.e. those for which $X_2 \setminus X_1$ consists of a single open cell, and the others. This is because the bipartite structure which can be put on a tree gives rise to a set of equations with just half the number of variables as in the general case, as can be seen by comparing theorems III.3 and III.5. We note that although this method always gives a Belyi function associated to D, there is no reason for it to give one defined over the smallest possible field. This question is also dealt with by Couveignes and Granboulan.

From now on we will suppose that there is a vertex at each end of every edge of the genus zero dessin D, so as to obtain clean Belyi functions. As mentioned at the end of part I, the Belyi function associated to any dessin of genus 0 is not well-defined, for if β is such a function, then β can be composed with any automorphism of $\mathbf{P}^1$, i.e. any element of $SL_2(\mathbf{C})$. In theorem III.3 we give a method for finding a Belyi function associated to any given genus zero dessin. Before doing so we note (cf. [SV]) that a rational clean Belyi function is easy to describe in terms of polynomials. Let $\beta(z) = A(z)/C(z)$ be a rational clean Belyi function. Then in particular $\beta(z) - 1 = \bigl(A(z) - C(z)\bigr)/C(z)$ must have roots of order exactly 2, so we must have $A(z) - C(z) = cB(z)^2$ for some polynomial $B(z) \in \mathbf{C}[z]$ having distinct roots and some constant c. As a converse we have:

Lemma III.2: *Let $A(z)$, $B(z)$ and $C(z) \in \mathbf{C}[z]$ be polynomials. Suppose that $B(z)$ has distinct roots, that $A(z) - C(z) = B(z)^2$, and that $AC' - CA' = \tilde{A}\tilde{C}B$ where for any polynomial $P(z) = \prod_i (z - a_i)^{n_i}$, we write $\tilde{P} = \prod_i (z - a_i)^{n_i - 1}$. Then $\beta(z) = A(z)/C(z)$ is a clean Belyi function.*

Proof: Set $\beta = A/C$. Then $\beta' = (AC' - CA')/C^2 = \tilde{A}\tilde{C}B/C^2$. So the roots of β' are given only by the (multiple) roots of A and of C, and the (simple) roots of B. The values of β at these roots are 0 (at the roots of A), ∞ (at the roots of C) and 1 (at the roots of B); moreover the ramification indices over 1 are all exactly 2 since B has distinct roots.   ◇

Let us now consider a clean genus zero dessin D.

Definition 7: The *valency* of a vertex of D is the number of edges coming out of it. The valency is a local property thus loops originating from the vertex are counted twice. The *valency* of an open cell is the number of edges bounding it, an edge being counted twice if the open cell lies on both sides of it.

If β is a clean rational Belyi function such that $D = \beta^{-1}([0,1])$, then the valency of each vertex (resp. open cell) of D is equal to the order of the corresponding zero (resp. pole) of β.

To a given dessin D, let us associate two valency lists. From now on, let $n = n_D$ denote the maximal valency of any vertex of D, and $m = m_D$ the maximal valency of any open cell. Let $V = \{u_1, \ldots, u_n\}$ be the vertex valency list where for $1 \leq i \leq n$, u_i is the number of vertices having valency i, and $C = \{v_1, \ldots, v_m\}$ be the open cell valency list, where for $1 \leq j \leq m$, v_j is the number of open cells of valency j. Note that there is only a finite number of genus 0 dessins having given valency lists V and C, for such a dessin must have e edges where $2e = \sum_i u_i + \sum_j v_j - 2$ by Euler's formula.

Let D be a genus zero dessin with e edges. We now begin the construction of the set of polynomial equations which will give a Belyi function associated to D. Let $V = \{u_1, \ldots, u_n\}$ and $C = \{v_1, \ldots, v_m\}$ be the valency lists of D. For $1 \leq i \leq n$, set

$$\tilde{P}_i(z) = z^{u_i} + C_{i,u_i-1} z^{u_i-1} + \cdots + C_{i,1} z + C_{i,0}$$

and for $1 \leq j \leq m$ set

$$\tilde{Q}_j(z) = z^{u_j} + D_{j,v_j-1} z^{v_j-1} + \cdots + D_{j,1} z + D_{j,0},$$

where the $C_{i,k}$ and the $D_{j,k}$ are indeterminates. Let the system of polynomials $\{\tilde{P}_i, \tilde{Q}_j\}$ be called $\tilde{R}_{V,C}$. Note that $\tilde{R}_{V,C}$ does not depend on the dessin D but only on its valency lists, which determine a finite number of dessins, all having e edges where $2e = \sum_i u_i + \sum_j v_j - 2$.

The aim of what follows is to show that there exist algebraic numbers $c_{i,k}$ and $d_{j,k}$ such that when the $C_{i,k}$ and the $D_{j,k}$ are replaced by these values, the rational function

$$\beta(z) = \frac{\prod_{i=1}^n \tilde{P}_i(z)^i}{\prod_{j=1}^m \tilde{Q}_j(z)^j}$$

becomes a Belyi function defined over $\overline{\mathbf{Q}}$ such that $D = \beta^{-1}([0,1])$.

The Belyi function β obtained in such a way will be defined only up to $SL_2(\mathbf{C})$. In order to fix a unique choice, we give new system of polynomials $R_{V,C}$ obtained from $\tilde{R}_{V,C}$ by specialization. The specialization consists in fixing three unknowns – which may be linear combinations of the indeterminates – to specific values. This can be done in any number of ways, and there are choices which have the advantage of minimizing the degree of the number field over which the Belyi function will be defined. We do not concern ourselves with this improvement here, but choose simply to set either a vertex or (the center of) an open cell of minimal valency to infinity. This is done in (i) of definition 8. Next, in (ii), we set one of the $C_{i,j}$ or the $D_{i,j}$ to the value 1 and another to 0, making sure that it is legitimate to do so.

Definition 8: The system $R_{V,C}$ of polynomials associated to the valency lists V and C is obtained from $\tilde{R}_{V,C}$ in three steps as follows.

(i) If the dessin has only one vertex, say of valency i_0, then $i_0 > 1$ by the assumption that D has a vertex at the end of every edge. Set

$$P_{i_0}(z) = \tilde{P}_{i_0}(z) - C_{i_0,1}z - C_{i_0,0} + z.$$

If the dessin has more than one vertex, choose an $i_0 \in \{1,\dots,n\}$ or a $j_0 \in \{1,\dots,m\}$ such that u_{i_0} or v_{j_0} is minimal in the set $\{u_i \mid 1 \le i \le n, u_i \ne 0\} \cup \{v_j \mid 1 \le j \le n, v_j \ne 0\}$. If an i_0 is chosen set

$$P_{i_0}(z) = \gamma(z^{u_{i_0}-1} + C_{i_0,u_{i_0}-2}z^{u_{i_0}-2} + \cdots + C_{i_0,1}z + C_{i_0,0}),$$

and if it is a j_0 set

$$Q_{j_0}(z) = \gamma(z^{v_{j_0}-1} + D_{j_0,v_{j_0}-2}z^{v_{j_0}-2} + \cdots + D_{j_0,1}z + D_{j_0,0}),$$

where γ is an indeterminate.

(ii) If the dessin has only one vertex, say of valency i_0, then the dessin possesses at least 2 open cells of valency 1, since such a dessin must consist of closed loops. Set

$$Q_1(z) = \tilde{Q}_1(z) - D_{1,1}z - D_{1,0} + z.$$

Now consider the case where the dessin has more than one vertex. If there exists any $i_1 \in \{1,\dots,n\}$ such that $u_{i_1} > 1$ and if an i_0 was chosen in (i), then $i_1 \ne i_0$ (resp. if there exists $j_1 \in \{1,\dots,m\}$ such that $v_{j_1} > 1$ and if a j_0 was chosen in (i), then $j_1 \ne j_0$) then set

$$P_{i_1}(z) = \tilde{P}_{i_1}(z) - C_{i_1,1}z - C_{i_1,0} + z$$

$$(\text{resp. } Q_{j_1}(z) = \tilde{Q}_{j_1}(z) - D_{j_1,1}z - D_{j_1,0} + z.)$$

If all non-zero u_i and v_j apart from the i_0 or j_0 chosen in (i) are equal to 1, then we can always choose a couple of one of the three forms (i_1,i_2), (i_1,j_1) or (j_1,j_2), in such a way that if an i_0 was chosen in (i) then i_1 and i_2 are different from i_0, and if a j_0 was chosen then j_1 and j_2 are different from j_0. If a couple of the type (i_1,i_2) is chosen set $P_{i_1}(z) = \tilde{P}_{i_1}(z) - C_{i_1,0}$ and $P_{i_2}(z) = \tilde{P}_{i_2}(z) - C_{i_2,0} + 1$. If a couple of type (i_1,j_1) is chosen set

$P_{i_1}(z) = \tilde{P}_{i_1}(z) - C_{i_1,0}$ and $Q_{j_1}(z) = \tilde{Q}_{j_1}(z) - D_{j_1,0} + 1$, and if a couple of type (j_1, j_2) is chosen set $Q_{j_1}(z) = \tilde{Q}_{j_1}(z) - D_{j_1,0}$ and $Q_{j_2}(z) = \tilde{Q}_{j_2}(z) - D_{j_2,0} + 1$.

(iii) For all $i \in \{1, \ldots, n\}$ and $j \in \{1, \ldots, m\}$ which were not chosen as an i_0, i_1, i_2, j_0, j_1 or j_2 as in (i) and (ii), set

$$P_i(z) = \tilde{P}_i(z)$$

and

$$Q_j(z) = \tilde{Q}_j(z).$$

Let the system of polynomials $R_{V,C}$ be given by the set $\{P_i, Q_j \mid 1 \le i \le n, 1 \le j \le m\}$.

Theorem III.3: *Let D be a genus zero dessin, assumed to have a vertex at each end of every edge. Let $V = \{u_1, \ldots, u_n\}$ and $C = \{v_1, \ldots, v_m\}$ be the valency lists of D, and let $R_{V,C} = \{P_i, Q_j\}$ be the associated polynomial system defined above. Let e be the number of edges of D, let $B_0, \ldots, B_{e-1}$ be indeterminates and set $B(z) = z^e + B_{e-1}z^{e-1} + \cdots + B_1 z + B_0$. Set $A(z) = \prod_{i=1}^{n} P_i(z)^i$ and $C(z) = \prod_{j=1}^{m} Q_j(z)^j$. Let $S_{V,C}$ be the system of polynomial equations given by comparing the coefficients on both sides of the equation*

$$A(z) - C(z) = \pm B(z)^2,$$

where the sign is positive or negative according to whether A or C has higher degree (by construction of $R_{V,C}$, their degrees are different). We have:

(i) For each solution s of $S_{V,C}$, let $\beta_s(z)$ be the rational function obtained from $A(z)/C(z)$ by substituting the values of the solution for the indeterminates. Then $\beta_s(z)$ is a rational clean Belyi function.

(ii) The dessins corresponding to the functions $\beta_s(z)$ are exactly the set of those having valency lists V and C. In particular, there exists at least one solution s of $S_{V,C}$ such that $D = \beta_s^{-1}([0,1])$.

(iii) The system $S_{V,C}$ admits only a finite number of solutions s. In particular, they are all defined over $\overline{\mathbb{Q}}$ and thus the same is true of the functions $\beta_s(z)$.

Proof: (i) This part is an immediate consequence of lemma III.2. Note that the sign in front of $B(z)^2$ is $+1$ exactly when an j_0 was chosen in (i) of the definition of $R_{V,C}$ and -1 when an i_0 was chosen.

(ii) This is an immediate consequence of the definition of valency given earlier and the remark immediately following this definition, relating the orders of the poles of β_s to the valencies of the open cells of D and those of the zeros to the valencies of the vertices.

(iii) Suppose that the system $S_{V,C}$ (which is a system with $2e$ equations and $2e$ indeterminates) admits an infinite number of solutions s. Then in particular there exists a dessin D' having the same valency lists V and C as D, such that an infinite number of solutions s give rise to Belyi functions $\beta_s(z)$ corresponding to D'. Now, either a vertex of D' of valency i_0 or an open cell of valency j_0 must be at ∞, and a vertex of valency i_1 or an open cell of valency j_1 must be at 0, according to the choices made in defining the system $R_{V,C}$, and the condition $C_{i_1,1} = 1$ means that the product of the vertices of valency i_1 is equal to 1. Clearly there are only a finite number of ways of realizing the dessin D' as the pre-image of a rational Belyi function under these conditions. In particular one such realization is given by an infinite number of Belyi functions $\beta_s(z)$, which is impossible by the Grothendieck correspondence. ◇

We now show how a similar but simpler system than $S_{V,C}$ can be obtained when the dessin is a tree.

Definition 9: A *tree* is a Grothendieck dessin $X_0 \subset X_1 \subset X_2$ of genus zero such that $X_2 \setminus X_1$ consists of exactly one open cell.

From the remark at the end of part I that $\beta^{-1}(\infty)$ gives a point in each open cell of the dessin corresponding to β, we see immediately that the Belyi function corresponding to a tree must be a polynomial. Such a polynomial has only two finite critical values, 0 and 1.

The following simplification for trees was described by Shabat (and is partially discussed in his article in this volume).

Definition 10: A polynomial $P \in \mathbf{C}[z]$ is said to be a *generalized Chebyshev polynomial* if there exist c_1 and $c_2 \in \mathbf{C}$ such that for all z_0 such that $P'(z_0) = 0$ we have either $P(z_0) = c_1$ or $P(z_0) = c_2$, i.e. P has at most 2 critical values. If the critical values of P are exactly $\{\pm 1\}$ we say that P is *normalized*.

Lemma III.4: *(i) Let $P(z)$ be a normalized generalized Chebyshev polynomial, and set $\beta(z) = 1 - P^2$. Then $\beta(z)$ is a clean Belyi polynomial and the dessin given by $\beta^{-1}([0,1])$ is a tree with ∞ in its open cell.*

(ii) Let T be a tree. Then there is a normalized generalized Chebyshev polynomial $P(z)$ such that setting $\beta(z) = 1 - P(z)^2$ we have $T = \beta^{-1}([0,1])$.

Proof: (i) If $\beta(z) = 1 - P^2(z)$ then β has only 0 and 1 as critical values. Thus β is clearly a Belyi function and since it has only one pole, $\beta^{-1}([0,1])$ must be a tree.

(ii) If T is a tree then there exists a rational Belyi function $\beta(z)$ such that $T = \beta^{-1}([0,1])$. Since T is a tree β has only one pole. Composing β with a suitable transformation in $SL_2(\mathbf{C})$ if necessary we may suppose the pole is at ∞ so β is a Belyi polynomial whose only critical values are at 0 and 1. Moreover because we assume that β is clean, we must have $\beta(z) - 1 = cQ(z)^2$ for some constant c and some polynomial Q having distinct roots. The critical points of β are the roots of Q and the critical points of Q. Moreover β can only have 0 and 1 as critical values, and 1 can only occur at the roots of Q, so at a critical point z_0 of Q which is not a root we must have $1 + cQ(z_0)^2 = 0$ so $Q(z_0) = \pm\sqrt{-1/c}$. Set $P(z) = \sqrt{-c}Q(z)$. Then $\beta(z) = 1 - P(z)^2$ and the critical values of P are ± 1.   ◇

The open cell valency list of a tree is particularly simple: there is only one open cell and its valency is twice the number of edges of the tree. Instead of using a vertex and an open cell valency list to describe the tree, we will describe it by two valency lists as follows. A bipartite structure on a tree is the assignation of a sign ± 1 to each vertex, in such a way that if a vertex is of one sign, every one of its neighbors is of the opposite sign. The bipartite structure is clearly unique up to global change of sign. From now on, let T be a tree with a bipartite structure, let n be the highest valency of any positive vertex and m the highest valency of any negative one. Let $V^+ = \{u_1, \ldots, u_n\}$ be the positive valency list, where u_i is the number of positive vertices having valency i, and $V^- = \{v_1, \ldots, v_m\}$ be the negative valency list, so v_j is the number of negative vertices having valency j. We will describe a set of polynomials R_{V^+,V^-} and a system of polynomial equations S_{V^+,V^-}, analogous to the sets $R_{V,C}$ and $S_{V,C}$ in theorem III.3, but smaller. We use identical notations as in the non-tree case in order to emphasize the similarity of the procedure.

For $1 \le i \le n$ set

$$\tilde{P}_i(z) = z^{u_i} + C_{i,u_i-1}z^{u_i-1} + \cdots + C_{i,1}z + C_{i,0}$$

and for $1 \le j \le m$ set

$$\tilde{Q}_j(z) = z^{v_j} + D_{j,v_j-1}z^{v_j-1} + \cdots + D_{j,1}z + D_{j,0},$$

where as earlier, the $C_{i,k}$ and the $D_{j,k}$ are indeterminates. Let $\tilde{R}_{V^+,V^-}$ be the set of polynomials $\{\tilde{P}_i, \tilde{Q}_j\}$; as before, this set only depends on the valency lists V^+ and V^- and therefore apply to a finite number of trees. We obtain a set of polynomials R_{V^+,V^-} from $\tilde{R}_{V^+,V^-}$ as follows. Choose an $i_0 \in \{1, \ldots, n\}$ such that $u_{i_0} \ne 0$ and set

$$P_{i_0}(z) = \tilde{P}_{i_0}(z) - C_{i,1}z - C_{i,0} + z.$$

For all $i \neq i_0$ set $P_i(z) = \tilde{P}_i(z)$ and for $1 \leq j \leq m$ set $Q_j(z) = \tilde{Q}_j(z)$. Let R_{V^+,V^-} be the set $\{P_i, Q_j\}$. Now we have a theorem for trees analogous to theorem III.3:

Theorem III.5: *Let T be a tree, assumed to have a vertex at the end of each edge, with a bipartite structure. Let $V^+ = \{u_1, \ldots, u_n\}$ and $V^- = \{v_1, \ldots, v_m\}$ be its positive and negative valency lists. Let*

$$P(z) = \prod_{j=1}^{m} Q_j(z)^j,$$

and let S_{V^+,V^-} be the set of polynomial equations obtained by comparing coefficients on both sides of the following equation:

$$P(z) - P(0) = \prod_{i=1}^{n} P_i(z)^i.$$

We have:

(i) For each solution s of S_{V^+,V^-}, let $P_s(z)$ be the normalized generalized Chebyshev polynomial given by replacing the indeterminates in the polynomial $\frac{2}{P(0)}P(z) - 1$ by the values of s, and let $\beta_s(z)$ be the polynomial obtained by replacing the indeterminates in the polynomial $1 - P_s(z)^2$ by the values of s. Then $\beta_s(z)$ is a clean Belyi polynomial.

(ii) The trees corresponding to the polynomials $\beta_s(z)$ are exactly the set of trees having valency lists V^+ and V^-.

(iii) The system S_{V^+,V^-} admits only a finite number of solutions, all defined over $\overline{\mathbb{Q}}$. In particular, all the $\beta_s(z)$ are defined over $\overline{\mathbb{Q}}$.

Proof: By construction, $P(z)$ has only two critical values, 0 and $P(0)$, so clearly $P_s(z)$ is a normalized generalized Chebyshev polynomial, and therefore $\beta_s(z)$ is a Belyi polynomial by lemma III.4. This proves (i). The proofs of (ii) and (iii) are identical to those in theorem III.3.   ◇

§3. The Gröbner basis algorithm

In order to explicitly obtain a rational Belyi function associated to a given genus zero dessin, it is necessary to be able to explicitly calculate all solutions to the set of equations $S_{V,C}$ or S_{V^+,V^-}. We do this using the Gröbner basis method (see [CLO] for a basic reference to this algorithm).

In order to apply this method, we need to impose an ordering on the indeterminates, which in turn imposes an ordering on the monomials in them. For instance, if the indeterminates are $x_1, \ldots, x_r$, we may put an ordering on the x_i via $x_i > x_j$ if and only if $i < j$. We put a lexicographic ordering on the monomials by decreeing that $x_i^a > x_j^b$ if and only if either $i < j$, or $i = j$ and $a > b$, and if A, B and C are monomials and $A < B$, then $AC < BC$.

Let S be an ideal in the polynomial ring $\mathbf{Q}[x_1, \ldots, x_r]$: in our case, the ideal generated by the system of equations $S_{V,C}$ or S_{V^+,V^-}. A Gröbner basis $\{g_1, \ldots, g_s\}$ of S with respect to the lexicographic ordering on the x_i has the following property: a set of representatives of $\mathbf{Q}[x_1, \ldots, x_r]/S$ is given by the set of power products in $x_1, \ldots, x_r$ which are not divisible by the leading (in the lexicographic sense) power product of any g_i (a power product is a monomial with coefficient equal to 1).

If, as in our case, the number of equations is equal to the number of indeterminates and the number of solutions to the system of equations generating S is finite, then there is only a finite number of power products in the x_i not divisible by the leading power product of any g_i. This implies that among the elements of the Gröbner basis $g_1, \ldots, g_s$ (with $s \geq r$), there are r of them, say $g_1, \ldots, g_r$ whose leading power products are of the form $x_1^{a_1}, \ldots, x_r^{a_r}$. In particular, g_r must be a polynomial in the single variable x_r. In our case, exactly one root of this polynomial belongs to the solution of $S_{V,C}$ which corresponds to the given dessin D. Now, the polynomial g_r may well be reducible. In that case, the set of solutions of $S_{V,C}$ corresponding to the set of dessins which are Galois conjugate to D come exactly from the irreducible factor of g_r one of whose roots corresponds to D itself. The solutions coming from roots of g_r which are not roots of this irreducible factor give other (complete Galois orbits of) dessins which are *not* Galois conjugate to D (see example 1 of IV for an example of this). In general, the system $S_{V,C}$ gives an ideal of dimension 0 of the polynomial ring, and the solutions corresponding to a Galois orbit of a given dessin correspond to one of the irreducible components of this ideal.

In order to apply the Gröbner basis method to the systems $S_{V,C}$ and S_{V^+,V^-} described in §2, we used the Maple package *grobner*. The routines in this package automatically select an ordering on the indeterminates. We have not yet found an example where the Gröbner basis given in this way does not have the form $\{g_1, \ldots, g_r\}$ where g_r is a polynomial in one of the indeterminates, and for $1 \leq i \leq r-1$, g_i is a polynomial in x_r and one other indeterminate, in which it is linear. We give a few examples in section IV. When this is true, the different solutions to the system $S_{V,C}$ (or S_{V^+,V^-}) are given by the roots of the final polynomial g_r.

IV. Examples of the method

We give here three basic examples of the procedure described in section III. Many more such examples are given in the articles by Birch, Couveignes-Granboulan and Malle in this volume. Note that in the examples given here, the Gröbner basis gave a Belyi function defined over the moduli field of the dessin which is also its field of definition in each case (see Couveignes-Granboulan for details on this question). We thus obtained an explicit minimal polynomial for the field of definition of the dessin. In the examples given here, the calculations were performed using Maple V, via the simple genus zero algorithm mentioned above and the Gröbner basis method.

Example 1: This is the tree mentioned in the introduction to this volume, whose Galois orbit contains only half of the trees having identical valency lists. Let T be the left-hand tree and T' the right-hand tree in the following diagram:

T has positive valency list $V^+ = \{5, 1, \ldots, 1\}$ (with 15 positive vertices of valency 1), and negative valency list $\{2, 3, 4, 5, 6\}$. In all there are exactly 24 trees having the same valency lists as T. Each one corresponds to a change in the ordering of the branches of T coming out of the central point; these define permutations of $\{2, 3, 4, 5, 6\}$ up to cyclic permutations.

The Gröbner basis method gives a minimal Belyi function as follows. We specialized by setting the unique positive vertex of valency 5 to 0 and the unique negative vertex of valency 6 to 1 (as usual, the open cell is located at ∞ so that our Belyi function will in fact be a polynomial). Let L be the splitting field of the following polynomial:

$$Q(z) = 104247\,z^{12} + 416988\,z^{11} + 977832\,z^{10} + 1716984\,z^9 + 2430621\,z^8 + 2818188\,z^7 + 2743316\,z^6 + 2259516\,z^5 + 1559145\,z^4 + 881776\,z^3 + 401604\,z^2 + 135828\,z + 26411.$$

For any root b_5 of $Q(z)$ (we call it b_5 because the negative vertex of valency

5 will be located at this root), define numbers b_2, b_3 and b_4 as follows:

$b_2 = \frac{-1}{90229675436255124}(11117953310160486933\, b_5^{11} +$
$74414217650153784975\, b_5^{10} +$
$224204971865186403387\, b_5^{9} + 439466725870053120081\, b_5^{8} +$
$649692977180346569502\, b_5^{7} + 770412410679635482950\, b_5^{6} +$
$739733459145142775770\, b_5^{5} + 574488699136179407930\, b_5^{4} +$
$359588050401471437497\, b_5^{3} + 176077260180110238271\, b_5^{2} +$
$60319246611391794719\, b_5 + 10806447247008193165).$

$b_3 = \frac{1}{11078199039673545 78}(696562804230787981773\, b_5^{11} +$
$34031585608293457345 59\, b_5^{10} +$
$857105422856926609963 5\, b_5^{9} + 1502111122258577231636 1\, b_5^{8} +$
$205034463061129556194 14\, b_5^{7} + 225754230655013470152 30\, b_5^{6} +$
$201681807516197272524 58\, b_5^{5} + 146453813399956012637 54\, b_5^{4} +$
$856583792213601487514 5\, b_5^{3} + 384805043080648582258 3\, b_5^{2} +$
$117317593846288720415 3\, b_5 + 18008042145995668020 1).$

$b_4 = \frac{1}{86887835605282712}(46002450933225137637\, b_5^{11} +$
$202981144281445544157\, b_5^{10} +$
$471365799526820237967\, b_5^{9} + 775670315522923479609\, b_5^{8} +$
$1004786466897370851324\, b_5^{7} + 104667181055507144740\, b_5^{6} +$
$879303815227739263954\, b_5^{5} + 598085687602105746722\, b_5^{4} +$
$324266105576407182307\, b_5^{3} + 129920504038754393755\, b_5^{2} +$
$32500268388779441943\, b_5 + 3375744126892136461).$

Set $P(z) = (z - b_2)^2(z - b_3)^3(z - b_4)^4(z - b_5)^5$. Then $P(z)$ is defined over L, and it is precisely the polynomial $P(z)$ of theorem III.5, b_2, b_3 and b_4 giving the positions of the negative vertices of valency 2, 3 and 4 respectively. Therefore for each root b_5 of $Q(z)$ we obtain a Belyi function by setting

$$\beta(z) = 1 - \left(\frac{2}{P(0)}P(z) - 1\right)^2.$$

It is easily verified that the 12 roots of $Q(z)$ give rise to 12 Belyi functions corresponding to non-identical trees. Therefore the Galois orbit of T consists in 12 trees and so the degree of its associated number field is 12. Thus, a primitive generator of the number field K_T is given by the root b_5 of $Q(z)$ such that the associated Belyi polynomial corresponds to T; this root is approximated by

$$b_5 \approx .07975979989 - .9494529866\, i,$$

the other values being given by

$$b_2 \approx .215145 + .535128299\,i, \quad b_3 \approx -.4121365 + .501616\,i,$$

$$b_3 \approx -.753923 - .244862\,i.$$

Recalling that $b_6 = 1$ and that the central point of T is located at 0 by choice of specialization, it is clear that these values give the right abstract tree since they give the correct ordering of the different branches around the central point.

The 24 possible orderings of the central branches of T can be expressed as permutations of $\{2, 3, 4, 5, 6\}$ starting with 2, i.e. permutations of $\{3, 4, 5, 6\}$. The 12 trees corresponding to the roots of $Q(z)$ turn out to correspond precisely to permutations of $\{3, 4, 5, 6\}$ by elements in A_4. This phenomenon appears to be quite mysterious. The other 12 trees having identical valency lists to those of T form a separate Galois orbit, that of the tree T' in the above diagram. The orbit is obtained from the splitting field of the polynomial

$$\tilde{Q}(z) = 104247\,z^{12} + 416988\,z^{11} + 977832\,z^{10} + 1717236\,z^9 + 2430117\,z^8 +$$
$$2818416\,z^7 + (8229940/3)\,z^6 + 2259416\,z^5 + 1559449\,z^4 + (2644796/3)\,z^3 +$$
$$401604\,z^2 +$$
$$135828\,z + 26411$$

and the three equations

$$b_2 = \tfrac{1}{4776142456713637719725 4}(20075431169583797779922436\,b_5^{11} +$$
$$7270622422973632991259 8919\,z^{10} + 13992259838714661704588 9469\,z^9 +$$
$$1899561691327833737051184 36\,z^8 + 20110250085593485855045 4541\,z^7 +$$
$$1606903049861110414356065 73\,z^6 + 893274272608492799060238 29\,z^5 +$$
$$2648381614760019254901525 7\,z^4 + 736018972350451724534604 7\,z^3 -$$
$$169379068386526162927840 14\,z^2 + 113683767612708972902943 72\,z -$$
$$32945489792102795802664 77).$$

$$b_3 = \tfrac{-1}{58640415718539663114406 3}(5342380868719752002936929 71\,z^{11} +$$
$$2257125852416175828475669710\,z^{10} + 504161475875776099376758 0908\,z^9 +$$
$$79457011107707709640290684 15\,z^8 + 9860654502732026849603165 379\,z^7 +$$
$$98204457505788041469993543 51\,z^6 + 78430330689790992734002262 51\,z^5 +$$
$$5024279762970053953329899659\,z^4 + 2541931899658532932729209056\,z^3 +$$
$$9145383977400159578095805\,z^2 + 17740867959699880018767590 0\,z +$$
$$1520874502563183976856956),$$

$$b_4 = \tfrac{1}{5519097949980203587238 24}(27325748166738582915470130 3\,z^{11} +$$
$$13119449374296035396118837 63\,z^{10} + 3222620979724854972171017811\,z^9 +$$

$54762709120445871058625287477\,z^8+724632199160487368844725634\,z^7+$

$7731471785759360608336468902\,z^6+6681776514588058634621916474\,z^5+$

$4679990153051251450192701318\,z^4+263761436606409165691870766\,7\,z^3+$

$1134731186643184171384287987\,z^2+32468016206070982087519915\,1\,z+$

$4386709016167355070643781\,1).$

We note that the division of the set of trees having same valency lists as T into two Galois orbits is not typical for this type of tree. A similar tree with negative valency list $\{1,2,3,4,5\}$ has a Galois orbit of order 24, as do those with negative valency lists $\{1,2,4,5,6\}$ and $\{2,3,4,5,7\}$, whereas the tree having negative valency list $\{1,2,3,4,6\}$ behaves like T. Another mysterious phenomenon pointed out to me by J-M. Couveignes is the remarkable similarity between the polynomials $Q(z)$ and $\tilde{Q}(z)$, whose three terms of highest degree and three terms of lowest degree are equal, and whose difference is divisible by $(X-1)^3$...

Example 2: We now consider an example of a dessin which is not a tree. Let D be given by

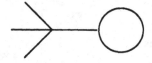

The valency lists of D are $V = \{3,0,1,1\}$ and $C = \{1,0,0,0,0,0,0,0,1\}$. There are in all three dessins having these valency lists; the other two are given by

We apply the Gröbner basis method by specializing the vertex of valency 4 to 0 and the vertex of valency 3 to 1. Then we find that the field of definition of D is of degree 3 and is given by

$$Q(z) = 147\,z^3 + 936\,z^2 + 1872\,z + 1120.$$

For any root r of $Q(z)$ set

$$N_r(z) = -459\,r^2 - 1260 - 1716\,r + 525\,z^2 + 350\,rz^2 +$$

$$567\,r^2 z + 2058\,rz + 1680\,z + 175\,z^3,$$

and

$$\gamma_r(z) = \frac{N_r(z)}{64(171477\,r^2 + 743823\,r + 740530)(-10\,z + 7\,r + 18)}.$$

Then we obtain a clean Belyi function for each root by setting

$$\beta_r(z) = -15882615z^4(z-1)^3\gamma_r(z).$$

A reconstruction from the dessin from this Belyi function as in III, §1 shows that the root of $Q(z)$ which corresponds to the dessin D is the real root, approximated by -1.093425511. The other two dessins are given by setting r to be the complex conjugate roots of $Q(z)$.

Example 3: We treat here the example given in [SV]. Let T be the tree with positive valency list $V^+ = \{1,1,1\}$ and negative valency list $V^- = \{2,2\}$ given by

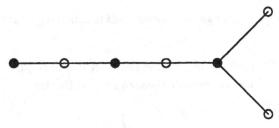

The system of equations R_{V^+, V^-} is given by

$$P_1(z) = z, \quad P_2(z) = z - 1 \text{ and } P_3(z) = z - C_{1,0}$$

and

$$Q_1(z) = z^2 + D_{1,1}z + D_{1,0} \text{ and } Q_2(z) = z^2 + D_{2,1}z + D_{2,0}.$$

There are exactly three solutions to these equations given as follows:

$$Q(z) = 25\,z^3 - 6\,z^2 - 6\,z - 2,$$

and $C_{1,0}$, $D_{1,1}$, $D_{2,1}$ and $D_{1,0}$ must take the values

$$2r, \quad 2/3 - (4/3)r, \quad -2/3 - (5/3)r, \quad \text{and} \quad 1/3 - (5/3)r^2 + (2/3)r$$

respectively. Thus if for a root r of $Q(z)$ we set

$$P_r(z) = \left(z^2 + (\frac{2}{3} - \frac{4}{3}r)z + (\frac{1}{3} + \frac{2}{3}r - \frac{5}{3}r^2)\right)\left(z^2 - (\frac{2}{3} + \frac{5}{3}r)z + r\right)^2,$$

we obtain a Belyi polynomial for each r by setting

$$\beta_r(z) = 1 - \left(\frac{2}{P_r(0)}P_r(z) - 1\right)^2.$$

The cubic equation $Q(z)$ has one real root r_0, and the Belyi function $\beta_{r_0}(z)$ corresponds to T. The remaining two complex conjugate roots of $Q(z)$ give the trees

(note that complex conjugation corresponds to reflecting a plane dessin over the real line).

Example 4: The calculation for the following dessin is quite complicated: it was performed by the number theory group in Bordeaux.

All dessins having identical valency lists to this dessin are in its Galois orbit, which has order 10. We give here a degree 10 polynomial whose roots describe the fields of definition of the orbit:

$Q(z) = z^{10} + 1482\,z^9 + 1689948\,z^8 + 890151444\,z^7 + 363946250304\,z^6 +$
$2267330869440\,z^5 - 1729356759663624\,z^4 + 75590803665798876\,z^3 -$

$1899051199966144224\,z^2 + 7231520112142277952\,z +$
634545639784165885776.

References

[B]   G. Belyi, On Galois extensions of a maximal cyclotomic field, Izv.
       Akad. Nauk SSSR, Ser. Mat. **43**:2 (1979), 269-276 (in Russian)
       [English transl.: Math. USSR Izv. **14** (1979), 247-256].

[C]   J-M. Couveignes, Calcul et rationnalité de fonctions de Belyi en genre
       0, *Annales de l'Institut Fourier* **44** (1), 1994.

[CLO] D. Cox, J. Little, D. O'Shea, Ideals, Varieties and Algorithms, Sprin-
       ger-Verlag, 1992.

[F]   O. Forster, Lectures on Riemann Surfaces, GRM 81, Springer-Verlag
       1981.

[G]   A. Grothendieck, *Esquisse d'un Programme*, Preprint 1985.

[JS]   G. Jones, D. Singerman, Theory of maps on orientable surfaces, Proc.
       London Math. Soc. (3) **37** (1978), 273-307.

[M]   B.H. Matzat, Konstruktive Galoistheorie, LNM 1284, Springer-Verlag,
       1985.

[Ma]   G. Malle, Polynomials with Galois groups $\mathrm{Aut}(M_{22})$, M_{22}, and
       $PSL_3(\mathbf{F_4}).2$ over $\mathbf{Q}$, *Math. Comp.* **51** (1988), 761-768.

[MV]   J. Malgoire and C. Voisin, Cartes Cellulaires, Cahiers Mathématiques
       de Montpellier No. 12, 1977.

[S]   G. Shabat, The Arithmetics of 1-, 2- and 3-edged Grothendieck
       dessins, Preprint IHES/M/91/75.

[SV]   G. Shabat and V. Voevodsky, Drawing Curves over Number Fields,
       The Grothendieck Festschrift, Vol. III. Birkhäuser, 1990.

*UA 741 du CNRS, Laboratoire de Mathématiques, Faculté des Sciences de
Besançon, 25000 Besançon, France

Dessins from a geometric point of view

Jean-Marc Couveignes* and Louis Granboulan**

Abstract

In this paper we study the topological aspects of dessins (via analytic description) with two distinct goals. Firstly we are interested in fields of definition and fields of moduli. We give a topological proof that there exist some dessins with no model defined over their field of moduli. This answers explicitly a question asked in [Har87]. Our second motivation is to collect practical and theoretical data for the explicit computation of covers given by some topological description, following ideas of Atkin [ASD71] Oesterlé and ourselves. This leads to a method for the computation of the linear space associated to a divisor on a given dessin.

§1. Introduction

This paper develops some practical applications of the archimedean analytic description of coverings through Puiseux series. In the second section, we recall a classical result due to Klein concerning the classification of genus zero Galois coverings, and related to the classification of regular polytopes. In the third section we give a review of many possible definitions of what a moduli field is. We do not claim to exhaust the list of various contradictory notions denoted by these words, but simply to avoid the frequent confusion about it. The fourth section is an illustration of what knowledge can be provided by local considerations at infinity. We show that such a study leads to interesting examples of coverings with strange rationality properties, which we can state by mere combinatorial considerations. In the fifth section we recall quite classical results related to the Legendre form of elliptic curves, which are useful in the next section. The sixth section consists in the analytic description of the linear systems associated with some divisors on the curve corresponding to a given dessin. It leads to a method for the explicit computation of an algebraic model. This provides us with an algorithmic correspondance between abstract dessins and explicit Belyi functions. We give quite general techniques. In the case where the genus of the dessin is small, the equations have a simple general form which helps make the method more beautiful. In view of that, it is convenient to have an a priori description of the equation of the curve and Belyi function, which is given by classical curve theory for low genera. We detail that in the seventh section.

The authors wish to thank Leila Schneps for many useful discussions and for the organization of the Luminy conference in April 1993, where we found the motivation for this work (especially the four talks given by Joseph Oesterlé).

Throughout this paper we represent the coverings as dessins. For definitions and motivations, the reader should read the article by Leila Schneps in this volume, and of course the introduction to this book. There are many possible combinatorial descriptions with such dessins. Let us illustrate this on a small example which we will consider throughout this paper every time we need to be more explicit. In our drawings, the points over 0 are denoted by a black bullet and the points over 1 correspond to the middles of the segments and to the extremities without bullets (unramified points over 1). This corresponds to Grothendieck's normalization. We ask that the ramification above 1 be equal to 1 or 2.

Let us consider the following genus zero and degree 3 dessin:

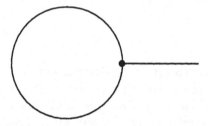

It has one vertex of multiplicity 3, corresponding to the totally ramified point over 0. There is one circular edge, corresponding to a point over 1 with ramification degree equal to 2, and one half-edge the extremity of which is an unramified point over 1. To finish, there are two faces. The inner one is unramified and the outer one is ramified of order 2.

Since the dessin is of degree 3, there are 3 flags which are drawn in the following picture

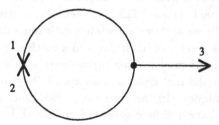

The monodromy of the dessin is given by the following three permutations of the flags which correspond to the elementary loops around 0, 1 and ∞.

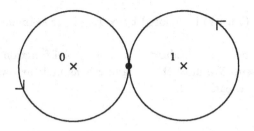

$$\sigma_0 = (1,2,3), \quad \sigma_1 = (1,2), \quad \sigma_\infty = (2,3).$$

The dessin itself is the preimage of the segment $[0,1]$ under the Belyi function. If we consider rather the preimage of the full real axis, we get a coloured triangulation of the sphere, consisting of three (grey) triangles oriented in the positive direction and sent by the Belyi function onto the upper half plane, and three (white) triangles, oriented in the inverse direction and lying above the lower half plane. This way, the dessin can be considered as a combinatorial covering of coloured triangulations.

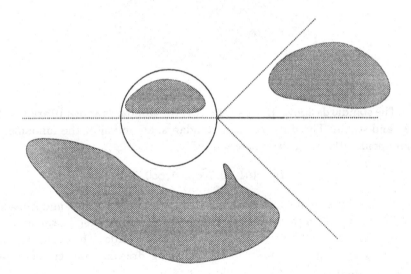

We now consider the elementary triangle $0, 1, \infty$ on the Riemann sphere. The middles of the three edges are -1, $1/2$ and 2. This splits the real axis into six open segments plus three points. The six open segments are called *standards* and we give each of them a name which will become clearer later. The segment $(0, 1/2)$ is denoted by $\vec{01}$; the segment $(1/2, 1)$ is denoted by $\vec{10}$, the segment $(1, 2)$ is denoted by $\vec{1\infty}$, the segment $(2, \infty)$ is denoted by $\vec{\infty 1}$, the segment $(\infty, -1)$ is denoted by $\vec{\infty 0}$ and the segment $(-1, 0)$ is denoted by $\vec{0\infty}$.

The preimages of these six standards under the Belyi function give 3×6 standards on the dessin. We draw these standards as little arrows and give an arbitrary number to each of them.

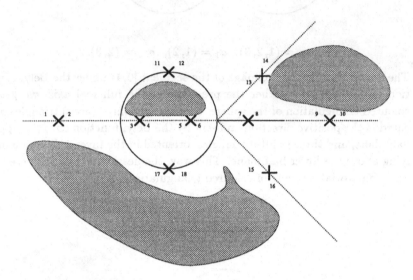

The standards above $\vec{01}$ are $\{7, 12, 18\}$, the standards above $\vec{10}$ are $\{11, 17, 8\}$ and so on. There is, on those standards, an action of the fundamental groupoid with "tangential base points"

$$\mathcal{B} = \{\vec{01}, \vec{10}, \vec{1\infty}, \vec{\infty 1}, \vec{\infty 0}, \vec{0\infty}\}$$

as defined by Deligne in [Del89] (see the appendix by Lochak and Emsalem to Ihara's article in this volume). This groupoid is generated by the paths $x_{\vec{v}}$ and $y_{\vec{v}}$ and $z_{\vec{v}}$ where $\vec{v}$ runs through the six standards. The paths $x_{\vec{01}} = x$, $y_{\vec{01}} = y$ and $x_{\vec{10}}$ are shown in the following drawing (we let the reader imagine what the other ones could be.)

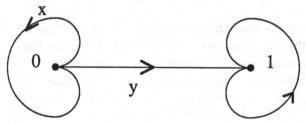

We also show $z_{0\bar{1}} = z$ below:

This action is given by the following maps:

$$x_{0\bar{1}} = (12, 18, 7),\ y_{0\bar{1}} = (7 \mapsto 8, 12 \mapsto 11, 18 \mapsto 17),\ y_{1\bar{0}} = y_{0\bar{1}}^{-1} \ldots$$

§2. Topological classification of genus zero covers

In this section, we recall a quite classical result first stated by Klein in its modern formulation [Kle13]. We need to introduce a certain number of Galois genus zero coverings of the sphere, corresponding to well-known dessins.

The first family corresponds to the dessins consisting of a star with e rays where e is a positive integer. A corresponding Belyi function is

$$y = f(x) = x^e$$

where e is the degree of the covering, totally ramified over 0 and ∞ and unramified elsewhere. We call these dessins C_e. Their topological Galois group is the cyclic group with e elements, C_e.

The second family corresponds to the polygon with $2e$ edges and admits the following Belyi function

$$-4y = x^e + x^{-e} - 2.$$

We call these dessins D_{2e}. Their topological Galois group is the dihedral group with $2e$ elements, D_{2e}.

We then have three coverings consisting of

- The tetrahedron which we call T of degree 12 and with Galois group the alternating permutation group on 4 letters A_4. A corresponding Belyi function is given by

$$yx^3(x^3 + 8)^3 = 2^6(x^3 - 1)^3.$$

- The octahedron O, of degree 24 with Galois group the full symmetric group on 4 letters S_4. A corresponding Belyi function is given by

$$y(x^8 + 14x^4 + 1)^3 = 2^2.3^3.x^4(x^4 - 1)^4.$$

- The icosahedron I, of degree 60 with Galois group the alternating permutation group on 5 letters A_5. A corresponding Belyi function is given by

$$y(x^{20} + 228x^{15} + 494x^{10} - 228x^5 + 1)^3 = x^5(x^{10} - 11x^5 - 1)^5.$$

We can now state Klein's theorem ([Kle13]), in which two coverings $\chi : \mathcal{C} \to \mathcal{D}$ and $\chi' : \mathcal{C}' \to \mathcal{D}'$ are said to be weakly isomorphic if there exist two isomorphisms c and d such that the following diagram commutes:

$$
\begin{array}{ccc}
\mathcal{C} & \xrightarrow{\ c\ } & \mathcal{C}' \\
\downarrow{\scriptstyle \chi} & & \downarrow{\scriptstyle \chi} \\
\mathcal{D} & \xrightarrow{\ d\ } & \mathcal{D}'
\end{array}
$$

They are *strongly* isomorphic if $\mathcal{D} = \mathcal{D}'$ and d can be chosen to be the identity.

Theorem 1: *Any algebraic Galois covering of the sphere is weakly isomorphic to one of the following: C_e or D_{2e} with $e \geq 1$, or T, O, or I.*

Proof: We first note that the Galois group G of such a covering G is a finite subgroup of $\mathrm{PGL}_2(\mathbf{C})$. Such subgroups are known to be isomorphic to one of the following: C_e, D_{2e} with $e \geq 2$, A_4, S_4, or A_5. The proof is quite elementary and uses the fact that a non-trivial element of finite order in $\mathrm{PGL}_2(\mathbf{C})$ has two fixed points ([Arm88, p 104]). If we call X the set of such fixed points, then G acts on X. One of the consequences of the proof is that there are 2 orbits if G is cyclic and 3 otherwise. The order of the stabilizer of a point in X just depends on its orbit. For each orbit $\mathcal{O}$ we denote by $(o_\mathcal{O}, s_\mathcal{O})$ the couple consisting of its cardinality and the order of the stabilizer of some element in $\mathcal{O}$. We list the values we obtain in each case:

- $(1, e), (1, e)$ for C_e.

- $(e, 2), (e, 2), (2, e)$ for D_{2e}.

- $(4, 3), (6, 2), (4, 3)$ for A_4.

- $(6,4)$, $(12,2)$, $(8,3)$ for S_4.

- $(12,5)$, $(30,2)$, $(20,3)$ for A_5.

It is clear that these fixed points are the ramification points of the covering with orders of ramification the orders of their stabilizer. This proves that either the covering is cyclic or there are exactly three singular values.

If the covering is cyclic, one can suppose that it is totally ramified over 0 and ∞ and that the single point above 0 is 0 and the single point above ∞ is ∞. We then get a function of the form $y = Ax^e$ which is clearly equivalent to the one we gave.

If there are three ramification values we can send them on 0, 1 and ∞ using the 3-transitivity of $\mathrm{PGL}_2(\mathbf{C})$. Note that we have put those three ramification values in some definite order in the above table. We respect this order in that we send the first one to 0, the second one to 1 and the third one to ∞.

Then, a strong isomorphism class of finite coverings is given by a subgroup of finite index of $\pi_1(\mathbf{P}_1(\mathbf{C}) - \{0, 1, \infty\}, b)$ where $b = 1/2$ is the base point. We choose the following basis of $\pi_1(\mathbf{P}_1(\mathbf{C}) - \{0, 1, \infty\}, b)$ that induces an isomorphism to the free group with two generators (σ_0, σ_1):

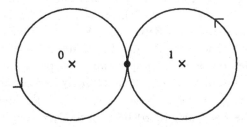

We write $\sigma_\infty^{-1} = \sigma_0 \sigma_1$. Now we can associate to G a subgroup g of π_1. Let us write G_0 for the covering in the list given above which has G as Galois group, and let g_0 be the corresponding subgroup. We prove that $g = g_0$.

Suppose for example that $G = A_4$. Both g_0 and g have index 12. They both contain σ_0^3, σ_1^2 and σ_∞^3 because of the ramification orders. The point now is that the subgroup generated by σ_0^3, σ_1^2 and σ_∞^3 is of finite index 12 (trivial from the classical presentation of A_4) so that $g = g_0 = \langle \sigma_0^3, \sigma_1^2, \sigma_\infty^3 \rangle$. The remaining cases are similar and reduce to the classical presentations of rotation groups.

Remark: Such a method no longer works for arbitrary genus. For example the following genus one dessin (where the opposite sides are identified) is Galois and the corresponding subgroup contains σ_0^4, σ_1^2 and σ_∞^4.

But the following dessin has the same property, and yet it is different. Indeed the subgroup generated by σ_0^4, σ_1^2 and σ_∞^4 is not of finite index.

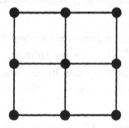

§3. Fields of definition, fields of moduli

In this section we simply recall a certain number of definitions in order to clarify our terminology for the rest of the paper. Let $\mathcal{D}$ be a dessin, that is, an isomorphism class over $\overline{\mathbf{Q}}$ of Belyi pairs. We recall that a Belyi pair is a made of a curve $\mathcal{C}$ defined over $\overline{\mathbf{Q}}$ and a function $\chi : \mathcal{C} \to \mathbf{P}_1(\overline{\mathbf{Q}})$ defined over $\overline{\mathbf{Q}}$ and unramified outside $\{0, 1, \infty\}$. Two Belyi pairs are said to be equivalent if the corresponding coverings are strongly isomorphic.

Let $\mathbf{K}$ be a number field and $\mathcal{C}$ a projective curve and ϕ a function on $\mathcal{C}$, defined over $\mathbf{K}$. If the Belyi pair $(\mathcal{C}, \phi)$ belongs to $\mathcal{D}$, we say that $\mathbf{K}$ is a *field of definition* of $\mathcal{D}$.

There is an action of Γ on the set of dessins. This action can be seen as the naive action on the coefficients of the equations of any Belyi pair. We call $\Gamma_{\mathcal{D}}$ the stabilizer of $\mathcal{D}$ and $\mathbf{K}_{\mathcal{D}}$ its fixed field. We call $\mathbf{K}_{\mathcal{D}}$ the moduli field of $\mathcal{D}$.

The moduli field is contained in any field of definition and is actually the intersection of all the possible fields of definition ([CH85]). It need not be a field of definition itself as we will show in section 4.

Note that we do not ask that the automorphisms of the covering (if any) be defined over the field of definition, which could have the effect of augmenting it. On the other hand, the field of moduli of $\mathcal{C}$ itself might be strictly smaller than the one of the dessin. For genus zero dessins, $\mathbf{P}_1(\mathbf{C})$ has $\mathbf{Q}$ as field of moduli but Lenstra proved that there exist genus zero dessins with arbitrary field of moduli (see the article by L. Schneps in this volume). To finish with the *distinguo* we should warn the reader that people studying modular forms over possibly non-congruence subgroups, usually consider structures that are somewhat richer than dessins. Following Birch (see his contribution to this volume), we define marked dessins to be dessins plus a fixed marked point over infinity. In this case, of course, the field of moduli might become bigger but it is more likely to be a field of definition (for example, it will always be one in genus zero).

In the case where the dessin has no automorphisms, it must admit a model over its field of moduli K_D by Weil's criterion ([Wei56]). In this case we note that the corresponding K_D-isomorphism class of curves is characteristic of the dessin. We will see an example of this in the next section.

§4. Galois action. Descending from C to R

In this section we illustrate the problems of fields of definition and descent on the toy example of descending from **C** to **R**. This is particularly interesting because we can give topological criteria for the descent. Further results on this subject can be found in [FD90]. Here, we are only interested in descent with extensions ramified over three points which thus can be chosen to be real, and which we take to be our favourite ones.

Let us denote by S_3 the sphere minus three points $P_1(C) - \{0, 1, \infty\}$ with base point $b = 1/2$ and the same basis as above for the π_1. A covering is thus given by two permutations a_0 and a_1 of the fibre over b, corresponding to the paths σ_0 and σ_1.

We write $M_{0,1,\infty}$ for the maximal extension of $R(t)$ unramified outside $\{0, 1, \infty\}$. We consider the following tower of extensions

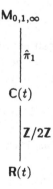

and the corresponding exact sequence of groups

$$1 \to \hat{\pi}_1 \to \mathcal{G} \to Z/2Z \to 1.$$

We recall that there exist two **R**-isomorphism classes of genus zero curves, the class of the straight line $\mathbf{P}_1(\mathbf{R})$ and the class of the plane curve given by the equation

$$x^2 + y^2 + z^2 = 0,$$

which we call $\tilde{\mathbf{P}}_1(\mathbf{R})$.

Given a dessin $\mathcal{D}$ by its monodromy (a_0, a_1), or equivalently, a triangulation of a surface, we ask three questions:

- Is the moduli field of $\mathcal{D}$ equal to **C** or **R**?

- If the moduli field is **R**, does the dessin admit a model over **R**?

- If a real genus zero dessin has no automorphisms, it admits a real model. Then can we say whether the underlying curve is $\mathbf{P}_1$ or $\tilde{\mathbf{P}}_1$?

We will give examples of all the possible situations and finish with an example of a real dessin (i.e. a dessin with real moduli field) with no real model.

To answer the first question we note that the outer action of $\mathbf{Z}/2\mathbf{Z}$ on $\hat{\pi}_1$ comes from an action on π_1 itself. Let τ denote the reflection of the plane induced by the unique non-trivial element $\tau \in \mathrm{Gal}(\mathbf{C}/\mathbf{R}) = \mathbf{Z}/2\mathbf{Z}$. This reflection is continuous and thus induces an involution of π_1. The images of σ_0, σ_1, σ_∞ are given by ${}^\tau\sigma_0 = \sigma_0^{-1}$, ${}^\tau\sigma_1 = \sigma_1^{-1}$, and ${}^\tau\sigma_\infty = \sigma_1\sigma_0$.

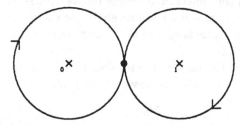

Now let $\chi : \mathcal{C} \to S_3$ be an algebraic covering of degree d and ${}^\tau\chi : {}^\tau\mathcal{C} \to S_3$ its conjugate under τ. There is a bijection induced by τ between the fibre of χ above b and the fibre of ${}^\tau\chi$ above b. Let $\{b_1, b_2, ..., b_d\}$ denote the points above b and $\{{}^\tau b_1, {}^\tau b_2, ..., {}^\tau b_d\}$ their images under τ. If σ is a closed path in π_1 and b_i a point above b on $\mathcal{C}$, then $\sigma(b_i)$ denotes the extremity of the lifted path on $\mathcal{C}$, with origin b_i. On the other hand, we can lift ${}^\tau\sigma$ onto ${}^\tau\mathcal{C}$, with origin ${}^\tau b_i$, so ${}^\tau\sigma({}^\tau b_i)$ is the extremity of the lifted path. Then ${}^\tau\sigma({}^\tau b_i) = {}^\tau(\sigma(b_i))$.

This means that the action of σ on the fibre $\chi^{-1}(b)$ is conjugated by τ to the action of ${}^\tau\sigma$ on the fibre ${}^\tau\chi^{-1}(b)$. Therefore, if χ was given by its monodromy (a_0, a_1), the monodromy of ${}^\tau\chi$ is (a_0^{-1}, a_1^{-1}) where a_0^{-1} and a_1^{-1} can be seen as permutations of $\{{}^\tau b_1, {}^\tau b_2, ..., {}^\tau b_d\}$ through the bijection with

$\{b_1, b_2, ..., b_d\}$ induced by τ. This gives an explicit description of the outer action of $\mathrm{Gal}(\mathbf{C}/\mathbf{R})$ in the above exact sequence.

Now, a dessin of degree d will be said to be real if and only if its field of moduli is $\mathbf{R}$. If (a_0, a_1) is its monodromy, this is just saying that there exists a permutation $\omega \in S_{\{b_1, b_2, ..., b_d\}}$ such that

$$\tau(a_0, a_1) = (a_0^{-1}, a_1^{-1}) = {}^{\omega}(a_0, a_1) = (\omega^{-1} a_0 \omega, \omega^{-1} a_1 \omega)$$

If this is the case, we note that ω belongs to the normalizer of $G = \langle a_0, a_1 \rangle$ in $S_{\{b_1, b_2, ..., b_d\}}$, and is defined up to an automorphism of $\mathcal{D}$ (we recall that the automorphism group of $\mathcal{D}$ is $a = \mathcal{Z}_{S_{\{b_1, b_2, ..., b_d\}}}(G)$, the centralizer of G in the full permutation group, see [Cou94]). Furthermore, since τ is an involution, we have $\omega^2 \in a$. Now, the dessin $\mathcal{D}$ admits a model over $\mathbf{R}$ if and only if ω can be chosen to satisfy Weil's cocycle condition

$$\omega^2 = 1.$$

Indeed, associated to ω, there is a morphism $H : \mathcal{C} \to {}^{\tau}\mathcal{C}$ such that the following diagram commutes

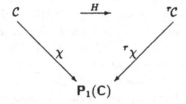

and H and ω are linked by the following identity

$$H(b_i) = {}^{\tau}(\omega(b_i)).$$

The cocycle condition on H for the existence of a real model is ${}^{\tau}HH = I$ which is immediately translated on ω as $\omega^2 = 1$.

We now come to the situation where the dessin $\mathcal{D}$ has no automorphisms. In this case ω is unique and ω^2 can only be equal to 1 and we have a model over $\mathbf{R}$ (here H is nothing but the identity)

and the action of τ extends to the real curve C in a way that makes the following diagram commute:

$$
\begin{array}{ccc}
C & \xrightarrow{\tau} & C \\
\downarrow{\scriptstyle\chi} & & \downarrow{\scriptstyle\chi} \\
P_1(C) & \xrightarrow{\tau} & P_1(C)
\end{array}
$$

The action of τ on the fibre above b is thus given by the formula

$$
{}^\tau b_i = \omega(b_i)
$$

and since ω conjugates a_0 and a_0^{-1}, it induces a permutation of the cycles of a_0 which gives the action of τ on the fibre $\chi^{-1}(0)$. In the same way we describe the Galois action on $\chi^{-1}(1)$ and $\chi^{-1}(\infty)$.

Suppose that among the cycles of σ_0 and σ_1 there is one which is fixed under the action of ω. Then, the corresponding point on C is real and thus C is isomorphic over R to the projective line $P_1(C)$.

To state the reciprocal assertion, we need to work a bit more. Suppose that χ is a real rational fraction: $\chi : P_1(R) \to P_1(R)$ associated to the dessin D. Let c be some connected component of the preimage of the open segment $(0, 1)$. Because χ is real and unramified over $(0, 1)$, c is either contained in R, or does not intersect it. If there exists such a c contained in R then its extremities are real thus proving the assertion that at least one point over $\{0, 1\}$ is real, and so the corresponding cycle must be fixed by ω. On the other hand, suppose that $\chi^{-1}((0, 1)) \cap R$ is empty. We note that $\chi^{-1}([0, 1]) \cap R$ cannot be empty because $\chi^{-1}([0, 1])$ is a connected non-empty subset of the plane which is invariant under the reflection τ. This again proves the desired statement.

We finish by stating

Theorem 2: *Let D be a dessin, given by its monodromy (a_0, a_1, a_∞). Then the field of moduli of D is R if and only if there exists some ω such that $a_0^{-1} = {}^\omega a_0$ and $a_1^{-1} = {}^\omega a_1$. In the latter case, the dessin admits a real model if and only if ω can be chosen so that $\omega^2 = 1$. If D is of genus zero and its automorphism group (the centralizer of $\langle a_0, a_1 \rangle$) is trivial, then the dessin admits a rational model over some real genus 0 curve. This curve is isomorphic to $P_1(R)$ if and only if the action of ω over the cycles of a_0 and a_1 has at least one fixed point.*

Examples

The rabbit is a real dessin with no automorphisms and admits a real model on the projective line.

To see this, we give numbers to the flags and compute the monodromy.

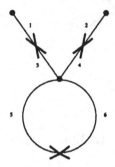

$$a_0 = (3,5,6,4), \quad a_1 = (1,3)(2,4)(5,6), \quad a_\infty = (4,2,6,3,1).$$

It is clear that there are no automorphisms. If a is a permutation which commutes with a_0, it must fix 1. But since it commutes with a_1 as well, it must fix 3 as well. Now, coming back to a_0 we see that a must be the identity. Furthermore we have

$$\omega = (3,4)(1,2)(5,6),$$

and check that the dessin is real.

The action of ω on the cycles of a_0 and a_1 fixes the cycle $(3,5,6,4)$ in a_0 and the cycle $(5,6)$ in a_1. This is more than enough to prove that the dessin has a real model on the projective line.

The rabbit with a lopped off left ear is a non-real dessin.

The monodromy is given by:

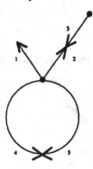

$$a_0 = (1, 4, 5, 2), \ a_1 = (2, 3)(4, 5), \ a_\infty = (5, 1, 2, 3).$$

Here there is no hope of finding an ω since such a permutation should fix 3 (from a_0) and 4 (from a_∞) and thus 2 and 5 as well (from a_1). This does not work.

The smiling rabbit obviously has an automorphism group of order 2.

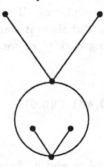

The rabbit with a lopped off left ear and a sidelong smirk on the right hand side is a real dessin with no non-trivial automorphisms and real model on the real curve $\tilde{P}_1$ with equation $x^2 + y^2 + z^2 = 0$.

Its monodromy is:

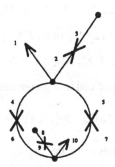

$$a_0 = (1,4,5,2)(9,6,7,10),\ a_1 = (2,3)(4,6)(8,9)(7,5),$$

$$a_\infty = (4,9,8,10,7)(3,5,6,1,2).$$

There are no non-trivial automorphisms (exercise) and there is a unique ω defined as

$$\omega = (1,10)(3,8)(9,2)(6,5)(4,7),$$

and none of the cycles of a_0 and a_1 are fixed by ω.

The double rabbit is a real dessin with no real model.

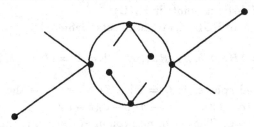

Its monodromy is:

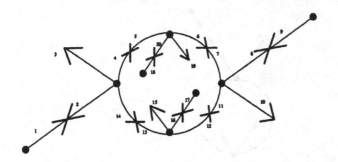

$$a_0 = (3, 2, 14, 4)(5, 20, 19, 6)(7, 11, 10, 8)(15, 13, 12, 16),$$

$$a_1 = (1, 2)(4, 5)(6, 7)(11, 12)(13, 14)(16, 17)(8, 9)(18, 20),$$

$$a_\infty = (18, 5, 14, 15, 16, 17, 12, 7, 19, 20)(4, 6, 8, 9, 10, 11, 13, 2, 1, 3).$$

There is an automorphism group of order 2 generated by

$$a = (3, 10)(8, 2)(9, 1)(14, 7)(6, 13)(5, 12)(11, 4)(20, 16)(15, 19)(17, 18).$$

The dessin is real for we can choose ω to be

$$\omega = (15, 3, 19, 10)(16, 2, 20, 8)(17, 1, 18, 9)(13, 4, 6, 11)(12, 14, 5, 7).$$

We could have chosen $a\omega$ instead. But $(a\omega)^2 = \omega^2$ is *not* the identity. This proves that our dessin although real, has no real model.

§5. Spheres minus four points

In this section we recall the basics about the Legendre form for elliptic curves. We are interested in building moduli spaces for spheres minus four points. To begin with, we define two different kinds of spheres minus four points. A non-coloured sphere minus four points is defined as a set of four distinct points $\{a, b, c, d\}$ on the complex projective line. A coloured sphere is a quadruplet of distinct points in $\mathbf{P}_1(\mathbf{C})$.

There are actions of $\mathrm{PGL}_2(\mathbf{C})$ on both sets, defined by

$$H\{a, b, c, d\} = \{Ha, Hb, Hc, Hd\}, \ H(a, b, c, d) = (Ha, Hb, Hc, Hd).$$

The set of coloured spheres is $\mathcal{C} = \mathbf{P}_1^4 - \mathcal{D}$ where $\mathcal{D}$ is the discriminant variety defined as $(a - b)(a - c)(a - d)(b - c)(b - d)(c - d) = 0$. The group $\mathcal{S}_4$ acts naturally on $\mathcal{C}$. The set of non-coloured spheres is the quotient $\mathcal{N}$ of $\mathcal{C}$ by $\mathcal{S}_4$.

We thus have a decolouration covering s_4 which is Galois with Galois group S_4.

$$\begin{array}{c} \mathcal{C} \\ \Big\downarrow {\scriptstyle s_4} \\ \mathcal{D} \end{array}$$

We define the classical cross-ratio function on $\mathcal{C}$

$$[a,b,c,d] = \lambda(a,b,c,d) = \frac{b-c}{b-a}\cdot\frac{d-a}{d-c}.$$

It is well known that two elements in $\mathcal{C}$ belong to the same $PGL_2(\mathbf{C})$-orbit if and only if λ takes the same value at those points.

We note that λ is invariant under the Klein group, seen as the subgroup V of S_4 generated by the permutations of type $(2,2)$. This subgroup is normal so that the covering splits in two. We note $\mathcal{H} = \mathcal{C}/V$, v the corresponding V-covering, and s_3 the S_3-covering of the lower part:

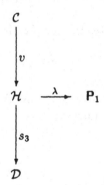

and we have the exact sequence

$$1 \to V \to S_4 \to S_3 \to 1.$$

It is tempting (although not particularly original...) to look at the action of S_3 on λ. It is given in the following list:

$$
\begin{array}{ll}
[[1,2]] & \lambda \mapsto 1-\lambda \\
[[1,3]] & \lambda \mapsto 1/\lambda \\
[[2,3]] & \lambda \mapsto \lambda/(\lambda-1) \\
[[1,2,3]] & \lambda \mapsto (\lambda-1)/\lambda \\
[[1,3,2]] & \lambda \mapsto 1/(1-\lambda)
\end{array}
$$

This action is killed by the function

$$J(\lambda) \stackrel{def}{=} 2^8 \frac{(\lambda^2 - \lambda + 1)^3}{\lambda^2(\lambda - 1)^2}$$

which defines a Galois covering with (strong) automorphism group S_3. Note the following amusing fact: J also admits a weak automorphism, namely

$$J(\frac{\lambda - 2}{2\lambda - 1}) = \frac{1728 J}{J - 1728} \qquad (1)$$

The linear fraction

$$\delta(\lambda) = \frac{\lambda - 2}{2\lambda - 1}$$

is the one which sends the triangle $(0, 1, \infty)$ to the triangle $(2, -1, 1/2)$. It is of order six. The linear fraction

$$\rho(J) = \frac{1728 J}{J - 1728}$$

is of order two and permutes the ramification locus of J. We have

$$\rho J = J\delta.$$

This will appear later on. We can draw the reciprocal image of $[0, 1728]$ under J and find the dessin below:

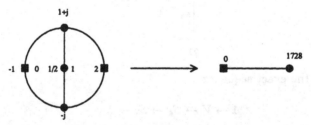

We set $J(a, b, c, d) = J(\lambda(a, b, c, d))$ and get a symmetric function of (a, b, c, d) defined over $\mathcal{D}$:

$$J(a, b, c, d) = 2^8 \cdot \frac{E_4^3}{\text{disc}(a, b, c, d)}$$

and

$$J - 1728 = 2^6 \cdot \frac{E_6^2}{\text{disc}(a, b, c, d)}, \qquad (2)$$

where $E_4 = 12\sigma_4 + \sigma_2^2 - 3\sigma_1\sigma_3$ and $E_6 = 72\sigma_2\sigma_4 - 2\sigma_2^3 + 9\sigma_1\sigma_2\sigma_3 - 27\sigma_3^2 - 27\sigma_4\sigma_1^2$, and $\mathrm{disc}(a, b, c, d) = \Delta$ denotes the discriminant. Those polynomials satisfy the expected invariance relations. For example, if $H(z) = (uz + v)/(wz + k)$ then we have

$$\frac{\Delta(H(a), H(b), H(c), H(d))}{\Delta(a, b, c, d)} = \frac{(uk - vw)^{12}}{(-u^4 + \sigma_1 u^3 w - \sigma_2 u^2 w^2 + \sigma_3 uw^3 - \sigma_4 w^4)^6}$$

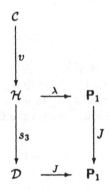

We note that the above commutative diagram is compatible with the Galois actions of S_3 on each side. It seems as well that the right hand side of this is incomplete (one level is lacking). In the sequel we try to see what can be done to complete this construction. We first remember of the existence of a Galois genus 0 extension of the sphere with group S_4. We build such an extension in the following way. Let $\mathcal{B}(x) = 1/4(x+1/x)^2 = 1+1/4(x-1/x)^2$ be the Galois function with automorphism group V, ramified over 0, 1, ∞.

We draw the corresponding dessin:

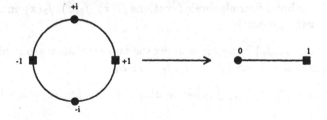

The composition $J \circ \mathcal{B}$ is a function ramified over 0, 1728, ∞ which defines the only genus zero S_4-extension of $\mathbf{P}_1$ ramified at those places (in *that* order).

To each value of x we associate the quadruplet $Q(x) = (x, -x, 1/x, -1/x)$ such that $\lambda(Q(x)) = [x, -x, 1/x, -1/x] = \mathcal{B}(x)$ and get the following commutative diagram:

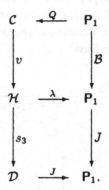

Note also that we can define $q(\lambda) = v(Q(x)) = v(x, -x, 1/x, -1/x)$ where x is any point such that $\mathcal{B}(x) = \lambda$. Such a point $q(\lambda)$ on $\mathcal{H}$ can be defined by its V-symmetric functions $\sigma_1 = 0$, $\sigma_2 = 2 - 4\lambda$, $\sigma_3 = 0$, $\sigma_4 = 1$, and $[x, -x, 1/x, -1/x] = \lambda$. This way we get the following commutative diagram

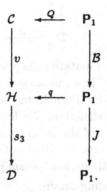

We would like to build four algebraic functions $f_1(x)$, $f_2(x)$, $f_3(x)$ and $f_4(x)$ with the following properties:

- The set $\{f_1, f_2, f_3, f_4\}$ is invariant under the automorphism group of $J \circ \mathcal{B}$, and this group $\mathcal{G}$ acts on $\{f_1, f_2, f_3, f_4\}$ like $\mathcal{S}_4$.

- The cross-ratio $[f_1, f_2, f_3, f_4]$ is (something like) $\lambda = [x, -x, 1/x, -1/x] = \mathcal{B}(x)$.

To do this, we write $\mathcal{S}_3^i$ for the stabilizer of i in $\mathcal{S}_4$ for $i \in \{1, 2, 3, 4\}$. The corresponding subextensions of $C(x)/C(j)$ are genus zero fields. We choose f_1 to be a generator of $C(x)^{\mathcal{S}_3^1}$. We then can choose $f_2(x) = f_1(-x)$, $f_3(x) = f_1(1/x)$ and $f_4(x) = f_1(-1/x)$. Note that the f_i are defined up to a linear transform on the left.

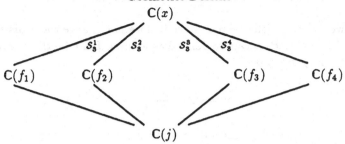

We can be more precise if we look for the minimal polynomial of f_1 with coefficients in $\mathbf{C}(j)$. To compute it, we just quotient the dessin corresponding to a cube by the group S_3^1 which can be seen as the stabilizer of one of the four diagonals of the cube.

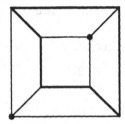

We thus get the following dessin.

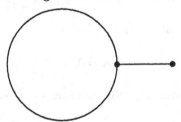

If we send the vertex of order three to zero and the one of order one to one, the corresponding Belyi function will be $X \mapsto Y$ such that

$$Y + 2^{14} \cdot X^3(X - 1) = 0.$$

In other words, we choose for f_i the four roots of the equation

$$j + 2^{14} \cdot f^3(f - 1) = 0. \tag{3}$$

On the other hand, the map $x \mapsto f_1$ is a Galois covering with group S_3. As we saw in the second section, such a covering must be equal to the classical J covering up to linear transforms L and R on both sides:

$$f_1(x) = L(J(R(x))).$$

We don't worry too much about L since it does not change the cross-ratio $[f_1, f_2, f_3, f_4]$. As for R, it can be defined as follows. Let r be the primitive 8-th root of unity given by

$$r = \sqrt{2} \cdot \frac{1+i}{2}$$

and let R be the linear transform defined by the matrix

$$R = \begin{pmatrix} 3 - r - 2r^2 + 4r^3 & 1 - r + 2r^2 \\ 3 - 2r - r^2 + 2r^3 & 2 + r - 2r^2 + 3r^3 \end{pmatrix}.$$

We set $f_1(x) = L(J(R(x)))$ and $f_2(x) = f_1(-x)$, $f_3(x) = f_1(1/x)$ and $f_4(x) = f_1(-1/x)$. Then it can be easily shown that the cross-ratio $[f_1, f_2, f_3, f_4]$ satisfies

$$[f_1(x), f_2(x), f_3(x), f_4(x)] = \delta(\mathcal{B}(x))$$

We also get the j-invariant thanks to (2)

$$J(f_1(x), f_2(x), f_3(x), f_4(x)) = J(f_1(x), f_1(-x), f_1(1/x), f_1(-1/x)) =$$

$$\rho(J(x, -x, 1/x, -1/x)) = \rho(j) = \frac{1728j}{j - 1728}.$$

The symmetric functions of the f_i are given by (3):

$$\sigma_1 = 1, \quad \sigma_2 = 0, \quad \sigma_3 = 0, \quad \sigma_4 = 2^{-14}j.$$

It is important not to confuse j, the invariant of $[x, -x, 1/x, -1/x]$, with $\rho(j)$, the invariant of the f_i.

We now define three maps. The first one, called D, from the j-space $\mathbf{P}_1(\mathbf{C})$ to the non-coloured space $\mathcal{D}$, is such that $D(j)$ is the point of $\mathcal{D}$ defined by its symmetric functions

$$\sigma_1 = 1, \sigma_2 = 0, \sigma_3 = 0, \sigma_4 = 2^{-14}j.$$

The second map, called H, from the λ-space $\mathbf{P}_1(\mathbf{C})$ to the half-coloured space $\mathcal{H}$, is such that $H(\lambda)$ is defined by its V-symmetric functions

$$\sigma_1 = 1, \sigma_2 = 0, \sigma_3 = 0, \sigma_4 = 2^{-14}J(\lambda) = 2^{-6}(\lambda^2 - \lambda + 1)^3/\lambda^2/(\lambda - 1)^2,$$

and the cross-ratio defined as

$$\delta(\lambda) = (\lambda - 2)/(2\lambda - 1).$$

The third map, called C, from the x-space $\mathbf{P}_1(\mathbf{C})$ to the coloured space $\mathcal{C}$, is such that $C(x)$ is defined by the quadruplet $(f_1(x), f_2(x), f_3(x), f_4(x))$ as above.

We then get the following commutative diagram in which the actions of S_4 as a Galois group on both sides are compatible with the arrows.

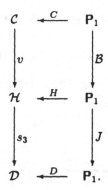

We have thus realized the covering of moduli spaces as a restriction of the covering of naive spaces. We finish by noting that $\lambda H = \delta$ and $JD = \rho$ which stresses the importance of (1). The functions (H, D) define something which is almost but not quite a section of (λ, J).

§6. Approximating dessins from Puiseux series

In this section we now come to the problem of computing explicitly some algebraic model for a given abstract dessin. In fact, we will do better: we will compute the linear space associated with any given divisor on the dessin. The result is given as Puiseux series. Of course, we must truncate the series and consider floating point coefficients if we want to work with finite memory and time. We show in the next section how to obtain some exact solution from such approximations. We consider the subgroup of $PGL_2(\mathbf{C})$ consisting of six linear transforms permuting 0, 1, and ∞. We describe it explicitly as follows:

$$H_{0\bar{1}}(\lambda) = \lambda = \lambda_{0\bar{1}}, \quad H_{0\bar{\infty}}(\lambda) = \frac{\lambda}{\lambda - 1} = \lambda_{0\bar{\infty}},$$

$$H_{1\bar{0}}(\lambda) = 1 - \lambda = \lambda_{1\bar{0}}, \quad H_{1\bar{\infty}}(\lambda) = \frac{\lambda - 1}{\lambda} = \lambda_{1\bar{\infty}},$$

$$H_{\infty\bar{0}}(\lambda) = \frac{1}{1 - \lambda} = \lambda_{\infty\bar{0}}, \quad H_{\infty\bar{1}}(\lambda) = \frac{1}{\lambda} = \lambda_{\infty\bar{1}}.$$

We note that for any standard $\vec{v}$ we have $H_{\vec{v}}(\vec{v}) = \vec{01} = (0, 1/2)$. Now let e be a positive integer. We build an e-th root of $\lambda_{\vec{v}}$ as follows. First let $\Lambda_{\vec{01},e}$ be defined for $\lambda_{\vec{01}} \in \mathbb{C} - \{\infty, 0\}$ as

$$\Lambda_{\vec{01},e}(\lambda_{\vec{01}}) = \lambda_{\vec{01}}^{1/e} = exp(2i\pi \mathrm{Log}(\lambda_{\vec{01}})e^{-1})$$

where Log is the principal determination of the logarithm. We then define the $\Lambda_{\vec{v},e}$ as

$$\Lambda_{\vec{v},e}(\lambda_{\vec{01}}) = \Lambda_{\vec{01},e}(H_{\vec{v}}(\lambda_{\vec{01}})) = \Lambda_{\vec{01},e}(\lambda_{\vec{v}}).$$

Now we define the domain $\mathcal{K}_{\vec{01}}$ to be the open circle of center 0 and radius 1 minus the segment $(-1, 0)$,

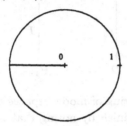

and similarly, $\mathcal{K}_{\vec{v}}$ is such that $H_{\vec{v}}(\mathcal{K}_{\vec{v}}) = \mathcal{K}_{\vec{01}}$. For example $\mathcal{K}_{\vec{0\infty}}$ is the half-plane $\Re(z) < 1/2$ minus the segment $(0, 1/2)$.

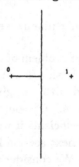

Note that there are two uniformizing parameters at any given point. For example, $\Lambda_{\vec{01},e}$ will be useful for analytic continuation from 0 to 1 and $\Lambda_{\vec{0\infty},e}$ will be useful for analytic continuation from 0 to ∞. The six domains of convergence form a covering of $\mathbf{P}_1$.

We consider a dessin $\mathcal{D}$ together with a Belyi function $\chi : \mathcal{C} \to \mathbf{P}_1 - \{0, 1, \infty\}$ for some algebraic curve $\mathcal{C}$, and a divisor D over $\mathcal{D}$, i.e. a divisor over the underlying curve $\mathcal{C}$ whose points all lie over $\{0, 1, \infty\}$. We write

$$D = \sum_i o_i P_i$$

and we write $\mathcal{L}(D)$ for the corresponding linear space. We will characterize it as the kernel of a certain operator built from some universal hermitian blocks. Let f be some function in $\mathcal{L}(D)$. Associated to each standard $\vec{v}$ above $\vec{01}$ there is a connected component of $\chi^{-1}(\mathcal{K}_{\vec{01}})$, and also an expansion of f as a series in $\Lambda_{\vec{01},e_i}$, where e_i is the ramification at the associated point P_i over 0.

$$f = \sum_{k \geq o_i} a_{\vec{v},k} \Lambda_{\vec{01},e_i}^k,$$

where o_i is the valence of f at P_i.

Similarly, we define uniformizing parameters and expansions of f at any standard of the dessin. We call $\mathbf{S}$ the set of all standards in the dessin. To a function $f \in \mathcal{L}(D)$ we associate the list of sequences of coefficients of its expansions at all standards

$$((a_{\vec{v},k})_k)_{\vec{v} \in \mathbf{S}}.$$

The sequence $(a_{\vec{v},k})_k$ is such that the associated entire series

$$\sum_k a_{\vec{v},k} X^k$$

is convergent on the open disk of radius one and is bounded outside any neighbourhood of $\{0,1\}$ in the disk. Such sequences form a linear space which we call $\mathbf{J}$. To each function $f \in \mathcal{L}(D)$ we associate a vector in $\mathbf{J}^{\mathbf{S}}$. This clearly induces an injection of linear spaces. We want to characterize its image as the kernel of a certain linear operator.

We now study the relations between the various expansions. The relations will be of three types. The first two types involve expansions at various standards related to the same point. The third one relates the expansions at two standards facing each other.

Let P_i be a point above 0 with ramification order e_i and $\vec{v}$ a standard at P_i. Let's say that $\vec{v}$ is over $\vec{01}$. We call $\vec{w} = x_{\vec{01}}(\vec{v})$ the next standard over $\vec{01}$ at P_i reached when turning counterclockwise.

A typical situation of that is in our example, the standards 7 and 12. We write the two corresponding expansions

$$f = \sum_{k \geq o_i} a_{\vec{v},k} \Lambda_{\vec{01},e_i}^k,$$

$$f = \sum_{k \geq o_i} a_{\vec{w},k} \Lambda_{\vec{01},e_i}^k,$$

where the coefficients are related by the obvious relations

$$a_{\vec{w},k} = \zeta_{e_i}^k \cdot a_{\vec{v},k} \tag{4}$$

where $\zeta_{e_i} = exp(2i\pi e_i^{-1})$ is the smallest primitive e_i-th root of unity. This relation simply expresses the monodromy of the logarithm.

We may think now of relating the expansion at $\vec{v}$ and the expansion at $\vec{u} = z_{0\vec{1}}(\vec{v})$, which is the first flag over $0\vec{\infty}$ met when turning counterclockwise.

$$f = \sum_{k \geq o_i} a_{\vec{v},k} \Lambda_{0\vec{1},e_i}^k,$$

$$f = \sum_{k \geq o_i} a_{\vec{u},k} \Lambda_{0\vec{\infty},e_i}^k.$$

This requires no more than expressing $\Lambda_{0\vec{\infty},e_i}$ from $\Lambda_{0\vec{1},e_i}$. Let $\xi_{e_i} = exp(i\pi e_i^{-1})$ be the smallest e_i-th root of -1; we find that

$$\Lambda_{0\vec{\infty},e_i} = \xi_{e_i} \cdot \frac{\Lambda_{0\vec{1},e_i}}{(1 - \lambda_{0\vec{1}})^{1/e_i}} = \xi_{e_i} \cdot \Lambda_{0\vec{1},e_i} \sum_{k \geq 0} \binom{e_i^{-1}}{k} \lambda_{0\vec{1}}^k \tag{5}$$

Now comes the only non-trivial type of relation, the one concerning for example the flags $\vec{v}$ and $\vec{t} = y_{0\vec{1}}(\vec{v})$. This time the two expansions are not over the same point since when $\vec{v}$ is over $0\vec{1}$ and concerns a point P_i over 0, on the contrary $\vec{t}$ is over $1\vec{0}$ and is attached to some point P_j above 1. We have the two corresponding expansions

$$f = \sum_{k \geq o_i} a_{\vec{v},k} \Lambda_{0\vec{1},e_i}^k,$$

$$f = \sum_{k \geq o_j} a_{\vec{t},k} \Lambda_{1\vec{0},e_j}^k.$$

Following Atkin [ASD71], we now equate these two expansions at some point x on the open segment $(\rho, \bar{\rho})$ where $\rho = exp(2i\pi/6)$ is a sixth root of unity. It is to be noted that for such an x, $1 - x = \bar{x}$.

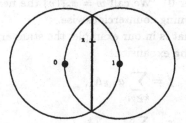

For convenience we adopt the following notation. Let $|a_{\vec{v},k}\rangle$ denote the infinite column vector of all coefficients in the expansion at $\vec{v}$, namely

$$|a_{\vec{v},k}\rangle = (a_{\vec{v},o_i}, a_{\vec{v},o_i+1}, a_{\vec{v},o_i+2}, ...)^{t}$$

where the tilde stands for transposition.

We also write $\langle x, o_i, e_i|$ for the infinite line vector

$$\langle x, o_i, e_i| = (\Lambda^{o_i}_{01,e_i}(x), \Lambda^{o_i+1}_{01,e_i}(x), \Lambda^{o_i+2}_{01,e_i}(x), ...)$$

Then the value taken by f at x is given by

$$f(x) = \langle x, o_i, e_i||a_{\tilde{v},k}\rangle$$

and the relation between the two expansions can be expressed for all x in the segment $(\rho, \bar{\rho})$ as

$$\langle x, o_i, e_i||a_{\tilde{v},k}\rangle = \langle 1 - x, o_j, e_j||a_{\tilde{i},k}\rangle = \langle \bar{x}, o_j, e_j||a_{\tilde{i},k}\rangle \qquad (6)$$

We write $|\bar{x}, o_i, e_i\rangle$ for the adjoint of $\langle x, o_i, e_i|$ and similarly, $|\bar{x}, o_j, e_j\rangle$ for the adjoint of $\langle x, o_j, e_j|$. We also define the operators v_{x,o_i,e_i}, v_{x,o_j,e_j} and c_{x,o_i,e_i,o_j,e_j} by

$$\begin{aligned}
v_{x,o_i,e_i} &= |\bar{x}, o_i, e_i\rangle\langle x, o_i, e_i| \\
v_{x,o_j,e_j} &= |\bar{x}, o_j, e_j\rangle\langle x, o_j, e_j| \\
c_{x,o_i,e_i,o_j,e_j} &= |\bar{x}, o_i, e_i\rangle\langle \bar{x}, o_j, e_j|.
\end{aligned}$$

We deduce from (6) that

$$\begin{pmatrix} v_{x,o_i,e_i} & -c_{x,o_i,e_i,o_j,e_j} \\ -c^*_{x,o_i,e_i,o_j,e_j} & \bar{v}_{x,o_j,e_j} \end{pmatrix} |a_{\tilde{v},k}\rangle \oplus |a_{\tilde{i},k}\rangle = 0 \qquad (7)$$

where $\bar{v}_{x,o_j,e_j}$ is the conjugate of v_{x,o_j,e_j}. This proves that the direct sum $|a_{\tilde{v},k}\rangle \oplus |a_{\tilde{i},k}\rangle$, obtained as the concatenation of the two column-vectors, is in the kernel of a given hermitian positive operator which we call

$$j_{x,o_i,e_i,o_j,e_j} = \begin{pmatrix} v_{x,o_i,e_i} & -c_{x,o_i,e_i,o_j,e_j} \\ -c^*_{x,o_i,e_i,o_j,e_j} & \bar{v}_{x,o_j,e_j} \end{pmatrix}.$$

We now choose a positive measure μ with non-finite support on $(\rho, \bar{\rho})$, such that μ is small enough around ρ and $\bar{\rho}$. For example we can choose it with non-finite compact support in $(\rho, \bar{\rho})$. This is safe enough but it may be even better to take $\mu(x)dx$, where $\mu(x)$ is a suitable power of $1 - |x|^2 = 1 - x + x^2$ or equivalently $J(x) = (1 - x + x^2)^3 x^{-2}(x - 1)^{-2}$. Then we integrate (7) over $(\rho, \bar{\rho})$. We define the integrals of the above operators:

$$V_{o_i,e_i} = \int v_{x,o_i,e_i} d\mu$$

$$V_{o_j,e_j} = \int v_{x,o_j,e_j} d\mu$$

$$C_{o_i,e_i,o_j,e_j} = \int c_{x,o_i,e_i,o_j,e_j} d\mu$$

$$J_{o_i,e_i,o_j,e_j} = \int j_{x,o_i,e_i,o_j,e_j} d\mu,$$

and we obtain

$$J_{o_i,e_i,o_j,e_j} |a_{\bar{v},k}) \oplus |a_{\bar{t},k}) = 0. \tag{8}$$

This finishes the characterization of $\mathcal{L}(D)$. We can collect all the relations in a blockwise matrix. The blocks are universal junction matrices, and the disposition of all the blocks reflects the topology of the dessin since it comes from the action of the fundamental groupoid on the standards. We note that all the entries of the junction operators are of the form

$$\int x^a (1 - x)^b d\mu$$

where a and b are rationals. We can think of expressing them with the beta function plus some hypergeometric functions.

Now, for the actual computation of a dessin we first choose a divisor on the dessin. For example, the Riemann-Hurwitz formula gives us a divisor which is in the canonical class, made up of ramification points. Indeed, if $\mathcal{D}$ is our dessin, we call P_i, Q_j, R_k the points above 0, 1, ∞ respectively, and p_i, q_j, r_k their multiplicities. If the dessin is clean, $q_j = 2$. We call $\mathcal{K}$ the following divisor, which is in the canonical class by the Riemann-Hurwitz formula:

$$\mathcal{K} = -\sum_i (P_i) + \sum_j (q_j - 1)(Q_j) - \sum_k (R_k).$$

If the genus is greater than or equal to 2 then the divisor $2\mathcal{K}$ is very ample. We compute the associated linear space $\mathcal{L}(2\mathcal{K})$ with enough accuracy as the kernel of the operator introduced above. To achieve this, we choose a given precision P and write down the junction matrices, truncated at rank P. We then build a blockwise matrix corresponding to the above divisor, from the truncated junction matrices, and compute its kernel. Actually, this matrix, being no more than an approximation, is not very likely to have a kernel. We just look for vectors with small images under this matrix, using the least-squares method. This provides us with an explicit, though approximate, description of the linear space $\mathcal{L}(2\mathcal{K})$ given by some base $(f_1, .., f_{\ell(2\mathcal{K})})$ where the f_i are Puiseux series with P terms. This linear space defines an embedding of the curve in a projective space. By looking for algebraic dependencies between the the f_i, we build an algebraic regular model $\mathcal{C}$ for the curve provide the genus is greater than or equal to 2 (if the genus is 0, a model of the curve is $\mathbf{P}_1$; if the genus is 1, we have some elliptic curve which can be determined by looking at any ample divisor, although there exist simpler techniques).

Now it remains to compute the Belyi function $\varphi(f_1, .., f_{\ell(2\mathcal{K})})$ from $\mathcal{C}$ to $\mathbf{P}_1$. We first compute the linear space associated to the divisor $-(\varphi) =$

$\sum_k r_k(R_k) - \sum_i p_i(P_i)$ in just the same way as above. It is of dimension 1, and we take a generator which we normalize with the conditions $\varphi(Q_j) = 1$. From all the Puiseux series expansions we have (both for φ and the f_i), we can express φ as an algebraic function of $(f_1, .., f_{\ell(2\kappa)})$.

Of course, all this is quite tedious, and we must develop sharper techniques depending on the genus of the curve as we will see in the next section for genus 0. Thanks to some *a priori* description, we can refine the approximation with an iterative method such as the one detailed below, which leads us to an exact algebraic solution over $\bar{\mathbf{Q}}$.

§7. Iterative ad hoc methods

In this section we describe iterative methods to compute genus zero dessins as rational functions from $\mathbf{P}_1(\mathbf{C})$ to $\mathbf{P}_1(\mathbf{C})$. We compute the positions α_i, β_i and γ_i of the points over 0, 1 and ∞. Algebraic methods are feasible in the case of relatively small dessins of low degree; numerous examples are given in the articles by Shabat, Malle, Birch in this volume. However it is possible to do the calculations much more efficiently via approximation and iteration.

Our method consists of three stages:

- Computing approximations of the positions of the points with ad hoc methods.

- Use an iterative algorithm to obtain the numeric convergence and compute the positions with hundreds of digits. We obtain a very good approximation of the Belyi function.

- Find the number field where the coefficients of the Belyi function are. We use a lattice reduction.

We work the whole time with the geometrical definition of the dessins: we try to compute some *canonical* positions of the vertices of a coloured triangulation of the curve. Thanks to this intuitive approach, we can occasionally help the computer with human intervention, if part of the solution seems obvious.

Finding the first approximation

The more general method uses Puiseux series, as stated in section 6, but this method needs memory and time, and cannot be used to analyse families of dessins.

In genus 0, we can use visual intuition and very quickly build an approximation of the Belyi function.

We say that two dessins are close if their combinatorial structure is close. For example, if we add one vertex to a dessin, the resulting dessin has a similar shape. This leads to a method of *growing families of dessins* where we add branches to an initial seed.

Here is for example a family of dessins, with the vertices in their correct positions. We see that the deformation caused by adding a point is small, so each time we can build an approximation of the positions of the vertices from the previous dessin.

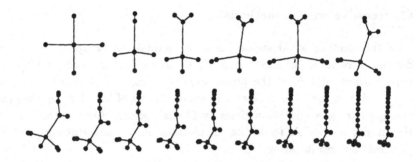

This stage needs our intuition, so we can place the new point near its future position, with visual considerations such as the regularity of the graph $\phi^{-1}([0,1])$ and the local symmetry around a vertex.

Numeric convergence

The second stage consists in solving some equations to find the position of the vertices with arbitrary precision.

We want to have a good system of equations, so as to really be able to compute the solutions with our computer. This system must be as simple and as small as possible and it must be stable, so we can converge to the exact solution even with rough approximations.

The case of trees

A dessin is called a "tree" if it is clean, of genus 0 and totally ramified over ∞. If the dessin is a tree, there is a system of polynomial equations in the coordinates of the vertices, that is easy to compute and easy to solve.

We follow the ideas of [Cou94].

If we write ϕ a Belyi function of degree d, it is a rational function with ramification points α_i of degree ν_i over 0 ($i \in A$; we denote by $N = |A|$ the number of vertices of the tree), ramification points β_i of degree 2 over 1

$(i \in B)$ and totally ramified over ∞. Hence we can write, if we put ∞ over ∞:

$$\phi = \frac{-1}{\lambda} \prod_{i \in A}(X - \alpha_i)^{\nu_i} = \frac{-1}{\lambda}\left(\prod_{i \in B}(X - \beta_i)\right)^2 + 1.$$

Since we fixed ∞ over ∞, the positions of the vertices are defined up to an affine transformation: we have 2 degrees of liberty. Now we have an equation in the positions of the vertices (α_i) and the segments (β_i):

$$\prod_{i \in A}(X - \alpha_i)^{\nu_i} = \left(\prod_{i \in B}(X - \beta_i)\right)^2 - \lambda \quad \text{which we write as:} \quad \Pi = Q^2 - \lambda.$$

(9)

This is the equation used by Atkin in [ASD71] and Shabat and Birch in this volume. This system could be solved with a Gröbner basis reduction algorithm, but the dessin should not be too large.

We want to build a better system of equations. We want to reduce the number of unknowns and obtain a system of equations independent of the (β_i).

We can factor the right-hand side of equation (9): $Q^2 - \lambda = (Q - \sqrt{\lambda})(Q + \sqrt{\lambda})$, so we can split the set of vertices A in two subsets A^+ and A^- – the blue and the red vertices – such that $i \in A^+$ if and only if $Q(\alpha_i) = +\sqrt{\lambda}$. We factor the left-hand part of equation (9):

$$\Pi^+ = \prod_{i \in A^+}(X - \alpha_i)^{\nu_i} = \left(Q + \sqrt{\lambda}\right)$$

$$\Pi^- = \prod_{i \in A^-}(X - \alpha_i)^{\nu_i} = \left(Q - \sqrt{\lambda}\right)$$

We use the notation:

$$\Theta = \prod_{i \in A}(X - \alpha_i)$$

$$\Sigma = \sum_{i \in A}\frac{\nu_i}{X - \alpha_i} \quad \Sigma^+ = \sum_{i \in A^+}\frac{\nu_i}{X - \alpha_i} \quad \Sigma^- = \sum_{i \in A^-}\frac{\nu_i}{X - \alpha_i}$$

$$\sigma = \Theta\Sigma \quad \sigma^+ = \Theta\Sigma^+ \quad \sigma^- = \Theta\Sigma^-$$

to differentiate equation (9):

$$\Pi\Sigma = 2QQ' \quad \text{i.e.} \quad \frac{\Pi}{\Theta}\sigma = 2QQ',$$

but since Q is prime to Π and to $\frac{\Pi}{\Theta}$, we deduce

$$\sigma = dQ \quad \text{i.e.} \quad \sigma^+ + \sigma^- = dQ$$

(10)

Now, we can compute $\sigma^+ - \sigma^-$, and we obtain an equation with the (α_i) and λ but without the (β_i).

$$\sigma^+ - \sigma^- = d\sqrt{\lambda} \tag{11}$$

because $(\sigma^+ - \sigma^-) - d\sqrt{\lambda}$ is a polynomial of degree less than $N-1$ with N distinct roots:

for i in A^+, $(\sigma^+ - \sigma^-)(\alpha_i) = \sigma^+(\alpha_i) = dQ(\alpha_i) = d\sqrt{\lambda}$

for i in A^-, $(\sigma^+ - \sigma^-)(\alpha_i) = -\sigma^-(\alpha_i) = -dQ(\alpha_i) = d\sqrt{\lambda}$.

If we add and substract equations (10) and (11), we obtain a system in the (α_i) that respects the coloration of the vertices:

$$2\sigma^+ = d\Pi^- \quad \text{and} \quad 2\sigma^- = d\Pi^+.$$

Let us define $\bar\nu_i = \nu_i$ for $i \in A^+$ and $\bar\nu_i = -\nu_i$ for $i \in A^-$. Equation (11) divided by Θ and with $U = 1/X$ is

$$\frac{d\sqrt{\lambda}U^{N-1}}{\prod_{i\in A}(1 - U\alpha_i)} = \sum_{i\in A^+}\frac{\nu_i}{1 - U\alpha_i} - \sum_{i\in A^-}\frac{\nu_i}{1 - U\alpha_i} = \sum_{i\in A}\frac{\bar\nu_i}{1 - U\alpha_i}.$$

Now, λ does not interfere with the terms of degree less than N; to eliminate λ, we write down the $N-1$ first terms of the Taylor expansion of this equation:

$$\forall 0 \le k \le N-2, \quad \sum_{i\in A}\bar\nu_i\alpha_i^k = 0. \tag{12}$$

The equation for $k = 0$ is trivial, so we have $N-2$ equations for N indeterminates. The set of solutions is invariant under affine transformations $(\alpha_i)_i \mapsto (A\alpha_i + B)_i$.

We add a few inequalities to the system, namely $\alpha_i \ne \alpha_j$ if $i \ne j$. This defines a smooth variety of dimension 2 in the space of dimension N. If we quotient this by the action of the group of affine transformations, we get a variety of dimension 0 in $\mathbf{P}_{N-2}(\mathbf{C})$. It is not necessarily a single point, not even necessarily irreducible over $\mathbf{Q}$, but one point must correspond to our Belyi function.

This proves that our system has a unique solution near our first approximation: the dessin we want to compute.

The case of dessins with all ramification orders even

If all the ramification orders of the dessin are even, we obtain equations similar to (12).

Let α_i, of ramification $2\nu_i$, denote the vertices and γ_i of ramification $2\mu_i$ the faces. Since all ramification indices are even, we can colour the vertices and the faces. We denote $\bar\nu_i$ and $\bar\mu_i$ the algebraic ramifications.

Then we have the system:

$$\forall i, \quad \forall 0 \le k \le \nu_i - 1, \qquad \sum_j \frac{\bar{\mu}_j}{(\gamma_j - \alpha_i)^k} = 0$$

$$\forall j, \quad \forall 0 \le k \le \mu_j - 1, \qquad \sum_i \frac{\bar{\nu}_i}{(\alpha_i - \gamma_j)^k} = 0.$$

Solving the system

The system (12) is a Vandermonde-like system. Let $\mathcal{A} = (\alpha_\infty, ..., \alpha_N)$ and the function $\mathcal{F}(\mathcal{A}) = (\sum_{j \in \mathcal{A}} \alpha_j^{\|})_{\|=\infty...\mathcal{N}-\epsilon}$ be such that our system is $\mathcal{F}(\mathcal{A}) = \iota$. Newton's algorithm for solving such equations begins with an approximation of the solution $\mathcal{A}_\iota$ and iterates the formula: $\mathcal{A}_{\backslash+\infty} = \mathcal{A}_\backslash - \mathcal{F}'_{\mathcal{A}_\backslash}{}^{-\infty} \mathcal{F}(\mathcal{A}_\backslash)$.

But $\mathcal{F}'_{\mathcal{A}_\backslash}$ is not invertible and $\mathcal{F}'_{\mathcal{A}_\backslash}{}^{-1}$ is defined up to an element of the kernel. We choose an $\mathcal{F}'_{\mathcal{A}_\backslash}{}^{-1}$ orthogonal to the kernel.

Now we have a method to solve $\mathcal{F}(\mathcal{A}) = \iota$, but we must be careful: the numeric representation forces us to work with $\mathcal{A}_\backslash \in \mathbf{C}^\mathcal{N}$, but we must be aware that $\mathcal{A}_\backslash$ is defined up to affine transformation.

We must normalize the $\mathcal{A}_\backslash$ to avoid a shift to infinity. We fix the center of the dessin at 0, and the scale of the dessin to a diameter of 1. The sum of the coordinates of the vertices is 0 and the maximal distance between two vertices is 1.

To handle large numbers, we use the PARI library.

Back to the algebraic point of view

The third stage uses the powerful lattice reduction tools to go from the geometrical point of view to the algebraic description: given very precise complex approximations of algebraic complex numbers, we build a lattice such that the shortest vector of this lattice gives the minimal polynomial. We use the method described in [LLL82].

The problem of finding a short vector in an integer lattice is hard, but efficient algorithms have been published in [LLL82], [Sch87] and [Sch88].

The previous stage allows us to compute these numbers to any desired precision. However, a priori we do not know what precision is necessary to find the exact solution with the lattice reduction method. We compute upper bounds for the degree and we guess the size of the coefficients of the polynomial.

The degree is lower than the number of "combinatorial" conjugates of the dessin i.e. the cardinal of the variety of dimension 0 in $\mathbf{P}_{N-2}(\mathbf{C})$, solution of our system. This number can be approximated by character formulae, see [Ser92].

Application to an example

Consider the dessin given by this graph, a tree with 10 vertices:

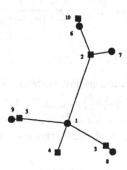

It is the 6th of the family of dessins shown above. We make the tree grow step by step, and then we compute the solution to 2000 digits.

The minimal polynomial of α_1/α_2 is the polynomial given here, of degree 24 and of discriminant $-1.2^{799}.3^{270}.5^{90}.7^{54}.N^2$ (N is a large number with no smaller factor than 127):

$$1216396531470080000\,x^{24} + 15167128532892096000\,x^{23}$$
$$+ 88567164003405619200\,x^{22} + 320465331330548463040\,x^{21}$$
$$+ 801926461469806116168\,x^{20} + 1468854325860309911334\,x^{19}$$
$$+ 2037128673503852027315\,x^{18} + 2189254042743982149456\,x^{17}$$
$$+ 1858352449953325455855\,x^{16} + 1271096908385844699688\,x^{15}$$
$$+ 717291487653207390204\,x^{14} + 342482003051130999024\,x^{13}$$
$$+ 140622333198259937516\,x^{12} + 49205805780202178532\,x^{11}$$
$$+ 13997991682162739850\,x^{10} + 2897517763455570160\,x^{9}$$
$$+ 284441186456050050\,x^{8} - 67794459856593624\,x^{7}$$
$$- 41017353384312340\,x^{6} - 10862737575891504\,x^{5}$$
$$- 1796582490031788\,x^{4} - 178029020920154\,x^{3}$$
$$- 7529198821413\,x^{2} + 14589968448\,x - 34245281017$$

Note the factorization of the leading coefficient

$$1216396531470080000 = 2^{11}.5^{4}.950309790211.$$

References

[Arm88] M.A. Armstrong. *Groups and Symmetry*. U.T.M. Springer-Verlag, 1988.

[ASD71] A.O.L. Atkin and H.P.F. Swinnerton-Dyer. Modular forms over non-congruence subgroups. In *Proceedings of symposia in pure mathematics*, number 19, AMS, 1971.

[CH85] K. Coombes and D. Harbater. Hurwitz families and arithmetic Galois groups. *Duke Math. J.* **52** (1985), 821-839.

[Cou94] J-M. Couveignes. Calcul et rationnalité de fonctions de Belyi en genre 0. *Ann. de l'Inst. Fourier* **44** (1), 1994.

[Del89] P. Deligne, Le groupe fondamental de la droite projective moins trois points. In *Galois groups over* **Q**, Y. Ihara, K. Ribet, J-P. Serre, eds. Math. Sci. Res. Inst. Publ. **16**, Springer-Verlag, New York, 1989.

[FD90] M. Fried and Pierre Dèbes. Rigidity and real residue class fields. *Acta Arith.* **LVI** (1990), 291-322.

[Har87] D. Harbater. Galois coverings of the arithmetic line, in *Springer Lect. Notes in Math.* **1240**, 165-195.

[Kle13] Felix Klein. Lectures on the Icosahedron and the Solution of Equations of the Fifth Degree. London, 1913.

[LLL82] A.K. Lenstra, H.W. Lenstra and L. Lovász. Factoring polynomials with rational coefficients. *Math. Ann.* **261** (1982), 515-534.

[Sch87] C-P. Schnorr. A hierarchy of polynomial time lattice basis reduction algorithms. *Th. Comp. Sci.* **53** (1987), 201-224.

[Sch88] C-P. Schnorr. A more efficient algorithm for lattice basis reduction. *J. Algorithms* **9** (1988), 47-62.

[Ser92] J-P. Serre. *Topics in Galois Theory*. Jones and Bartlett, 1992.

[Wei56] André Weil. The field of definition of a variety. *Amer. J. Math* **78** (1956), 509-524.

*Membre de l'Option Recherche du Corps des ingénieurs de l'Armement
UMR d'Algorithmique Arithmétique de Bordeaux, Université de Bordeaux, et Groupe de recherche en complexité et cryptographie, DMI, Ecole Normale Supérieure, France
**Groupe de recherche en complexité et cryptographie, DMI, Ecole Normale Supérieure, France

Maps, hypermaps and triangle groups

Gareth Jones* and David Singerman*

§1. Introduction

It is now widely known that maps on connected oriented surfaces are parametrized by the transitive permutation representations (or conjugacy classes of subgroups) of Grothendieck's cartographic group [G]

$$C_2^+ = \langle \rho_0, \rho_1, \rho_2 \mid \rho_1^2 = \rho_0\rho_1\rho_2 = 1 \rangle;$$

the general theory can be found in [JS1, MV], for example, and the basic idea can be traced back at least as far as Heffter's work [He1, He2] in the last century on map-colourings. Similarly, an idea of Tutte [T] can be used to show that maps on surfaces which may be non-orientable or with boundary correspond to the group

$$C_2 = \langle \sigma_0, \sigma_1, \sigma_2 \mid \sigma_i^2 = (\sigma_2\sigma_0)^2 = 1 \rangle,$$

which contains C_2^+ as the "even subgroup" of index 2 generated by

$$\rho_0 = \sigma_1\sigma_2, \quad \rho_1 = \sigma_2\sigma_0 \quad \text{and} \quad \rho_2 = \sigma_0\sigma_1$$

(see [BS, Jo1, V1, V2] for example). If we restrict to triangular maps (as in [Jo2]), then the corresponding groups are

$$T_2 = \langle \sigma_0, \sigma_1, \sigma_2 \mid \sigma_i^2 = (\sigma_2\sigma_0)^2 = (\sigma_0\sigma_1)^3 = 1 \rangle$$

and its even subgroup

$$T_2^+ = \langle \rho_0, \rho_1, \rho_2 \mid \rho_1^2 = \rho_2^3 = \rho_0\rho_1\rho_2 = 1 \rangle.$$

Similarly, this theory can be extended to parameterize hypermaps (which include all maps) by using the hypercartographic group

$$\mathcal{H}_2 = \langle \sigma_0, \sigma_1, \sigma_2 \mid \sigma_i^2 = 1 \rangle$$

and (in the oriented case) the corresponding oriented hypercartographic group [Cor, CoMa]

$$\mathcal{H}_2^+ = \langle \rho_0, \rho_1, \rho_2 \mid \rho_0\rho_1\rho_2 = 1 \rangle.$$

The importance of this last category is illustrated by Belyĭ's Theorem [B], which implies that a complex algebraic curve X is defined over the field $\overline{\mathbb{Q}}$

of algebraic numbers if and only if it is uniformised by a subgroup of finite index in a triangle group Δ [CW, Wo]; each conjugacy class of such subgroups corresponds, via the natural epimorphism $\mathcal{H}_2^+ \to \Delta$, to an oriented hypermap, which may be regarded as a tessellation of X corresponding to the choice of a Belyĭ function $\beta : X \to \mathbf{P}^1(\mathbf{C}) = \mathbf{C} \cup \{\infty\}$ (a finite covering of the complex projective line, or Riemann sphere, unbranched outside $\{0, 1, \infty\}$). In particular, maps on X correspond to clean Belyĭ functions, those for which the monodromy permutation at 1 is a product of disjoint 2-cycles.

In the first part of this paper we will outline this theory of maps and hypermaps, which is explained in detail in [JS1], [BS], [CS] and [IS]. We will identify the six triangle groups defined above with certain subgroups of the extended modular group $\Pi = PGL_2(\mathbf{Z})$; this enables us to realise the associated maps, triangulations and hypermaps as quotients of universal tessellations of the hyperbolic plane, and hence as tessellations of Riemann surfaces or Klein surfaces. In the second part we will show how various functors, such as barycentric subdivision, are induced by homomorphisms between these groups; as pointed out to us by Alexander Zvonkin, many of these functors correspond to transformations of Belyĭ functions. Similarly, we will show how the automorphisms of these groups induce operations on the objects in the associated categories; for example, we will see that Π has an outer automorphism which transforms the triangulation corresponding to the Fermat curve $x^n + y^n = 1$, with Belyĭ function x^n, into the torus map $\{3,6\}_{n,0}$ described by Coxeter and Moser in [CoMo].

§2. Oriented maps

The simplest case to consider is $\mathcal{C}_2^+$. The permutation representations of this group correspond to oriented maps on surfaces without boundary. (Throughout this paper, representations will be assumed to be *transitive*, while surfaces, maps and hypermaps will always be *connected*.)

Given a map $\mathcal{M}$ on an oriented surface $\mathcal{S}$, we can define permutations r_i of the set Ω of darts (directed edges) of $\mathcal{M}$ in the following way. We let r_1 denote the permutation which interchanges the two darts on the same edge, and we let r_0 be the permutation which rotates each dart around the vertex to which it points, using the orientation of $\mathcal{S}$ to send it to the next dart at that vertex (see Fig. 1).

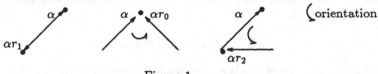

Figure 1

We then find that the permutation $r_2 = (r_0 r_1)^{-1}$ rotates the darts around faces, again following the orientation of S. Clearly $r_1^2 = r_0 r_1 r_2 = 1$, so we obtain a permutation representation $\rho_i \mapsto r_i$ of C_2^+ on Ω, transitive since S is connected.

Now we want *every* transitive permutation representation of C_2^+ to correspond to a map, so we need to allow ρ_1 to have fixed points on Ω. We do this by allowing $\mathcal{M}$ to have *free edges*, that is, edges with only one vertex and one dart. The involution r_1 then fixes each such dart.

Figure 2

As an example, consider the star-map $\mathcal{M}_n$ in Figure 2 (see also [SV]). This is a map of genus 0 with one vertex (at 0), one face (centred at ∞) and n free edges (from 0 to the n-th roots of 1). Here $r_0 = (12 \ldots n), r_1 = (1)(2) \ldots (n)$ and $r_2 = (n \ldots 21)$. This map is an n-sheeted covering of $\mathcal{M}_1$, corresponding to the Belyĭ function $\mathbf{P}^1(\mathbf{C}) \to \mathbf{P}^1(\mathbf{C})$, $z \mapsto z^n$. This is branched at 0 and ∞ only, with trivial branching at 1; the permutations r_0, r_1 and r_2 describe the monodromy permutations at $0, 1$ and ∞.

We have described how oriented maps give rise to permutation representations of C_2^+. We now want to show that each such representation arises from a map. First note that each map determines a conjugacy class of subgroups M of index $|\Omega|$ in C_2^+, the stabilisers of the darts. These subgroups are called *map subgroups* in [JS1] (and *Borel subgroups* in [SV]), and can be used to reconstruct the map. We briefly describe the idea, which is explained in detail in [JS1]. In fact, in that paper we worked with the cartographic group which describes maps of type (m, n) (or $\{n, m\}$ in the notation of [CoMo]), that is, maps for which m is the least common multiple of the vertex valencies, and n is the least common multiple of the face valencies (we allow $m, n = \infty$). This group is the triangle group

$$\Delta = (m, 2, n)$$
$$= \langle \rho_0, \rho_1, \rho_2 \mid \rho_0^m = \rho_1^2 = \rho_2^n = \rho_0 \rho_1 \rho_2 = 1 \rangle.$$

It leaves invariant the "universal map" $\hat{\mathcal{M}}(m, n)$ of type (m, n), consisting of regular n-gons with m meeting at each vertex. This map lies on a simply-connected Riemann surface $\mathcal{U} = \mathcal{U}(m, n)$, where

$$\mathcal{U}(m,n) = \begin{cases} \text{hyperbolic plane} & \text{if } \frac{1}{m} + \frac{1}{n} < \frac{1}{2}, \\[2mm] \text{Euclidean plane} & \text{if } \frac{1}{m} + \frac{1}{n} = \frac{1}{2}, \\[2mm] \text{Riemann sphere} & \text{if } \frac{1}{m} + \frac{1}{n} > \frac{1}{2}. \end{cases}$$

Given a permutation representation of Δ on any set Ω, the stabiliser of a point is a subgroup M of index $|\Omega|$ in Δ; since M leaves $\hat{\mathcal{M}}(m, n)$ invariant, we can form the quotient map $\mathcal{M} = \hat{\mathcal{M}}(m, n)/M$ which lies on the Riemann surface $\mathcal{U}/M$. This is the map we are seeking, in the sense that the permutations r_i described above give the original representation of Δ.

Now every representation of C_2^+ can be factored through Δ for some m, n, via the obvious epimorphism $\theta : C_2^+ \to \Delta$, so we can pull M back to obtain the required representation of C_2^+ on the cosets of $\theta^{-1}(M)$. Thus the representations of C_2^+ parametrize oriented maps without boundary.

If we restrict attention to triangular maps, then $n = 3$ and $m = \infty$ (there is no restriction on vertex valencies, so the relation $\rho_0^m = 1$ is vacuous). Thus the analogue T_2^+ of C_2^+ for oriented triangulations is the triangle group $\Delta = (\infty, 2, 3)$, which can be identified with the modular group

$$\Gamma = PSL_2(\mathbf{Z}) = SL_2(\mathbf{Z})/\{\pm I\} = \{ \pm \begin{pmatrix} a & b \\ c & d \end{pmatrix} \mid a, b, c, d \in \mathbf{Z}, \ ad - bc = 1 \},$$

as in [Jo2]. To do this, we use the presentation

$$\Gamma = \langle S_0, S_1, S_2 \mid S_1^2 = S_2^3 = S_0 S_1 S_2 = 1 \rangle,$$

where

$$S_0 = \pm \begin{pmatrix} 1 & 1 \\ 0 & 1 \end{pmatrix}, \quad S_1 = \pm \begin{pmatrix} 0 & -1 \\ 1 & 0 \end{pmatrix}, \quad S_2 = \pm \begin{pmatrix} 0 & 1 \\ -1 & 1 \end{pmatrix};$$

the isomorphism $\rho_i \mapsto S_i$ ($i = 0, 1, 2$) then identifies T_2^+ with Δ. The corresponding universal triangulation $\hat{T} = \hat{\mathcal{M}}(\infty, 3)$ is the Farey map [JSW] illustrated in Figure 3. The vertex set is the extended set of rationals $\hat{\mathbf{Q}} = \mathbf{Q} \cup \{\infty\}$; vertices a/b and c/d (in reduced form) are joined by an edge if and only if $ad - bc = \pm 1$; each edge lies in two triangles, and these are the faces of $\hat{T}$. In Figure 3 we give two illustrations of $\hat{T}$, the first in the upper half-plane $\mathcal{U} = \mathcal{U}(\infty, 3)$ and the second in a hemisphere of the Riemann sphere.

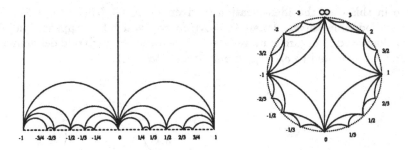

Figure 3 : the universal triangulation $\hat{T}$

Now Γ acts on $\overline{\mathcal{U}} = \mathcal{U} \cup \hat{\mathbf{Q}}$ by

$$\pm \begin{pmatrix} a & b \\ c & d \end{pmatrix} : z \mapsto \frac{az + b}{cz + d};$$

Figure 4 shows a fundamental region F for Γ on $\mathcal{U}$, together with the actions of its generators. This action of Γ preserves $\hat{T}$; indeed, Γ is the orientation-preserving automorphism group of $\hat{T}$, permuting its darts regularly (notice how F corresponds to the dart directed along the edge 0∞ towards ∞). Each subgroup $M \leq T_2^+ = \Gamma$ has an induced action on $\hat{T}$, and the conjugacy class of M in Γ corresponds to the quotient triangulation $\hat{T}/M$ of the compactification $\overline{\mathcal{U}}/M$ of the Riemann surface $\mathcal{U}/M$ (here $\omega = e^{2\pi i/3}$ and $\rho = e^{2\pi i/6}$).

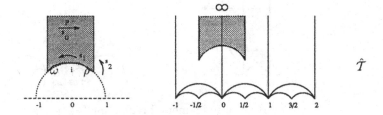

Figure 4 : a fundamental region for Γ

In a similar way, *all* oriented maps without boundary are parametrized by conjugacy classes of subgroups of $\mathcal{C}_2^+ = (\infty, 2, \infty)$, which can be identified with the subgroup $\Gamma_0(2)$ of index 3 in Γ represented by the matrices $\pm \begin{pmatrix} a & b \\ c & d \end{pmatrix}$ with c even. This has a presentation

$$\Gamma_0(2) = \langle U_0, U_1, U_2 \mid U_1^2 = U_0 U_1 U_2 = 1 \rangle,$$

where

$$U_0 = \pm \begin{pmatrix} -1 & 0 \\ 2 & -1 \end{pmatrix}, \quad U_1 = \pm \begin{pmatrix} -1 & 1 \\ -2 & 1 \end{pmatrix}, \quad U_2 = \pm \begin{pmatrix} 1 & 1 \\ 0 & 1 \end{pmatrix},$$

so in this case the identification is given by $\rho_i \mapsto U_i$ ($i = 0, 1, 2$). The corresponding universal map $\hat{\mathcal{M}} = \hat{\mathcal{M}}(\infty, \infty)$ (which first appeared in [Si2]) is illustrated in Figure 5; its vertices are the rationals with odd denominator, and a/b is joined to c/d if and only if $ad - bc = \pm 1$.

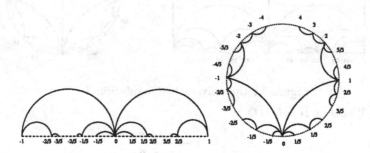

Figure 5 : the universal map $\hat{\mathcal{M}}$

$\Gamma_0(2)$ acts as the group of orientation-preserving automorphisms of $\hat{\mathcal{M}}$; Figure 6 shows a fundamental region (corresponding to the dart along the edge 01 towards 0) and the action of the generators. As in the case of triangulations, each subgroup $M \leq \Gamma_0(2)$ corresponds to the map $\hat{\mathcal{M}}/M$.

Figure 6 : a fundamental region for $\Gamma_0(2)$

§3. Maps on arbitrary surfaces

A similar theory applies to the group C_2, except that map subgroups which do not lie in C_2^+ correspond to maps on non-orientable surfaces or on surfaces with boundary. This is explained in detail in [BS] (see also [Jo1] and [T]). Our approach is to consider the action of C_2 on the flags (called *blades* in [BS]), which are represented as "half-darts" as in Figure 7. If the surface S is without boundary we define three permutations s_0, s_1, s_2 of the set Ω of flags as in Figure 8:

Figure 7   Figure 8

Thus s_i changes the i-dimensional component of each flag, leaving the other two components unchanged. Since

$$s_0^2 = s_1^2 = s_2^2 = (s_2 s_0)^2 = 1,$$

we obtain a permutation representation $\sigma_i \mapsto s_i$ of C_2 in which each s_i is a product of 2-cycles. As before, we need to allow σ_i to have fixed points in Ω. This occurs if S has non-empty boundary, as shown in Figure 9 (in the first diagram, the edge is a free edge):

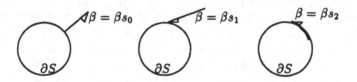

Figure 9

For example, Figure 10 shows a map on a Möbius band, with $s_0 = (1\,2)(3\,4), s_1 = (1)(3)(2\,4), s_2 = (1\,4)(2\,3)$.

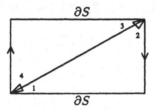

Figure 10

In this way, every map on a surface (which may be non-orientable or with boundary) gives rise to a permutation representation of C_2, which is transitive if and only if the map is connected.

As in the case of C_2^+, we can reconstruct the map from a transitive permutation representation by using map subgroups. Instead of the triangle group $\Delta = (m, 2, n)$ we use the extended triangle group

$$\Delta = [m, 2, n]$$
$$= \langle \sigma_0, \sigma_1, \sigma_2 \mid \sigma_i^2 = (\sigma_1 \sigma_2)^m = (\sigma_2 \sigma_0)^2 = (\sigma_0 \sigma_1)^n = 1 \rangle.$$

This group, generated by reflections in the sides of a triangle with angles $\pi/m, \pi/2, \pi/n$, leaves the universal map $\hat{\mathcal{M}}(m,n)$ invariant. Given a transitive representation of Δ on a set Ω, the stabiliser of a point is a subgroup M of index $|\Omega|$ in Δ. The required map is $\hat{\mathcal{M}}(m,n)/M$, which lies on the Klein surface $\mathcal{U}/M$. This surface has a boundary if and only if M contains reflections (conjugates of some σ_i), or equivalently, the images s_i of σ_i have fixed points. The number of boundary components can be computed from the permutation representation by using a theorem of Hoare [Ho]. The orientability of $\mathcal{U}/M$ can be determined from the Schreier coset graph of M in Δ by deleting all loops corresponding to fixed points of σ_0, σ_1 and σ_2: then the surface is orientable if and only if the resulting graph is bipartite [HS]. Finally the genus can be computed from the Riemann-Hurwitz or Euler-Poincaré formula.

If the map subgroup M is not contained in the even subgroup $\Delta^+ = (m, 2, n)$ of Δ, so that $\mathcal{U}/M$ is non-orientable or with boundary, then we can form the subgroup $M^+ = M \cap \Delta^+$. The Riemann surface $\mathcal{U}/M^+$, the complex double of $\mathcal{U}/M$, carries the map $\hat{\mathcal{M}}(m,n)/M^+$ which is the canonical double of $\hat{\mathcal{M}}(m,n)/M$. As $\mathcal{U}/M = (\mathcal{U}/M^+)/(M/M^+)$, the Klein surface $\mathcal{U}/M$ is the quotient of a Riemann surface by an anticonformal involution. Similarly $\hat{\mathcal{M}}(m,n)/M = (\hat{\mathcal{M}}(m,n)/M^+)/(M/M^+)$, so every map on a non-orientable surface or a surface with boundary can be obtained from a map on a Riemann surface by taking the quotient with respect to an anticonformal involution.

As in the case of $\mathcal{C}_2^+$, we can use epimorphisms $\mathcal{C}_2 \to \Delta$ to pull map subgroups back to $\mathcal{C}_2 = [\infty, 2, \infty]$, so maps on connected surfaces (possibly with boundary) are parametrized by conjugacy classes of subgroups of $\mathcal{C}_2$, and similarly, triangulations correspond to conjugacy classes of subgroups of $\mathcal{T}_2 = [\infty, 2, 3]$. We can identify $\mathcal{T}_2$ with the extended modular group

$$\Pi = PGL_2(\mathbf{Z}) = GL_2(\mathbf{Z})/\{\pm I\} = \{ \pm \begin{pmatrix} a\,b \\ c\,d \end{pmatrix} \mid a, b, c, d \in \mathbf{Z}, \ ad - bc = \pm 1 \},$$

which acts on $\mathcal{U}$ by

$$\pm \begin{pmatrix} a\,b \\ c\,d \end{pmatrix} : z \mapsto \begin{cases} \frac{az+b}{cz+d} & \text{if } ad - bc = 1; \\[2mm] \frac{a\bar{z}+b}{c\bar{z}+d} & \text{if } ad - bc = -1, \end{cases}$$

and we can identify $\mathcal{C}_2$ with the subgroup $\Pi_0(2)$ of index 3 in Π consisting of the above transformations with c even. To do this, we use the presentations

$$\Pi = \langle R_0, R_1, R_2 \mid R_i^2 = (R_2 R_0)^2 = (R_0 R_1)^3 = 1 \rangle,$$

where

$$R_0 = \pm \begin{pmatrix} 0 & 1 \\ 1 & 0 \end{pmatrix}, \quad R_1 = \pm \begin{pmatrix} -1 & 1 \\ 0 & 1 \end{pmatrix}, \quad R_2 = \pm \begin{pmatrix} -1 & 0 \\ 0 & 1 \end{pmatrix},$$

and

$$\Pi_0(2) = \langle T_0, T_1, T_2 \mid T_i^2 = (T_2 T_0)^2 = 1 \rangle,$$

where

$$T_0 = R_1 = \pm \begin{pmatrix} -1 & 1 \\ 0 & 1 \end{pmatrix}, \quad T_1 = R_2 = \pm \begin{pmatrix} -1 & 0 \\ 0 & 1 \end{pmatrix},$$

$$T_2 = R_2^{R_1 R_0} = \pm \begin{pmatrix} 1 & 0 \\ 2 & -1 \end{pmatrix}.$$

Figures 11 and 12 show fundamental regions for these two groups, together with the actions of their generators. As in the orientable cases, each conjugacy class of subgroups M of $\mathcal{T}_2$ or of $\mathcal{C}_2$ then corresponds to the appropriate quotient triangulation or map (here $\rho = e^{2\pi i/6}$ and $\sigma = (1+i)/2$).

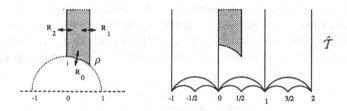

Figure 11 : a fundamental region for Π

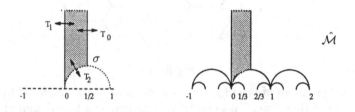

Figure 12 : a fundamental region for $\Pi_0(2)$

§4. Hypermaps

A theory of oriented hypermaps was introduced by Cori in [Cor]; for a comprehensive recent survey see [CoMa]. Geometrically, hypermaps arise as embeddings of hypergraphs in oriented surfaces without boundary; the basic components are termed hypervertices, hyperedges and hyperfaces. Algebraically, hypermaps correspond to transitive permutation representations of the hypercartographic group $\mathcal{H}_2^+$, in the same way as oriented maps are obtained from C_2^+. The only essential difference is that, in a hypermap, a hyperedge may be incident with an arbitrary (positive) number of hyper-vertices and hyperfaces, so we delete the relation $\rho_1^2 = 1$ from C_2^+ to obtain the triangle group $\mathcal{H}_2^+ = (\infty, \infty, \infty)$, a free group of rank 2. Otherwise, the theories of oriented maps and hypermaps are similar; for details, see [CS]. A theory of hypermaps on non-orientable surfaces or surfaces with boundary has recently been developed in [IS], extending the work in [BS]. Such a hypermap corresponds to a transitive permutation representation of $\mathcal{H}_2 = [\infty, \infty, \infty]$ on the set Ω of flags, and can be obtained from an oriented hypermap by factoring out an anticonformal involution.

A hypermap $\mathcal{H}$ has type (p, q, r) if p, q and r are the least common multiples of the valencies of the hypervertices, hyperedges and hyperfaces (the cycle-lengths of the permutations $s_1 s_2, s_2 s_0, s_0 s_1$); thus a map is simply a hypermap with $q \le 2$. For example, Figure 13 shows a hypermap of type $(3, 3, 3)$ on a torus, with hypervertices, hyperedges and hyperfaces represented by the regions labelled $0, 1, 2$ respectively; the underlying hypergraph is the Fano plane, the projective plane $\mathbf{P}^2(\mathbf{Z}_2)$ of order 2, with seven points and seven lines corresponding to the hypervertices and hyperedges [Si1].

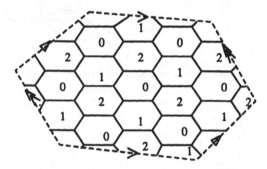

Figure 13

Each hypermap of type (p, q, r) is a quotient of the universal hypermap $\hat{\mathcal{H}}(p, q, r)$ of that type. This can be represented as a tessellation of

$$\mathcal{U} = \mathcal{U}(p, q, r) = \begin{cases} \text{hyperbolic plane} & \text{if } \frac{1}{p} + \frac{1}{q} + \frac{1}{r} < 1, \\ \text{Euclidean plane} & \text{if } \frac{1}{p} + \frac{1}{q} + \frac{1}{r} = 1, \\ \text{Riemann sphere} & \text{if } \frac{1}{p} + \frac{1}{q} + \frac{1}{r} > 1. \end{cases}$$

The hypervertices, hyperedges and hyperfaces, respectively labelled $0, 1$ and 2, are $2p$-gons, $2q$-gons and $2r$-gons, one of each meeting at every vertex of the tessellation; the vertices and edges of the tessellation form a Cayley graph for the extended triangle group

$$\Delta = [p, q, r] = \langle \sigma_0, \sigma_1, \sigma_2 \mid \sigma_i^2 = (\sigma_1 \sigma_2)^p = (\sigma_2 \sigma_0)^q = (\sigma_0 \sigma_1)^r = 1 \rangle,$$

which acts on $\mathcal{U}$ as the automorphism group of $\hat{\mathcal{H}}(p,q,r)$.

For example, the universal hypermap $\hat{\mathcal{H}} = \hat{\mathcal{H}}(\infty,\infty,\infty)$ is shown in Figure 14: $\hat{\mathcal{H}}$ is the dual of the universal triangulation $\hat{\mathcal{T}} = \hat{\mathcal{M}}(\infty,3)$ in Figure 3, with hypervertices, hyperedges and hyperfaces corresponding to vertices $a/b \in \hat{\mathbb{Q}}$ with a even and b odd, a and b both odd, or a odd and b even. (These are the equivalence classes $[0], [1]$ and $[\infty]$ containing $0, 1$ and ∞, where we define $a/b \equiv c/d$ if $a \equiv c$ and $b \equiv d \bmod(2)$.)

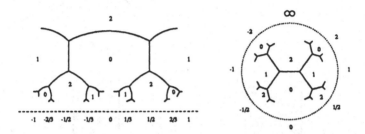

Figure 14 : the universal hypermap $\hat{\mathcal{H}}$

The orientation-preserving automorphism group $\mathcal{H}_2^+ = (\infty,\infty,\infty)$ of $\hat{\mathcal{H}}$ can be identified with the principal congruence subgroup $\Gamma(2)$ of level 2 in Γ, the normal subgroup of index 6 represented by the matrices $\pm\left(\begin{smallmatrix} a & b \\ c & d \end{smallmatrix}\right)$ with b, c both even. Similarly, the full automorphism group $\mathcal{H}_2 = [\infty,\infty,\infty]$ can be identified with $\Pi(2)$, the corresponding congruence subgroup of Π.

These identifications use the following presentations:

$$\Pi(2) = \langle V_0, V_1, V_2 \mid V_i^2 = 1 \rangle,$$

where

$$V_0 = T_1^{T_0} = \pm\begin{pmatrix} -1 & 2 \\ 0 & 1 \end{pmatrix}, \quad V_1 = T_1 = \pm\begin{pmatrix} -1 & 0 \\ 0 & 1 \end{pmatrix},$$

$$V_2 = T_2 = \pm\begin{pmatrix} 1 & 0 \\ 2 & -1 \end{pmatrix},$$

and

$$\Gamma(2) = \langle W_0, W_1, W_2 \mid W_0 W_1 W_2 = 1 \rangle,$$

where

$$W_0 = V_1 V_2 = \pm\begin{pmatrix} -1 & 0 \\ 2 & -1 \end{pmatrix}, \quad W_1 = V_2 V_0 = \pm\begin{pmatrix} -1 & 2 \\ -2 & 3 \end{pmatrix},$$

$$W_2 = V_0 V_1 = \pm\begin{pmatrix} 1 & 2 \\ 0 & 1 \end{pmatrix}.$$

Figures 15 and 16 show fundamental regions for $\Pi(2)$ and $\Gamma(2)$, together with the actions of their generators.

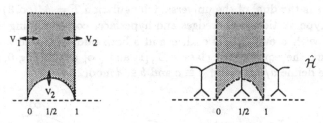

Figure 15 : a fundamental region for $\Pi(2)$

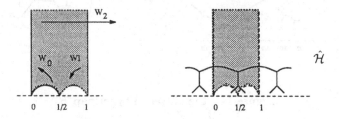

Figure 16 : a fundamental region for $\Gamma(2)$

§5. Hypermaps and Belyĭ functions

The connections between Belyĭ functions and triangle groups are very clearly explained in [CW]. If X is a complex algebraic curve, then it follows from Belyĭ's Theorem [B] that the following are equivalent:

a) X is defined over the field $\overline{\mathbf{Q}}$ of algebraic numbers;

b) there is a Belyĭ function $\beta : X \to \mathbf{P}^1(\mathbf{C})$, that is, a finite covering, unbranched outside $\{\,0, 1, \infty\,\}$;

c) $X \cong \mathcal{U}/H$ for some subgroup H of finite index in a triangle group $\Delta = (p, q, r)$, where p, q, r are finite;

d) $X \cong \overline{\mathcal{U}}/M$ for some subgroup M of finite index in $\mathcal{H}_2^+ = \Gamma(2)$, where $\overline{\mathcal{U}}$ denotes $\mathcal{U} \cup \hat{\mathbf{Q}}$.

(The subgroup M in (d) is simply the inverse image of H in (c) under the obvious epimorphism $\Gamma(2) = (\infty, \infty, \infty) \to (p, q, r)$.) When these conditions are satisfied, the situation is summarised by means of the following commutative diagram, in which the horizontal arrows denote the natural projections, and λ is the classical λ-function, automorphic with respect to $\Gamma(2)$ (see §6.7 of [JS2], for example); this is extended to $\overline{\mathcal{U}}$ so that the fibres over the points $0, 1, \infty \in \mathbf{P}^1(\mathbf{C})$ are the equivalence classes $[0], [1], [\infty] \subseteq \hat{\mathbf{Q}}$ corresponding to the sets of hypervertices, hyperedges and hyperfaces of $\hat{\mathcal{H}}$

in Figure 14.

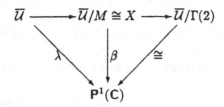

The choice of a Belyĭ function β then determines a finite hypermap on X, namely $\hat{\mathcal{H}}/M$, which is obtained by lifting the trivial oriented hypermap $\hat{\mathcal{H}}/\Gamma(2)$ on $\overline{U}/\Gamma(2) \cong \mathbf{P}^1(\mathbf{C})$ via β to X. Conversely, each finite oriented hypermap determines a conjugacy class of subgroups M of finite index in $\Gamma(2)$, and hence gives rise to a pair (X, β) where X is the algebraic curve uniformised by M and β is the Belyĭ function on X given by the above diagram. Thus finite oriented hypermaps can be regarded as the natural combinatorial models of algebraic curves defined over $\overline{\mathbf{Q}}$; in particular, they provide a faithful representation of the Galois group of $\overline{\mathbf{Q}}$ over $\mathbf{Q}$ (see the article by L. Schneps in this volume for this action on maps of genus 0).

§6. Categories associated with triangle groups

We have seen that hypermaps (including maps and triangulations) correspond to permutation representations of certain triangle groups Δ. In general, Δ is an extended triangle group

$$[p, q, r] = \langle \sigma_0, \sigma_1, \sigma_2 \mid \sigma_i^2 = (\sigma_1\sigma_2)^p = (\sigma_2\sigma_0)^q = (\sigma_0\sigma_1)^r = 1 \rangle,$$

but for oriented surfaces without boundary, we take Δ to be the even subgroup

$$(p, q, r) = \langle \rho_0, \rho_1, \rho_2 \mid \rho_0^p = \rho_1^q = \rho_2^r = \rho_0\rho_1\rho_2 = 1 \rangle$$

of index 2 in $[p, q, r]$, where

$$\rho_0 = \sigma_1\sigma_2, \qquad \rho_1 = \sigma_2\sigma_0, \qquad \rho_2 = \sigma_0\sigma_1.$$

(We allow $p, q, r = \infty$, corresponding to vacuous relations.)

The transitive permutation representations of Δ form a category $\mathbf{C}(\Delta)$, in which the objects Ω correspond to hypermaps of the appropriate type; the morphisms $\Omega_1 \to \Omega_2$ are the functions which commute with the actions of Δ, and these correspond to coverings of hypermaps. Each object $\Omega \in \mathbf{C}(\Delta)$ is isomorphic to the set $M\backslash\Delta$ of cosets of a point-stabiliser $M = \Delta_\alpha \le \Delta$, where $\alpha \in \Omega$; morphisms correspond to inclusions of stabilisers, and the automorphism group $\mathrm{Aut}(\Omega)$ of Ω (within $\mathbf{C}(\Omega)$) is isomorphic to $N_\Delta(M)/M$. The most symmetric objects are the regular, or Galois, objects,

128   Gareth Jones and David Singerman

those for which M is a normal subgroup, so that $\operatorname{Aut}(\Omega) \cong \Delta/M$. These are important since every (finite) object Ω is the quotient of some (finite) regular object $\tilde{\Omega}$ by a subgroup $A \le \operatorname{Aut}(\tilde{\Omega})$ [JS1, Corollary 6.8]: one can take $\tilde{\Omega}$ to correspond to the kernel $\tilde{M} = \cap_{g \in \Delta} M^g$ of the action of Δ on Ω, with A corresponding to $M/\tilde{M}$.

We are particularly interested in the following groups Δ and their associated categories:

Triangle group Δ	Category $C(\Delta)$	Objects Ω
$\mathcal{T}_2 = [\infty, 2, 3] = \Pi$	**T**	triangulations;
$\mathcal{T}_2^+ = (\infty, 2, 3) = \Gamma$	**T$^+$**	oriented triangulations;
$\mathcal{C}_2 = [\infty, 2, \infty] = \Pi_0(2)$	**M**	maps;
$\mathcal{C}_2^+ = (\infty, 2, \infty) = \Gamma_0(2)$	**M$^+$**	oriented maps;
$\mathcal{H}_2 = [\infty, \infty, \infty] = \Pi(2)$	**H**	hypermaps;
$\mathcal{H}_2^+ = (\infty, \infty, \infty) = \Gamma(2)$	**H$^+$**	oriented hypermaps.

(In the cases **T**, **M** and **H** we allow surfaces with boundary; in the oriented cases **T$^+$**, **M$^+$** and **H$^+$** we do not.)

We have identified each of the six groups Δ in this table with a certain subgroup of the extended modular group $\Pi = PGL_2(\mathbf{Z})$, so that the action of Π on $\mathcal{U}$ allows us to realise each object $\Omega \in C(\Delta)$ as a hypermap on a Riemann surface or Klein surface, the compactification of $\mathcal{U}/M$ where $M = \Delta_\alpha$, $\alpha \in \Omega$. The inclusions between these six groups are shown in Figure 17, where edge-labels denote indices of subgroups, small dots denote subgroups of Π and large ones normal subgroups of Π.

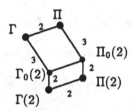

Figure 17

We will now show how homomorphisms $\Delta_1 \to \Delta_2$ between these groups Δ induce functors between the corresponding categories; in particular, inclusions $\Delta_1 \le \Delta_2$ and automorphisms $\Delta \to \Delta$ have useful interpretations.

§7. Functors induced by homomorphisms

We first consider functors induced by inclusions. Suppose that we have an inclusion $\Delta_1 \le \Delta_2$ between any two groups. Each object $\Omega_1 \in C(\Delta_1)$ has the form $\Omega_1 \cong M \backslash \Delta_1$ for some conjugacy class of subgroups $M \le \Delta_1$;

these all lie in a single conjugacy class in Δ_2, so they correspond to a unique object $\Omega_2 \cong M \backslash \Delta_2 \in C(\Delta_2)$. Note that $|\Omega_2| = |\Delta_2 : M| = |\Delta_2 : \Delta_1|.|\Delta_1 : M| = |\Delta_2 : \Delta_1|.|\Omega_1|$. Inclusions between subgroups $M \leq \Delta_1$ give rise to inclusions in Δ_2, so morphisms in $C(\Delta_1)$ induce morphisms in $C(\Delta_2)$, and we have a functor $C(\Delta_1) \to C(\Delta_2)$. If Ω_2 is regular in $C(\Delta_2)$ then M is normal in Δ_2, and hence normal in Δ_1, so Ω_1 is regular in $C(\Delta_1)$; as we shall see, the converse is not generally true.

We now give some examples of such functors.

(1) *Even subgroups.* Consider the inclusion

$$\Delta_1 = (p,q,r) < [p,q,r] = \Delta_2,$$

for instance $\Gamma < \Pi$, $\Gamma_0(2) < \Pi_0(2)$ or $\Gamma(2) < \Pi(2)$. The corresponding functor $F : C(\Delta_1) \to C(\Delta_2)$ simply forgets the orientation of each oriented object. For example:

(a) The group $\Delta_1 = \Gamma_0(2)$, a free product $C_2 * C_\infty$, has three normal subgroups M with Δ_1/M isomorphic to the alternating group A_5 ([Ha], [DJ]), so there are three regular oriented maps $\mathcal{M}$ with rotation group A_5: these are the icosahedron, the dodecahedron and the great dodecahedron (a map of type $(5,5)$ and genus 4 [Cox,§6.2]). Each M is normal in $\Delta_2 = \Pi_0(2)$, so each $\mathcal{M}$ is regular in $C(\Delta_2) = M$, that is, $\mathcal{M}$ is *reflexible* [CoMo, §8.1], meaning that it has an additional orientation-reversing automorphism.

(b) The regular oriented maps $\mathcal{M}_1 = \{4,4\}_{2,1}$ and $\mathcal{M}_2 = \{4,4\}_{1,2}$, described in §8.2 of [CoMo] and shown in Figure 18, correspond to the two normal subgroups $M_1, M_2 \lhd \Delta_1 = (4,2,4)$ with $\Delta_1/M_i \cong AGL_1(5)$ [JS1].

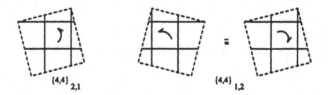

Figure 18

Neither subgroup is normal in $\Delta_2 = [4,2,4]$, so although these maps are regular in M^+, as *oriented* maps, they are not regular in M, that is, they are *chiral* rather than reflexible. In fact, $\mathcal{M}_1$ and $\mathcal{M}_2$ are conjugate in Δ_2, so $\mathcal{M}_1 \cong \mathcal{M}_2$ in M: they are isomorphic as unoriented maps, but not as oriented maps, each being the mirror-image of the other. These are examples of regular embeddings of the complete graph K_5; regular and reflexible embeddings of all complete graphs are classified in [JJ].

(2) *The Walsh and Vince functors.* Consider the inclusions

$$\Pi(2) < \Pi_0(2), \qquad \Gamma(2) < \Gamma_0(2),$$

both of index 2. The subgroups $\Delta_1 = \Pi(2)$ and $\Gamma(2)$ correspond to the hypermap categories **H** and **H⁺**, while $\Delta_2 = \Pi_0(2)$ and $\Gamma_0(2)$ correspond to the map categories **M** and **M⁺**. Walsh [Wa] showed how to represent each oriented hypermap $\mathcal{H} \in$ **H⁺** as a bipartite map $\mathcal{M} = W(\mathcal{H})$ on the same surface: black and white vertices of $\mathcal{M}$ correspond to hypervertices and hyperedges of $\mathcal{H}$, and edges of $\mathcal{M}$ correspond to incident hypervertex-hyperedge pairs, while faces of $\mathcal{M}$ correspond to hyperfaces of $\mathcal{H}$ (see Figure 19). Conversely, every bipartite oriented map represents a hypermap.

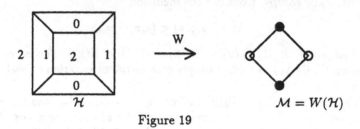

Figure 19

This functor W is induced by the inclusion $\Gamma(2) < \Gamma_0(2)$. If we regard each finite hypermap $\mathcal{H} \in$ **H⁺** as corresponding to a Belyĭ function β on an algebraic curve X defined over $\hat{\mathbf{Q}}$, then applying W to $\mathcal{H}$ corresponds to replacing β with the clean Belyĭ function $\gamma = 4\beta(1 - \beta)$ obtained from the $\Gamma_0(2)$-automorphic function $4\lambda(1 - \lambda)$:

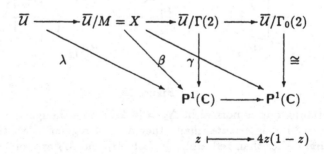

There is an obvious extension of W to **H**, induced by $\Pi(2) < \Pi_0(2)$. As shown by Corn and Singerman [CS], if $\mathcal{M}$ is regular then so is $\mathcal{H}$, whereas if $\mathcal{H}$ is regular then $\mathcal{M}$ is edge-transitive but not necessarily regular (see Figure 20, for example).

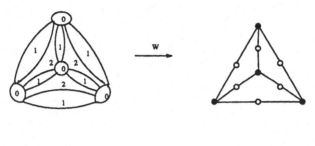

$\mathcal{H}$ regular $\mathcal{M}$ not regular

Figure 20

A similar functor V from hypermaps to maps has been introduced by Vince [V1, §3]: each hyperface is contracted to a point, giving a map $V(\mathcal{H})$ whose faces (coloured alternately black and white) correspond to the hypervertices and hyperedges of $\mathcal{H}$; this is the dual map of $W(\mathcal{H})$, obtained by composing the inclusion $\Pi(2) < \Pi_0(2)$ with the automorphism of $\Pi_0(2)$ transposing T_0 and T_2.

(3) *Stellations.* Consider the inclusions

$$\Pi_0(2) < \Pi, \qquad \Gamma_0(2) < \Gamma,$$

both of index 3. The subgroups $\Delta_1 = \Pi_0(2)$ and $\Gamma_0(2)$ correspond to the map categories **M** and $\mathbf{M^+}$, while $\Delta_2 = \Pi$ and Γ correspond to the categories **T** and $\mathbf{T^+}$ of triangulations.

Given a map $\mathcal{M}$, the stellation $S(\mathcal{M})$ is a triangulation of the same surface: each k-gonal face of $\mathcal{M}$ is divided into k triangular faces by joining a new vertex in its interior to each of the k incident vertices of $\mathcal{M}$, as in Figure 21:

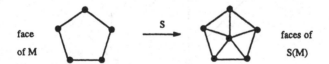

face   faces of
of M   S(M)

Figure 21

This functor $S : \mathbf{M} \to \mathbf{T}$, $\mathbf{M^+} \to \mathbf{T^+}$ is induced by the inclusions $\Pi_0(2) < \Pi$, $\Gamma_0(2) < \Gamma$. In the oriented case, it corresponds to replacing a clean Belyĭ function γ (induced by $\lambda_0 = 4\lambda(1 - \lambda)$) with the Belyĭ function $\delta = 27\gamma^2/(4-\gamma)^3$ induced by $1/J : \overline{\mathcal{U}} \to \mathbf{P^1(C)}$, where J is the classical modular function [JS2, Ch.6]

$$J = \frac{(4 - \lambda_0)^3}{27\lambda_0^2} = \frac{4(1 - \lambda + \lambda^2)^3}{27\lambda^2(1 - \lambda)^2}.$$

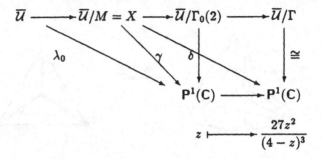

Not only is the Belyĭ function δ clean, but also the monodromy permutation over ∞ consists of 3-cycles, corresponding to the face-centres of $S(\mathcal{M})$.

(4) *Compositions.* The inclusions

$$\Pi(2) < \Pi, \qquad \Gamma(2) < \Gamma,$$

both of index 6, can be regarded as compositions $\Pi(2) < \Pi_0(2) < \Pi$, $\Gamma(2) < \Gamma_0(2) < \Gamma$, and so the functors $C : \mathbf{H} \to \mathbf{T}$ and $\mathbf{H}^+ \to \mathbf{T}^+$ they induce are the compositions of the functors W and S (Figure 22):

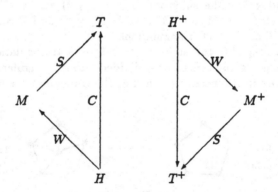

Figure 22

Applying C to a hypermap $\mathcal{H} \in \mathbf{H}$, we obtain the dual of the underlying trivalent map; this triangulation is a tripartite map, the three sets of vertices corresponding to the hypervertices, hyperedges and hyperfaces of $\mathcal{H}$. In the oriented case, C corresponds to replacing a Belyĭ function β (induced by λ) with the Belyĭ function $\delta = 27\beta^2(1 - \beta)^2/4(1 - \beta + \beta^2)^3$ induced by

$$1/J = 27\lambda^2(1-\lambda)^2/4(1-\lambda+\lambda^2)^2.$$

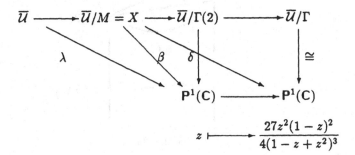

$$z \longmapsto \frac{27z^2(1-z)^2}{4(1-z+z^2)^3}$$

Figure 23 shows how the octahedron arises as $C(\mathcal{H})$ for a hypermap $\mathcal{H}$.

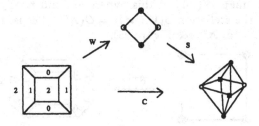

Figure 23

The next examples of functors are induced by homomorphisms which are not inclusions.

(5) The epimorphism

$$\theta : \Pi(2) = [\infty, \infty, \infty] \to [\infty, 2, \infty] = \Pi_0(2),$$

which sends each V_i to T_i, induces the functor $\mathbf{M} \to \mathbf{H}$ which regards each map $\mathcal{M}$, corresponding to some $M \le \Pi_0(2)$, as a hypermap (corresponding to $\theta^{-1}(M) \le \Pi(2)$). Similarly, the epimorphism

$$\Pi_0(2) = [\infty, 2, \infty] \to [\infty, 2, 3] = \Pi,$$

which sends each T_i to R_i, allows us to regard each triangulation as a map.

(6) *Barycentric subdivision*. This functor $B : \mathbf{M} \to \mathbf{T}$ is induced by the epimorphism $\theta : \Pi(2) \to \Pi_0(2)$ in the previous example, together with the inclusion $\Pi(2) < \Pi$ in example (4) (see Figure 24). Each $\mathcal{M} \in \mathbf{M}$ corresponds to some $M \le \Pi_0(2)$, and its barycentric subdivision $B(\mathcal{M})$ is the triangulation corresponding to the inclusion $\theta^{-1}(M) \le \Pi$:

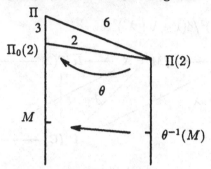

Figure 24

Firstly, if $\mathcal{M}$ corresponds to $M \leq \Pi_0(2)$, then $\theta^{-1}(M) \leq \Pi(2)$ corresponds to $\mathcal{M}$ regarded as a hypermap; next, $\theta^{-1}(M) \leq \Pi_0(2)$ corresponds to the Walsh bipartite map $W(\mathcal{M})$ of this hypermap, and finally $\theta^{-1}(M) \leq \Pi$ corresponds to the stellation $S(W(\mathcal{M})) = C(\mathcal{M})$, that is, the barycentric subdivision $B(\mathcal{M})$ of $\mathcal{M}$ (see Figure 25).

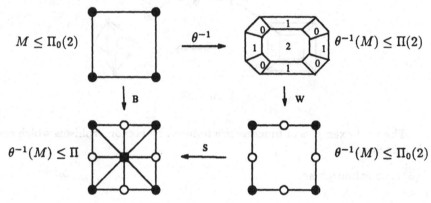

Figure 25

Since $\theta^{-1}(M) \leq \Pi(2)$, $B(\mathcal{M})$ covers the triangulation (of a disc) in Figure 26 corresponding to $\Pi(2) \triangleleft \Pi$:

Figure 26

§8. Automorphisms

For each $\Delta = [p, q, r]$ or (p, q, r), the automorphism group $\mathrm{Aut}(\Delta)$ permutes the conjugacy classes of subgroups $M \leq \Delta$, preserving inclusions, so it acts on the category $C(\Delta)$. The inner automorphism group $\mathrm{Inn}(\Delta)$ acts trivially, so there is an induced action of the outer automorphism class group

$$\mathrm{Out}(\Delta) = \mathrm{Aut}(\Delta)/\mathrm{Inn}(\Delta)$$

on $C(\Delta)$ (Léger considers this situation more generally in [Lé]). We shall examine the "operations" induced by $\mathrm{Out}(\Delta)$ on $C(\Delta)$ for the six subgroups Δ considered earlier. In each case, it is straightforward to determine $\mathrm{Out}(\Delta)$ from a decomposition of Δ as a free product (possibly with amalgamation), using the structure theorems for free products [LS]; the details can be found in [Ja1, Jo3, JT1] for example.

(1) $\Delta = \Pi = \langle R_0, R_1, R_2 \mid R_i^2 = (R_2 R_0)^2 = (R_0 R_1)^3 = 1 \rangle$.

This group is the amalgamated free product $V_4 *_{C_2} S_3$ of the Klein four-group $\langle R_0, R_2 \rangle \cong V_4$ and $\langle R_0, R_1 \rangle \cong S_3$, with amalgamated subgroup $\langle R_0 \rangle \cong C_2$. Dyer [Dy], correcting the error in [HR], showed that $\mathrm{Out}(\Pi) \cong C_2$, generated by the class containing the automorphism

$$\alpha : R_0 \mapsto R_0, \quad R_1 \mapsto R_1, \quad R_2 \mapsto R_0 R_2$$

which fixes $\langle R_0, R_1 \rangle$ and acts on $\langle R_0, R_2 \rangle$ by transposing R_2 and $R_0 R_2$. This cannot be inner, since it does not preserve the normal subgroup $\Gamma = \langle R_0 R_1, R_1 R_2 \rangle$. The effect of α on congruence subgroups of small index is shown in Figure 27 (see [JT2]). Small dots denote normal subgroups of Π and large ones characteristic subgroups of Π. We have $\Pi/N(12) \simeq S_4 \times S_4$.

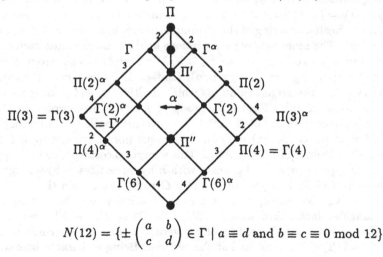

$$N(12) = \{\pm \begin{pmatrix} a & b \\ c & d \end{pmatrix} \in \Gamma \mid a \equiv d \text{ and } b \equiv c \equiv 0 \bmod 12\}$$

Figure 27

In [JT1] it was shown that α induces the following operation on triangulations $T \in C(\Pi) = \mathbf{T}$: cut the surface along each edge of T, and then rejoin the two incident faces along their common edge with the *opposite* orientation, as in Figure 28. For example, the k triangles surrounding a vertex of valency k become an annulus or a Möbius band as k is even or odd (see Figure 29).

This operation (which, by abuse of notation, we will also denote by α) preserves edges and faces, but transposes vertices and Petrie polygons (closed zig-zag paths). The underlying surface is not generally preserved: both its orientability and its Euler characteristic may change.

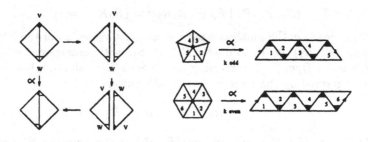

Figure 28   Figure 29

Application. The Fermat curve $x^n + y^n = 1$ is a Riemann surface X_n of genus $g = (n-1)(n-2)/2$. The Belyĭ function $(x,y) \mapsto x^n$ represents X_n as an n^2-sheeted covering of the Riemann sphere, branched over the points $0, 1$ and ∞. The monodromy permutations at these three points each consist of disjoint cycles of length n, so $X_n \cong \mathcal{U}/M$ where M is a subgroup of index n^2 in the triangle group $\Delta = (n,n,n)$. (In fact M is the commutator subgroup of Δ [Wo]: the transformations which multiply x and y independently by n-th roots of unity form a subgroup of $\mathrm{Aut}(X_n)$ isomorphic to $C_n \times C_n$, corresponding to $\Delta/M \cong \Delta^{ab}$.) Now M lifts back to a subgroup M_n of index n^2 in $(\infty, \infty, \infty) = \Gamma(2)$, the unique normal subgroup in $\Gamma(2)$ with quotient isomorphic to $C_n \times C_n$. This subgroup corresponds to a hypermap $\mathcal{H}_n$ of type (n,n,n) and genus g with n hypervertices, n hyperedges and n hyperfaces. By applying the functor C to $\mathcal{H}_n$ we obtain the triangulation $\mathcal{F}_n$ of X_n, consisting of $3n$ vertices (of valency $2n$), $3n^2$ edges and $2n^2$ triangular faces, discussed in [Wo, §6]. Thus $\mathcal{F}_n$ (which corresponds to $M_n \leq \Pi$) is an n^2-sheeted regular covering of the map $\mathcal{F}_1$ (corresponding to $M_1 = \Gamma(2) \leq \Pi$), branched at the vertices. Being a characteristic subgroup of $\Gamma(2)$, which is itself normal in Π, M_n is normal in Π and hence $\mathcal{F}_n$ is reflexible. (This triangulation $\mathcal{F}_n$ also arises in topological graph theory [Wh, §10.3] as a minimum-genus imbedding of the complete tripartite graph $K_{n,n,n}$.)

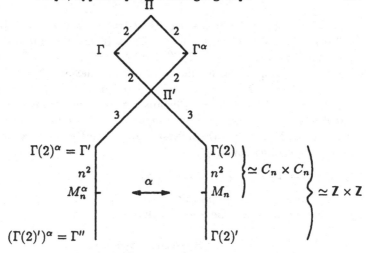

Figure 30

Under α, $\Gamma(2)$ is mapped to the derived group $\Gamma' = \Gamma(2)^\alpha$ of Γ, and so M_n is mapped to the unique normal subgroup $M_n^\alpha \triangleleft \Gamma'$ with $\Gamma'/M_n^\alpha \cong C_n \times C_n$ (see Figure 30). This subgroup M_n^α corresponds to the reflexible torus map $\mathcal{T}_n = \{3,6\}_{n,0}$ described in §8.4 of [CoMo], so α transposes the maps $\mathcal{F}_n$ and $\mathcal{T}_n$, as shown in Figure 31. (Note that $\mathcal{F}_3 \cong \mathcal{T}_3$, since $M_3^\alpha = M_3$.)

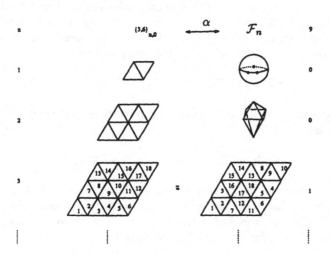

| n | n^2-fold unbranched covering of $3,6_{n,0}$ | n^2-fold branched covering of $\mathcal{F}_1$ | $(n-1)(n-2)/2$ |

Figure 31

For each n, $\mathcal{T}_n$ is the quotient of the 6-valent triangulation $\{3,6\}$ of the Euclidean plane $\mathbf{R}^2$, corresponding to $\Gamma'' \leq \Pi$, by the subgroup M_n^α/Γ'' of index $12n^2$ in $\mathrm{Aut}(\{3,6\}) = \Pi/\Gamma'' \cong [6,2,3]$. Hence each Fermat triangulation $\mathcal{F}_n$ is the quotient of the infinite triangulation $\{3,6\}^\alpha$, corresponding to $(\Gamma'')^\alpha = \Gamma(2)' \leq \Pi$, by the subgroup $M_n/\Gamma(2)'$ of index $12n^2$ in $\mathrm{Aut}(\{3,6\}^\alpha) = \Pi/\Gamma(2)' \cong [6,2,3]$. Just as $\{3,6\}$ is the inverse limit of the torus maps $\mathcal{T}_n$ ($n \in \mathbf{N}$), with coverings $\mathcal{T}_n \to \mathcal{T}_m$ induced by factors $m|n$, one can regard $\{3,6\}^\alpha$ as the inverse limit of the Fermat triangulations $\mathcal{F}_n$, with corresponding coverings.

(2) $\Delta = \Gamma = \langle S_0, S_1, S_2 \mid S_1^2 = S_2^3 = S_0 S_1 S_2 = 1 \rangle$.

In this case, Γ is a free product $C_2 * C_3$, and so $\mathrm{Out}(\Gamma) \cong C_2$, generated by the class containing the automorphism

$$\beta : S_1 \mapsto S_1, \quad S_2 \mapsto S_2^{-1}$$

which fixes the factor $\langle S_1 \rangle \cong C_2$ and inverts $\langle S_2 \rangle \cong C_3$. This is induced by conjugation by the generator R_1 of Π (in fact, $\mathrm{Aut}(\Gamma) \cong \Pi$). The corresponding operation on $\mathbf{C}(\Gamma) = \mathbf{T}^+$ simply reverses the orientation of each oriented triangulation.

(3) $\Delta = \Pi_0(2) = \langle T_0, T_1, T_2 \mid T_i^2 = (T_2 T_0)^2 = 1 \rangle$.

This is the free product of $\langle T_0, T_2 \rangle \cong V_4$ and $\langle T_1 \rangle \cong C_2$. The group $\mathrm{Out}(\Pi_0(2)) \cong S_3$ is induced by the 3! permutations of the three involutions T_0, T_2 and $T_0 T_2$ in the factor V_4. The corresponding operations on maps $\mathcal{M} \in \mathbf{M}$ have been studied by Wilson [Wi], Lins [L], Jones and Thornton [JT1], and Léger and Terrasson [LT]. The automorphism

$$\delta : T_0 \mapsto T_2, \quad T_1 \mapsto T_1, \quad T_2 \mapsto T_0$$

induces the duality operation, interchanging vertices and faces, while the automorphism

$$\alpha : T_0 \mapsto T_0, \quad T_1 \mapsto T_1, \quad T_2 \mapsto T_0 T_2$$

has the same effect as the corresponding automorphism α of Π: it cuts a map along each edge and then rejoins faces with the reverse orientation, thus interchanging vertices and Petrie polygons. These two automorphisms generate $\mathrm{Out}(\Pi_0(2))$; for example, the involution

$$\alpha \delta \alpha = \delta \alpha \delta : T_0 \mapsto T_0 T_2, \quad T_1 \mapsto T_1, \quad T_2 \mapsto T_2$$

interchanges faces and Petrie polygons, preserving vertices and edges.

As an illustration, if we apply these six operations to the octahedron $\mathcal{M}$, as in Figure 32, we find that $\mathcal{M}^\delta$ is the cube, $\mathcal{M}^\alpha$ and $\mathcal{M}^{\alpha\delta}$ are the torus maps $\{3,6\}_{2,0}$ and $\{6,3\}_{2,0}$, while $\mathcal{M}^{\delta\alpha}$ and $\mathcal{M}^{\delta\alpha\delta}$ are the nonorientable maps $\{4,6\}_3$ and $\{6,4\}_3$ of characteristic -2 in Table 8 of [CoMo].

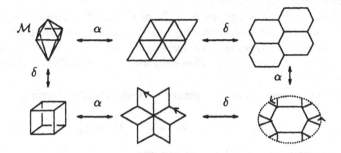

Figure 32

This orbit of length 6 shows that $\mathrm{Out}(\Pi_0(2))$ acts faithfully on **M**. As another example, the tetrahedron lies in an orbit of length 3, containing the maps on the projective plane obtained by antipodal identification of the cube and the octahedron (Figure 33).

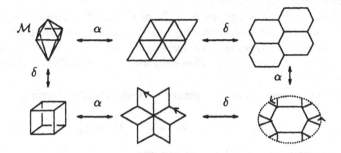

Figure 33

(4) $\Delta = \Gamma_0(2) = \langle U_0, U_1, U_2 \mid U_1^2 = U_0 U_1 U_2 = 1 \rangle$.

This is the free product of $\langle U_1 \rangle \cong C_2$ and $\langle U_2 \rangle \cong C_\infty$. It therefore follows easily from the structure theorems for free products [LS] that $\mathrm{Out}(\Gamma_0(2)) \cong V_4$, generated by the classes containing the automorphisms

$$\beta : U_1 \mapsto U_1, \quad U_2 \mapsto U_2^{-1}$$

(which reverses the orientation of each map $\mathcal{M} \in \mathbf{M}^+$) and the duality operation

$$\delta : U_1 \mapsto U_1, \quad U_2 \mapsto U_0 = U_2^{-1} U_1.$$

(5) $\Delta = \Pi(2) = \langle V_0, V_1, V_2 \mid V_i^2 = 1 \rangle$.

This is the free product of three groups $\langle V_i \rangle \cong C_2$. As shown by James [Ja1] (see also [Jo3]), $\mathrm{Out}(\Pi(2)) \cong PGL_2(\mathbf{Z})$: $\mathrm{Aut}(\Pi(2))$ preserves $\Gamma(2)$, the unique torsion-free subgroup of index 2 in $\Pi(2)$, and acts on $\Gamma(2)^{\mathrm{ab}} \cong \mathbf{Z} \times \mathbf{Z}$ as $GL_2(\mathbf{Z})$; now $\mathrm{Inn}(\Gamma(2))$ acts trivially on $\Gamma(2)^{\mathrm{ab}}$, while $V_1(\in \Pi(2) \setminus \Gamma(2))$ inverts the free generators W_0, W_2 of $\Gamma(2)$ and hence induces $-I$ on $\Gamma(2)^{\mathrm{ab}}$;

thus $\mathrm{Out}(\Pi(2)) \cong GL_2(\mathbf{Z})/\{\pm I\} = PGL_2(\mathbf{Z})$. James showed that this group acts faithfully on $\mathbf{C}(\Pi(2)) = \mathbf{H}$, so in the case of hypermaps we have an *infinite* group of operations.

There is an obvious subgroup $S \cong S_3$ of $\mathrm{Out}(\Pi(2))$, induced by the 3! permutations of the generating involutions V_i; the corresponding operations on $\mathbf{H}$ simply rename hypervertices, hyperedges and hyperfaces, leaving the underlying surface unchanged. This subgroup was described (for oriented hypermaps) in [M] by Machì.

$\mathrm{Out}(\Pi(2))$ is generated by S together with the class containing the automorphism

$$\gamma : V_0 \mapsto V_0^{V_2}, \quad V_1 \mapsto V_1, \quad V_2 \mapsto V_2.$$

This is the restriction to $\Pi(2)$ of the automorphism

$$\delta\alpha\delta : T_0 \mapsto T_0 T_2, \quad T_1 \mapsto T_1, \quad T_2 \mapsto T_2$$

of $\Pi_0(2)$; it corresponds to applying the operation α to the 2-face-coloured Vince map $V(\mathcal{H}) = W(\mathcal{H})^\delta$ of a hypermap $\mathcal{H}$:

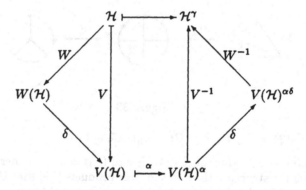

If τ_{ij} denotes the automorphism in S transposing V_i and V_j, and if ρ_0, ρ_1 and ρ_2 are the classes in $\mathrm{Out}(\Pi(2))$ corresponding to $\tau_{02}, \tau_{12}{}^{\gamma\tau_{01}}$ and γ, then

$$\mathrm{Out}(\Pi(2)) = \langle \rho_0, \rho_1, \rho_2 \mid \rho_i^2 = (\rho_2\rho_0)^2 = (\rho_0\rho_1)^3 = 1 \rangle,$$

so we have an isomorphism $\mathrm{Out}(\Pi(2)) \to \Pi$, $\rho_i \mapsto R_i$. Thus, as pointed out in [Jo3], each hypermap $\mathcal{H}$ determines a subgroup $H \leq \Pi$, corresponding to the operations leaving $\mathcal{H}$ invariant, and hence determines a triangulation $\mathcal{T}$, corresponding to $H \leq \Pi$. (Strictly speaking, this isomorphism is not canonical, so $\mathcal{H}$ actually determines a *pair* of triangulations $\mathcal{T}, \mathcal{T}^\alpha$.) The flags of $\mathcal{T}$ (or of $\mathcal{T}^\alpha$) correspond to the images of $\mathcal{H}$ under the various operations on hypermaps, so if $\mathcal{H}$ is finite then so is $\mathcal{T}$.

(6) $\Delta = \Gamma(2) = \langle W_0, W_1, W_2 \mid W_0 W_1 W_2 = 1 \rangle.$

This is a free group $F_2 = C_\infty * C_\infty$ of rank 2, so $\mathrm{Out}(\Gamma(2))$ is isomorphic to $GL_2(\mathbf{Z})$, represented faithfully on $\Gamma(2)^{\mathrm{ab}} \cong \mathbf{Z} \times \mathbf{Z}$. It is generated by the classes corresponding to the restrictions to $\Gamma(2)$ of the automorphisms τ_{ij} and γ of $\Pi(2)$, together with conjugation by V_1 (which generates the kernel of the epimorphism $\mathrm{Out}(\Gamma(2)) \cong GL_2(\mathbf{Z}) \to PGL_2(\mathbf{Z}) \cong \mathrm{Out}(\Pi(2))$, and induces orientation-reversal on $\mathbf{H}^+$).

As an illustration, Hall [Ha] has shown that F_2 has 19 normal subgroups M with $F_2/M \cong A_5$, corresponding to the 19 orbits of $\mathrm{Aut}(A_5) \cong S_5$ on generating pairs for A_5. It follows that there are (up to isomorphism) 19 regular oriented hypermaps $\mathcal{H}$ with rotation-group $\mathrm{Aut}(\mathcal{H}) \cong A_5$. These are described in [d'AJ]; they are all reflexible, and form 6 orbits under S:

(a)   an S-orbit of 6 hypermaps of genus 0 (and type a permutation of $(2,3,5)$), including the icosahedron and dodecahedron;

(b)   an S-orbit of 3 hypermaps of genus 4 (and type a permutation of $(2,5,5)$), including the great dodecahedron;

(c)   an S-orbit of 3 hypermaps of genus 5 (and type a permutation of $(3,3,5)$);

(d)   an S-orbit of 3 hypermaps of genus 9 (and type a permutation of $(3,5,5)$); the two generators of order 5 are conjugate in A_5;

(e)   an S-orbit of 3 hypermaps of genus 9 (and type a permutation of $(3,5,5)$); the two generators of order 5 are not conjugate in A_5;

(f)   a single S-invariant hypermap of genus 13 and type $(5,5,5)$.

Now $\mathrm{Aut}(F_2)$ permutes these 19 normal subgroups M, and Neumann and Neumann [NN] have shown that it has two orbits, of lengths 9 and 10, distinguished by the order of the commutator of the two generators (either 3 or 5 respectively). It follows that the 19 hypermaps form two orbits under $\mathrm{Out}(\Gamma(2))$, one consisting of the S-orbits (b), (c) and (e), the other (a), (d) and (f).

§9. Higher dimensions

The n-dimensional analogue of the cartographic group $\mathcal{C}_2$ is, as conjectured by Shabat and Voevodsky in [SV], the Coxeter group

$$\mathcal{C}_n = \langle \sigma_0, \ldots, \sigma_n \mid \sigma_i^2 = (\sigma_i \sigma_j)^2 = 1 \quad (|i - j| > 1) \rangle.$$

This group, together with its associated combinatorial and topological structures, has been studied by Vince [V1, V2]; James [Ja2] has shown that if $n > 2$ then the associated group of operations is $\mathrm{Out}(\mathcal{C}_n) \cong D_4$. Higher-dimensional analogues of maps have also been studied by Costa [Cos1, Cos2] and Dress [Dr].

References

[B]   G.V.Belyĭ, On Galois extensions of a maximal cyclotomic field, *Izv. Akad. Nauk SSSR*, **43** (1979), 269–276 (Russian); *Math. USSR Izvestiya* **14** (1980), 247–256 (English transl.).

[BS]   R.P.Bryant and D.Singerman, Foundations of the theory of maps on surfaces with boundary, *Quarterly J. Math. Oxford*, (2) **36** (1985), 17–41.

[CW]   P.Beazley Cohen and J.Wolfart, Dessins de Grothendieck et variétés de Shimura, *C.R.Acad.Sci.Paris*, **315** Sér. I (1992), 1025–1028.

[Cor]   R.Cori, Un code pour les graphes planaires et ses applications, *Astérisque* **27** (1975).

[CoMa] R.Cori and A.Machì, Maps, hypermaps and their automorphisms: a survey I, II, III, *Expositiones Math.* **10** (1992), 403–427, 429–447, 449–467.

[CS]   D.Corn and D.Singerman, Regular hypermaps, *European J. Combinatorics*, **9** (1988), 337–351.

[Cos1] A.F.Costa, On manifolds admitting regular combinatorial maps, *Rend. Circ. Mat. Palermo* **24** (1990), 327–335.

[Cos2] A.F.Costa, Locally regular coloured graphs, *J. Geometry* **43** (1992), 57–74.

[Cox]   H.S.M.Coxeter, *Regular Polytopes*, Macmillan, New York, 1963.

[CoMo] H.S.M. Coxeter and W.O.J. Moser, *Generators and Relations for Discrete Groups*, Springer-Verlag, Berlin-Heidelberg-New York, 1965.

[d'AJ] A.J.Breda d'Azevedo and G.A.Jones, Regular Platonic hypermaps, preprint, Southampton, 1993.

[DJ]   M.L.N.Downs and G.A.Jones, Enumerating regular objects with a given automorphism group, *Discrete Math.*, **64** (1987), 299–302.

[Dr]   A.W.M.Dress, Regular polytopes and equivariant tessellations from a combinatorial point of view, in *Algebraic Topology Göttingen 1984*, Lecture Notes in Mathematics **1172**, Springer-Verlag, Berlin, 1985, 56–72.

[Dy]   J.L.Dyer, Automorphism sequences of integer unimodular groups, *Illinois J. Math.*, **22** (1978), 1–30.

[G]   A.Grothendieck, Esquisse d'un programme, preprint, 1984.

[Ha]   P.Hall, The Eulerian functions of a group, *Quarterly J. Math. Oxford*, **7** (1936).

[He1] L.Heffter, Über das Problem der Nachbargebiete, *Math. Ann.*, **38** (1891), 477–508.

[He2] L.Heffter, Über metacyklische Gruppen und Nachbarconfigurationen, *Math. Ann.*, **50** (1898), 261–268.

[Ho] A.H.M.Hoare, Subgroups of N.E.C. groups and finite permutation groups, *Quarterly J. Math. Oxford*, (2) **41** (1990), 45–59.

[HS] A.H.M.Hoare and D.Singerman, The orientability of subgroups of plane groups, in *Groups–St.Andrews 1981* (eds. C.M.Campbell and E.F.Robertson), London Math. Soc. Lecture Notes **71**, 1982, 221–227.

[HR] L-K.Hua and I.Reiner, Automorphisms of the unimodular group, *Trans. Amer. Math. Soc.* **72** (1952), 467–473.

[IS] M.Izquierdo and D.Singerman, Hypermaps on surfaces with boundary, preprint, Southampton, 1993.

[Ja1] L.D.James, Operations on hypermaps, and outer automorphisms, *European J. Combinatorics*, **9** (1988), 551–560.

[Ja2] L.D.James, Complexes and Coxeter groups—operations and outer automorphisms, *J.Algebra*, **113** (1988), 339–345.

[JJ] L.D.James and G.A.Jones, Regular orientable imbeddings of complete graphs, *J. Combinatorial Theory (B)*, **39** (1985), 353–367.

[Jo1] G.A.Jones, Graph imbeddings, groups, and Riemann surfaces, in *Algebraic Methods in Graph Theory, Szeged 1978* (eds. L.Lovász and V.T. Sós), Colloq. Math. Soc. János Bolyai **25**, North-Holland, Amsterdam, 1978, 297–311.

[Jo2] G.A.Jones, Triangular maps and non-congruence subgroups of the modular group, *Bull. London Math. Soc.*, **11** (1979), 117–123.

[Jo3] G.A.Jones, Operations on maps and hypermaps, in *Sém. Lotharingien de Combinatoire* (ed. G.Nicoletti), Bologna **13** (1985), 45–55.

[JS1] G.A.Jones and D.Singerman, Theory of maps on orientable surfaces, *Proc. London Math. Soc.*, (3) **37** (1978), 273–307.

[JS2] G.A.Jones and D.Singerman, *Complex Functions, an Algebraic and Geometric Viewpoint*, Cambridge University Press, Cambridge, 1987.

[JSW] G.A.Jones, D.Singerman and K.Wicks, The modular group and generalized Farey graphs, in *Groups—St. Andrews, 1989* (eds. C. M. Campbell and E.F.Robertson), London Math. Soc. Lecture Notes **160** (1991), 316–338.

[JT1] G.A.Jones and J.S.Thornton, Operations on maps, and outer automorphisms, *J. Combinatorial. Theory (B)*, **35** (1983), 93–103.

[JT2]   G.A.Jones and J.S.Thornton, Automorphisms and congruence sub-
         groups of the extended modular group, *J. London Math. Soc.*, (2)
         34 (1986), 26–40.

[Lé]   C.Léger, Opérations sur les cartes et métamorphoses de la catégorie
         des *G*-ensembles, *Revista Mat. Univ. Complutense Madrid* **4** (1991),
         45–51.

[LT]   C.Léger and J-C.Terrasson, L'action du groupe symétrique *S*3
         sur l'ensemble des classes d'isomorphisme de pavages de surface
         fermée, *C. R. Acad. Sci. Paris*, **302** Sér. I (1986), 39–42.

[L]   S.Lins, Graph-encoded maps, *J. Combinatorial Theory (B)*, **32**
         (1982), 171–181.

[LS]   R.C.Lyndon and P.E.Schupp, *Combinatorial Group Theory*, Sprin-
         ger-Verlag, Berlin-Heidelberg-New York, 1977.

[M]   A.Machì, On the complexity of a hypermap, *Discrete Math.*, **42**
         (1982), 221–226.

[MV]   J.Malgoire and C.Voisin, Cartes cellulaires, *Cahiers Mathématiques
         de Montpellier*, **12** (1977).

[NN]   B.H.Neumann and H.Neumann, Zwei Klassen charakteristischer
         Untergruppen und ihre Faktorgruppen, *Math. Nachr.* **4** (1951),
         106–125.

[SV]   G. Shabat and V. A. Voevodsky, Drawing curves over number
         fields, in *Grothendieck Festschrift III* (ed. P.Cartier *et al.*), Progress
         in Math. **88**, Birkhäuser, 1990, 199–227.

[Si1]   D.Singerman, Klein's Riemann surface of genus 3 and regular im-
         beddings of finite projective planes, *Bull. London Math. Soc.*, **18**
         (1986), 364–370.

[Si2]   D.Singerman, Universal tessellations, *Revista Mat. Univ. Com-
         plutense Madrid* **1** (1988), 111–123.

[T]   W.T.Tutte, What is a map? in *New Directions in Graph Theory*
         (ed. F. Harary), Academic Press, New York, 1973, 309–325.

[V1]   A.Vince, Combinatorial maps, *J. Combinatorial Theory (B)*, **34**
         (1983), 1–21.

[V2]   A.Vince, Regular combinatorial maps, *J. Combinatorial Theory
         (B)*, **35** (1983), 256–277.

[Wa]   T.R.S.Walsh, Hypermaps versus bipartite maps, *J. Combinatorial
         Theory (B)*, **18** (1975), 155–163.

[Wh]   A.T.White, *Graphs, Groups and Surfaces*, North-Holland, Ams-
         terdam, 1984.

[Wi]   S.E.Wilson, Operators over regular maps, *Pacific J. Math., 81* (1979), 559–568.

[Wo]   J.Wolfart, Mirror-invariant triangulations of Riemann surfaces, triangle groups and Grothendieck dessins: variations on a theme of Belyi, preprint, Frankfurt, 1992.

*Department of Mathematics, University of Southampton, Southampton SO9 5NH, United Kingdom

Fields of Definition of Some
Three Point Ramified Field Extensions

Gunter Malle*
Mathematisches Institut und IWR
Im Neuenheimer Feld 368, D – 69120 Heidelberg

1 Introduction

In this paper we collect some computational data on fields of definition of three point ramified field extensions of $\bar{\mathbb{Q}}(t)$ of degree $n \leq 13$. No general results on such fields of definition, except for upper bounds on their degree and restrictions on possible ramified primes, are known at present. A more precise description of these fields would probably lead to results in the inverse problem of Galois theory, i.e., the question whether every finite group occurs as geometric Galois group over the field $\mathbb{Q}(t)$. The data presented here already gave hints for the possible validity of a translation theorem. This could then be proved to hold in general in [4].

Without any further knowledge one would expect the Galois group of a field of definition to be the full symmetric group. Our lists contain a number of examples where the Galois group is strictly smaller, although no known result predicts this.

There have been some previous investigations in this direction, for example by Atkin and Swinnerton-Dyer [1], who were interested in fields of definition attached to factor groups of the modular group $SL_2(\mathbb{Z})$, and by Thompson [8], who considered a particular example in the context of the Guralnick–Thompson conjecture on groups of genus zero.

2 Background

We are interested in the fields of definition of finite field extensions K of $\mathbb{C}(t)$ ramified over the three points $\{0, 1, \infty\}$. There is a natural bijection

* The author gratefully acknowledges financial support by the Deutsche Forschungsgemeinschaft

between these extensions and the finite unramified coverings of the Riemann sphere $\mathbb{P}^1(\mathbb{C})$ with the three points $\{0, 1, \infty\}$ removed, which in turn are in bijection with the set of dessins d'enfant (see [3*] and [6*] in this volume). The latter are classified by conjugacy classes of subgroups Δ of finite index of the fundamental group

$$\pi_1 = \langle \gamma_1, \gamma_2, \gamma_3 \mid \gamma_1 \gamma_2 \gamma_3 = 1 \rangle$$

of $\mathbb{P}^1(\mathbb{C}) \backslash \{0, 1, \infty\}$ (cf. [6*]). Any such subgroup Δ of index n gives rise to a transitive permutation representation ϕ of π_1 into S_n. The Galois group of the Galois closure of $K/\mathbb{C}(t)$ is then isomorphic to the image of π_1 under this homomorphism.

It is well known that the same picture applies to finite extensions K of $\bar{\mathbb{Q}}(t)$ ramified only over $\{0, 1, \infty\}$ (see for example [5], I, §4, Satz 3). Any such extension $K/\bar{\mathbb{Q}}(t)$ is already defined over a number field k of finite degree over $\mathbb{Q}$. It is these fields that we are interested in.

Let N denote the Galois closure of $K/\bar{\mathbb{Q}}(t)$. The homomorphism $\phi : \pi_1 \to S_n$ maps the generators γ_i onto generators σ_i of the group $G = \phi(\pi_1) \leq S_n$, which is isomorphic to the Galois group $\mathrm{Gal}(N/\bar{\mathbb{Q}}(t))$. Conversely, for any triple $(\sigma_1, \sigma_2, \sigma_3)$ of elements generating $G \leq S_n$ with $\sigma_1\sigma_2\sigma_3 = 1$, there exists a homomorphism from π_1 to S_n mapping the γ_i to σ_i. Hence any such triple $(\sigma_1, \sigma_2, \sigma_3)$ gives rise to a field extension $K/\bar{\mathbb{Q}}(t)$. Triples which are conjugate in S_n yield the same extension. More precisely the field extensions of $\bar{\mathbb{Q}}(t)$ of degree n unramified outside $\{0, 1, \infty\}$ and having group $G \leq S_n$ are parametrized by sets $\Sigma_G(C_1, C_2, C_3) :=$

$$\{\sigma_1, \sigma_2, \sigma_3 \mid \sigma_i \in C_i, \sigma_1\sigma_2\sigma_3 = 1, G = \langle \sigma_1, \sigma_2, \sigma_3 \rangle\} \bmod S_n,$$

where the C_i run over the conjugacy classes of S_n meeting G. This is called the Hurwitz classification [5]. The field extension corresponding to $\sigma = (\sigma_1, \sigma_2, \sigma_3)$ is denoted by K_σ. It is easily seen that if two extensions $K/\bar{\mathbb{Q}}(t)$, $K'/\bar{\mathbb{Q}}(t)$ are algebraically conjugate, then they are classified by the same class triple $\mathbf{C} = (C_1, C_2, C_3)$ of the same subgroup G of S_n. So we get:

Proposition A: *An upper bound for the degree of the field of definition k_σ of $K_\sigma/\bar{\mathbb{Q}}(t)$ is given by the number $n_G(\mathbf{C}) := |\Sigma_G(\mathbf{C})|$.*

Beckmann [2] has shown the following:

Proposition B: *For the choice $\{0, 1, \infty\}$ of ramification points, only prime divisors of the group order $|G|$ can ramify in $k_\sigma/\mathbb{Q}$.*

The reality type of k_σ can be read off from the tuple σ, as was noted by Serre [7]:

Proposition C: *The action of complex conjugation on the fields of definition induces an action*

$$\rho : [\sigma_1, \sigma_2, \sigma_3] \mapsto [\sigma_1^{-1}, \sigma_2^{-1}, \sigma_2 \sigma_3^{-1} \sigma_2^{-1}]$$

on the corresponding classes modulo conjugation $[\sigma]$ of element triples.

In the case where the class vector **C** admits symmetries, a little bit more is known. In those cases, either there is a non-trivial action of geometric automorphisms on $\Sigma_G(C_1, C_2, C_3)$, which forces $k/\mathbb{Q}$ to be imprimitive, or the class vector can be obtained by a Galois translation from another class triple [5]. In some additional cases, the translation by a non-Galois extension also leads to information on the field of definition (see [4]).

Apart from that, the precise nature of k is at present a mystery.

3 The calculations

The calculation divides up into several different parts. First, for given degree n the triples σ of elements modulo conjugation of S_n with product 1 must be determined. Then the groups generated by these triples must be identified. This step was done using a C–program written by B. Przywara.

Next, the corresponding field extensions have to be constructed. This is done as described for example in [5], using Gröbner basis techniques. The occurring systems of non-linear equations were solved using a Pascal program written by the author. The construction of the field extension immediately yields a generating polynomial $f(x)$ for the field of definition $k_\sigma/\mathbb{Q}$. In general, this polynomial is very badly conditioned, i.e., the order generated by a root of f has large index in the maximal order. The program system KANT II was used to calculate the field discriminant of k_σ from f and to obtain a small generating equation. We are grateful for having obtained access to the Kant II system prior to its publication. This program also determines the number of real embeddings r_1 of k_σ. Of course, this can also be deduced without the explicit knowledge of k_σ from the purely group theoretical data using Proposition C. Independently, the field discriminant was also computed using the computer algebra system PARI.

The Galois group of the polynomial f and hence of the Galois closure of $k_\sigma/\mathbb{Q}$ was determined from the decomposition types modulo small primes in the case that it was the symmetric group. For groups of degree at most

150   Gunter Malle

seven, the MAPLE Galois group program was used, and for degree 8 a program by Helmut Geyer for the determination of the Galois groups of degree 8 polynomials gave the answer.

4 The tables

The tables contain the data for field extensions of degree at most thirteen. However, the following restrictions have been made:
- Only extensions of genus zero are considered, since only for those a good and sufficiently mechanical algorithm for the computation of fields of definition is available.
- Only primitive permutation groups G were treated, since non-primitive groups correspond to extensions containing a proper subextension, and so can be built together from smaller primitive ones.
- If a structure constant n_G equals one, then certainly the extension is defined over $\mathbb{Q}$, so these have as well been omitted in the tables.

In addition to the above, the tables were further restricted as follows:
- For degree $n = 9$, only class vectors $\mathbf{C}$ containing at least one class of involutions are retained.
- For $10 \leq n \leq 13$, we have only included those class vectors $\mathbf{C}$ containing one class of involutions and one class of elements of order three. In other words, the corresponding Galois groups are factor groups of the modular group $\mathrm{SL}_2(\mathbb{Z})$ modulo (in general: non-congruence) subgroups. Such extensions had already been studied in [1], 2.3 and 4.2. Contrary to the expectation put forward by these authors, we find examples where primes not dividing n or the element orders in the three classes ramify in the field of definition.

Thus the fields of definition considered here correspond to genus zero dessins of degree at most 13.

The tables are organized as follows. For each degree n we give a separate table for each primitive permutation group G having a class triple $\mathbf{C}$ satisfying the above restrictions. The first column contains the description of the class vector in terms of triples of partitions of n. (Trivial parts of partitions are suppressed.) The second counts the number $n_G(\mathbf{C})$ of classes of triples generating G modulo conjugation in S_n. In the third, we give the Galois group of the Galois closure of the field of definition k. Note that by the Hurwitz classification it is a subgroup of the symmetric group $S_{n(\mathbf{C})}$. The fourth column contains the field discriminant $d(k)$. By Proposition B of Beckmann, $d(k)$ is only divisible by prime divisors of $|G|$. In the case that

neither KANT II nor PARI were able to compute the field discriminant, a *
is printed instead. The fifth column shows the number r_1 of real embeddings
of k. In the last column we give comments ("geo", "tr") if the structure of
the Galois group is restricted by the action of geometric automorphisms [5]
or due to translation arguments [4].

In the second set of tables, we display generating equations for the fields
of definition occurring in the previous tables. These are ordered by their
degree d and then by their discriminant. The polynomials are given by the
descending sequence of their coefficients, the highest $a_d = 1$ being omitted.
These tables only cover fields of degree at most 9, since for larger fields the
coefficients become to unwieldy. For field degrees 3 and 4 we have included
an additional column counting the number of occurrences of this particular
field of definition in the first set of tables.

Remark: The fields of definition found here also give a large number of
examples for number fields ramified over only a few very small primes. It is
not clear how such examples could be effectively constructed otherwise.

Degree 5, $G = S_5$

C	$n(C)$	Gal	$d(k)$	r_1
3 − 4 − 3 2	2	S_2	$2^3.3$	2

Degree 6, $G = S_6$

C	$n_G(C)$	Gal	$d(k)$	r_1	remarks
2^2 − 3 2 − 6	3	S_3	$-2^2.3^3$	1	
3 − 3 2 − 6	2	S_2	-2^2	0	
3^2 − 3 2 − 3 2	3	S_3	$-2^2.3^3$	1	as 2^2-32-6
4 − 4 − 4 2	2	S_2	$-2^2.5$	0	
4 − 4 2 − 3 2	4	S_4	$-2^6.3^3.5$	2	
4 − 5 − 3 2	4	S_4	$2^6.3^3.5^2$	0	
4 2 − 3 2 − 3 2	4	$2 \wr S_2$	$-2^6.3^3.5$	2	geo
5 − 3 2 − 3 2	7	$(2 \wr S_3)^+ \times 1$	$2^6.3^4.5^4$	2	geo, as 2^3-5-32

$G = A_6$

C	$n_G(C)$	Gal	$d(k)$	r_1	remarks
2^2 − 4 2 − 5	2	S_2	-3.5	0	
2^2 − 5 − 5	2	S_2	-3.5	0	as 2^2-42-5
3 − 4 2 − 5	2	S_2	$2^3.5$	2	

Degree 7, $G = S_7$

C	$n_G(\mathbf{C})$	Gal	$d(k)$	r_1	remarks
$2^2 - 6 - 6$	4	S_4	$2^4.3^3.7$	0	as 2^3-42-6
$2^2 - 6 - 5\,2$	3	S_3	$-2^2.5.7$	1	
$2^2 - 6 - 4\,3$	3	S_3	$-2^2.3^3.7$	1	
$2^2 - 5\,2 - 5\,2$	3	S_3	$-2^2.5.7$	1	as $2^3-42-52$
$2^2 - 5\,2 - 4\,3$	2	S_2	$3.5.7$	2	
$2^2 - 4\,3 - 4\,3$	2	S_2	$2^2.7$	2	as $2^3-42-43$
$2^3 - 4\,2 - 6$	4	S_4	$2^4.3^3.7$	0	
$2^3 - 4\,2 - 5\,2$	3	S_3	$-2^2.5.7$	1	
$2^3 - 4\,2 - 4\,3$	2	S_2	$2^2.7$	2	
$2^3 - 5 - 6$	3	S_3	$-3^3.5$	1	
$2^3 - 5 - 4\,3$	2	S_2	$-2^2.5$	0	
$2^3 - 3\,2 - 7$	3	S_3	$-2^2.3.7^2$	1	
$3 - 6 - 5\,2$	2	S_2	$2^2.3.5$	2	
$3 - 6 - 4\,3$	2	S_2	2^3	2	
$3 - 5\,2 - 4\,3$	2	S_2	$2^3.3.5$	2	
$3^2 - 4 - 6$	2	S_2	$-2^3.3$	0	
$3^2 - 4 - 5\,2$	3	S_3	$-2^3.5^2$	1	
$3^2 - 3\,2 - 6$	6	A_6	$2^6.3^6.7^2$	2	
$3^2 - 3\,2 - 5\,2$	4	S_4	$-2^2.3^3.5^2.7^2$	2	
$3^2 - 3\,2 - 4\,3$	4	S_4	$-2^6.3^3.7^2$	2	
$4 - 4\,2 - 6$	4	$2 \wr S_2$	$2^6.3.5^2.7$	0	
$4 - 4\,2 - 5\,2$	4	S_4	$-2^6.5^2.7$	2	
$4 - 4\,2 - 4\,3$	4	S_4	$-2^6.3^3.5.7$	2	
$4 - 5 - 5\,2$	2	S_2	2^3	2	
$4 - 5 - 4\,3$	3	S_3	$2^2.3^3.5^2$	3	
$4 - 3\,2 - 7$	3	S_3	$-2^3.3^3.7^2$	1	
$4 - 3\,2^2 - 6$	3	S_3	$-2^3.3^3.5$	1	
$4 - 3\,2^2 - 4\,3$	2	S_2	$2^2.3.5$	2	
$4\,2 - 3\,2 - 6$	12	S_{12}	$-2^{19}.3^{12}.5^2.7^5$	2	
$4\,2 - 3\,2 - 5\,2$	9	S_9	$-2^{14}.3^8.5^5.7^5$	3	
$4\,2 - 3\,2 - 4\,3$	9	S_9	$-2^{16}.3^6.5.7^5$	3	
$5 - 3\,2 - 6$	6	S_6	$2^6.3^6.5^3.7^2$	2	
$5 - 3\,2 - 5\,2$	7	S_7	$-2^6.3^7.5^3.7^2$	1	
$5 - 3\,2 - 4\,3$	5	S_5	$-2^8.3^3.5^3.7^2$	3	
$3\,2 - 3\,2 - 7$	6	$2 \wr S_3$	$-2^6.3^3.7^4$	0	geo
	+3	S_3	$-2^2.3.7^2$	1	as 2^3-32-7
$3\,2 - 3\,2^2 - 6$	6	S_6	$-2^8.3^6.5.7^2$	4	
$3\,2 - 3\,2^2 - 5\,2$	4	S_4	$-2^2.3^3.5^2.7^2$	2	
$3\,2 - 3\,2^2 - 4\,3$	3	S_3	$-2^3.5.7^2$	1	

Degree 7, $G = A_7$

C	$n_G(C)$	Gal	$d(k)$	r_1	remarks
$2^2 - 4\,2 - 7$	2	S_2	3.7	2	
$2^2 - 5 - 7$	2	S_2	3.7	2	
$2^2 - 3\,2^2 - 7$	2	S_2	3.7	2	
$3 - 4\,2 - 7$	2	S_2	-2.7	0	
$3^2 - 4\,2 - 4\,2$	4	S_4	$2^4.3^3.7$	0	from 2^8-42-6
$3^2 - 4\,2 - 5$	5	S_5	$2^6.3^3.5^3.7$	1	
$3^2 - 4\,2 - 3\,2^2$	3	S_3	$-2^3.3^3.7$	1	
$3^2 - 5 - 5$	3	S_3	$-3^3.5$	1	from 2^3-5-6
$3^2 - 5 - 3\,2^2$	3	S_3	$2^2.3^3.5^2$	3	
$4\,2 - 4\,2 - 4\,2$	12	$S_3 \wr 2$	$2^{22}.3^9.5^3.7^3$	0	geo
$4\,2 - 4\,2 - 5$	8	$2 \wr S_4$	$-2^{12}.3^6.5^5.7$	2	geo
	$+3$	S_3	$-2^2.5.7$	1	from $2^8-42-52$
$4\,2 - 4\,2 - 3\,2^2$	4	$2 \wr S_2$	$2^6.3.5^2.7$	0	geo
	$+2$	S_2	$2^2.7$	2	from $2^3-42-43$
$4\,2 - 5 - 5$	4	$2 \wr S_2$	$2^6.3^3.5.7$	0	geo
$4\,2 - 5 - 3\,2^2$	4	$S_3 \times 1$	$-2^3.3.5^2.7$	1	
$4\,2 - 3\,2^2 - 3\,2^2$	2	S_2	3.5.7	2	geo
$5 - 5 - 3\,2^2$	2	S_2	$-2^2.5$	0	geo
	$+2$	S_2	$2^2.3$	2	from 2^3-5-43

$G = L_2(7)$

C	$n_G(C)$	Gal	$d(k)$	r_1	remarks
$2^2 - 3^2 - 7$	2	S_2	-7	0	
$2^2 - 4\,2 - 7$	2	S_2	-7	0	
$3^2 - 3^2 - 4\,2$	4	$2 \wr S_2$	$-2^6.3^2.7$	2	geo
$3^2 - 4\,2 - 4\,2$	2	S_2	$2^2.7$	2	geo
$4\,2 - 4\,2 - 4\,2$	4	$2 \wr S_2$	$-2^6.7$	2	geo

Degree 8, $G = S_8$

C	$n_G(\mathbf{C})$	Gal	$d(k)$	r_1	remarks
$2^2 - 6 - 8$	2	S_2	$2^2.7$	2	
$2^2 - 3^2\,2 - 8$	2	S_2	-2^2	0	
$2^2 - 8 - 5\,2$	5	S_5	$2^9.3.5^3.7$	1	
$2^2 - 8 - 4\,3$	5	S_5	$2^9.3^4.7$	1	
$2^3 - 4\,2 - 8$	6	S_6	$2^{11}.3.5^2.7$	2	
$2^3 - 4\,2^2 - 7$	3	S_3	$-2^3.3.7^2$	1	
$2^3 - 4\,2^2 - 5\,3$	2	S_2	$2^3.5$	2	
$2^3 - 4^2 - 5\,2$	3	S_3	$-2^3.5^2$	1	
$2^3 - 4^2 - 4\,3$	2	S_2	2^3	2	
$2^3 - 5 - 8$	2	S_2	$2^3.3.5$	2	
$2^3 - 3\,2^2 - 8$	6	S_6	$2^9.3^4.5^2$	2	
$2^3 - 6 - 6\,2$	3	S_3	$-2^3.3^3.7$	1	
$2^3 - 6 - 7$	5	S_5	$2^3.3^3.7^4$	1	
$2^3 - 6 - 5\,3$	4	S_4	$-2^3.3^3.5^3.7$	2	
$2^3 - 3^2\,2 - 6\,2$	2	S_2	$2^3.3$	2	
$2^3 - 3^2\,2 - 7$	3	S_3	$-2^3.5.7^2$	1	
$2^3 - 3^2\,2 - 5\,3$	2	S_2	$2^3.3$	2	
$2^3 - 6\,2 - 5\,2$	7	S_7	$-2^{12}.3^3.5^5.7$	1	
$2^3 - 6\,2 - 4\,3$	5	S_5	$2^{10}.3^4.7$	1	
$2^3 - 7 - 5\,2$	9	S_9	$2^{12}.3.5^6.7^6$	1	
$2^3 - 7 - 4\,3$	9	S_9	$2^{14}.3^6.5.7^6$	1	
$2^3 - 5\,2 - 5\,3$	5	S_5	$2^{10}.3^2.5.7$	1	
$2^3 - 4\,3 - 5\,3$	5	S_5	$-2^{10}.3.5^3.7$	3	
$2^4 - 6 - 4\,3$	2	S_2	-3	0	
$2^4 - 5\,2 - 5\,2$	3	S_3	$-2^3.5^2$	1	as $2^3 - 4^2 - 52$
$2^4 - 4\,3 - 4\,3$	2	S_2	2^3	2	as $2^3 - 4^2 - 43$
$3 - 8 - 5\,2$	2	S_2	$-2^2.5$	0	
$3 - 8 - 4\,3$	2	S_2	$-2^3.3$	0	
$3^2 - 4\,2^2 - 6$	5	S_5	$-2^6.3^4.7^2$	3	
$3^2 - 4\,2^2 - 5\,2$	4	S_4	$-2^5.3^3.5^2.7^2$	2	
$3^2 - 4\,2^2 - 4\,3$	7	S_7	$2^{12}.3^7.7^2$	3	
$3^2 - 3\,2 - 8$	6	S_6	$-2^{14}.3^4.7^2$	0	
$3^2 - 6 - 6$	3	S_3	$-2^3.3^3.7$	1	as $2^3 - 6 - 62$
$3^2 - 6 - 3^2\,2$	4	S_4	$-2^5.3^3.7^2$	2	
$3^2 - 6 - 5\,2$	9	S_9	$2^{16}.3^6.5^6.7^3$	1	
$3^2 - 6 - 4\,3$	6	S_6	$2^8.3^6.7^3$	2	
$3^2 - 3^2\,2 - 3^2\,2$	2	S_2	$2^3.3$	2	as $2^3 - 3^2 2 - 62$

C	$n_G(\mathbf{C})$	Gal	$d(k)$	r_1	remarks
$3^2 - 3^2\,2 - 5\,2$	8	S_8	$-2^{10}.3^4.5^5.7^2$	2	
$3^2 - 3^2\,2 - 4\,3$	4	S_4	$-2^5.3^3.7^2$	2	
$3^2 - 5\,2 - 5\,2$	6	$2\wr S_3$	$-2^{10}.5^3.7^4$	0	geo
	+7	S_7	$-2^{12}.3^3.5^5.7$	1	as $2^8-6\,2-5\,2$
	+1	1	1	1	as $2^4-6-5\,2$
	+1	1	1	1	as $2^4-3^2\,2-5\,2$
$3^2 - 5\,2 - 4\,3$	12	$S_6\wr 2$	$2^{20}.3^{11}.5^9.7^5$	4	
$3^2 - 4\,3 - 4\,3$	6	$(2\wr S_3)^+$	$2^{12}.3^2.7^4$	2	geo
	+5	S_5	$2^{10}.3^4.7$	1	as $2^8-6\,2-4\,3$
	+2	S_2	-3	0	as $2^4-6-4\,3$
$4 - 4\,2 - 8$	3	S_3	$-2^2.5.7$	1	
$4 - 4\,2^2 - 7$	3	S_3	$-2^2.5.7^2$	1	
$4 - 4\,2^2 - 5\,3$	3	S_3	$2^2.3^3.5^2$	3	
$4 - 4^2 - 5\,2$	3	S_3	$-2^3.5^2$	1	
$4 - 3\,2^2 - 8$	3	S_3	$-3^3.5$	1	
$4 - 6 - 6\,2$	2	S_2	$3.5.7$	2	
$4 - 6 - 5\,3$	3	S_3	$2^3.3^3.5^2.7$	3	
$4 - 3^2\,2 - 6\,2$	2	S_2	-3.5	0	
$4 - 3^2\,2 - 7$	3	S_3	$-2^3.5.7^2$	1	
$4 - 6\,2 - 5\,2$	4	$S_3 \times 1$	$-2^3.3.5^2.7$	1	
$4 - 6\,2 - 4\,3$	6	S_6	$2^9.3^7.5^2.7$	2	
$4 - 7 - 5\,2$	4	S_4	$2^3.5^2.7^2$	0	
$4 - 7 - 4\,3$	4	S_4	$2^5.3^3.5.7^2$	0	
$4 - 5\,2 - 5\,3$	4	S_4	$-2^3.3^3.5^2.7$	2	
$4 - 4\,3 - 5\,3$	4	S_4	$-2^5.3^3.5^2.7$	2	
$4\,2 - 4\,2^2 - 6$	9	S_9	$-2^{12}.3^{10}.5.7^5$	3	
$4\,2 - 4\,2^2 - 3^2\,2$	4	S_4	$-2^6.3^3.5$	2	
$4\,2 - 4\,2^2 - 5\,2$	13	S_{13}	$-2^{27}.3^6.5^{10}.7^5$	3	
$4\,2 - 4\,2^2 - 4\,3$	12	S_{12}	$-2^{27}.3^6.5.7^5$	2	
$4\,2 - 3\,2 - 8$	12	S_{12}	$-2^{35}.3^{10}.5^3.7^5$	2	
$4\,2 - 6 - 6$	6	$2\wr S_3$	$2^6.3^6.5^3.7^4$	2	geo
$4\,2 - 6 - 3^2\,2$	9	S_9	$-2^{12}.3^6.5^3.7^5$	3	
$4\,2 - 6 - 5\,2$	18	S_{18}	$*$	2	
$4\,2 - 6 - 4\,3$	18	S_{18}	$*$	4	
$4\,2 - 3^2\,2 - 3^2\,2$	4	$2\wr S_2$	$2^6.3^2.5$	0	geo
$4\,2 - 3^2\,2 - 5\,2$	11	S_{11}	$2^{24}.3^4.5^6.7^5$	3	
$4\,2 - 3^2\,2 - 4\,3$	11	S_{11}	$-2^{23}.3^{10}.5^4.7^5$	5	
$4\,2 - 5\,2 - 5\,2$	32	$\leq 2\wr S_{16}$	$*$	2	geo

156 Gunter Malle

C	$n_G(C)$	Gal	$d(k)$	r_1	remarks
$4\,2 - 5\,2 - 4\,3$	28	S_{28}	$*$	4	
$4\,2 - 4\,3 - 4\,3$	24	$\le 2 \wr S_{12}$	$*$	6	geo
$4\,2^2 - 5 - 6$	4	S_4	$-2^5.3^3.5^2.7^2$	2	
$4\,2^2 - 5 - 3^2\,2$	3	S_3	$2^2.3^3.5^2$	3	
$4\,2^2 - 5 - 5\,2$	6	S_6	$2^{12}.3^3.5^4.7^2$	2	
$4\,2^2 - 5 - 4\,3$	7	S_7	$2^{10}.3^3.5^4.7^2$	3	
$4\,2^2 - 3\,2 - 7$	9	S_9	$-2^{16}.3^8.5.7^6$	3	
$4\,2^2 - 3\,2 - 5\,3$	5	S_5	$2^{11}.3^3.5^3$	1	
$4\,2^2 - 3\,2^2 - 6$	6	S_6	$2^6.3^6.5.7^2$	2	
$4\,2^2 - 3\,2^2 - 3^2\,2$	4	S_4	$-2^6.3^3.5$	2	
$4\,2^2 - 3\,2^2 - 5\,2$	11	S_{11}	$2^{23}.3^8.5^9.7^2$	3	
$4\,2^2 - 3\,2^2 - 4\,3$	6	S_6	$2^{12}.3^3.5.7^2$	2	
$4^2 - 3\,2 - 6$	4	S_4	$-2^6.3^3.7^2$	2	
$4^2 - 3\,2 - 3^2\,2$	2	S_2	$2^3.3$	2	
$4^2 - 3\,2 - 5\,2$	6	A_6	$2^{14}.3^4.5^4.7^2$	2	
$4^2 - 3\,2 - 4\,3$	6	S_6	$-2^{14}.3^4.7^2$	0	
$5 - 3\,2 - 8$	4	S_4	$2^{11}.3^3.5.7^2$	0	
$5 - 6 - 3^2\,2$	5	S_5	$2^6.3^4.5^3.7^2$	1	
$5 - 6 - 5\,2$	4	S_4	$2^5.3^3.5.7^2$	0	
$5 - 6 - 4\,3$	6	S_6	$-2^{12}.3^6.5^4.7^2$	0	
$5 - 3^2\,2 - 5\,2$	5	S_5	$2^5.3^3.5^2.7^2$	1	
$5 - 3^2\,2 - 4\,3$	6	S_6	$-2^{12}.3^7.5^3.7^2$	4	
$5 - 5\,2 - 5\,2$	8	$2 \wr S_4$	$-2^9.3^6.5^4.7^4$	2	geo
	$+3$	S_3	$-2^3.5^2$	1	as $2^4-52-52$
$5 - 5\,2 - 4\,3$	14	$\le S_{14}$	$2^{33}.3^{11}.5^9.7^4$	2	
$5 - 4\,3 - 4\,3$	6	$(2 \wr S_3)^+$	$2^6.3^4.5^4.7^4$	2	geo
	$+1$	1	1	1	as $2^4-52-43$
$3\,2 - 3\,2^2 - 8$	12	S_{12}	$2^{33}.3^6.5^3.7^2$	0	
$3\,2 - 6 - 6\,2$	10	S_{10}	$2^{17}.3^9.5.7^5$	2	
$3\,2 - 6 - 7$	8	S_8	$2^{15}.3^7.7^6$	0	
$3\,2 - 6 - 5\,3$	8	S_8	$-2^{17}.3^6.5^5.7^5$	2	
$3\,2 - 3^2\,2 - 6\,2$	6	S_6	$2^{11}.3^6.5$	2	
$3\,2 - 3^2\,2 - 7$	9	S_9	$-2^{15}.3^7.5^2.7^6$	3	
$3\,2 - 3^2\,2 - 5\,3$	5	S_5	$2^{11}.3^4.5^2$	1	
$3\,2 - 6\,2 - 5\,2$	14	S_{14}	$2^{32}.3^{13}.5^9.7^5$	2	
$3\,2 - 6\,2 - 4\,3$	14	S_{14}	$-2^{35}.3^{14}.5^2.7^5$	4	
$3\,2 - 7 - 5\,2$	17	S_{17}	$*$	3	
$3\,2 - 7 - 4\,3$	17	S_{17}	$*$	3	

C	$n_G(C)$	Gal	$d(k)$	r_1	remarks
3 2 − 5 2 − 5 3	15	S_{15}	*	5	
3 2 − 4 3 − 5 3	11	S_{11}	$2^{29}.3^4.5^7.7^5$	3	
3 2^2 − 6 − 6	10	$2 \wr S_5$	$2^{12}.3^9.5^2.7^6$	2	geo
	+2	S_2	−3	0	as $2^4−6−43$
3 2^2 − 6 − 3^2 2	6	S_6	$-2^6.3^6.5^3.7^2$	4	
3 2^2 − 6 − 5 2	18	S_{18}	*	0	
3 2^2 − 6 − 4 3	18	S_{18}	*	0	
3 2^2 − 3^2 2 − 3^2 2	2	S_2	-2^2	0	geo
3 2^2 − 3^2 2 − 5 2	7	S_7	$2^{11}.3^6.5^2.7^2$	3	
3 2^2 − 3^2 2 − 4 3	9	S_9	$-2^{20}.3^6.5^3.7^2$	3	
3 2^2 − 5 2 − 5 2	22	$\leq 2 \wr S_{11}$	*	4	geo
	+1	1	1	1	as $2^4−52−43$
3 2^2 − 5 2 − 4 3	25	S_{25}	*	5	
3 2^2 − 4 3 − 4 3	18	$\leq 2 \wr S_9$	*	2	geo
	+2	S_2	2^3	2	as $2^4−43−43$

Degree 8, $G = A_8$

C	$n_G(C)$	Gal	$d(k)$	r_1	remarks
2^2 − 6 2 − 7	2	S_2	$-2^3.7$	0	
2^2 − 6 2 − 5 3	3	S_3	$-2^3.3^3.5$	1	
2^2 − 7 − 7	3	S_3	$-2^3.3.7^2$	1	from $2^8−42^2−7$
2^2 − 7 − 5 3	4	$2 \wr S_2$	$-2^3.3^3.5^2.7^2$	2	
2^2 − 5 3 − 5 3	3	$S_2 \times 1$	$2^3.5$	2	$2^8−42^2−53,2^4−42−53$
3 − 6 2 − 7	2	S_2	3.7	2	
3 − 6 2 − 5 3	2	S_2	3.5.7	2	
3 − 7 − 5 3	2	S_2	5	2	
3^2 − 3^2 − 5 3	2	S_2	3.7	2	geo
3^2 − 4 2 − 6 2	5	S_5	$2^9.3^4.7$	1	
3^2 − 4 2 − 7	6	S_6	$2^{12}.3^3.5.7^4$	2	
3^2 − 4 2 − 5 3	8	S_8	$-2^{15}.3^6.5^6.7^2$	2	
3^2 − 4^2 − 3 2^2	2	S_2	−3	0	
3^2 − 5 − 6 2	5	S_5	$2^4.3^4.5^3.7$	1	
3^2 − 5 − 7	3	S_3	$-2^2.3^3.5.7$	1	
3^2 − 5 − 5 3	3	S_3	$-2^2.3^3.5.7$	1	
3^2 − 3 2^2 − 6 2	6	S_6	$2^{10}.3^6.7$	2	
3^2 − 3 2^2 − 7	11	S_{11}	$-2^{14}.3^{10}.5.7^7$	1	
3^2 − 3 2^2 − 5 3	7	S_7	$-2^{10}.3^5.5^5.7$	1	

C	$n_G(C)$	Gal	$d(k)$	r_1	remarks
$4\,2 - 4\,2 - 4^2$	6	S_6	$2^{11}.3.5^2.7$	2	from $2^3{-}42{-}8$
$4\,2 - 4\,2 - 6\,2$	16	$\leq 2\wr S_8$	$*$	2	geo
$4\,2 - 4\,2 - 7$	18	$\leq 2\wr S_9$	$*$	0	geo
$4\,2 - 4\,2 - 5\,3$	15	$2\wr S_7 \times 1$	$2^{28}.3^7.5^{10}.7^4$	2	geo, as $2^4{-}42{-}53$
$4\,2 - 4^2 - 5$	3	S_3	$-2^3.3.5^2.7$	1	
$4\,2 - 4^2 - 3\,2^2$	6	S_6	$2^{11}.3^3.5^2.7$	2	
$4\,2 - 5 - 6\,2$	7	S_7	$-2^{13}.3^7.5^4.7^2$	1	
$4\,2 - 5 - 7$	6	S_6	$2^{13}.3^3.5^3.7^4$	2	
$4\,2 - 5 - 5\,3$	8	$S_4 \wr 2$	$2^{16}.3^5.5^5.7^2$	4	
$4\,2 - 3\,2^2 - 6\,2$	15	S_{15}	$2^{35}.3^{15}.5^6.7^2$	3	
$4\,2 - 3\,2^2 - 7$	22	S_{22}	$*$	2	
$4\,2 - 3\,2^2 - 5\,3$	14	$S_7 \wr 2$	$2^{35}.3^{10}.5^8.7^2$	6	
$4^2 - 5 - 5$	2	S_2	$2^3.3.5$	2	from $2^3{-}5{-}8$
$4^2 - 5 - 3\,2^2$	4	S_4	$2^6.3^3.5^2$	0	
$4^2 - 3\,2^2 - 3\,2^2$	6	S_6	$2^9.3^4.5^2$	2	from $2^3{-}32^2{-}8$
$5 - 5 - 6\,2$	2	S_2	7	2	geo
$5 - 5 - 5\,3$	3	$S_2 \times 1$	3.7	2	geo, as $2^4{-}5{-}53$
$5 - 3\,2^2 - 6\,2$	6	S_6	$2^{10}.3^6.5^3.7$	2	
$5 - 3\,2^2 - 7$	9	S_9	$2^{12}.3^8.5^5.7^6$	1	
$5 - 3\,2^2 - 5\,3$	5	S_5	$2^8.3^3.5^2.7$	1	
$3\,2^2 - 3\,2^2 - 6\,2$	12	$2\wr A_6$	$-2^{20}.3^{12}.5^5.7$	2	geo
$3\,2^2 - 3\,2^2 - 7$	17	$\leq 2\wr S_8 \times 1$	$*$	4	geo, as $2^4{-}32^2{-}7$
$3\,2^2 - 3\,2^2 - 5\,3$	9	$2\wr S_4 \times 1$	$-2^{16}.3^2.5^4.7$	2	geo, as $2^4{-}32^2{-}53$

$$G = \mathrm{Hol}(Z_2^3)$$

C	$n_G(C)$	Gal	$d(k)$	r_1	remarks
$2^2 - 4^2 - 7$	2	S_2	-7	0	
$2^2 - 6\,2 - 7$	2	S_2	-7	0	
$2^2 - 7 - 7$	2	S_2	-7	0	as $2^4{-}42{-}7$
$2^4 - 4\,2 - 7$	2	S_2	-7	0	
$3^2 - 4\,2 - 4^2$	2	S_2	$2^2.7$	2	
$3^2 - 4\,2 - 6\,2$	4	$2\wr S_2$	$-2^6.3^2.7$	2	
$3^2 - 4\,2 - 7$	4	V_4	$2^6.7^2$	0	
$4\,2 - 4\,2 - 4^2$	4	$2\wr S_2$	$-2^6.7$	2	geo
$4\,2 - 4\,2 - 6\,2$	2	S_2	$2^2.7$	2	geo
$4\,2 - 4\,2 - 7$	2	S_2	-7	0	as $2^4{-}42{-}7$

$G = \mathrm{PGL}_2(7)$

C	$n_G(C)$	Gal	$d(k)$	r_1
$2^3 - 3^2 - 8$	2	S_2	2^3	2

$G = L_2(7)$

C	$n_G(C)$	Gal	$d(k)$	r_1	remarks
$3^2 - 3^2 - 4^2$	2	S_2	2^3	2	from 2^3-3^2-8

$G = 2^3 : 7 : 3$

C	$n_G(C)$	Gal	$d(k)$	r_1	remarks
$3^2 - 3^2 - 6\,2$	4	$2 \wr S_2$	$3^3.7$	0	geo
$3^2 - 3^2 - 7$	4	V_4	$3^2.7^2$	0	geo

Degree 9, $G = S_9$

C	$n_G(C)$	Gal	$d(k)$	r_1	remarks
$2^2 - 6\,3 - 8$	5	S_5	$2^9.3^6$	1	
$2^2 - 6\,3 - 7\,2$	4	S_4	$-2^2.3^5.7^2$	2	
$2^2 - 6\,3 - 5\,4$	3	S_3	$2^2.3^5.5$	3	
$2^2 - 8 - 8$	6	S_6	$-2^{10}.3^5.7$	0	as 2^4-42^2-8
$2^2 - 8 - 7\,2$	5	S_5	$2^9.3^6.7^2$	1	
$2^2 - 8 - 5\,4$	5	S_5	$2^9.3^5.5^3$	1	
$2^2 - 7\,2 - 7\,2$	5	S_5	$2^4.3^5.7^3$	1	as 2^4-42^2-72
$2^2 - 7\,2 - 5\,4$	4	S_4	$-2^2.3^5.5^2.7^2$	2	
$2^2 - 5\,4 - 5\,4$	3	S_3	$2^2.3^5.5$	3	as 2^4-42^2-54
$2^3 - 4\,2^2 - 9$	9	S_9	$-2^7.3^{18}.5.7$	3	
$2^3 - 4^2 - 6\,3$	3	S_3	$-2^3.3^4$	1	
$2^3 - 4^2 - 8$	6	S_6	$-2^{11}.3^4.5^2$	0	
$2^3 - 4^2 - 7\,2$	5	S_5	$2^5.3^4.5.7^3$	1	
$2^3 - 4^2 - 5\,4$	4	S_4	$-2^5.3^4.5^2$	2	
$2^3 - 3^2\,2 - 9$	10	S_{10}	$-2^9.3^{18}.5^3$	0	
$2^3 - 6 - 9$	3	S_3	$-2^3.3^3.7$	1	
$2^3 - 3\,2^3 - 9$	3	S_3	$-3^3.5$	1	
$2^3 - 6\,2 - 6\,3$	9	S_9	$2^{14}.3^{12}.7$	1	
$2^3 - 6\,2 - 8$	14	S_{14}	$-2^{33}.3^{15}.5^2.7^3$	0	

C	$n_G(\mathbf{C})$	Gal	$d(k)$	r_1	remarks
$2^3 - 6\,2 - 7\,2$	13	S_{13}	$*$	1	
$2^3 - 6\,2 - 5\,4$	11	S_{11}	$2^{16}.3^{12}.5^8.7$	3	
$2^3 - 6\,3 - 7$	7	S_7	$-2^{10}.3^4.5.7^4$	1	
$2^3 - 6\,3 - 5\,2^2$	4	S_4	$-2^2.3^4.5^3.7$	2	
$2^3 - 6\,3 - 4\,3\,2$	6	S_6	$2^6.3^8.7$	2	
$2^3 - 6\,3 - 5\,3$	10	S_{10}	$2^{10}.3^{11}.5^7$	2	
$2^3 - 7 - 8$	10	S_{10}	$2^{25}.3^5.7^6$	2	
$2^3 - 7 - 7\,2$	6	$S_5 \times 1$	$2^{10}.3^5.7^2$	1	tr $2^3 - 3^3 - 7\,2$
$2^3 - 7 - 5\,4$	7	S_7	$-2^{12}.3^4.5^5.7^4$	1	
$2^3 - 8 - 5\,2^2$	6	S_6	$2^9.3^4.5^4.7^2$	2	
$2^3 - 8 - 4\,3\,2$	12	S_{12}	$-2^{30}.3^{11}.5^2.7^2$	2	
$2^3 - 8 - 5\,3$	16	S_{16}	$*$	0	
$2^3 - 9 - 5\,2$	10	S_{10}	$2^{14}.3^{18}.5^7.7$	2	
$2^3 - 9 - 4\,3$	10	S_{10}	$2^{14}.3^{18}.5^3.7$	2	
$2^3 - 5\,2^2 - 7\,2$	6	S_6	$2^4.3^4.5^3.7^4$	2	
$2^3 - 5\,2^2 - 5\,4$	4	$S_3 \times 1$	$-3^4.5.7$	1	tr $2^3 - 3^3 - 5\,4$
$2^3 - 4\,3\,2 - 7\,2$	9	$S_8 \times 1$	$-2^4.3^9.5.7^6$	2	
$2^3 - 4\,3\,2 - 5\,4$	7	S_7	$2^8.3^8.5^4.7$	3	
$2^3 - 7\,2 - 5\,3$	10	S_{10}	$2^{12}.3^9.5^8.7^6$	2	
$2^3 - 5\,3 - 5\,4$	8	$A_7 \times 1$	$2^{12}.3^8.5^2$	3	
$2^4 - 4\,2^2 - 8$	6	S_6	$-2^{10}.3^5.7$	0	
$2^4 - 4\,2^2 - 7\,2$	5	S_5	$2^4.3^5.7^3$	1	
$2^4 - 4\,2^2 - 5\,4$	3	S_3	$2^2.3^5.5$	3	
$2^4 - 3^2\,2 - 6\,3$	3	S_3	-3^5	1	
$2^4 - 3^2\,2 - 8$	6	S_6	$-2^{11}.3^6.5^2$	0	
$2^4 - 3^2\,2 - 7\,2$	4	S_4	$-2^2.3^5.7^2$	2	
$2^4 - 3^2\,2 - 5\,4$	3	S_3	-3^5	1	
$2^4 - 6 - 6\,3$	2	S_2	-7	0	
$2^4 - 6 - 8$	4	S_4	$2^8.3^3.7$	0	
$2^4 - 6 - 5\,4$	2	S_2	-5.7	0	
$2^4 - 6\,3 - 5\,2$	5	S_5	$2^3.3^5.5^3.7$	1	
$2^4 - 6\,3 - 4\,3$	4	S_4	$-2^4.3^5.7$	2	
$2^4 - 8 - 5\,2$	9	S_9	$2^{22}.3^5.5^6.7$	1	
$2^4 - 8 - 4\,3$	9	S_9	$2^{22}.3^9.5^2.7$	1	
$2^4 - 5\,2 - 7\,2$	5	S_5	$2^5.3^5.5^3.7^2$	1	
$2^4 - 5\,2 - 5\,4$	5	S_5	$2^4.3^5.5^2.7$	1	
$2^4 - 4\,3 - 7\,2$	4	S_4	$-2^2.3^5.7^2$	2	
$2^4 - 4\,3 - 5\,4$	4	S_4	$-2^5.3^5.5.7$	2	

$G = A_9$

C	$n_G(C)$	Gal	$d(k)$	r_1	remarks
$2^2 - 4^2 - 9$	3	S_3	-3^5	1	
$2^2 - 6\,2 - 9$	6	S_6	$2^5.3^{11}.7$	2	
$2^2 - 7 - 9$	3	S_3	$2^3.3^5.7$	3	
$2^2 - 9 - 5\,2^2$	3	S_3	$3^5.5.7$	3	
$2^2 - 9 - 4\,3\,2$	6	S_6	$2^4.3^{11}.7$	2	
$2^2 - 9 - 5\,3$	6	$S_3 \wr 2$	$-2^3.3^{11}.5^4$	0	
$2^4 - 4\,2 - 9$	3	S_3	$-3^4.5.7$	1	
$2^4 - 4^2 - 6\,2$	4	A_4	$2^6.3^4$	0	
$2^4 - 4^2 - 7$	4	A_4	$2^6.7^2$	0	
$2^4 - 4^2 - 4\,3\,2$	3	S_3	$-2^2.3^4$	1	
$2^4 - 4^2 - 5\,3$	5	A_5	$2^6.3^4.5^2$	1	
$2^4 - 3\,2^2 - 9$	6	S_6	$2^4.3^{10}.5$	2	
$2^4 - 6\,2 - 6\,2$	4	A_4	$2^6.3^4$	0	as 2^4-4^2-62
$2^4 - 6\,2 - 7$	6	S_6	$2^6.3^7.7^3$	2	
$2^4 - 6\,2 - 5\,2^2$	3	S_3	$-3^4.5.7$	1	
$2^4 - 6\,2 - 4\,3\,2$	6	S_6	$2^7.3^8.7$	2	
$2^4 - 6\,2 - 5\,3$	8	S_8	$2^6.3^9.5^5.7$	0	
$2^4 - 7 - 7$	4	A_4	$2^6.7^2$	0	as 2^4-4^2-7
$2^4 - 7 - 4\,3\,2$	4	S_4	$2^5.3^4.7^2$	0	
$2^4 - 7 - 5\,3$	8	S_8	$2^6.3^7.5^6.7^5$	0	
$2^4 - 5\,2^2 - 4\,3\,2$	3	S_3	$2^2.3^4.5$	3	
$2^4 - 5\,2^2 - 5\,3$	3	S_3	$-3^4.5^2.7$	1	
$2^4 - 4\,3\,2 - 4\,3\,2$	3	S_3	$-2^2.3^4$	1	as 2^4-4^2-432
$2^4 - 4\,3\,2 - 5\,3$	4	S_4	$-2^5.3^4.5.7$	2	
$2^4 - 5\,3 - 5\,3$	7	$A_5 \times 1^2$	$2^6.3^4.5^2$	1	geo, as 2^4-4^2-53

$G = \text{Hol}(Z_3^2)$

C	$n_G(C)$	Gal	$d(k)$	r_1
$2^3 - 3^3 - 8$	2	S_2	-2^3	0
$2^3 - 6\,2 - 8$	2	S_2	-2^3	0

$G = \Gamma L_2(8)$

C	$n_G(C)$	Gal	$d(k)$	r_1	remarks
$2^4 - 3^2 - 9$	2	S_2	-3	0	
$2^4 - 6\,2 - 6\,2$	8	$2.S_4$	$2^8.3^8.7^2$	0	geo

$G = L_2(8)$

C	$n_G(C)$	Gal	$d(k)$	r_1
$2^4 - 7 - 7$	2	1^2	1	2

$G = G_{144}$

C	$n_G(C)$	Gal	$d(k)$	r_1
$2^3 - 4^2 - 8$	2	S_2	-2^3	0

Degree 10, $G = S_{10}$

C	$n_G(C)$	Gal	$d(k)$	r_1
$2^3 - 3^3 - 10$	5	F_{20}	$2^4.5^5$	1

$G = A_{10}$

C	$n_G(C)$	Gal	$d(k)$	r_1
$2^4 - 3^3 - 8\,2$	3	S_3	$-2^3.5^2$	1
$2^4 - 3^3 - 9$	6	S_6	$-3^8.5^2.7$	0
$2^4 - 3^3 - 6\,4$	3	S_3	$-2^2.3.5^2$	1
$2^4 - 3^3 - 7\,3$	3	S_3	$-5^2.7$	1

Degree 11, $G = S_{11}$

C	$n_G(C)$	Gal	$d(k)$	r_1
$2^5 - 3^3 - 10$	6	S_6	$-5^2.7^2.11$	0
$2^5 - 3^3 - 9\,2$	3	S_3	$-3^4.11$	1
$2^5 - 3^3 - 8\,3$	2	S_2	$2^3.11$	2
$2^5 - 3^3 - 7\,4$	3	S_3	$-2^2.11$	1
$2^5 - 3^3 - 6\,5$	2	S_2	-11	0

$G = A_{11}$

C	$n_G(\mathbf{C})$	Gal	$d(k)$	r_1
$2^4 - 3^3 - 11$	8	S_8	$-5^2.7^3.11^6$	2

$G = \mathrm{L}_2(11)$

C	$n_G(\mathbf{C})$	Gal	$d(k)$	r_1
$2^4 - 3^3 - 11$	2	S_2	-11	0

Degree 12, $G = S_{12}$

C	$n_G(\mathbf{C})$	Gal	$d(k)$	r_1
$2^5 - 3^3 - 12$	9	S_9	$2^{10}.3^9.5^2.7^4.11$	1
$2^5 - 3^4 - 9\ 2$	3	S_3	$-2^2.3^4$	1
$2^5 - 3^4 - 8\ 3$	2	S_2	-2^3	0
$2^5 - 3^4 - 7\ 4$	2	S_2	-7	0
$2^5 - 3^4 - 6\ 5$	2	S_2	$2^2.3$	2
$2^5 - 3^4 - 7\ 3\ 2$	2	S_2	$2^2.7$	2

$G = A_{12}$

C	$n_G(\mathbf{C})$	Gal	$d(k)$	r_1
$2^4 - 3^4 - 11$	3	S_3	$-2^2.11^2$	1
$2^4 - 3^4 - 7\ 5$	2	S_2	$2^2.7$	2

$G = M_{12}$

C	$n_G(\mathbf{C})$	Gal	$d(k)$	r_1
$2^4 - 3^4 - 10\ 2$	2	S_2	$-2^2.5$	0
$2^4 - 3^4 - 11$	2	S_2	-11	0
$2^6 - 3^3 - 11$	2	S_2	-11	0

$G = \mathrm{PGL}_2(11)$

C	$n_G(\mathbf{C})$	Gal	$d(k)$	r_1
$2^5 - 3^4 - 10$	2	S_2	5	2

Degree 13, $G = S_{13}$

C	$n_G(\mathbf{C})$	Gal	$d(k)$	r_1
$2^5 - 3^4 - 12$	20	S_{20}	*	0
$2^5 - 3^4 - 11\ 2$	10	S_{10}	*	0
$2^5 - 3^4 - 10\ 3$	10	S_{10}	*	0
$2^5 - 3^4 - 9\ 4$	9	A_9	$2^6.3^{12}.13^6$	1
$2^5 - 3^4 - 8\ 5$	7	D_7	$-2^9.13^6$	1
$2^5 - 3^4 - 7\ 6$	8	S_8	$-2^4.3^5.7^3.13^6$	2

$G = A_{13}$

C	$n_G(\mathbf{C})$	Gal	$d(k)$	r_1
$2^4 - 3^4 - 13$	10	S_{10}	$-2^2.5^2.7^2.13^8$	0
$2^6 - 3^3 - 13$	5	S_5	$7.11.13^3$	1
$2^6 - 3^4 - 8\ 4$	4	S_4	$2^5.3^3.13$	0
$2^6 - 3^4 - 9\ 3$	4	S_4	$2^2.3^4.13$	0
$2^6 - 3^4 - 10\ 2$	6	S_6	$-3^3.5^5.13$	0
$2^6 - 3^4 - 11$	6	S_6	$-2^4.3^4.7.11^4$	0
$2^6 - 3^4 - 6\ 4\ 3$	2	S_2	13	2
$2^6 - 3^4 - 8\ 3\ 2$	3	S_3	$-2^3.13$	1
$2^6 - 3^4 - 7\ 4\ 2$	3	S_3	$-2^2.7.13$	1
$2^6 - 3^4 - 6\ 5\ 2$	2	S_2	-3.13	0
$2^6 - 3^4 - 7\ 5$	4	S_4	$2^2.3^3.5.7.13$	0

$G = \mathrm{L}_3(3)$

C	$n_G(\mathbf{C})$	Gal	$d(k)$	r_1
$2^4 - 3^4 - 13$	4	$2 \wr S_2$	13^3	0

field degree 3

$d(k)$	a_2, a_1, a_0		$d(k)$	a_2, a_1, a_0	
$-2^2.3^3$	$0,0,2$	2	$-3^4.5.7$	$0,-12,19$	3
$-2^2.3^4$	$0,-3,4$	3	$3^5.5.7$	$0,-27,-51$	1
$-2^3.3^4$	$0,-3,-10$	1	$-5^2.7$	$1,2,3$	1
-3^5	$0,0,3$	3	$-2^3.3.5^2.7$	$-1,2,-38$	3
$-3^3.5$	$0,3,-1$	4	$2^3.3^3.5^2.7$	$0,-45,110$	1
$-2^3.3^3.5$	$3,6,-2$	2	$-3^4.5^2.7$	$0,15,40$	1
$2^2.3^4.5$	$0,-12,-14$	1	$-2^2.3.7^2$	$-1,5,1$	2
$2^2.3^5.5$	$0,-18,12$	3	$-2^3.3.7^2$	$1,-2,6$	2
$-2^3.5^2$	$1,2,-2$	6	$-2^3.3^3.7^2$	$0,-21,70$	1
$-2^2.3.5^2$	$-1,-3,-3$	1	$-2^2.5.7^2$	$-1,5,-13$	1
$2^2.3^3.5^2$	$0,-15,20$	4	$-2^3.5.7^2$	$1,5,-15$	3
$-2^2.3^3.7$	$0,-6,12$	1	$-2^2.11$	$-1,1,1$	1
$-2^3.3^3.7$	$0,-6,-16$	4	$-3^4.11$	$0,6,-1$	1
$2^3.3^5.7$	$0,-27,30$	1	$-2^2.11^2$	$1,4,-2$	1
$-2^2.5.7$	$0,2,2$	5	$-2^3.13$	$0,-1,2$	1
$-2^2.3^3.5.7$	$0,-3,12$	2	$-2^2.7.13$	$0,4,2$	1

field degree 4

$d(k)$	a_3, a_2, a_1, a_0		$d(k)$	a_3, a_2, a_1, a_0	
$2^6.3^4$	$2,6,6,3$	2	$-2^3.3^3.5^3.7$	$-1,6,9,-9$	1
$-2^6.3^3.5$	$2,6,0,-6[a]$	3	$-2^2.3^4.5^3.7$	$0,0,10,-15$	1
$-2^6.3^3.5$	$2,-3,-4,-2[b]$	1	$2^6.7^2$	$2,2,0,2[e]$	2
$2^6.3^2.5$	$-2,-4,8,10$	1	$2^6.7^2$	$-2,1,0,14[f]$	1
$2^6.3^3.5^2$	$2,3,4,20[c]$	1	$3^2.7^2$	$-1,-1,-2,4$	1
$2^6.3^3.5^2$	$0,6,-16,12[d]$	1	$-2^5.3^3.7^2$	$0,-3,-6,-6$	2
$-2^5.3^4.5^2$	$0,-9,-24,-18$	1	$-2^6.3^3.7^2$	$0,-6,-8,-24$	2
$-2^6.7$	$2,1,2,1$	2	$2^5.3^4.7^2$	$0,9,34,30$	1
$-2^6.3^2.7$	$2,-1,-8,-8$	2	$-2^2.3^5.7^2$	$0,-6,14,-12[g]$	1
$3^3.7$	$-1,3,-1,1$	1	$-2^2.3^5.7^2$	$-1,0,16,-4$	2
$2^4.3^3.7$	$2,0,0,3$	3	$2^5.3^3.5.7^2$	$0,15,6,18$	2
$2^8.3^3.7$	$0,6,4,3$	1	$2^{11}.3^3.5.7^2$	$0,36,176,942$	1
$-2^4.3^5.7$	$0,3,14,-3$	1	$2^3.5^2.7^2$	$1,-3,0,8$	1
$-2^6.3^3.5.7$	$2,3,14,7$	1	$-2^2.3^3.5^2.7^2$	$1,-9,-29,-14$	2
$2^6.3^3.5.7$	$-8,30,-32,16$	1	$-2^3.3^3.5^2.7^2$	$0,18,0,-24$	1
$-2^5.3^4.5.7$	$0,-9,-16,-6$	1	$-2^5.3^3.5^2.7^2$	$2,-9,60,60$	2
$-2^5.3^5.5.7$	$-2,0,30,-33$	1	$-2^2.3^5.5^2.7^2$	$-1,9,-35,-70$	1
$-2^6.5^2.7$	$2,-2,-6,-5$	1	$2^5.3^3.13$	$0,-3,2,6$	1
$2^6.3.5^2.7$	$0,2,0,21$	2	$2^2.3^4.13$	$1,3,-1,2$	1
$-2^3.3^3.5^2.7$	$1,-3,-6,6$	1	$2^2.3^3.5.7.13$	$1,9,-5,4$	1
$-2^5.3^3.5^2.7$	$-2,0,-20,-20$	1	13^3	$-1,2,4,3$	1

[a] $= S_4$, [b] $= 2 \wr S_2$, [c] $= 4-5-32$, [d] $= 4^2-5-32^2$, [e] $= A_4$, [f] $= V_4$,
[g] $= 2^2 - 63 - 72$.

field degree 5

$d(k)$	$a_4,\ldots,a_0$	$d(k)$	$a_4,\ldots,a_0$
$2^9.3^6$	$-1,-2,6,-6,6$	$2^4.3^4.5^3.7$	$-3,9,-15,21,-21$
$2^6.3^4.5^2$	$0,-2,4,-6,4$	$2^3.3^5.5^3.7$	$-1,8,-4,7,37$
$2^{11}.3^4.5^2$	$1,-8,0,18,-30$	$-2^6.3^4.7^2$	$-1,-5,7,-7,7$
$2^{11}.3^3.5^3$	$-1,-6,30,-45,45$	$2^{10}.3^5.7^2$	$-1,-8,-20,2,-2$
$2^9.3^5.5^3$	$1,14,-10,5,-35$	$2^9.3^6.7^2$	$-2,16,-8,37,-14$
$2^4.5^5$	$0,0,0,0,-2$	$2^5.3^3.5^2.7^2$	$1,5,-5,-2,2$
$2^9.3^4.7$	$-1,4,-8,17,-5$	$-2^8.3^3.5^3.7^2$	$-1,0,36,-72,24$
$2^{10}.3^4.7$	$1,-8,-4,14,14$	$2^6.3^4.5^3.7^2$	$-1,-2,22,-49,49$
$2^{10}.3^2.5.7$	$2,6,4,3,-2$	$2^5.3^5.5^3.7^2$	$2,11,-12,-96,-96$
$2^8.3^3.5^2.7$	$0,-4,-8,12,-16$	$2^4.3^5.7^3$	$-1,-10,24,-15,3$
$2^4.3^5.5^2.7$	$-1,-7,5,7,7$	$2^5.3^4.5.7^3$	$1,-3,-3,6,18$
$2^9.3.5^3.7$	$1,-6,10,25,-55$	$2^3.3^3.7^4$	$1,-1,4,-8,5$
$-2^{10}.3.5^3.7$	$-2,-6,12,-9,10$	$7.11.13^3$	$1,1,5,2,1$
$2^6.3^3.5^3.7$	$1,2,-2,-20,20$		

field degree 6

$d(k)$	$a_5,\ldots,a_0$	$d(k)$	$a_5,\ldots,a_0$
$2^{11}.3^6.5$	$0,-3,-8,-3,0,5$	$2^6.3^6.5.7^2$	$3,-3,-18,-21,-15,-5$
$2^4.3^{10}.5$	$0,3,-4,9,-6,1$	$2^6.3^6.5^3.7^2$	$0,-9,4,42,-60,20$
$2^9.3^4.5^2$	$2,-1,-6,-2,4,-1$	$-2^6.3^6.5^3.7^2$	$0,-3,6,-12,-24,-8$
$-2^{11}.3^4.5^2$	$2,3,0,-2,0,2$	$-2^{12}.3^7.5^3.7^2$	$0,-36,-64,24,192,64$
$-2^{11}.3^6.5^2$	$0,0,-8,9,0,18$	$2^{12}.3^3.5^4.7^2$	$-2,-5,20,25,-62,19$
$2^6.3^4.5^4$	$0,3,10,3,0,1$	$2^9.3^4.5^4.7^2$	$-2,1,-6,26,20,-1$
$-2^3.3^{11}.5^4$	$0,-9,-18,54,36,96$	$2^{14}.3^4.5^4.7^2$	$0,0,-120,-150,0,2000$
$-2^{10}.3^5.7$	$2,-2,-4,2,4,4$	$-2^{12}.3^6.5^4.7^2$	$0,18,-140,333,-360,166$
$2^{10}.3^6.7$	$0,3,4,3,0,-1$	$2^8.3^6.7^3$	$0,9,4,15,18,-10$
$2^6.3^8.7$	$0,6,-2,3,-6,2$	$2^6.3^7.7^3$	$-3,3,-4,12,-6,-6$
$2^7.3^8.7$	$0,0,-6,-3,0,-2$	$2^{12}.3^2.7^4$	$0,-10,0,31,0,-36$
$2^4.3^{11}.7$	$-3,6,-16,15,-21,16$	$-2^6.3^3.7^4$	$2,8,18,24,18,6$
$2^5.3^{11}.7$	$3,-3,-17,-12,30,46$	$2^{12}.3^3.5.7^4$	$2,-9,4,4,0,28$
$2^{11}.3.5^2.7$	$2,1,4,5,6,3$	$-2^{10}.5^3.7^4$	$0,-13,0,40,0,20$
$2^{11}.3^3.5^2.7$	$0,6,0,6,12,2$	$2^{13}.3^3.5^3.7^4$	$2,11,36,-54,60,-6$
$2^9.3^7.5^2.7$	$0,-3,2,12,-24,-38$	$2^4.3^4.5^3.7^4$	$0,-2,2,-12,8,-12$
$-3^8.5^2.7$	$-3,6,-3,3,-3,2$	$2^6.3^6.5^3.7^4$	$-3,18,14,63,147,56$
$2^{10}.3^6.5^3.7$	$0,-3,8,0,0,20$	$2^6.3^4.5^4.7^4$	$0,-3,0,3,0,-36$
$-2^{14}.3^4.7^2$	$0,6,16,18,0,4$	$-5^2.7^2.11$	$3,5,5,0,-1,2$
$2^6.3^6.7^2$	$0,3,-4,0,0,-4$	$-2^4.3^4.7.11^4$	$3,9,11,30,18,46$
$2^{12}.3^3.5.7^2$	$0,-2,0,-3,-4,2$	$-3^3.5^5.13$	$2,0,0,5,0,5$
$-2^8.3^6.5.7^2$	$0,-9,-4,15,12,5$		

field degree 7

$d(k)$	$a_6, \ldots, a_0$
$2^{12}.3^8.5^2$	$0, 0, 0, -3, 12, -2, -12$
$2^8.3^8.5^4.7$	$1, 3, -19, 11, -45, 35, -7$
$-2^{12}.3^3.5^5.7$	$2, 2, 10, 10, 4, 28, -12$
$-2^{10}.3^5.5^5.7$	$-3, 10, -30, 40, -48, 84, -60$
$2^{12}.3^7.7^2$	$-1, -3, -1, -11, 3, -11, -1$
$2^{11}.3^6.5^2.7^2$	$-3, 3, 1, -12, 24, -12, -12$
$-2^6.3^7.5^3.7^2$	$1, -6, -2, 16, 0, -8, 10$
$2^{10}.3^3.5^4.7^2$	$0, -6, 4, 8, -4, -14, -4$
$-2^{13}.3^7.5^4.7^2$	$-1, 6, -26, 26, 30, -140, 140$
$-2^{10}.3^4.5.7^4$	$-2, 3, 7, -22, 36, -30, 10$
$-2^{12}.3^4.5^5.7^4$	$-2, 2, 10, -15, -138, -204, 954$
$-2^9.13^6$	$-2, -2, -4, 3, -2, -6, -4$

field degree 8

$d(k)$	$a_7, \ldots, a_0$
$-2^{16}.3^2.5^4.7$	$4, 6, 4, -13, -28, 10, 24, -3$
$-2^{12}.3^6.5^5.7$	$0, 8, 0, 18, 0, -4, 0, -35$
$2^6.3^9.5^5.7$	$0, 9, 5, 30, 0, 65, -15, 45$
$2^8.3^8.7^2$	$0, 0, 0, -3, 0, 12, -18, 9$
$-2^{10}.3^4.5^5.7^2$	$-1, 0, -3, 0, 2, 8, -10, 4$
$2^{16}.3^5.5^5.7^2$	$-2, -9, 38, -46, -42, 181, -242, 111$
$-2^{15}.3^6.5^6.7^2$	$0, -10, -20, 60, 200, 370, 300, -325$
$-2^9.3^6.5^4.7^4$	$2, -9, -10, 34, 12, 0, 90, -30$
$-2^{17}.3^6.5^5.7^5$	$-2, -20, 104, -496, 1364, -328, -1792, -1828$
$2^6.3^7.5^6.7^5$	$-3, 23, -36, 75, -291, 53, 522, 256$
$2^{15}.3^7.7^6$	$0, 3, -10, -9, -24, -34, 108, 138$
$-2^4.3^9.5.7^6$	$-2, 10, -5, 1, 58, -68, 91, -32$
$-5^2.7^3.11^6$	$0, 2, -3, 10, -14, 14, -8, 1$
$-2^4.3^5.7^3.13^6$	$-1, -10, 1, 55, 0, -120, -6, 96$

field degree 9

$d(k)$	$a_8,\ldots,a_0$
$2^{14}.3^{12}.7$	$-3,12,-22,42,-48,56,-36,24,-4$
$-2^7.3^{18}.5.7$	$0,0,-9,9,0,-6,9,0,-2$
$2^{22}.3^9.5^2.7$	$-2,8,-8,20,-8,68,-32,38,-4$
$2^{22}.3^5.5^6.7$	$-1,-12,12,116,-116,-440,360,540,-540$
$-2^{20}.3^6.5^3.7^2$	$-1,-4,-16,-16,-16,16,24,36,12$
$2^{16}.3^6.5^6.7^3$	$3,14,32,76,120,140,160,100,100$
$-2^{12}.3^{10}.5.7^5$	$3,-6,-10,69,135,-88,-312,-108,76$
$-2^{12}.3^6.5^3.7^5$	$1,-4,-16,-17,27,30,-30,-60,60$
$-2^{14}.3^8.5^5.7^5$	$3,-15,-85,-135,87,251,-1215,-2670,-1070$
$-2^{16}.3^6.5.7^5$	$3,2,-2,6,38,54,30,-9,-3$
$2^{14}.3^6.5.7^6$	$4,7,4,-17,-56,-23,4,-32,-36$
$-2^{16}.3^8.5.7^6$	$1,-19,5,99,-249,15,1167,-768,132$
$-2^{15}.3^7.5^2.7^6$	$-2,-2,-20,46,82,-20,-200,-200,300$
$2^{12}.3^8.5^5.7^6$	$2,-16,-32,36,60,420,1800,3375,2250$
$2^{12}.3.5^6.7^6$	$2,3,9,-10,69,168,-8,-64,80$
$2^{10}.3^9.5^2.7^4.11$	$3,-12,-42,30,216,234,36,-9,-9$
$2^6.3^{12}.13^6$	$3,-6,-30,0,90,86,-30,-9,87$

References

[1]　A.O.L. ATKIN, H.P.F. SWINNERTON-DYER, Modular forms on non-congruence subgroups, in: Combinatorics, *Proc. Symp. Pure Math.* **XIX**, AMS 1971.

[2]　S. BECKMANN, Ramified primes in the field of moduli of branched coverings of curves, *J. Algebra* **125** (1989), 236–255.

[3*]　J. BIRCH, Noncongruence subgroups, covers and drawings, this volume.

[4]　G. MALLE, Genus zero translates of three point ramified Galois extensions, *Manuscripta Math.* **71** (1991), 97–111.

[5]　B.H. MATZAT, Konstruktive Galoistheorie, Springer Lecture Notes 1284, Berlin - Heidelberg - New York 1987.

[6*]　L. SCHNEPS, Dessins d'enfant on the Riemann sphere, this volume.

[7]　J.-P. SERRE, Groupes de Galois sur ℚ, *Astérisque* **161–162**, 73–85.

[8]　J.G. THOMPSON, Groups of genus zero and certain rational functions, in: Groups–Canberra, Springer Lecture Notes 1456, pp.185–190.

On the classification of plane trees by their Galois orbits

George B. Shabat*

Contents

Abstract

We consider a particular case of the Grothendieck correspondence between arithmetical and combinatorial-topological classes of objects. Our combinatorial-topological objects are plane trees (up to isotopy). The arithmetical ones are what we call the *generalized Chebyshev polynomials*, i.e. the polynomials in one variable with only two critical values. We consider such polynomials with complex coefficients up to affine equivalence and claim that all the equivalence classes have representatives with algebraic coefficients. The Grothendieck correspondence is established by assigning to each generalized Chebyshev polynomial its *critical tree*, i.e. the preimage of the segment joining its critical values; this correspondence turns out to be 1-1 in a suitable sense.

This construction allows one to define the action of the *absolute Galois group* $\mathrm{Gal}(\overline{\mathbf{Q}}/\mathbf{Q})$ on the isotopy classes of plane trees. Considering this action we

1) group the isotopy classes of plane trees into some finite combinatorial classes (of *bicoloured valency*), called valency classes; the Galois orbits of trees are contained in these classes and often (perhaps "generally" in some sense?) coincide with them.

2) suggest some rough classification of the plane trees in terms of this action. Basically, we single out three *special* types of trees and call all the others *general*. The corresponding arithmetic turns out to be *cyclotomic* in two special cases and *mysterious* (to the author) in the one that remains. We prove the finiteness theorem which claims that there is only a finite number of different valency classes of bicoloured trees of general type of a given size.

§0. Introduction

At the present stage of realization of Grothendieck's program it is easier to state problems than to describe solutions. For example:

– To find a direct way from combinatorial-topological to arithmetical structure without transcendental transit;

– To understand the action of the absolute Galois group $\mathrm{Gal}(\overline{\mathbf{Q}}/\mathbf{Q})$ on the dessins;

– To construct reduction of dessins modulo primes;

– To study and understand dessins-labeled decompositions of moduli-like spaces;

– To investigate connections of dessins d'enfant theory with other (sometimes apparently distant) subjects.

The case of plane trees is a doubly particular case of Grothendieck's dessins d'enfants: the genus of the surface where the graph is drawn is zero and the complement of it consists of a single cell. However, the above problems turn out to be non-trivial even in this case and we find the situation quite fascinating since the considered objects are very classical and the results tend to be extremely explicit.

The main concepts and results of the paper are joint with Nikolai Adrianov. They open some approach to the gradual understanding of the Grothendieck correspondence in the case of trees according to the degrees of number fields of their definition.

Though the material of the paper is completely motivated by the *Esquisse d'un Programme* [Gro], the exposition is self-contained. The proofs and details are to be found in [AdrSh].

I am indebted to many people whose interest and mathematical assistance were invaluable during the years of isolation of *Esquisse*'ical activities. My enthusiasm concerning plane trees was initiated by Vladimir Voevodsky; numerous examples were treated and re-treated by participants of my Moscow University Seminar, especially by Nikolai Adrianov. Professional computer-assisted calculations performed by Jean Betréma and Alexander Zvonkin, by Leila Schneps [Schn], and by Jean-Marc Couveignes [Couv] were inspiring both mathematically and morally. Discussions with C. Soulé, J. Oesterlé, C. Itzykson, P. Cohen and R. Penner helped me to acquire understanding of perspectives and relations with other mathematical and physical topics. I would like to thank Leila Schneps for her role in organizing the Luminy conference, and her energy in accumulating the efforts of individuals in the development of Grothendieck's program.

§1. Definitions and notation

1.1. *Plane bicoloured trees*

1.1.0. A set $T \subset \mathbf{C}$ together with two finite subsets of it $V_\pm \subset T$ is called a *plane bicoloured tree* if it satisfies the following conditions:

(1) The complement $T \setminus (V_+ \cup V_-)$ is a disjoint union of *open edges*, homeomorphic to open real intervals;

(2) Any open edge is bounded exactly by one point of V_+ and one point of V_-.

1.1.1. Two plane bicoloured trees $(T, V_\pm)$ and $(T', V'_\pm)$ are called *isotopic* if there exists a family of homeomorphisms

$$h_t : \mathbf{C} \to \mathbf{C} \quad (t \in [0, 1])$$

constituting a continous function

$$h : \mathbf{C} \times [0, 1] \to \mathbf{C},$$

such that h_0 is the identity, $h_1(T) = T'$ and $h_1(V_\pm) = V'_\pm$.

1.1.2. For a bicoloured plane tree $(T, V_\pm)$ we define two valency functions $v_\pm : V_\pm \to \mathbf{N}$, assigning to every vertex the number of open edges of T bounded by it.

1.2. *Generalized Chebyshev Polynomials*

1.2.0 Definition: Let k be an arbitrary algebraically closed field. A non-constant polynomial in one variable $P(z) \in k[z] \setminus k$ is called a *generalized Chebyshev polynomial* if it has no more than two critical values, i.e. there exist values $c_+, c_- \in k$ such that for any $a \in k$ such that $\frac{dP}{dz}(a) = 0$, we have $P(a) \in \{c_\pm\}$.

1.2.1 The classical Chebyshev polynomials

$$T_n(z) = \cos(n \cdot \arccos(z))$$

(cf. [Cheb]) provide standard examples of complex polynomials with critical values ± 1. These examples explain the term "generalized Chebyshev polynomial"; though they are defined in trigonometric terms, they have integer coefficients and thus can be considered over an arbitrary field.

1.2.2. Example: the polynomial z^n for $n \geq 2$ has just one critical value, so it is a generalized Chebyshev polynomial.

1.2.3 We say that the two polynomials $P(z)$ and $AP(az + b) + B$ (with $A, B, a, b \in k$ and $Aa \neq 0$) are *affinely equivalent*. A generalized Chebyshev polynomial remains generalized Chebyshev under this kind of affine transformation (its critical values are of course transformed).

1.3. *Critical trees of complex generalized Chebyshev polynomials*

1.3.0. Construction theorem: *Let P be a complex generalized Chebyshev polynomial with critical values* $c_\pm$. *Then the set*

$$T = P^{-1}[c_-, c_+]$$

with the subsets $V_\pm = P^{-1}(c_\pm)$ *is a bicoloured plane tree, called the* critical tree *of the polynomial P.*

1.3.1. Theorem: *Association of a critical tree to a generalized Chebyshev polynomial establishes a bijective correspondence between the classes of affine equivalence of generalized Chebyshev polynomials and isotopy classes of plane bicoloured trees.*

1.3.2. A striking consequence of this theorem is that *any plane tree can be continuously transformed into its "true" shape.* See [BeZv].

1.3.3. Any embedding of the field $\overline{\mathbb{Q}}$ of algebraic numbers into the field $\mathbb{C}$ of complex numbers induces the embedding of the corresponding sets of generalized Chebyshev polynomials. Indeed we have:

Theorem: *Any embedding of* $\overline{\mathbb{Q}}$ *into* $\mathbb{C}$ *induces a bijective correspondence between the classes of affine equivalence of generalized Chebyshev polynomials over* $\overline{\mathbb{Q}}$ *and* $\mathbb{C}$.

§2. Galois orbits of plane trees

2.1. Fix an algebraically closed field k and a subgroup $\Gamma \subset \mathrm{Aut}(k)$. We assume that Γ acts on the polynomials in $k[z]$ by acting on the coefficients. It is immediate that this action preserves the generalized Chebyshev polynomials.

2.2. Theorem: *The action of a group Γ on the set of generalized Chebyshev polynomials over k induces its action on the set of classes under affine equivalence of such polynomials.*

2.3. Lemma: *Suppose we are given two groups Γ and G, acting on the same set X, and an action of the group Γ on the group G. Suppose that these actions satisfy*

$$\gamma(g \cdot x) = \gamma(g) \cdot \gamma(x)$$

for all $\gamma \in \Gamma$, $g \in G$, $x \in X$. Then the action of the group Γ on the orbit set $X \setminus G$ defined by the formula

$$\gamma([x])_G = [\gamma(x)]_G$$

is well-defined.

Theorem 2.2 is proved by applying lemma 2.3 where X is the set of generalized Chebyshev polynomials over k and G is the group of affine transformations of these polynomials.

§3. Bicoloured valency classes of plane trees

3.0. The results of 1.3.3 and 2.2 define the action of $\mathrm{Gal}(\overline{\mathbf{Q}}/\mathbf{Q})$ on the set of isotopy classes of bicoloured plane trees.

3.1. Given any bicoloured plane tree $(T, V_\pm)$ with the valency functions $v_\pm : V_\pm \to \mathbf{N}$, we call its *bicoloured valency class* the set of (isotopy classes of) bicoloured plane trees with the isomorphic pair of valency functions.

3.2. Theorem: *The Galois orbit of a bicoloured plane tree lies in its bicoloured valency class.*

This follows from the system of equations defining the generalized Chebyshev polynomial P corresponding to the tree:

$$P(z) - c_\pm = \lambda_\pm \prod_{s \in V_\pm} (z - s)^{v_\pm(s)}.$$

Since these equations define the bicoloured valencies as multiplicities of polynomial equations, these valencies are clearly Galois-invariant.

§4. Special and general trees

In these diagrams, circles denote negatively coloured vertices and squares positive ones. The valencies are written inside them.

4.1. A tree is said to be of *branchless type* if all its valencies do not exceed 2. It has the following form, with a finite (odd or even) number of edges e, and is denoted by I_e:

I_e:

4.2. A tree is of *diameter ≤ 4 type* if any of its vertices can be reached from any other one passing along no more than 4 edges. Apart from several degenerate cases, a tree of this type has a unique *centre* (which we assume to be of positive type) from which any *terminal* vertex (i.e. vertex of valency 1) can be reached passing along no more than 2 edges.

The diameter ≤ 4 trees, denoted by $IV_{m_1\dots m_e}$ look as follows:

$IV_{m_1,\dots,m_e}$: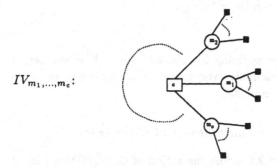

They are completely determined by the *cyclically ordered* valencies of the negative (non-central and non-terminal) vertices.

4.3. A tree is of *3-valenced diameter 6 type*, denoted by VI_{mcn}, if it is as in the following diagram:

VI_{mcn}: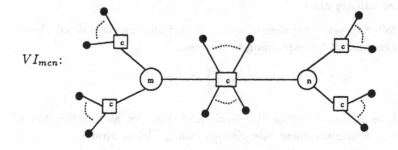

4.4. The tree is of *general type* if it does not belong to one of the special types 4.1 - 4.3 above (and does not become such after switching of signs).

4.5 Arithmetics of special trees

4.5.1. Using the trigonometric definition of the classical Chebyshev polynomials (see 1.2.1) one easily shows that the corresponding arithmetic is cyclotomic.

4.5.2. We explain briefly why the arithmetic of the 3-valenced diameter 6 type is also cyclotomic. The equations 3.2 for the tree VI_{mcn} after putting the vertices with valencies m and n to 0 and 1, take the form

$$P - c_- = z^m(1 - z)^n Q(z),$$

$$P - c_+ = R(z)^c.$$

Differentiating these equations with respect to z, factoring and taking into account the mutual coprimeness of z, $1 - z$, $Q(z)$ and $R(z)$, we see that

$$\frac{dR}{dz} = cz^{m-1}(1 - z)^{n-1}.$$

Thus we obtain

$$R(w) = \gamma_{mcn} + \int_0^w z^{m-1}(1 - z)^{n-1} dz,$$

where γ_{mcn} is some integration constant. The condition $R(0)^c = R(1)^c$ gives the cyclotomic equation for this constant. Taking the appropriate roots of unity we obtain all the representatives of the bicoloured valency classes VI_{mcn}; they differ only in the number of edges going "up" or "down" from the central vertex.

4.5.3. The arithmetic of diameter 4 trees seems quite interesting but definitely is not well understood for the time being.

(a) The fields of definition of these trees "generally" appear to have symmetric Galois group. But nobody, perhaps, can describe these fields.

(b) Leila Schneps has found an example IV_{23456} (see frontispiece) of the bicoloured valency class which is *not* a Galois orbit (see [Schn, III, Example 1]).

(c) The author and N. Adrianov have found two infinite series of such examples of the form IV_{mmmnn}, the smallest one is IV_{55566}. Combinatorially these series are not distinguished from the general trees of this form by any visible reason. The ability to "see" the splitting of Galois orbits of this sort would imply some progress in understanding the Grothendieck correspondence.

§5. Finiteness theorem

Theorem: *For any positive integer R there exists only a finite number of bicoloured valency classes of general type of size R.*

Sketch of Proof: Let us introduce the notation

$$\#\#S = \sum_{x \in S} \frac{1}{\#\mathrm{Aut}(x)},$$

and for a finite set V and a function $v : V \to N$ where N is an arbitrary set, we write

$$\binom{V}{v} = \frac{(\#V)!}{\prod_{n \ldots N}[\#(v^{-1}n)]!}$$

where $0! = 1$ as usual. This notation is a set-theoretic version of the polynomial coefficients' notation:

$$\binom{q_1 + q_2 + \cdots}{q_1 q_2 \cdots} = \frac{(q_1 + q_2 + \cdots)!}{q_1! q_2! \cdots}.$$

Counting formula: For a pair of finite sets $V_\pm$ and a pair of functions $v_\pm : V_\pm \to \mathsf{N}$, we write $[V_\pm, v_\pm]$ for the set of isomorphism classes of plane trees with the corresponding valency functions. This set is non-empty only if

$$\sum_{s \in V_+} v_+(s) = \sum_{s \in V_-} v_-(s). \tag{1}$$

With this notation, we have:

$$\#\#[V_\pm, v_\pm] = \frac{\binom{V_+}{v_+}\binom{V_-}{v_-}}{\#V_+ \#V_-}. \tag{2}$$

This formula can be deduced from known results (cf. [HarPal]). The proof is written out in [AdrSh]. As a corollary of (2) we obtain the

Main inequality: *If a pair of valency distributions $v_\pm : V_\pm \to \mathsf{N}$ satisfying (1) has no more than R realizations as plane trees, then we have the inequality*

$$\binom{V_+}{v_+}\binom{V_-}{v_-} \le R \#V_+ \#V_-$$

The finiteness theorem follows just from a formal analysis of this inequality.

References

[AdrSh] N. Adrianov, G. Shabat, Plane trees and Generalized Chebyshev
         Polynomials, to appear.

[BeZv]   J. Bétréma, A. Zvonkin, Une vraie forme d'arbre planaire, in TAP-
         SOFT '93: Theory and Practice of Software Development. (M.C.
         Gaudel, J.-P. Jouannaud eds). Lecture Notes in Computer Science
         no 668, Springer, 1993, 599-612

[Cheb]   P.I. Chebyshev, Théorie des mechanismes connus sous le nom de
         parallélogrammes, Mémoires présentées á l'Acad. Imp. des Sci.
         de St. Petersbourg par divers savants, VII (1854), 539-568.

[Couv]   J-M. Couveignes, Calcul et rationnalité de fonctions de Belyi en
         genre 0, to appear in *Ann. de l'Inst. Fourier.*

[Gro]   A. Grothendieck, Esquisse d'un Programme, preprint 1984.

[HarPal] F. Harary, E.M. Palmer, Graphical enumeration, Academic Press,
         New York-London, 1973.

[Schn]   L. Schneps, Dessins d'enfants on the Riemann sphere, this volume.

*Institute of New Technologies, 10 N. Radishchevskaya St. Moscow 109004,
Russia

Triangulations

Michel Bauer et Claude Itzykson*

Table des matières

Introduction

Ces notes traitent de diverses questions liées à la triangulation des surfaces.

Ce sujet, à l'origine de la partie la plus élémentaire de la topologie, se révèle d'une richesse insoupçonnée. D'une part physiciens et mathématiciens en explorant la "gravité quantique" à deux dimensions en ont tiré des informations sur l'espace des modules des courbes algébriques. D'autre part un théorème de Belyi établit une équivalence entre courbes arithmétiques et recouvrements finis de la droite projective ramifiés aux images réciproques de trois points (0, 1 et ∞) de sorte qu'une décomposition de la sphère de Riemann en deux triangles peut être relevée en une triangulation caractéristique de la courbe. Il s'ensuit que le groupe de Galois $\mathrm{Gal}(\bar{\mathbf{Q}}/\mathbf{Q})$ agit sur les triangulations (des surfaces orientables compactes) dont les sommets adjacents portent deux valeurs distinctes parmi trois possibles.

La combinatoire, la théorie des groupes à divers titres, la topologie et l'arithmétique semblent inextricablement mêlées.

On se bornera ici à décrire les aspects les plus élémentaires, accessibles aux auteurs, en mettant plutôt l'accent sur les points qui leur sont obscurs.

Que la théorie des groupes (finis ou plus généralement discrets) soit liée aux décompositions cellulaires trouve son origine dans l'interprétation graphique d'une présentation par générateurs et relations qui date au moins de Cayley (cf. le chapitre 8 du livre de Coxeter et Moser). Une décomposition cellulaire d'une surface compacte orientable est engendrée par un "groupe cartographique" (orienté) selon la terminologie de Grothendieck dans son "Esquisse d'un programme". Notre collègue J.-M. Drouffe en avait aussi donné la construction, qu'on pourrait généraliser en dimension supérieure. Ce groupe apparaît aussi dans des travaux d'analyse combinatoire consacrés aux hypergraphes. Nous allons voir qu'il permet de donner des preuves élémentaires de propriétés combinatoires simples.

Il est peut être plus intéressant de souligner combien de questions restent sans réponse, les plus fascinantes étant de nature arithmétique. Ce sont aussi celles devant lesquelles les physiciens se trouvent malheureusement les plus démunis. On pourra consulter à ce sujet les travaux de Voevodsky et Shabat, et pour les aspects les plus fondamentaux le livre "Gal($\bar{\mathbf{Q}}/\mathbf{Q}$)" où sont rassemblées les contributions des théoriciens des nombres difficiles à déchiffrer pour les non-spécialistes.

Pour donner une idée de notre ignorance citons ici le problème relatif au théorème d'existence fondamental de Riemann dans son mémoire sur les fonctions Abeliennes: une variété analytique unidimensionnelle compacte est une variété algébrique.

Si la donnée des ramifications d'une telle surface au dessus de trois points 0, 1, ∞, de la droite projective sur $\mathbf{C}$ détermine effectivement la structure complexe et si on dispose dès l'abord d'une fonction rationnelle sur la surface Σ donnée par l'application $\beta : \Sigma \longrightarrow CP_1$, il n'y a à notre connaissance aucune méthode constructive générale pour engendrer le corps des fonctions rationnelles sur Σ. Ceci rend la bijection (de Belyi),

$$(\Sigma, \beta) \; \underset{(2)}{\overset{(1)}{\underset{\longleftarrow}{\longrightarrow}}} \; \text{décomposition cellulaire finie}$$

hautement non triviale dans la direction (2) au sens algébrique voire arithmétique. Tout au plus comme le font Voevodsky et Shabat peut on chercher à accumuler des exemples. A fortiori les questions de corps de définition et d'action Galoisienne sont elles hors de portée en dehors de quelques généralités.

Comme on l'a indiqué, l'étude détaillée des décompositions cellulaires est par ailleurs intéressante, en ce qu'elle se transporte en une décomposition de l'espace des modules des classes d'isomorphisme des courbes algébriques à n points marqués, espace de dimension $6g - 6 + 2n$ (sur $\mathbf{R}$). On évoquera brièvement dans une dernière section, quelques résultats dus à Harer, Zagier, Penner, Witten et Kontsevich, en se bornant à la partie algébrique, la plus familière aux auteurs.

On s'est efforcé de trouver un compromis entre la terminologie des physiciens et celle des mathématiciens et entre le français et le franglais. En particulier on appellera *groupe cartographique* (groupe cartographique orienté selon Grothendieck) tout groupe discret qui admet une présentation en termes de deux générateurs dont l'un est une involution (complétée le cas échéant par des relations). Un tel groupe est un quotient naturel du groupe fondamental de la droite projective (sur C) privée de trois points. De même, on appellera *carte* (finie) la donnée d'un ensemble (fini) de *brins* (drapeaux orientés selon Grothendieck) muni d'une action transitive d'un groupe cartographique, équipé de sa présentation, tel que l'involution agisse sans point fixe. Ceci revient à dire que la classe d'équivalence de l'involution n'intersecte pas le stabilisateur B de l'un quelconque des brins. Il s'ensuit pour G fini que son ordre et le nombre de brins (qui en est un sous multiple) sont pairs. Si le stabilisateur B est réduit à l'identité ou parlera de carte *régulière* (par référence aux solides réguliers). Sauf indication contraire nous nous limiterons aux groupes et cartes finis.

Ces notes sont en partie le fruit d'une collaboration avec P. Cohen et J. Wolfart d'une part, J.-B. Zuber et P. Di Francesco d'autre part. J.-M. Luck nous a prêté main forte pour la partie numérique. Nous les remercions tous vivement.

§1. Groupe cartographique

On considère un ensemble fini de cellules bidimensionnelles polygonales (le nombre de côtés est aussi supposé fini) tel que les arêtes soient identifiées par paires (qui peuvent appartenir à la même cellule). La surface obtenue est compacte pour toute topologie raisonnable. Nous supposerons toujours qu'elle est connexe. Comment s'assurer qu'elle est orientable ? Chaque cellule peut être orientée, ce qui conduit à un ordre cyclique des sommets. On demande alors que les arêtes identifiées portent une orientation opposée. Ces données, que nous appelons par abus de langage une carte, ont été diversement dénommées dans des travaux récents. Dans les modèles de matrices des physiciens, elle apparaît comme "graphes de Feynmann à double arêtes (ou propagateurs) orientés" dans un développement perturbatif ('t Hooft), ou encore "graphes épais" (Penner).i Les sommets où ces arêtes dédoublées se rencontrent héritent d'un ordre cyclique, tandis que les faces correspondent aux cycles formés par les lignes orientées:

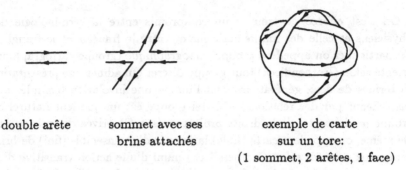

double arête   sommet avec ses   exemple de carte
                      brins attachés   sur un tore:
                                               (1 sommet, 2 arêtes, 1 face)

Figure 1: Conventions graphiques

A chaque carte on associe une carte duale obtenue en échangeant les
sommets et les faces. Soit S le nombre de sommets, A celui des arêtes (après
identification) et F celui des faces. Coupant chaque arête en deux demi
arêtes chacune attachée à un sommet (ces sommets pouvant être distincts
ou confondus), nous obtenons $2A$ brins (par dualité on obtient ainsi les
doubles arêtes chacune attachée à une face). Si S_v et F_v désignent les
nombres respectifs de sommets et de faces de valence v, on a les relations

$$2A = \Sigma \, v \, S_v = \Sigma \, v \, F_v$$
$$S = \Sigma \, S_v \qquad F = \Sigma \, F_v$$

Ainsi le nombre de sommets ou de faces de valence impaire est-il pair.

Définissons deux éléments du groupe Σ_{2A} des permutations des $2A$
brins, supposés indexés, en utilisant la représentation flêchée des physiciens.
Chaque arête entrante en un sommet est suivie d'une arête sortante. La
permutation σ_0 résultante a pour décomposition cyclique une représentation
des sommets avec leur valence. De même chaque brin est associé à un
partenaire pour reconstituer une double arête définissant une involution σ_1
possédant A cycles de longueur 2. Enfin la décomposition cyclique de la
permutation σ_2 telle que

$$\sigma_0 \sigma_1 \sigma_2 = \text{id}.$$

code les faces avec leur valences. Le groupe cartographique G ainsi engendré
agit transitivement sur la carte connexe qui peut être identifiée avec l'espace
quotient G/B où B est le stabilisateur d'un brin (changer de brin conjugue
B dans G), G étant muni de sa présentation $\{\sigma_0, \sigma_1, \sigma_2\}$. Un élément de G
qui agit trivialement sur les brins est forcément l'identité , ce qui impose
des restrictions sur B. En effet, si G est un groupe fini à trois générateurs
σ_2, σ_1, σ_0 vérifiant entre autres les relations $\sigma_1^2 = \sigma_0 \sigma_1 \sigma_2 = 1$, $\sigma_1 \neq 1$ et
B est un sous groupe de G, on peut essayer d'associer une carte à G/B,
dont les éléments sont identifiés aux demi-arêtes. Les arêtes sont les orbites
de l'action de σ_1, les faces les orbites de l'action de σ_2 et les sommets les

orbites de l'action de σ_0. Ceci reconstruit bien une carte si et seulement si les orbites de σ_1 ont toujours deux éléments, c'est-à-dire si et seulement si σ_1 agit sans points fixes sur G/B. Ceci signifie que la classe de conjugaison de σ_1 dans G ne rencontre pas B. Dans ce cas, il est facile de vérifier que le groupe cartographique de la carte associée à G/B est le quotient de G par le plus grand sous groupe invariant de G contenu dans B (ce sous groupe invariant est l'intersection des conjugués de B dans G). Il y a donc une correspondance biunivoque entre les cartes de groupe cartographique G muni de sa présentation $\{\sigma_0, \sigma_1, \sigma_2\}$ et les sous groupes de G disjoints de la classe de conjugaison de σ_1 et ne contenant aucun sous groupe invariant non trivial de G.

La caractéristique d'Euler d'une carte

$$\chi = S - A + F$$

est paire. En effet comme sous groupe de Σ_{2A} le groupe cartographique G possède une représentation alternée unidimensionnelle: $g \longrightarrow \rho(g) = (-1)^{\#\text{cycles pairs}}$. Si S_+ (F_+) désignent le nombre de sommets (faces) de valence paire, on a d'après ce qui précède

$$\begin{aligned} \rho(\sigma_0) &= (-1)^{S_+} = (-)^S \\ \rho(\sigma_1) &= (-1)^A \\ \rho(\sigma_2) &= (-1)^{F_+} = (-1)^F \end{aligned}$$

et la relation $\sigma_0\sigma_1\sigma_2 = \text{id.}$ entraîne $(-1)^\chi = 1$.

En fait χ est non seulement pair mais inférieur ou égal à 2 comme l'argument suivant (datant au moins de Poincaré) l'indique. Notons n_i, $i = 0, 1, 2$, l'ordre de σ_i. Alors $n_1 = 2$ et n_0 (n_2) est le plus petit commun multiple des valences des sommets (des faces). Considérons l'espace vectoriel V des 1-formes harmoniques, c'est-à-dire des fonctions φ définies sur les brins (à valeurs réelles) telles que pour $i = 0, 1, 2$, et tout brin x

$$\sum_{0 \leq k \leq n_i - 1} \varphi\left(\sigma_i^k x\right) = 0$$

Evidemment $\dim V \leq A$. Les S (F) relations $i = 0$ $(i = 2)$ satisfont une unique relation en raison de la connexité, de sorte que

$$\dim_{\mathbf{R}} V = A - (S - 1) - (F - 1) = 2 - \chi = 2g \geq 0$$

ce qui prouve l'assertion et définit le genre $g \geq 0$.

Les inégalités

$$2A \leq n_0 S \qquad 2A \leq n_2 F$$

entraînent

$$\chi \geq 2A \left(\frac{1}{n_0} + \frac{1}{2} + \frac{1}{n_2} - 1 \right)$$

Ainsi

   (i) si $\frac{1}{n_0} + \frac{1}{2} + \frac{1}{n_2} > 1$ le genre est nul

   (ii) si $\frac{1}{n_0} + \frac{1}{2} + \frac{1}{n_2} = 1$ le genre est 0 ou 1

Lorsque les inégalités précédentes sont des égalités on dira que la carte est *semi-régulière* (tous les sommets et toutes les faces ont même valence). Il est clair qu'une carte régulière (cas où le stabilisateur d'un brin est réduit à l'identité) est semi-régulière mais la réciproque n'est pas toujours vraie.

Le groupe G défini ci-dessus s'identifie au groupe cartographique orienté de Grothendieck qui agit sur les drapeaux formés d'un sommet d'une arête incidente et d'une face bordant l'arête, muni d'une relation d'orientation. On définit trois involutions renversant l'orientation échangeant un élément (sommet, arête ou face) de chaque drapeau. Les produits par paires engendrent le groupe cartographique orienté. Cependant ces définitions doivent être modifiées si une arête est incidente sur le même sommet à ses deux extrémités, s'il existe des sommets de valence un, ou enfin si une arête n'est incidente que sur une seule face, inconvénient que ne possède pas la définition précédente.

Le groupe cartographique G ne doit pas être confondu avec le *groupe de symétrie H* de la carte. Ce dernier est défini comme le centralisateur de G dans Σ_{2A}, c'est-à-dire l'ensemble des permutations des $2A$ brins qui commutent avec l'action de G.

Proposition 1. L'ordre du groupe de symétrie H divise $2A$ et est isomorphe (mais non identique) au groupe cartographique si et seulement si la carte est régulière.

Soit x un brin générique et faisons agir G et H respectivement à gauche et à droite de telle sorte que la commutativité s'écrive

$$(gx)h = g(xh) \qquad g \in G, \quad h \in H$$

La transitivité de G entraîne que les orbites de H ont toutes le même nombre d'éléments égal à l'ordre de H de sorte que

$$2A = |H| \times \#(\text{orbites})$$

Si la carte est régulière, l'ensemble des brins s'identifie au groupe cartographique (en tant qu'ensemble) par son action à gauche qui commute avec l'action de G à droite donc H contient un groupe isomorphe à G, et d'après ce qui précède, H est isomorphe à G. Finalement si H est isomorphe à G son ordre est à la fois un diviseur et un multiple de $2A$, donc $|H| = |G| = 2A$ ce qui n'est possible pour une action transitive de G que si G agit sans point fixe et la carte est régulière.

En genre zéro les cartes régulières (ou même semi-régulières) correspondent aux solides platoniciens. L'inégalité $\frac{1}{n_0} + \frac{1}{n_2} > \frac{1}{2}$ se récrit $(n_0 - 2)(n_2 - 2) < 4$ ce qui donne lieu à la table (avec $G \sim H$)

(a) groupe diédral d'ordre $2n$

   $n_0 = 2$, $n_2 = n \geq 2$, $S = A = n$, $F = 2$

   $n_0 = n$, $n_2 = 2$, $A = F = n$, $S = 2$

(b) groupe tétraédral A_4 d'ordre 12

   $n_0 = n_2 = 3$, $S = F = 4$, $A = 6$

(c) groupe actaédral S_4 d'ordre 24

   $n_0 = 4$, $n_2 = 3$, $S = 6$, $A = 12$, $F = 8$

   $n_0 = 3$, $n_2 = 4$, $S = 8$, $A = 12$, $F = 6$

(d) groupe icosaédral A_5 d'ordre 60

   $n_0 = 5$, $n_2 = 3$, $S = 12$, $A = 30$, $F = 20$

   $n_0 = 3$, $n_2 = 5$, $S + 10$, $A = 30$, $F = 12$

Dans les cas (b), (c), (d) le groupe G admet un système de deux générateurs l'un d'ordre 2 l'autre d'ordre $3 = \inf(n_0, n_2)$.

En genre 1 l'égalité $(n_0 - 2)(n_2 - 2) = 4$ correspond aux quotients des réseaux réguliers

$$
\begin{array}{ll}
(a) & \begin{array}{lll} n_0 = 3 & n_2 = 6 & \text{hexagonal} \\ n_0 = 6 & n_2 = 3 & \text{triangulaire} \end{array} \left.\vphantom{\begin{array}{l} a \\ a \end{array}}\right\} \\
(b) & n_0 = n_2 = 4 \quad \text{carré}
\end{array}
$$

Les cas les plus intéressants sont relatifs à un genre g supérieur à 1. Pour une carte régulière

$$g - 1 = A\left(\frac{1}{2} - \frac{1}{n_0} - \frac{1}{n_2}\right)$$

Nous retrouvons l'inégalité de Hurwitz en demandant que le groupe de symétrie H soit d'ordre $2A$ aussi grand que possible, c'est-à-dire à g fixé en minimisant la quantité positive $\frac{1}{2} - \frac{1}{n_0} - \frac{1}{n_2}$ ou encore n_0 et n_2, chacun d'eux devant être supérieur ou égal à 3 (pour que $g > 1$). Les valeurs $(3,3)$, $(3,4)$, $(3,5)$ et $(3,6)$ correspondant à $g \leq 1$, on trouve que $(3,7)$ est l'extremum cherché (correspondant soit à $n_0 = 3$, $n_2 = 7$ soit à la carte duale $n_0 = 7$, $n_2 = 3$). Dans ces conditions $|G| = 2A$ et l'on en tire

Proposition 2. Pour g supérieur à 1, l'ordre du groupe de symétrie d'une carte régulière est inférieur ou égal à $84(g - 1)$.

Quand cette borne est atteinte on parle de carte régulière de Hurwitz et de groupe (cartographique ou de symétrie) de Hurwitz engendré par trois générateurs de produit unité et d'ordre 2,3,7 respectivement.

Pour une carte de Hurwitz on a S(ou F) $= 28(g - 1)$, $A = 48(g - 1)$, F(ou S) $= 12(g - 1)$. On sait que cette borne n'est pas toujours atteinte. Par exemple en genre 2 l'ordre du plus grand groupe de symétrie est 48 (il s'agit du produit direct $\mathbb{Z}/2\mathbb{Z} \times S_4{}^*$). En revanche en genre 3 on a le célèbre groupe de Hurwitz $PSL_2(\mathsf{F}_7) \sim L_3(\mathsf{F}_2)$ d'ordre 168 (pour une revue des groupes de Hurwitz voir M. Conder).

En revenant au cas d'une carte arbitraire soit S_v (F_v) le nombre de sommets (de faces) de valence v. Dans des modèles de matrices chaque carte est comptée avec un poids de la forme $\frac{1}{|H|}$ où l'ordre du groupe symétrie apparaît sous la forme

$$|H| = \prod_v \frac{v^{S_v} S_v!}{\nu_S}$$

expression dans laquelle l'entier positif ν_S est défini de la manière suivante. Ayant indexé tous les brins par $2A$ symboles, on compte le nombre ν_S de façons distinctes de les joindre de manière à produire la carte donnée (sans indexation). En d'autres termes ν_S est le nombre d'involutions $\sigma_1 \in \Sigma_{2A}$ telles que, σ_0 étant donné, ces deux permutations engendrent des groupes G isomorphes mais distincts. Deux tels choix σ_1 et σ_1' ayant même décomposition cyclique (2^A) sont conjugués dans Σ_{2A}. Il existe donc une permutation k commutant à σ_0 telle que $\sigma_1' = k^{-1}\sigma_1 k$. Si $\mathrm{Com}(\sigma_0)$ désigne le commutant de σ_0 dans Σ_{2A} et $\mathrm{Com}(\sigma_0, \sigma_1)$ – isomorphe à H – le sous groupe de ces éléments qui commutent aussi avec σ_1, on a

$$\nu_S = \frac{|\mathrm{Com}(\sigma_0)|}{|\mathrm{Com}(\sigma_0, \sigma_1)|}$$

La classe de σ_0 étant $\prod_v (v)^{S_v}$ l'ordre de son commutant est

$$|\mathrm{Com}(\sigma_0)| = \prod_v v^{S_v} S_v!$$

tandis que $|\mathrm{Com}(\sigma_0, \sigma_1)| = |H|$ ce qui justifie l'expression ci-dessus. Bien entendu on a pour une définition adéquate la formule duale

$$|H| = \prod_v \frac{v^{F_v} F_v}{\nu_F}$$

---

* Correspondant à une carte formée de 16 triangles se rencontrant par 8 en 6 sommets (avec 24 arêtes) ou par dualité de 6 octogones se rencontrant par triplets en 16 sommets, double recouvrement de la sphère ramifié aux sommets d'un octaèdre.

Le groupe de symétrie H agit linéairement sur l'espace vectoriel V des 1-formes harmoniques par

$$\varphi \longrightarrow \varphi^h \quad : \qquad \varphi^h(x) = \varphi(xh)$$

où l'on vérifie sans peine que φ^h est encore harmonique. Il s'ensuit que V est un H-module (en général réductible). Dans le cas, probablement très spécial, où en tensorisant par $\mathbf{C}$, V se décomposerait en deux représentations irréductibles conjuguées de H, on en concluerait[*] que $g\|\,|H|$. C'est ce qui se produit par exemple pour la carte de Hurwitz de $PSL_2(\mathbf{F}_7)$, mais g étant premier avec $g - 1$, ceci ne peut avoir lieu pour d'autres cartes de Hurwitz que si g est un diviseur de 84 ce qui majorerait l'ordre du groupe de symétrie correspondant par $84 \times 83 = 6.972$.

Il existe une description intrinsèque du groupe de symétrie H d'une carte. Marquons un brin x, soit $B_x \subset G$ son groupe d'isotropie dans le groupe cartographique, et considérons le normalisateur $\mathrm{Norm}(B_x) \subset G$, c'est-à-dire le sous groupe des $g \in G$ tels que $g\,B_x g^{-1} = B_x$. Il est clair que B_x est un sous groupe invariant de $\mathrm{Norm}(B_x)$.

Proposition 3. Le groupe de symétrie H est isomorphe à

$$\mathrm{Norm}(B_x)\,/\,B_x$$

Si la carte est régulière $B = \{\mathrm{id}\}$, $\mathrm{Norm}(B) = G$ et nous retrouvons l'isomorphisme $H \sim G$.

La preuve se fait en trois étapes:

(i) Si x et $y = xh$ sont des brins de la même orbite du groupe de symétrie, alors $B_x = B_y$. En effet $g \in B_x$ implique que $g(xh) = (gx)h = xh$, donc $B_x \subset B_y$ et de même $B_y \subset B_x$.

(ii) H privé de l'identité agit sans point fixe sur les brins comme conséquence de la transitivité de G (voir plus haut).

(iii) Soient deux brins x et y tels que $B_x = B_y = B$. Par l'hypothèse de transitivité, il existe $\gamma \in G$ tel que $y = \gamma x$ et $\gamma B \gamma^{-1} = B$ donc $\gamma \in \mathrm{Norm}(B)$. Fixant x et γ, pour un brin générique z, il existe $g \in G$ tel que $z = gx$.

L'application h des brins dans les brins

$$h \quad : \qquad z \longrightarrow zh = g\gamma x$$

---

[*] La dimension d'une représentation irréductible (sur $\mathbf{C}$) divise l'ordre d'un groupe fini (cf. Serre).

est bien définie en vertu de la propriété de γ. Pour tout $g' \in G$

$$(g'z)\,h = (g'g)\,\gamma x = g'(zh)$$

En identifiant l'ensemble des brins avec le quotient G/B il s'ensuit que $(gB)h = gB\gamma = g\gamma B$ ce qui donne l'application $h : gB \longrightarrow g\gamma B$ telle que

$$g' \circ h = h \circ g' \; : \quad gB \longrightarrow g'g\gamma B$$

Nous avons ainsi un homomorphisme $\mathrm{Norm}(B) \longrightarrow H$ de noyau B. Cet homomorphisme est surjectif puisque étant donné $h \in H$ on peut trouver $\gamma \in \mathrm{Norm}(B)$ tel que $y = xh = \gamma x$ ce qui complète la preuve.

On dira qu'une carte est un *recouvrement* d'une autre si les conditions suivantes sont satisfaites. Soit $\hat{G}(\hat{\sigma}_0, \hat{\sigma}_1, \hat{\sigma}_2)$ le groupe cartographique de la première carte agissant sur un ensemble de $2\hat{A}$ brins $\hat{B}$ et $G(\sigma_0, \sigma_1, \sigma_2)$, $\mathcal{B}$ les quantités analogues pour la seconde. On demande qu'il existe une application surjective $f : \hat{B} \to \mathcal{B}$ telle que $f(\hat{\sigma}_i \hat{x}) = \sigma_i f(\hat{x})$.

Comme on l'a remarqué plus haut, un élément de G qui agit trivialement sur tous les brins est forcément l'identité. Comme f est surjective, la définition d'un recouvrement impose que toute relation entre $\hat{\sigma}_0, \hat{\sigma}_1, \hat{\sigma}_2$ dans $\hat{G}$ est vraie pour $\sigma_0, \sigma_1, \sigma_2$ dans G, ce qui signifie que G est un quotient de $\hat{G}$. On peut représenter $\hat{B}$ comme $\hat{G}/\hat{B}$ et $\mathcal{B}$ comme G/B. On peut alors décrire plus explicitement les recouvrements. Partant de $\hat{G}$ un groupe cartographique, G un groupe quotient de $\hat{G}$ (où l'on note p la projection $\hat{G} \xrightarrow{p} G \longrightarrow 1$) tel que $p(\hat{\sigma}_1) \neq e$, et B un sous groupe de G tel que $\mathcal{B} = G/B$ soit une carte de groupe cartographique G, les recouvrements de $\mathcal{B}$ de groupe cartographique $\hat{G}$ sont indexés par les sous groupes de $p^{-1}(B)$. Notant $\hat{B}$ un tel sous groupe, $\hat{B} = \hat{G}/\hat{B}$ recouvre $\mathcal{B}$. La preuve est élémentaire. On prouve d'abord que tous les sous groupes de $p^{-1}(B)$ ont les propriétés voulues pour que $\hat{G}/\hat{B}$ soit bien une carte de groupe cartographique $\hat{G}$. On montre ensuite que le diagramme

$$\begin{array}{ccc} \hat{G} & \xrightarrow{\;p\;} & G \\ \downarrow & & \downarrow \\ \hat{G}/\hat{B} & & G/B \end{array}$$

se complète naturellement en un diagramme commutatif

$$\begin{array}{ccc} \hat{G} & \xrightarrow{\;p\;} & G \\ \downarrow & & \downarrow \\ \hat{G}/\hat{B} & \xrightarrow{\;f\;} & G/B \end{array}$$

où f a les propriétés d'un recouvrement.

Un cas particulier est le suivant. On considère une carte et son groupe cartographique $G(\sigma_0, \sigma_1, \sigma_2)$, on prend pour $\hat{\mathcal{B}}$ le groupe G lui-même en tant qu'ensemble. Alors G agit sur $\hat{\mathcal{B}}$ par multiplication à gauche. On prend $\hat{\sigma}_i = \sigma_i$ qui est une involution sans point fixe sur $\hat{\mathcal{B}}$. On choisit un brin x de la carte initiale et on définit l'application $\hat{\mathcal{B}} \longrightarrow \mathcal{B}$ par $\left(g \in \hat{\mathcal{B}} \right) \longrightarrow (gx \in \mathcal{B})$. Toutes les fibres ont pour cardinal $|B_x|$ et on vérifie que les propriétés de recouvrement sont satisfaites, ainsi

Proposition 4. Toute carte possède un recouvrement par une carte régulière admettant comme groupe de symétrie le groupe cartographique de la carte initiale.

En raison de la symétrie de la carte régulière, cette construction ne dépend pas essentiellement du choix du brin x dans la carte initiale. Si on conserve la notation n_0, n_2 pour les plus petits communs multiples des valences des sommets et des faces de la carte initiale et si on pose $n = |B_x|$ de sorte que $|G| = 2An$, la carte régulière qui la recouvre a pour caractéristique d'Euler

$$\chi' = 2An \left(\frac{1}{n_0} - \frac{1}{2} + \frac{1}{n_2} \right) \leq n \, \chi$$

donc un genre g' tel que

$$g' \geq 1 + n(g-1)$$

inégalité intéressante si $g > 1$. Il va sans dire que n_0 et n_2, les ordres de σ_0 et σ_2, divisent $|G| = 2An$.

Les groupes cartographiques (finis), c'est-à-dire ceux qui admettent une présentation en termes de deux générateurs dont l'un est une involution, sont très nombreux. * Nous n'avons pas fait une étude exhaustive de cette question (qui le mériterait). Mentionnons à titre d'exemples les quotients finis du groupe modulaire $PSL_2(\mathbf{Z})$ et les groupes symétriques Σ_n. Les premiers parce que le groupe modulaire admet une présentation en termes de deux générateurs dont l'un est une involution, l'autre d'ordre trois, les seconds parce qu'ils sont engendrés par un cycle $\sigma_0 = (1, 2, ..., n)$ d'ordre n et une transposition $\sigma_1 = (1, 2)$ tandis que $\sigma_2 = (\sigma_0 \sigma_1)^{-1} = (2)(n, n-1, ..., 3, \hat{2}, 1)$ est d'ordre $(n-1)$. La carte régulière correspondante est telle que $S = \frac{n!}{n}$, $A = \frac{n!}{2}$, $F = \frac{n!}{n-1}$ (avec F échangé avec S pour la carte duale) et son genre est

$$g = 1 + \frac{(n-2)!}{4} \left[n^2 - 5n + 2 \right]$$

---

* De fait tous les groups simples finis ont une présentation en termes de deux générateurs, voir les articles de Steinberg et d'Asbacher-Guralnick mentionnés dans la liste des références.

qui vaut 0 pour $n \leq 4$ [*] tandis que pour $n > 4$, le genre croît très rapidement. [†] Un autre exemple intéressant est fourni pour p premier supérieur à 3 par le groupe affine modulo p agissant sur la droite affine L_p identifiée avec le corps F_p, correspondant aux transformations

$$\sigma_2 \qquad u \longrightarrow au + b, \qquad a \in \mathsf{F}_p^*, \qquad u, b \in \mathsf{F}_p$$

où F_p^* est le groupe multiplicatif cyclique de F_p de générateur η (c'est-à-dire que $p - 1$ est la plus petite puissance telle que $\eta^{p-1} = 1 \bmod p$). On peut prendre comme générateurs

$$\sigma_0 \qquad u \longrightarrow \eta u \qquad\qquad \sigma_0^{p-1} = 1$$

$$\sigma_1 \qquad u \longrightarrow -u - 1 \qquad \sigma_1^2 = 1$$

d'où

$$\sigma_2 = -\eta^{-1} u - 1 \qquad
\begin{array}{ll}
p \equiv 1 \bmod 4 & \sigma_2^{p-1} = 1 \\[2mm]
p \equiv -1 \bmod 4 & \sigma_2^{\frac{p-1}{2}} = 1
\end{array}$$

Une carte régulière associée aura $S = p$, $A = \frac{p(p-1)}{2}$ et suivant que $p \equiv 1(4)$, $F = p$ (il s'agit de $p - 1$-gones) ou $p \equiv -1(4)$, $F = 2p$ (qui sont des $\frac{p-1}{2}$-gones).

Le genre est alors donné par

$$g - 1 = \begin{cases} \dfrac{p(p - 5)}{4} & \text{si} \quad p \equiv 1(4) \\[4mm] \dfrac{p(p - 7)}{4} & \text{si} \quad p \equiv -1(4) \end{cases}$$

Comme il y a p sommets et $\frac{p(p-1)}{2}$ arêtes, la carte correspond au graphe complet sur p points complété par les faces (on vérifie que chaque arête est incidente sur deux sommets distincts sinon il existerait un entier k tel que $\sigma_1 = \sigma_0^k$, ce qui est impossible). On en déduit que le graphe complet à p sommets (p premier supérieur à 3) peut être tracé sans intersections sur une surface compacte orientable de genre g donné par la formule précédente. Ceci est à comparer avec le problème classique de coloriage des cartes (ou son dual) selon lequel (Ringel (1974)) en genre $g > 0$ pour colorier toute carte de sorte que des faces adjacentes portent des couleurs distinctes, il

---

[*] Les cas n=2 et 3 correspondent respectivement aux cartes ⬭ et △ et sont identifiables aux cas diédraux D_2 et D_6.

[†] Pour l'étude de ce cas voir pages 42-43 du livre *Topics in Galois Theory* de Serre mentionné dans la liste des références.

suffit de n couleurs où l'entier n, nombre chromatique, est la partie entière de

$$\frac{7 + \sqrt{1 + 48g}}{2}$$

Cette partie entière est en tous cas une borne supérieure. De manière équivalente, le graphe complet à n sommets peut être tracé sur une surface (orientable) de genre $g \geq \frac{(n-3)(n-4)}{12}$ (inégalité de Heawood). En effet, les faces étant au moins triangulaires $2A = n(n-1) \geq 3F$ de sorte que la formule d'Euler entraîne $2g - 2 = \frac{n(n-1)}{2} - n - F \geq \frac{n(n-1)}{6} - n$, qui est l'inégalité précédente. Le théorème de coloriage en genre positif est alors que l'inégalité peut être saturée, en ce sens qu'on peut trouver $g = \gamma(n) = \text{Inf}\left\{r \in \mathbf{Z}, r \geq \frac{(n-3)(n-4)}{12}\right\}$.

En genre 1 le nombre chromatique est 7, ce qui s'identifie au cas précédent quand $p = 7$. Montrons comment pour $p = 5, 7$ la construction précédente se présente en termes d'une courbe elliptique arithmétique.

Soit d'abord $p = 5 \equiv 1(4)$. Pour tracer sur un tore le graphe complet (auto-dual) ayant 5 sommets, 10 arêtes et 5 faces carrées, nous considérons les sommets du réseau carré bi-dimensionnel comme l'ensemble des entiers de Gauss $\mathbf{Z}\left(\sqrt{-1}\right)$. L'idéal premier engendré par $\tau = 2 + \sqrt{-1}$ de norme 5 ($\tau^2 - 4\tau + 5 = 0$) correspond à un sous réseau et le quotient $\mathbf{Z}\left(\sqrt{-1}\right)/\tau\mathbf{Z}\left(\sqrt{-1}\right)$ (qui s'identifie au corps $\mathbf{F}_5$) contient 5 sommets et 5 carrés du pavage initial. Si nous identifions les arêtes avec celles du réseau initial, chaque sommet (chaque face) est joint au quatre autres comme dans la figure suivante:

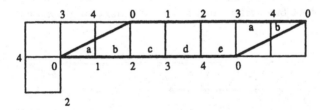

Figure 2a: Carte correspondant au groupe affine sur $\mathbf{F}_5$

Nous avons ainsi le graphe complet dessiné sur la courbe elliptique d'invariant modulaire $j(\tau/5) = j\left(\sqrt{-1}\right) = 12^3$. De plus le groupe de symétrie de $\mathbf{Z}\left(\sqrt{-1}\right)$ engendré par les translations et rotations multiples d'un quart de tour devient un groupe de symétrie de la configuration ($\sqrt{-1}\ \tau = 2\tau - 5$, $\sqrt{-1}\ 5 = 5\tau - 10$) et son action est celle du groupe affine sur la droite sur $\mathbf{F}_5$. Si on représente les brins par une paire ($x \in$

$\mathbf{Z}\left(\sqrt{-1}\right)$ mod τ, $\left(\sqrt{-1}\right)^{a}$, a mod 4) deux générateurs possibles sont

$$\sigma_0\left(x,\ \sqrt{-1}^{a}\right) = \left(x, \sqrt{-1}^{a+1}\right)$$
$$\sigma_1\left(x, \sqrt{-1}^{a}\right) = \left(x + \sqrt{-1}^{a},\ \sqrt{-1}^{a+2}\right)$$

On vérifie les relations

$$\sigma_0^4 = \sigma_1^2 = (\sigma_0\sigma_1)^4 = \left(\sigma_0^2\sigma_1\right)^5 = \sigma_1\sigma_0^2\sigma_1\sigma_0^3\sigma_1\sigma_0 = 1$$

impliquant que $\sigma_2^4 = 1$ (comme il se doit) et que $\sigma_0^2\sigma_1$ engendre un sous groupe cyclique invariant d'ordre 5

$$\sigma_1^{-1}\left(\sigma_0^2\sigma_1\right)\sigma_1 = \left(\sigma_0^2\sigma_1\right)^4, \quad \sigma_0^{-1}\left(\sigma_0^2\sigma_1\right)\sigma_0 = \left(\sigma_0^2\sigma_1\right)^2.$$

Tout élément du groupe s'écrit alors

$$g_{\alpha\beta} = \left(\sigma_0^2\sigma_1\right)^{\alpha}\sigma_0^{\beta} \qquad \alpha \text{ mod } 5, \qquad \beta \text{ mod } 4$$

avec la loi de multiplication

$$g_{\alpha_1\beta_1}g_{\alpha_2\beta_2} = g_{\alpha_1+3^{\beta_1}\alpha_2,\beta_1+\beta_2},$$

qui est bien celle du groupe affine puisque 3 est un générateur de F_5^*.

Dans le cas $p = 7$, $g = 1$ on peut reprendre l'exercice utilisant l'identification des sommets du réseau triangulaire avec les entiers de Eisenstein $\mathbf{Z}(\omega)$, $\omega = \exp 2\sqrt{-1}\frac{\pi}{6}$ de la forme $m+n\omega$ de norme m^2+n^2+mn. L'idéal premier est engendré par $\tau = 2 + \omega$, de norme 7 ($\tau^2 - 5\tau + 7 = 0$).

Le réseau triangulaire modulo l'idéal engendré par τ contient 7 sommets, 21 arêtes et 14 faces triangulaires correspondant au graphe complet sur 7 points:

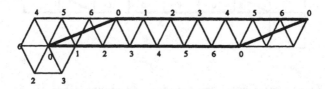

Figure 2b: Carte correspondant au groupe affine sur F_7

Comme précédemment les brins sont des paires ($x \in \mathbf{Z}(\omega)$ mod τ, ω^a, a mod 6) et l'action du groupe de symétrie du réseau est engendrée par

$$\sigma_0\left(x,\omega^a\right) = \left(x,\omega^{a+1}\right)$$
$$\sigma_1\left(x,\omega^a\right) = \left(x + \omega^a,\ \omega^{a+3}\right)$$

On vérifie que

$$\sigma_0^6 = \sigma_1^2 = (\sigma_0\sigma_1)^3 = (\sigma_0^3\sigma_1)^7 = 1$$

Comme

$$\sigma_1^{-1}\left(\sigma_0^3\sigma_1\right)\sigma_1 = \left(\sigma_0^3\sigma_1\right)^6$$
$$\sigma_0^{-1}\left(\sigma_0^3\sigma_1\right)\sigma_0 = \left(\sigma_0^3\sigma_1\right)^5$$

l'élément $\sigma_0^3\sigma_1$ engendre un sous groupe cyclique invariant d'ordre 7. Un élément générique du groupe est représenté sous la forme

$$g_{\alpha,\beta} = \left(\sigma_0^3\sigma_1\right)^\alpha \sigma_0^\beta \quad , \quad \alpha \bmod 7, \quad \beta \bmod 6$$

avec

$$g_{\alpha_1,\beta_1}g_{\alpha_2,\beta_2} = g_{\alpha_1+3^{\beta_1}\alpha_2,\beta_1+\beta_2}$$

qui est le groupe affine sur $\mathbf{F}_7$, 3 étant aussi dans ce cas un générateur de $\mathbf{F}_7^*$. La courbe elliptique correspondante a pour invariant modulaire $j\left(\frac{\tau}{7}\right) = j(\omega) = 0$.

La décomposition barycentrique d'une carte conduit à une triangulation dont les sommets correspondent aux sommets, milieux d'arêtes et milieux des faces. Pour cette triangulation les nombres de sommets, d'arêtes et de faces triangulaires deviennent

$$S' = S + A + F$$
$$A' = 6A$$
$$F' = \Sigma\, 2vF_v = 4A$$

et le genre est invariant. Nous l'appelerons carte (ou triangulation) dérivée. Soit G' son groupe cartographique. Les sommets de la triangulation peuvent être classés en trois familles du type S, A ou F. En identifiant les sommets de chaque classe on obtient la carte régulière à 3 sommets, 3 arêtes et 2 faces (triangulation élémentaire d'une sphère) de groupe cartographique et de symétrie Σ_3. Ceci engendre un homomorphisme surjectif $G' \longrightarrow \Sigma_3 \longrightarrow 1$ de noyau Γ (et correspond à la présentation de Belyi). On se convainc que Γ est engendré par les éléments $\tilde\sigma_0 = \sigma_0'^2$ et $\tilde\sigma_1 = \sigma_1'\sigma_0'^2\sigma_1'$. Ces derniers engendrent bien un sous groupe invariant Γ dans G' puisque

$$\sigma_0'^{-1}\tilde\sigma_0\sigma_0' = \tilde\sigma_0$$
$$\sigma_1'^{-1}\tilde\sigma_0\sigma_1' = \tilde\sigma_1$$
$$\sigma_0'^{-1}\tilde\sigma_1\sigma_0' = \tilde\sigma_0^{-1}\tilde\sigma_1^{-1}$$
$$\sigma_1'^{-1}\tilde\sigma_1\sigma_1' = \tilde\sigma_0$$

où on a utilisé la relation $(\sigma_0'\sigma_1')^3 = 1$ de la triangulation, et les images t_0, t_1 de σ_0', σ_1' dans l'application $G' \longrightarrow G'/\Gamma$ vérifient

$$t_0^2 = t_1^2 = (t_0t_1)^3 = 1$$

qui sont les relations génératrices de Σ_3 (en tant que groupe de Weyl de l'algèbre de Lie sl_3). Enfin, on s'assure que ni t_0 ni t_1 ne sont identiquement l'identité.

Les brins de la carte originale forment une sous-famille de ceux de la triangulation, stable pour l'action de $\tilde{\sigma}_0$ et $\tilde{\sigma}_1$, action qui devient celle respectivement de σ_0 et σ_1. On en déduit la suite exacte d'homomorphismes

$$1 \longrightarrow \Gamma \longrightarrow G' \longrightarrow \Sigma_3 \longrightarrow 1$$
$$\downarrow$$
$$G$$
$$\downarrow$$
$$1$$

En général Γ n'est pas isomorphe à G comme le montre l'exemple de genre zéro à 2 sommets, 1 arête et 1 face pour lequel $|G \sim Z/2Z| = 2$, tandis que la carte dérivée, qui n'est pas régulière, possède 4 sommets, 6 arêtes et 4 faces et un groupe cartographique à 24 éléments, cependant que $|\Gamma \sim Z/2Z \times Z/2Z| = 4$. Cette carte dérivée s'obtient à partir de la carte tétraédrale de même nombre de sommets, arêtes et faces, en procédant à la transformation dite "flip" (correspondant à l'échange de diagonale dans un quadrilatère) affectant deux faces du tétraèdre. On passe ainsi d'un groupe cartographique d'ordre 12 (A_4) à G' d'ordre 24 (qui à la différence de A_4 possède un élément d'ordre 4).

Les triangulations dérivées d'une carte ne sont pas les plus générales parmi celles où il est possible de partager les sommets en trois classes, les sommets d'un triangle appartenant à chacune d'entre elles. C'est cette classe plus générale qui intervient dans le théorème de Belyi. Cependant une telle carte admettra une triangulation dérivée barycentrique.

§2. Structure complexe

Aux données combinatoires d'une carte on associe une surface de Riemann compacte. Il s'agit d'un recouvrement fini de la sphère de Riemann qui n'est ramifié qu'aux images inverses de 0, 1, ∞ (correspondant respectivement aux sommets, "milieux" d'arêtes et "milieux" de faces). Ayant fait choix d'un point base arbitraire (hors de ces trois points), les générateurs c_0, c_1, c_∞ du groupe fondamental $\pi_1(\mathbb{C}P_1 - \{0, 1, \infty\})$ correspondent à des circuits élémentaires autour de 0, 1, ∞ et satisfaisant à la relation

$$c_0 c_1 c_\infty = \mathrm{id}.$$

agissent sur les fibres de recouvrement comme σ_0, σ_1, σ_2 respectivement sur les brins de la carte. Notons π_1 pour faire bref, le groupe fondamental ci-dessus. On a donc un homomorphisme surjectif

$$\pi_1 \longrightarrow G \longrightarrow 1$$

dont le noyau est engendré par $c_0^{n_0}$, c_1^2, $c_\infty^{n_2}$ et les autres relations satisfaites par les σ_i.

La droite projective réelle coupe la sphère marquée en 0, 1, ∞ en deux triangles. Notons Σ le recouvrement ramifié défini ci-dessus et β la projection sur la sphère de Riemann. L'image inverse par β de la triangulation de la sphère devient la triangulation dérivée de la carte. En particulier, les arêtes initiales s'identifient à $\beta^{-1}[0,1]$ tandis que $\beta^{-1}[1,\infty]$ dessine sur Σ la carte duale. De même β^{-1} permet de relever la structure complexe qui transforme Σ en une surface de Riemann compacte, revêtement à $2A$ feuillets de la sphère de Riemann ramifiée en $\beta^{-1}(0,1,\infty)$. La projection β définit, au moyen de la coordonnée x sur la sphère, un élément du corps rationnel sur Σ qui prend aux sommets, milieux d'arêtes et milieux de faces de la carte qu'on imagine tracée sur Σ, les valeurs 0, 1, ∞ respectivement qui sont des valeurs critiques. En un sommet (milieu de face) de valence v, la fonction x s'exprime comme $(t - t_s)^v \left(\frac{1}{(t-t_f)^v} \right)$ en fonction d'un paramètre uniformisant local, à une racine v-ième de l'unité près (si $v = 1$, il n'y a pas ramification). Dans cette construction il est clair que toute symétrie de la carte provient d'un automorphisme de la surface de Riemann laissant la fonction β invariante.

La surface Σ peut elle même être considérée comme la "surface de Riemann" d'une seconde fonction rationnelle y prenant une valeur donnée (par exemple ∞) en $2A$ points de Σ qui se projettent en $2A$ points distincts de $CP_1 - \{0,1,\infty\}$ (qu'on aura tout intérêt à prendre ou rationnels ou algébriques). Dans ces conditions y, bien que n'étant pas entièrement spécifié, satisfaira une équation polynomiale de degré $2A$ à coefficients polynomiaux en x définissant une carte algébrique. On note que au dessus du point 1 la fonction x possède A points de ramification d'ordre 2.

Donnons une description équivalente de la structure complexe de Σ. Choisissons par exemple un recouvrement simplement connexe de Σ moins $\beta^{-1}\{0,1,\infty\}$ par un plan hyperbolique $\mathcal{H}$ en identifiant $CP_1 - \{0,1,\infty\}$ à $\mathcal{H}/\Gamma(2)$ où $\Gamma(2)$ est le sous groupe de niveau 2 du groupe modulaire (le groupe modulaire étant le quotient $PSL_2(\mathbf{Z})$ et le sous groupe $\Gamma(2) \subset PSL_2(\mathbf{Z})$, étant formé des matrices $\begin{pmatrix} a & b \\ c & d \end{pmatrix} \equiv I$ modulo 2, quotienté par son centre $\pm I$). On représentera alors $\Sigma - \beta^{-1}\{0,1,\infty\}$ par le quotient $\mathcal{H}/\Gamma$, où le sous-groupe que nous notons encore Γ est d'indice fini dans $\Gamma(2)$ (précisément d'indice $2A$). Soit x la fonction qui applique $\mathcal{H}/\Gamma(2)$ sur $CP_1 - \{0,1,\infty\}$ normalisée en demandant qu'elle prenne les valeurs 0, 1,

∞ aux trois points paraboliques. La projection β est alors donnée par
$M \in \mathcal{H}/\Gamma \longrightarrow x(M)$. Enfin l'action du groupe cartographique sur les $2A$
brins s'identifiera à celle de $\Gamma(2)$ sur $\Gamma(2)/\Gamma$. En d'autres termes, il existera
des homomorphismes surjectifs $\Gamma(2) \longrightarrow G \longrightarrow 1^*$, tels que $\Gamma(2)/\Gamma \sim G/B$.
En un sens, on peut alors considérer Σ comme une courbe modulaire.

On peut remplacer dans cette construction $\Gamma(2)$ par un groupe
triangulaire Δ de signature $n_0, n_1 = 2, n_2$ et un sous groupe $\Gamma \subset \Delta$ d'indice
fini $(2A)$ dans Δ. Le quotient $\mathcal{H}/\Gamma$ apparaît alors comme un modèle, en
général singulier de Σ, Γ n'étant pas nécessairement fuchsien (c'est-à-dire
agissant avec des points fixes).

En se référant à la remarque qui clôt la section précédente, si on dispose
en général d'une application $\beta : \Sigma \longrightarrow \mathbb{C}P_1$, ramifiée en $0, 1, \infty$, on pourra
toujours lui associer l'application composée $\beta' = f \circ \beta$ en passant à la
triangulation dérivée appliquant respectivement les sommets $x = 0, 1, \infty$
sur $f(x) = 0$, les "milieux" d'arêtes $x = -1, \frac{1}{2}, 2$ sur $f(x) = 1$ et les
"milieux" de faces $x = \frac{1 \pm i\sqrt{3}}{2}$ sur $f(x) = \infty$ soit en respectant les notations
classiques de Klein

$$f(x) : 1 - f(x) : 1 = 27x^2(1-x)^2 : \left(2x^3 - 3x^2 - 3x + 2\right)^2 : 4\left(x^2 - x + 1\right)^3$$

En d'autres termes, on considère la variable x comme le birapport de 4
valeurs distinctes et f comme l'invariant sous l'action du groupe de Klein à
6 éléments, image des permutations:

$$SL_2\left(F_2\right) \sim PSL_2(\mathbb{Z})/\Gamma(2)$$

Bien entendu les ordres de ramification de β' au dessus de $f(x) = 1$ sont
tous égaux à deux.

§3. Structure arithmétique

C'est à ce point qu'un physicien commence à perdre pied. Le résultat
le plus important – à la vérité pour l'instant le seul de ce domaine – est
dû à Belyi comme indiqué dans l'introduction. Nous allons esquisser (une
partie de) la démonstration pour son intérêt constructif dans un sens, en

---

* Le noyau de cet homomorphisme est Γ', le plus grand sous groupe invariant de
$\Gamma(2)$ d'indice fini contenu dans Γ. D'où $G \sim \Gamma(2)/\Gamma'$, $B \sim \Gamma/\Gamma'$. Enfin le groupe
d'automorphisme est le quotient du normalisateur de Γ dans $\Gamma(2)$ par Γ. On notera
au passage une conséquence du théorème de Nielsen-Schreier :tout sous groupe d'indice
fini j du groupe $\Gamma(2)$ (groupe libre à deux générateurs) est un groupe libre à $j + 1$
générateurs.

regrettant de ne pouvoir en dire autant dans l'autre sens. Reprenant les termes de l'auteur

Théorème 1 (Belyi). Une courbe algébrique régulière et complète sur un corps de caractéristique nulle est définie sur $\bar{\mathbf{Q}}$ si et seulement si on peut la présenter comme revêtement de $\mathbf{P}_1$ ramifié aux images inverses de 0, 1, ∞. La restriction à 0, 1, ∞ n'est pas essentielle puisque le groupe des automorphismes (algébriques) de $\mathbf{P}_1$ est transitif sur les triplets de points. Soit donc une courbe X sur $\bar{\mathbf{Q}}$ (dite alors arithmétique) et t un élément du corps rationnel sur la courbe dont U désigne l'ensemble fini des valeurs critiques (qui peuvent contenir le point à l'infini) de sorte que U (ou $U \backslash \{\infty\}$) soit dans $\bar{\mathbf{Q}}$. Cette donnée permet de considérer X comme recouvrement de $\mathbf{P}_1$ ramifié aux images inverses de U. Il s'agit maintenant de passer de cet ensemble à $(0, 1, \infty)$. Belyi procède en deux temps. Par une application polynomiale on passe de U à V tel que V (ou $V \backslash \{\infty\}$) soit dans $\mathbf{Q}$ puis par une seconde application polynomiale de V à $\{0, 1, \infty\}$.

Soit h_1 le polynôme minimal à coefficients rationnels qui s'annule sur U (ou $U \backslash \{\infty\}$). L'application $X \longrightarrow h_1 \circ t$ est ramifiée aux images inverses de l'origine, des valeurs critiques du polynôme $\{h_1(\xi) \mid h_1'(\xi) = 0\}$ et peut être de l'infini. Soit h_2 un second polynôme minimal à coefficients rationnels qui s'annule aux valeurs critiques $h_1(\xi) \in \bar{\mathbf{Q}}$. On a $\deg h_2 \leq \text{degree } h_1 - 1$ par un raisonnement classique. Les valeurs critiques de l'application $h_2 \circ h_1 \circ t$ sont en ∞, 0, l'image de 0 par h_2 (qui est rationnelle) et les valeurs critiques de h_2 en nombre strictement inférieur à celles de h_1. Poursuivant de la sorte, en un nombre d'itérations inférieur ou égal à degré(h_1) on obtient un élément $f \in \bar{\mathbf{Q}}(X)$, $f = h_\ell \circ h_{\ell-1} \circ \ldots \circ h_1 \circ t$ n'ayant que des valeurs critiques rationnelles et (peut être) l'∞. Soit $V \subset \mathbf{Q}$ cet ensemble fini de points. Dans un second temps, il s'agit de trouver un polynôme g (à coefficients rationnels) de valeurs critiques 0, 1, ∞ et tel que $g(V) \subset \{0, 1, \infty\}$. Si l'on parvient à construire un tel polynôme g, l'application $g \circ f$ donnera la présentation cherchée de X comme recouvrement de $\mathbf{P}$, non ramifié en dehors des images inverses de $\{0, 1, \infty\}$. Supposons qu'on dispose d'un polynôme g_1, ayant des propriétés analogues à g mais relativement à un sous ensemble $V_1 \subset \mathbf{Q}$ tel que $V = V_1 \cup \{\xi\}$. L'élément $g_1 \circ f \in \bar{\mathbf{Q}}(X)$ aura comme valeurs de ramification finies $(0, 1, g_1(\xi))$. Si l'on peut trouver un polynôme à coefficients rationnels g_2 admettant $\{0, 1, \infty\}$ comme seules valeurs critiques et tel que l'image de 0, 1, $s = g_1(\xi)$ appartienne à l'ensemble $\{0, 1\}$, le polynôme g cherché sera alors $g = g_2 \circ g_1$. Répétant autant de fois qu'il le faut cette construction on voit qu'on se ramène au cas ou l'ensemble V est réduit à trois points que par un automorphisme de P_1, on peut supposer de la forme 0, 1, $s = \frac{m}{m+n}$, m, n entiers positifs et à la construction d'un polynôme à coefficients rationnels, dont la dérivée s'annule en 0 et 1 et qui envoie 0, 1 et s sur l'ensemble 0, 1. Un tel polynôme p est

$$p(x) = \frac{(m+n)^{m+n}}{m^m n^n} x^m (1-x)^n$$

En 0 ou 1 ce polynôme s'annule tandis qu'en $s = \frac{m}{n+m}$ l'unique point distinct de 0, 1 ou sa dérivée s'annule, sa valeur est 1.

La réciproque, à savoir que si X est un recouvrement (fini) $\mathbf{P}$, non ramifié en dehors des images inverses de $\{0, 1, \infty\}$, alors X peut être défini sur $\bar{\mathbf{Q}}$ est une propriété de rigidité qui repose sur un critère de Weil. On a vu ci-dessus que le théorème d'existence de Riemann définit X comme surface de Riemann compacte en fait comme une courbe algébrique (complète). Si σ est un automorphisme de $\mathbf{C}$ laissant $\mathbf{Q}$ fixe, son action sur X définit un autre recouvrement de $\mathbf{P}_1$ ramifié aux images inverses de 0, 1, ∞ de même degré (de même nombre de feuillets). Pour que X^σ soit une surface de Riemann compacte (connexe) il ne peut y avoir qu'un choix fini de σ tel que X^σ soit non isomorphe à X (il vient au même de dire que chaque tel recouvrement de degré n correspond à un sous groupe d'indice n dans $\pi_1(\mathbf{C}P_1 \backslash \{0, 1, \infty\})$ et qu'il n'y a qu'un nombre fini de tels sous groupes). En d'autres termes pour presque tous les σ, $X^\sigma \equiv X$. Le critère de Weil revient alors à dire que dans cette situation X peut être défini sur une extension finie de $\mathbf{Q}$.

Nous ne sommes pas sûrs de bien comprendre cette preuve. Mais en examinant des cas concrets (ou la donnée est l'action finie de $\pi_1(\mathbf{C}P_1 \backslash \{0, 1\infty\})$ sur les feuillets de recouvrements) et "solubles" (ce qui est rare) on se convainc qu'on est amené à résoudre des cascades d'équations à coefficients rationnels.

Donnons d'abord quelques exemples. Peut être le plus frappant est-il celui présenté par Oesterlé décrivant les conjugués d'un dessin d'enfant (au sens propre).

Sur une surface de genre zéro supposons qu'on dispose d'une application

$$\begin{array}{ccc} \beta: & y & \longrightarrow & x \\ \Sigma \sim & \mathbf{P}_1 & \longrightarrow & \mathbf{P}_1 \end{array}$$

dont les points de ramification appartiennent à l'ensemble $\beta^{-1}(0, 1, \infty)$. Il s'agit de construire l'application rationnelle

$$x = \frac{P(y)}{Q(y)}$$

où P et Q sont premiers entre eux (sur $\bar{\mathbf{Q}}$), y apparaissant comme un générateur du corps des fonctions rationnelles sur Σ. Quitte à substituer à y une transformée homographique, on peut supposer $\deg P = \deg Q = n$, degré du recouvrement, normaliser Q à la forme $y^n + \dots$ et s'assurer que la valeur $y = \infty$ correspond à un point régulier d'image rationnelle – par exemple -1 – (et l'on pourra choisir $\left(\frac{dx}{dy^{-1}} \Big|_{y^{-1}=0} = 1 \right)$ ce qui revient à écrire

$$\frac{P}{Q} = \frac{-y^n + ay^{n-1} + \dots}{y^n + b\, y^{n-1} + \dots} \quad , \quad a + b = 1 \ .$$

De la sorte il reste $2n - 1$ coefficients à déterminer. Puisque seules les préimages de 0, 1, ∞ sont des points de ramification, il s'ensuit que $P'Q - Q'\, P = 0$ entraîne $QP(Q - P) = 0$, c'est-à-dire, comme $Q\,P$ et $P - Q$ sont premiers entre eux, que l'on peut trouver trois entiers positifs $\alpha_0, \alpha_1, \alpha_\infty$, et un polynôme S tels que

$$P^{\alpha_0}(Q - P)^{\alpha_1} Q^{\alpha_\infty} = (P'Q - Q'P)\, S$$

D'ailleurs on pourra se limiter à $\alpha_0 + \alpha_1 + \alpha_\infty \leq 2n - 1$. En effet factorisons (sur $\mathbb{C}$) $P'Q - Q'P$, de degré inférieur ou égal à $2n - 1$, en produit de trois polynômes $\rho_0 \rho_1 \rho_\infty$ tels que les racines de ρ_0 (respectivement ρ_1, ρ_∞) annulent P (respectivement $Q - P$, Q). La multiplicité d'une racine de ρ_i est majorée par $\deg \rho_i$ de sorte qu'on peut choisir $\alpha_i \leq \deg \rho_i$ et $\alpha_0 + \alpha_1 + \alpha_\infty \leq 2n - 1$ impliquant un nombre fini de triplets $\{\alpha_i\}$ possibles.

Effectuant la division euclidienne de $P^{\alpha_0}(Q - P)^{\alpha_1} Q^{\alpha_\infty}$ par $P'Q - Q'P$, opération rationnelle dans le corps engendré sur $\mathbb{Q}$ par les coefficients, on obtient un polynôme en y de degré inférieur ou égal à $2n - 2$. L'annulation de ses $2n - 1$ coefficients fournit ainsi pour tout triplet $\{\alpha_i\}$, $2n - 1$ conditions algébriques qui ont au plus un nombre fini de solutions correspondant aux applications de Belyi de genre zéro et de degré n, comme on peut le voir en bornant le nombre de décompositions d'une surface à valences des sommets et des faces données. Celles ayant un ordre de ramification 2 en $\beta^{-1}(1)$ correspondant aux triangulations dérivées décrites ci-dessus.

Donnons l'exemple classique (Klein) de la décomposition tétraédrale de la sphère, carte régulière et de la carte dérivée en douze paires de triangles. Dans la normalisation choisie $y = \infty$ est l'un des points de ramification – on pourrait y remédier par une transformation homographique – et correspond à un "milieu de face", envoyé sur $x = 1$, tandis qu'on envoie les sommets en $x = 0$ et les milieux d'arêtes à l'infini.

Posant

$$s = y\left(y^3 + 8\right) \ , \quad a = -y^6 + 20y^3 + 8 \ , \quad f = \left(y^3 - 1\right)$$

tels que $-s^3 + a^2 + 64f^3 = 0$, on définit

$$\beta : \quad y \longrightarrow x = \frac{P(y)}{Q(y)} = \frac{s^3}{a^2}$$

et on vérifie que

$$P(Q - P)Q = -\frac{1}{3}(P'Q - Q'P)\, s\, a\, f$$

Le groupe cartographique identique au groupe de symétrie est le groupe alterné à 12 éléments A_4 et la courbe est définie sur $\mathbf{Q}$.

Avec les mêmes conventions, les cartes régulières de groupe S_2, S_3, S_4 correspondent respectivement aux applications, définies sur $\mathbf{Q}$:

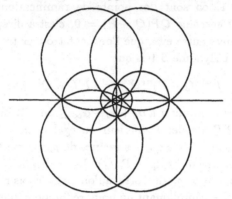

Figure 3: Carte régulière de groupe cartographique S_4 (cube ou octaèdre)

$$y \quad \longrightarrow \quad x = y^2$$

$$y \quad \longrightarrow \quad x = \left(\frac{y^3 - 1}{y^3 + 1}\right)^2$$

$$y \quad \longrightarrow \quad x = \frac{1}{64}\frac{\left[\left(48y^4 - 8y^2 + 3\right)\left(16y^4 + 40y^2 + 1\right)\right]^3}{\left[y\left(4y^2 - 1\right)\left(16y^4 + 8y^2 + 9\right)\left(144y^4 + 8y^2 + 1\right)\right]^2}$$

Il existe aussi une construction "transcendante" du générateur du corps des fonctions rationnelles dans le cas des cartes régulières de la sphère (correspondant aux solides réguliers).

Reprenons d'abord l'exemple du tétraèdre régulier ci-dessus et l'application de Belyi

$$x = \frac{y^3 \left(y^3 + 8\right)^3}{\left(y^6 - 20y^3 - 8\right)^2}$$

Une image inverse du demi plan Im $x > 0$ est un triangle dans le plan y dont les côtés peuvent être des arcs de cercles (correspondant à la subdivision barycentrique du tétraèdre). Une application de Schwarz

$$t = \int_0^x \frac{\mathrm{d}z}{z^{1-\frac{1}{3}}(1-z)^{1-\frac{1}{6}}}$$

transforme le demi plan Im $x > 0$ dans le sixième d'un triangle équilatéral d'angles au sommets $\frac{\pi}{2}$, $\frac{\pi}{3}$, $\frac{\pi}{6}$. La fonction inverse $y(t)$ satisfait à

$$\left(6\frac{dy}{dt}\right)^2 = 1 - y^3$$

qui est l'équation affine de la courbe elliptique F^3 (la courbe de Fermat de degré 3 et de genre 1) double recouvrement de la sphère ramifiée aux sommets du tétraèdre et le corps cherché $\mathbf{Q}(y(t))$ où y est à normalisation près, la fonction de Weierstrass sur cette courbe. La construction s'interprète en "mettant à plat" le tétraèdre régulier comme un ensemble de triangles équilatéraux avec identification des côtés. La fonction de Belyi cherchée est alors une fonction doublement périodique dont la dérivée s'annule aux sommets, prenant 2 fois chaque valeur dans une cellule élémentaire du groupe de translation (rapport modulaire $e^{2i\pi/6}$) : à translation près, c'est la fonction de Weierstrass.

On peut répéter une construction analogue pour l'octaèdre où l'on trouve que le corps rationnel cherché est un sous corps de celui des fonctions méromorphes sur une courbe de genre 4, triple recouvrement de la sphère ramifié aux images inverses des 6 sommets d'un octaèdre régulier, d'équation affine

$$4\left(12\frac{dy}{dt}\right)^3 = y^6 - 20y^3 - 8$$

(on posera $v = 12\frac{dy}{dt}$).

Enfin le cas de l'icosaèdre donne comme générateur une fonction méromorphe sur une surface de genre 25, sextuple recouvrement de la sphère ramifié aux images inverses des sommets de l'icosaèdre.

A notre connaissance l'exemple le plus simple de carte ayant des conjugués galoisiens est une troncation du "dessin d'enfant" d'Oesterlé. Il correspond au symbole féminin et possède deux conjugués:

Figure 4: Le symbole féminin et ses deux conjugués

Ce sont des cartes à deux faces, l'une a un côté et l'autre neuf. On cherche donc une fraction rationnelle de la forme $\frac{s(z)}{z+u}$ où $s(z)$ est de degré 10, et $-u$

le centre de la face à un côté, le centre de l'autre face étant à l'∞. Fixer le sommet d'ordre 4 à être à l'origine rigidifie le dessin aux dilatations près. Alors $s(z)$ est de la forme $\ell z^4 (z+t)^3 p(z)$ où $-t$ est la position du sommet d'ordre 3 et où p est un polynôme unitaire de degré 3 dont les racines sont les positions des trois sommets d'ordre 1. Demander que $\frac{s(z)}{z+u} - 1$ soit de la forme $\frac{\ell q(z)^2}{z+u}$ où les zéros du polynôme unitaire q sont les milieux d'arêtes amène à un problème d'élimination. Il faut choisir ℓ, t, u, les trois coefficients de p et les cinq de q de telle sorte que le polynôme

$$\ell z^4 (z+t)^3 p(z) - (z+u) - \ell q(z)^2$$

soit identiquement nul. Les coefficients dominants se compensent, donc il reste dix équations homogènes en onze variables. On montre alors que les coefficients inconnus sont polynomiaux en t et u, et que t et u sont liés par la relation de degré 3

$$t^3 + 9t^2 u + 99 t u^2 + 105 u^3 = 0$$

dont les trois solutions sont

$$\frac{t}{u} = -3 - \frac{6}{\alpha} + 4\alpha, \; -3 - \frac{6j}{\alpha} + 4j^2 \alpha, \; -3 - \frac{6j^2}{\alpha} + 4j\alpha$$

où $j = \exp \frac{2i\pi}{3}$ et α est la racine cubique réelle de 6.

La première solution est réelle, et pour des raisons de symétrie, elle est donc associée au symbole féminin original. En général, il est difficile sauf en traçant numériquement l'image réciproque du segment $[0,1]$ pour chaque solution, de voir quel dessin conjugué est associé à une solution donnée des équations algébriques satisfaites par la fonction de Belyi. Les deux conjugués du symbole féminin illustrent déjà ce dilemme:

Figure 5: Le symbole féminin et ses conjugués comme image inverse des applications de Belyi

En plus de ceux qui figurent dans la contribution d'Oesterlé on trouvera d'autres exemples dans les articles de Shabat et Voevodsky (voir ci-dessous) et Wolfart.

Les considérations qui précèdent sont de nature trop élémentaires pour éclairer l'action de $\mathrm{Gal}(\bar{\mathbf{Q}}/\mathbf{Q})$ sur les cartes et la liste des problèmes ouverts est énorme. Citons parmi ceux-ci:

1. Caractériser les courbes correspondant aux cartes régulières pour des familles aussi larges que possibles. Nous ne connaissons même pas la solution pour les groupes de permutations.

2. Quelles sont les cartes dont la courbe est définie sur $\mathbf{Q}$? Examiner les réductions modulo un nombre premier.

3. Pour un genre positif, dans un modèle affine $P(x,y) = 0$ où x est la fonction de Belyi d'une carte y étant une seconde fonction rationnelle sur la courbe (et donc $P(x,y)$ de degré $n = 2A$ en y) chercher, par un choix approprié de y, à minorer le degré en x.

4. Caractériser la représentation du groupe d'automorphisme sur les formes harmoniques discrètes.

Cette énumération pourrait se poursuivre longtemps ou se résumer en une seule question: Quelles sont les propriétés arithmétiques susceptibles d'être lues sur une carte et quelles sont les cartes associées par $\mathrm{Gal}(\bar{\mathbf{Q}}/\mathbf{Q})$?

Pour répondre partiellement à la question 3, donnons une construction, celle du *diviseur canonique* – ou plutôt d'un représentant de la classe d'équivalence correspondante – susceptible de fournir des renseignements utiles.

Soit x la fonction rationnelle correspondant à l'application de Belyi $\Sigma \longrightarrow \mathbf{P}_1$ et $\mathcal{S}$, $\mathcal{A}$ et $\mathcal{F}$ les images réciproques par β des valeurs critiques 0, 1, ∞. S'il s'agit d'une triangulation dérivant d'une carte, ce sont les points correspondants respectivement aux sommets, arêtes et faces de la carte initiale.

La différentielle $\mathrm{d}x$ est alors méromorphe et son diviseur est un représentant de la classe canonique K

$$\mathrm{div}(\mathrm{d}x) = \sum_{s \in \mathcal{S}} (v(s) - 1)s + \sum_{a \in \mathcal{A}} (v(a) - 1)a - \sum_{f \in \mathcal{F}} (v(f) + 1)f$$

Soit $L(K)$ l'espace vectoriel (a priori sur $\mathbf{C}$ ou plutôt sur $\bar{\mathbf{Q}}$) des fonctions rationnelles sur Σ telles que

$$y \in L(K) \Longleftrightarrow \mathrm{div}(y) + \mathrm{div}(\mathrm{d}x) \geq 0$$

Ce sont des fonctions dont les zéros appartiennent à $\mathcal{F}$, les pôles à $\mathcal{A} \cup \mathcal{S}$ et telles en outre qu'en tout point $f \in \mathcal{F}$ y a un zéro d'ordre au moins $v(f)+1$, en tout $a \in \mathcal{A}$ un pôle d'ordre au plus $v(a) - 1$, enfin en tout $s \in \mathcal{S}$ un pôle d'ordre au plus $v(s) - 1$.

Le théorème de Riemann Roch implique qu'en genre positif

$$\ell(K) = \dim L(K) = g$$

et que pour tout $y \in L(K)$, $y\,dx$ différentielle holomorphe possède $2g - 2$ zéros.

Considérons d'abord le cas particulier de genre 1. Il s'ensuit que $L(K)$ est unidimensionnel. En d'autres termes une telle fonction y est unique à un facteur constant près et de plus $y\,dx = dt$ est une différentielle holomorphe sans zéro. Dans ces conditions les inégalités précédentes deviennent des égalités, c'est dire que

$$\operatorname{div}(y) = \sum_{f \in \mathcal{F}}(v(f) + 1)f - \sum_{a \in \mathcal{A}}(v(a) - 1)a - \sum_{s \in \mathcal{S}}(v(s) - 1)s$$

Dans le cas d'une triangulation dérivée $v(a) = 2$ pour tout a, par conséquent

$$\text{nombre de zéros de } y = \sum_{f \in \mathcal{F}}(v(f) + 1) = F + 2A$$

$$\text{nombre de poles de } y = A + \sum_{s \in \mathcal{S}}(v(s) - 1) = 3A - S$$

Ces quantités sont bien égales puisqu'en genre 1 on a $S + F = A$. Il s'ensuit – toujours d'après Riemann – que les paires x, y, éléments de $\bar{\mathbf{Q}}(\Sigma)$, satisfont à une relation polynomiale

$$P(x, y) = 0$$

telle que

$$\deg_x P = 2A + F$$
$$\deg_y P = 2A$$

donnant un plongement $\Sigma \longrightarrow \mathbf{P}_1 \times \mathbf{P}_1$.

On adaptera ces raisonnements sur les degrés si la triangulation de Belyi est arbitraire auquel cas on aura

$$\deg_y P = \deg(\beta) = \sum_{f \in \mathcal{F}} v(f) = \sum_{s \in \mathcal{S}} v(s)$$
$$\deg_x P = |\mathcal{F}| + \deg(\beta)$$

Dans le cas général $(g \geq 1)$ on aura défini par le procédé précédent un espace vectoriel $L(K)$ de g fonctions rationnelles sur Σ linéairement indépendantes. L'une quelconque prend chaque valeur un nombre égal à celui de ses pôles (comptés avec leur multiplicité), c'est-à-dire

$$\deg(y) \leq \sum_{a \in \mathcal{A}}(v(a) - 1) + \sum_{s \in \mathcal{S}}(v(s) - 1)$$

Pour une triangulation dérivée on a donc

$$\deg(y) \leq 3A - S$$

Comme $A - S - F = 2(g - 1)$ ceci est équivalent à

$$\deg(y) \leq F + 2A + 2(g - 1)$$

l'inégalité étant remplacée par une inégalité en genre 1.

Comme précédemment on en conclut une relation polynomiale

$$P(x, y) = 0$$

quel que soit $y \in L(K)$ avec

$$\begin{aligned} \deg_x P &\leq F + 2A + 2(g - 1) \\ \deg_y P &= 2A \end{aligned}$$

Ajoutons quelques remarques:

(i) Il y a disymétrie de ces expressions entre sommets et faces, qu'on pourra compenser en considérant les relations entre y et x^{-1}.

(ii) Les raisonnements précédents ne permettent pas d'affirmer qu'une paire x, y engendre effectivement le corps $\bar{Q}(x, y)$. C'est cependant génériquement très vraisemblable. (iii) Le groupe d'automorphisme de la carte agit sur $L(K)$. Resterait à examiner de plus près cette action. (iv) On pourrait examiner plus généralement $L(nK)$.

Quoi qu'il en soit la construction qu'on vient de décrire fournit un moyen effectif de relier les données combinatoires et analytiques en permettant d'aborder dans certains cas l'analyse du corps $\bar{Q}(\Sigma)$ et son éventuelle restriction, un corps de nombres, extension finie de $\mathbf{Q}$.

A titre d'illustration on va analyser un exemple dû à Shabat et Voevodsky en genre 1 qui conduit au corps de définition $\mathbf{Q}\left(\sqrt{7}\right)$ et donc à un unique conjugué. Soit la carte tracée sur un tore formée de 8 faces (triangulaires) de 12 arêtes et de quatre sommets groupés en deux paires $\left(s_1^+, s_1^-\right)$ chacun de valence 7 et $\left(s_2^+, s_2^-\right)$ chacun de valence 5. Le degré de l'application de Belyi est donc 24. Soit x la fonction correspondante qui prend la valeur zéro en $s_1^\pm$ points critiques d'ordre 7 et $s_2^\pm$ points critiques d'ordre 5, la valeur 1 aux douze milieux d'arêtes (indiqués par des cercles sur la figure 6a) points critiques d'ordre 2, enfin la valeur ∞ aux huit milieux de faces points critiques d'ordre 3 (puisque les faces sont des triangles).

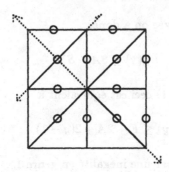

Figure 6a: La triangulation d'un tore considérée par Shabat et Voevodsky

Superficiellement la figure admet une symétrie $\mathbb{Z}/2\mathbb{Z} \times \mathbb{Z}/2\mathbb{Z}$ indiquée schématiquement par des axes de symétrie, l'un d'eux échangeant les sommets s_1^+ et s_1^-, l'autre les sommets s_2^+ et s_2^-. Plus précisément, il s'agit de deux anti-automorphismes de la surface de Riemann dont le produit est une involution qui échange les faces $f_i^+ \longleftrightarrow f_i^-$, remarque qui va se révéler essentielle dans la suite. D'ailleurs on vérifie sans peine que le groupe d'automorphisme de la carte définie comme ci-dessus se réduit à cette involution. Montrons que ces données permettent de construire une fonction méromorphe sur la surface de Riemann correspondante de degré deux comme on s'y attend pour une courbe elliptique. En effet soit $y \, dx$ l'unique différentielle holomorphe à facteur près.

La fonction y (de degré 32) a des pôles d'ordre 6 en $s_1^\pm$, d'ordre 4 en $s_2^\pm$ et des pôles simples aux douze milieux d'arêtes a_i tandis qu'elle a des zéros quadruples aux huit milieux de faces f_i. Il s'ensuite que la combinaison

$$z = x^5(x-1)^3 y^6$$

est à un facteur près la fonction cherchée ayant deux pôles simples en $s_1^\pm$ et deux zéros simples en $s_2^\pm$. Le théorème de Riemann-Hurwitz implique alors que z a quatre points critiques (où dz, ou si l'on préfère $\frac{dz}{y \, dx}$, s'annule simplement). Pour le voir il suffit de décomposer la sphère où z prend ses valeurs en deux polygônes à n cotés dont les sommets sont ces points critiques, de relever cette décomposition sur le tore en quatre polygônes à n côtés possédant $2n$ arêtes et n sommets. Comme la caractéristique d'Euler s'annule $n - 2n + 4 = 0$ et $n = 4$. Une autre façon de le voir consiste à remarquer que sur le tore les formes différentielles ont autant de zéros que de pôles. Or dz a deux pôles doubles donc quatre zéros. Ces quatre points sont insensibles à la normalisation (multiplicative) de z. Comme nous l'avons remarqué, les automorphismes d'une carte proviennent d'automorphismes de la courbe arithmétique associée. Nous notons $\tilde{\mu}$ l'automorphisme du tore associé à l'automorphisme μ de la carte. Il est facile de vérifier que

μ laisse 4 arêtes fixes (en échangeant les deux demi-arêtes de ces quatre arêtes). Donc $\tilde{\mu}$ a la même propriété, et possède un point fixe sur le milieu de ces quatre arêtes. Mais les automorphismes d'un tore ont soit zéro soit quatre points fixes. En effet si t est un paramètre uniformisant, un automorphisme est soit une translation (sans point fixe) soit la composée d'une translation et de l'inversion $t \longrightarrow -t$ (quatre points fixes car $2t = 0$ a quatre solutions). Choisissant l'origine en un point fixe de $\tilde{\mu}$ on a donc $\tilde{\mu}(t) = -t$, et $dt \longrightarrow -dt$. Par définition d'un automorphisme x (donc dx) est invariant. En conséquence $y \longrightarrow -y$ donc z est invariant. On a alors le résultat général suivant: sur une surface de Riemann les points fixes d'un automorphisme non trivial sont points critiques de toutes les fonctions méromorphes invariantes par cet automorphisme. Dans notre cas particulier la démonstration est triviale. Les fonctions invariantes dans $\tilde{\mu}$ sont les fonctions de t^2, donc leur différentielle à un zéro au point fixe. On en déduit que les quatre points critiques de la fonction invariante z sont les points critiques de la fonction invariante x situées sur les quatre arêtes globalement invariantes. Donc les points critiques de z sont les quatre milieux d'arêtes a_1, a_2, a_3, a_4.

En suivant Shabat et Voevodsky cherchons maintenant à reconstruire x comme fonction rationnelle de z. Définissons $\phi = \prod_{1 \le i \le 4} \left(z - z\left(f_i^{\pm}\right)\right)$ polynôme unitaire de degré 4. La fonction ϕ^3/z^5 se comporte comme z^7 pour z grand et comme z^{-5} pour z tendant vers zéro. Le produit $x\phi^3/z^5$ est donc une constante puisque sans pôle, et l'on vérifie alors que x a bien des pôles d'ordre trois aux milieux des faces. Posons

$$\frac{\rho}{x} = \frac{\phi^3}{z^5} \qquad (*)$$

où ρ est une constante. On constate qu'il s'agit bien d'une fonction de degré 24. Il faut alors s'assurer que $x - 1$ a des zéros doubles aux douze milieux d'arêtes parmi lesquels a_1, a_2, a_3, a_4 où z a des points critiques simples. On posera

$$\alpha = \prod_{1 \le i \le 4} \left(z - z\left(a_i\right)\right)$$

et l'équation de la courbe cherchée pourra s'écrire

$$u^2 = \alpha(z)$$

tandis que

$$\rho\left(\frac{1}{x} - 1\right) = \frac{\alpha\beta^2}{z^5} \qquad (**)$$

où β est à son tour un polynôme du quatrième degré dont les zéros sont les images par paires sur la sphère z des huit milieux d'arêtes restants. Reste à exprimer l'égalité des expressions de x tirée de (*) et (**). Il est plus simple d'égaler les dérivées par rapport à z de ces deux équations ce qui fait disparaître l'inconnue ρ. Il vient ainsi

$$\phi^2 \left[3z\phi' - 5\phi\right] = \beta \left[z \left(\alpha'\beta + 2\alpha\beta'\right) - 5\alpha\beta\right]$$

où ϕ, α, β sont trois polynômes unitaires de degré 4 en z. Chaque membre est factorisé en produit d'un polynôme de degré 4 et un de degré 8. On constate que les termes de plus haut degré sont compatibles avec les égalités

$$3z\phi' - 5\phi = 7\beta$$
$$7\phi^2 = z\left(\alpha'\beta + 2\alpha\beta'\right) - 5\alpha\beta$$

Ou encore en éliminant β

$$49\phi^2 = \left(z\alpha' - 5\alpha\right)\left(3z\phi' - 5\phi\right) + 2\alpha\left(3z^2\phi'' - 2z\phi'\right)$$

Les inconnues sont les huit coefficients des deux polynômes unitaires α et ϕ tandis que nous obtenons en apparence huit équations en identifiant les deux membres de l'équation précédente où les coefficients de z^8 sont égaux. Cependant l'une des équations est conséquence des sept autres ce qui reflète le fait que l'échelle de z est arbitraire. Ceci n'affecte pas la valeur de l'invariant modulaire. On trouve pour équation de la courbe avec $k = \pm\sqrt{7}$

$$u^2 = z^4 + 156\ z^3 + (17150 + 4256\ k)z^2$$
$$+(4452140 + 1690304\ k)z - (279416375 + 10544000\ k)$$

soit pour l'invariant modulaire deux valeurs positives qui encadrent 12^3

$$j = 4(114302 + 43141\ k) = 4(8 - 3\ k)^2(2 + k)^6(10 + 3\ k)^3$$

On confirme bien que le corps de définition de la carte initiale est $\mathbf{Q}\left(\sqrt{7}\right)$ et qu'elle n'admet qu'une conjuguée:

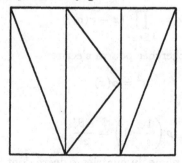

Figure 6b: La triangulation conjuguée à 6a

Resterait à associer la valeur positive, $\sqrt{7}$ disons, à l'une des deux cartes, ce que nous n'avons pu faire que numériquement. Nous avons commencé par tracer l'image réciproque du segment $[0, 1]$ par l'application de Belyi sur la sphère de la fonction z (Figure 7) avant de remonter sur le tore.

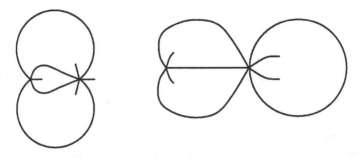

Figure 7: *L'image réciproque du segment* $[0, 1]$ *par l'application de Belyi dans l'exemple de Shabat et Voevodsky dans la variable* z

Le dessin original est associé à $k = -\sqrt{7}$, et son conjugué à $k = +\sqrt{7}$.

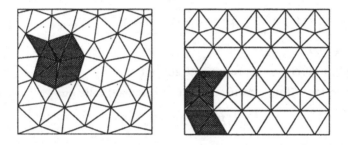

Figure 8: *La triangulation de Shabat et Voevodsky relevée sur le recouvrement universel:* $k = -\sqrt{7}$, $k = +\sqrt{7}$

Parmi les faibles informations dont on dispose sur les cartes qui se déduisent par action du groupe de Galois figure le fait que les nombres S_v et F_v de sommets et de faces de valence v sont les mêmes pour tout v (et bien sûr le nombre d'arêtes est conservé). Ceci suggère la définition suivante. On dira que deux cartes sont *associées* si elles ont les mêmes valeurs de S_v et F_v pour tout v – sans qu'elles soient nécessairement connexes. Or on a le résultat suivant de théorie des groupes – facile à vérifier –. Soit G un groupe fini d'ordre $|G|$, $\tilde{G}$ son dual, ensemble des représentations linéaires irréductibles (sur $\mathbf{C}$) C_1, C_2, C_3 trois classes de G

Lemme. Le nombre N_{C_1,C_2,C_3} de triplets g_1, g_2, g_3 avec $g_i \in C_i$ tels que $g_1 g_2 g_3 = 1$ est donné par

$$N_{C_1,C_2,C_3} = \frac{|C_1||C_2||C_3|}{|G|} \sum_{r \in \tilde{G}} \frac{\chi^r(C_1)\,\chi^r(C_2)\,\chi^r(C_3)}{\dim_r}$$

où $\dim_r$ est la dimension de la représentation r et $\chi^r(C)$ la valeur du caractère correspondant sur la classe C de cardinal $|C|$.

La donnée des S_v, F_v tels que $2A = \sum_v v S_v = \sum_v v F_v$ nous fournit trois classes de Σ_{2A} et un ensemble de cartes associées. Le nombre de triplets σ_0, σ_1, σ_2 avec $\sigma_0 \in \prod_v (v)^{S_v}$, $\sigma_1 \in (2)^A$ et $\sigma_2 \in \prod_v (v)^{F_v}$ (avec la notation standard des classes du groupe des permutations) tels que $\sigma_2 \sigma_1 \sigma_0 = 1^*$ est égal à la somme sur ces cartes associées (c'est-à-dire de tels triplets à *équivalence près dans* Σ_{2A}) des entiers $\frac{|\Sigma_{2A}|}{|H|}$ où H est le groupe d'automorphismes.

On en déduit l'égalité

$$F(\{S_v, F_v\}) \equiv \frac{(2A-1)!!}{\prod_v v^{S_v + F_v} S_v! F_v!} \times$$

$$\sum_{r \in \tilde{\Sigma}_{2A}} \frac{\chi^r\left(\prod_v (v)^{S_v}\right)\chi^r\left((2)^A\right)\chi^r\left(\prod_v(v)^{F_v}\right)}{\dim_r} = \sum_{\text{cartes associées}} \frac{1}{|H_{\text{carte}}|}$$

qui conduit au critère de rationalité suivant

Proposition. Etant donnée une carte (connexe) telle que

$$|H|F(\{S_v, F_v\}) = 1$$

alors le corps de définition est $\mathbf{Q}$.

Ce critère est évidemment à rapprocher (mais est moins intrinsèque) du critère de rigidité énoncé dans l'exposé de Serre cité en référence (qui ne s'intéresse en outre qu'au cas régulier où $|H|$ est l'ordre du groupe cartographique).

Exemple: Le tore $S = 1$, $A = 3$, $F = 2$ de genre 1:

---

* Dans certains exemples nous avons inversé l'ordre $\sigma_2 \sigma_1 \sigma_0 = 1$ (au lieu de $\sigma_0 \sigma_1 \sigma_2 = 1$), ceci n'affecte évidemment pas la nature des résultats.

$$\sigma_0 = (1, 4, 2, 5, 3, 6)$$

$$\sigma_1 = (1, 5)(2, 6)(3, 4) = \sigma_0^3$$

$$\sigma_2 = (1, 2, 3)(4, 5, 6) = \sigma_0^2$$

le groupe cartographique est abélien d'ordre 6, de généra-teur σ_0, isomorphe á $\mathbf{Z}/6\mathbf{Z}$.

Les permutations qui commutent avec $\sigma_0, \sigma_1, \sigma_2$, forment un groupe isomorphe (la carte est régulière): $|H| = 6$. En se servant de la table des caractères de Σ_6 (F.D. Murnaghan, "The Theory of Group Representations" Dover, New York (1963), page 146) on trouve que la quantité à évaluer est

$$\frac{6 \times 5!!}{6^1 1! \, 3^2 2!} \cdot \left(1 + 1 - \frac{1}{5} - \frac{1}{5} - \frac{2}{10} - \frac{2}{10} \right) = 1$$

ce qui est bien ce qu'on attend pour la courbe elliptique d'invariant modulaire $j = 0$ "définie" sur $\mathbf{Q}$.

De même on vérifie que la carte de genre zéro représentée sur la Figure 9 est définie sur $\mathbf{Q}$.

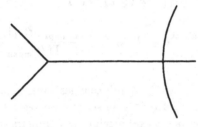

Figure 9: Le dessin d'enfant "guillotiné"

On a tracé sur cette figure l'image inverse par la fonction de Belyi du segment $[0, 1]$.

Un cas très particulier où l'on saurait sans trop de peine évaluer $F(\{S_v\}, \{F_v\})$ est celui où la carte n'a qu'un sommet ou qu'une face (c'est-à-dire tous les $S_v = 0$ sauf $S_{2A} = 1$, ou tous les $F_v = 0$ sauf $F_{2A} = 1$).

Dans ce cas, comme parmi les représentations de Σ_n seules celles de la forme

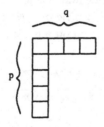

ont un caractère qui ne s'annule pas identiquement sur le cycle (n) on peut utiliser le résultat que

$$\sum_{p+q=n-1} (-1)^p y^p z^q \chi^{p,q} \left(\prod_v (v)^{k_v}\right) = \frac{\prod_v (z^v - y^v)^{k_v}}{z-y}$$

qui donne

$$\dim_{p,q} = \binom{n-1}{p} = \binom{n-1}{q}$$

$$\chi^{p,q}((n)) = (-1)^p$$

et si $n = 2A$

$$\chi^{p,q}\left((2)^A\right) = (-1)^{\left[\frac{p+1}{2}\right]} \binom{A-1}{\left[\frac{p}{2}\right]}$$

où le crochet désigne la partie entière, résultats qu'on aura l'occasion d'utiliser plus loin.

Plus généralement

$$\sum_{\substack{\Sigma\, vS_v = \Sigma\, vF_v = 2A}} \prod_v (t_v)^{S_v} (t'_v)^{F_v} F(\{S_v\}\{F_v\})$$

$$= \sideset{}{'}\sum_{\substack{\ell_1 > \ell_2 > \dots > \ell_{2A} \geq 0 \\ \Sigma\, \ell_k = A(2A+1)}} (-1)^{\frac{A(A-1)}{2}} \mathrm{ch}_{\ell.}(\theta)\mathrm{ch}_{\ell.}(\theta') \frac{\prod_{\ell\,\mathrm{impair}} \ell!! \prod_{\ell\,\mathrm{pair}}(\ell-1)!!}{\prod(\ell_{\mathrm{impair}} - \ell_{\mathrm{pair}})}$$

Les indices $\ell_1 > \ell_2 > \dots > \ell_{2A} \geq 0$ indexent les tableaux d'Young ayant dans la ligne i, $f_i = \ell_i + i - 2A$ boîtes, la somme est restreinte aux termes tels que le nombre des ℓ pairs est égal à celui des ℓ impairs (c'est-à-dire A); enfin les fonctions de Schur généralisées (caractères du groupe linéaire) sont exprimées en fonction des variables $\theta_v = \frac{t_v}{v}$ (ou $\theta'_v = \frac{t'_v}{v}$) par

$$\mathrm{ch}_{\ell.}(\theta.) = \det p_{f_i+i-j}(\theta.)|_{1\leq i,j\leq 2A}$$
$$\sum_0^\infty z^f p_f(\theta) = \exp \sum_1^\infty z^n \theta_n .$$

Il est en fait possible de trouver une fonction génératrice des quantités $F_C(\{S_v\}, \{F_v\})$ où la somme des $\frac{1}{|H|}$ ne porte que sur les cartes associées connexes à l'aide d'une intégrale matricielle. Soit dM la mesure de Lebesgue sur l'ensemble des matrices hermitiennes $N \times N$ et Λ une matrice définie positive. Dans le régime $\Lambda \longrightarrow \infty$ on a (grâce à une remarque d'I. Kostov) le développement asymptotique

$$\ell\mathrm{n}\left(\frac{\int \mathrm{d}M \exp\left(-\frac{1}{2}\mathrm{tr}\, M\Lambda M\Lambda + \sum_1^\infty g_k \mathrm{tr} \frac{M^k}{k}\right)}{\int \mathrm{d}M \exp\left(-\frac{1}{2}\mathrm{tr}\, M\Lambda M\Lambda\right)}\right) =$$

$$\sum F_C(\{S_v\}, \{F_v\}) g_1^{S_1} g_2^{S_2} \cdots (\mathrm{tr}\,\Lambda^{-1})^{F_1} (\mathrm{tr}\,\Lambda^{-2})^{F_2} \dots$$

Hélas à la différence des problèmes traités dans la section suivante, on ne dispose pour l'instant d'aucune méthode efficace pour étudier cette fonction génératrice en dehors de l'expression donnée ci-dessus.

§4. Décompositions cellulaires de l'espace des modules

A. La caractéristique virtuelle

Jusqu'ici nous avons considéré chaque carte séparément. Il est possible d'utiliser des familles de cartes pour donner une décomposition cellulaire de l'espace des modules des courbes. La technique des intégrales matricielles permet alors d'obtenir des résultats concernant la topologie de cet espace de modules. Nous donnons brièvement dans cette section et la suivante deux exemples frappants.

Soit $\Gamma_{g,n}$ le groupe modulaire correspondant à des surfaces de Riemann à n points marqués, quotient discret du groupe des difféomorphismes par ceux qui sont homotopes à l'identité. Il est possible de trouver un espace contractible $\mathcal{T}$ sur lequel agit ce groupe et on appelle caractéristique d'Euler virtuelle $\chi(\Gamma_{g,n})$ le quotient de la caractéristique d'Euler de $\mathcal{T}/\Gamma'$, où $\Gamma' \subset \Gamma_{g,n}$ est un sous groupe d'indice fini $[\Gamma_{g,n} : \Gamma']$ agissant sans point fixe sur $\mathcal{T}$, par l'indice en question. On se limitera au cas $n > 0$, $2g - 2 + n > 0$ et on posera d'abord $n = 1$.

Selon Harer et Zagier pour $g > 0$

$$\chi(\Gamma_{g,1}) = \sum_{r=2g}^{6g-3} (-1)^{r-1} \frac{\lambda_g(r)}{2r}$$

où l'entier $\lambda_g(r)$ est obtenu de la manière suivante. On considère un polygône à $2r$ côtés et on compte le nombre $\lambda_g(r)$ d'appariements distincts des côtés conduisant à une surface de genre g orientable (l'unique face correspond au point marqué) tels que la carte obtenue ne contiennent que des sommets de valence supérieure à deux. Ces cartes sont donc telles que

$$2A = 2r = \sum_{v \geq 3} v S_v \geq 3S$$

d'où l'on tire $r \leq 6g - 3$ comme indiqué ci-dessus, l'autre borne $r \geq 2g$ étant évidente. Pour obtenir la quantité voulue passons par l'intermédiaire des intégrales matricielles. Soit $\mathcal{M}$ l'espace vectoriel des matrices hermitiennes à N lignes et colonnes, $\dim_{\mathbf{R}} \mathcal{M} = N^2$ et si M désigne un élément générique soit

$$dM = \prod_{1 \leq i \leq N} dM_{ii} \prod_{i < j} d\,\mathrm{Re}\,M_{ij}\,d\,\mathrm{Im}\,M_{ij}$$

la mesure de Lebesgue sur $\mathcal{M}$ invariante par l'action adjointe du groupe unitaire $U(N)$. Considérons alors une partition $\underline{\nu}$ de l'entier $2r$, c'est-à-dire $\underline{\nu} \equiv \{\nu_1, ..., \nu_{2r} \mid \Sigma\, j\, \nu_j = 2r\}$ et posons

$$t_{\underline{\nu}}(M) = \prod_j \left(\operatorname{tr} M^j\right)^{\nu_j}$$

D'après les règles perturbatives de Feynman la valeur moyenne

$$\langle t_{\underline{\nu}} \rangle = \frac{\int dM\, e^{-\frac{1}{2}\operatorname{Tr} M^2} t_{\underline{\nu}}(M)}{\int dM\, e^{-\frac{1}{2}\operatorname{Tr} M^2}}$$

s'obtient comme somme sur des produits de cartes connexes ayant au total ν_j sommets de valence j chacune affectée d'un poids égal à N^F (F nombre de faces).

Chaque carte (connexe ou pas) est décrite par un groupe cartographique engendré par deux permutations (de $2r$ objets) σ_0 et σ_1, avec $\sigma_1^2 = 1$ et $\sigma_0 \in \underline{\nu}$, en identifiant les partitions avec les classes de conjugaison du groupe de permutations correspondant, Σ_{2r}, dont on notera $\underline{\Sigma}_{2r}$ l'ensemble des classes. Rappelons que le nombre d'éléments de la classe $\underline{\nu}$ est

$$|\underline{\nu}| = \frac{(2r)!}{\prod_j j^{\nu_j} \nu_j!}$$

Pour un élément $\sigma \in \Sigma_{2r}$ on écrira parfois $[\sigma]$ pour désigner sa classe au lieu $\underline{\nu}(\sigma)$. Cela étant, ayant choisi un représentant $\sigma_0 \in \underline{\nu}$, ce qui revient à indexer les $2r$ matrices M figurant dans $t_{\underline{\nu}}(M)$, les règles perturbatives s'expriment sous la forme

$$\langle t_{\underline{\nu}} \rangle = \sum_{\underline{\mu} \in \underline{\Sigma}_{2r}} \sum_{\substack{\sigma_1 \in [2^r] \\ [\sigma_0\sigma_1] \in \underline{\mu}}} N^{\Sigma \mu_i}$$

où l'on somme sur les permutations $\sigma_1 \in [2^r]$ pondérées par $N^{\Sigma \mu_i}$, μ_i étant le nombre de cycles de longueur i de $\sigma_0\sigma_1$, c'est-à-dire le nombre de faces de valence i.

Introduisons les caractères χ^Y (Y un tableau d'Young à $2r$ boîtes) du groupe Σ_{2r} qui satisfont aux relations d'orthogonalité et de complétude

$$\sum_Y \chi^Y(\sigma)\chi^Y(\tau) = \frac{(2r)!}{|[\sigma]|}\delta_{[\sigma],[\tau]}$$

$$\sum_{\tau \in \Sigma_{2r}} \chi^Y(\tau)\chi^Y(\sigma\tau) = (2r)!\delta_{YY'}\frac{\chi^Y(\sigma)}{\chi^Y([1^{2r}])}$$

où $\chi^Y\left(\left[1^{2r}\right]\right)$ est bien évidemment la dimension de la représentation indexée par Y. Un calcul immédiat donne

$$\langle t_{\underline{\nu}} \rangle = \sum_{\underline{\mu} \in \Sigma_{2r}} \langle t_{\underline{\nu}} \rangle_{\underline{\mu}}$$

où

$$\frac{\langle t_{\underline{\nu}} \rangle_{\underline{\mu}}}{|\underline{\mu}| N^{\Sigma \mu_i}} \equiv \frac{\langle t_{\underline{\mu}} \rangle_{\underline{\nu}}}{|\underline{\nu}| \, N^{\Sigma \nu_i}} = \frac{|[2^r]|}{(2r)!} \sum_Y \frac{\chi^Y\left([2^r]\right)\chi^Y(\underline{\mu})\chi^Y(\underline{\nu})}{\chi^Y\left([1^{2r}]\right)}$$

La symétrie de cette expression reflète l'échange des cartes avec leurs duales (c'est-à-dire l'échange des sommets avec les faces).

Revenons à $\chi\left(\Gamma_{g,1}\right)$. Par dualité on voit que le calcul de $\lambda_g(r)$ revient à l'évaluation de $\langle t_{[2r]} \rangle \equiv \langle \text{tr } M^{2r} \rangle$ qui n'introduit que des cartes connexes. Plus précisément

$$\lambda_g(r) = \sum_{\substack{\underline{\mu},\mu_1=\mu_2=0 \\ \Sigma\mu_i=r+1-2g}} \frac{\langle t_{[2r]} \rangle_{\underline{\mu}}}{N^{\Sigma\mu_i}}$$

La condition $\Sigma\mu_i = \{$nombre de faces de la carte duale connexe à un seul sommet et r arêtes$\} = r+1-2g$, exprime qu'on obtient une surface de genre g, tandis que $\mu_1 = \mu_2 = 0$ reflète la restriction introduite dans la définition de $\lambda_g(r)$. Cette restriction rend malaisé le calcul de $\lambda_g(r)$ sous cette forme et justifie des développements combinatoires fort élaborés dans l'article de Harer et Zagier.

Si on introduit une quantité intermédiaire $\epsilon_r(g)$ obtenue en levant cette restriction, il est facile en revanche de voir à l'aide des expressions données ci-dessus en termes de caractères que

$$\sum_{g \le \left[\frac{r}{2}\right]} N^{1+r-2g} \epsilon_g(r) \;=\; \langle t_{[2r]} \rangle$$

$$= \frac{(2r-1)!!}{2} \text{coeff. de } y^{r+1} \text{ dans } \left(\frac{1+y}{1-y}\right)^N$$

Il est cependant possible en jouant de nouveau sur la dualité de trouver une façon élégante d'évaluer $\chi\left(\Gamma_{g,1}\right)$. Observons qu'on peut encore écrire

$$\lambda_g(r) = \sum_{\substack{\underline{\mu},\mu_1=\mu_2=0 \\ \Sigma\mu_i=r+1-2g}} \frac{\langle t_{\underline{\mu}} \rangle_{[2r]}}{N} \frac{2r}{\prod_j j^{\mu_j}\mu_j!}$$

Soit en clair

$$\frac{\lambda_g(r)}{2r} = \frac{1}{N} \sum_{\substack{\mu_3,\mu_4,\ldots \\ \Sigma 3\mu_3 + 4\mu_4 + \ldots = 2r \\ \Sigma\mu_i = r + 1 - 2g}} \left\langle \frac{1}{\mu_3!}\left(\frac{\operatorname{tr} M^3}{3}\right)^{\mu_3} \frac{1}{\mu_4!}\left(\frac{\operatorname{tr} M^4}{4}\right)^{\mu_4} \ldots \right\rangle$$

Les conditions de sommation impliquent que

$$\mu_3 + 2\mu_4 + 3\mu_5 + \ldots = 4g - 2$$

et justifient la remarque suivante due à Penner. Soit

$$Z(N,x) = \left\langle \exp\left(-\sum_{k \geq 3} x^{k-2}\frac{M^k}{k}\right) \right\rangle$$

alors

$$\phi(N,x) = \log Z(N,x)$$

admet un développement perturbatif formel en contributions de cartes connexes qui peut se réorganiser en une série de puissances de N

$$\phi(N,x) = \sum_{s \geq 1} \phi_s(x)N^s$$

avec

$$\phi_1(x) = \sum_{g \geq 1} x^{4g-2}\left(\sum_{r=2g}^{6g-3}(-1)^{r+1}\frac{\lambda_g(r)}{2n}\right)$$

$$= \sum_{g} x^{4g-2}\chi(\Gamma_{g,1})$$

Bien entendu l'intégrale intervenant dans $Z(N,x)$ est formelle. Pour lui donner un sens remarquons que la représentation intégrale de la fonction Γ d'Euler est équivalente pour x réel positif tendant vers zéro à

$$\frac{(ex^2)^{\frac{1}{x^2}}}{(2\pi x^2)^{1/2}}\Gamma\left(\frac{1}{x^2}\right) = \frac{\int_{-\infty}^{1/x} dm \exp\left(-\sum_{k \geq 2}\frac{m^k x^{k-2}}{k}\right)}{\int_{-\infty}^{\infty} dm \exp\left(-\frac{m^2}{2}\right)}$$

A des termes exponentiellement petits près, le développement asymptotique pour $x \longrightarrow \infty$ (formule de Stirling) n'est rien d'autre que le développement perturbatif de cette intégrale où pour chaque terme, on pourra étendre l'intervalle d'intégration à $(-\infty, +\infty)$. C'est le sens qu'on va donner à l'expression matricielle ci-dessus après avoir effectué l'intégrale "angulaire" compacte réduisant $Z(N,x)$ à une intégrale sur les valeurs propres λ_i chacune limitée à une intervalle semi infini $(-\infty, \frac{1}{x})$ sans modifier la série asymptotique perturbative. De la sorte nous obtenons avec la notation $\Delta(\lambda) \equiv \prod_{1 \leq i < j \leq N} (\lambda_i - \lambda_j)$

$$Z(N,x) = \frac{1}{(2\pi)^{N/2} \prod_1^N p!} \int \prod_1^N \left(d\lambda_i \theta(1 - \lambda_i x)(1 - \lambda_i x)^{\frac{1}{x^2}} e^{\frac{\lambda_i}{x}} \right) \Delta(\lambda)^2$$

où $\theta(u)$ est la fonction saut d'Heaviside, $\theta(u) = 1$ pour $u > 0$, $\theta(u) = 0$ pour $u < 0$. La représentation intégrale de la fonction Γ donnée ci-dessus nous fournit

$$Z(N,x) = \left[\frac{(ex^2)^{\frac{1}{x^2}}}{\sqrt{2\pi}} \Gamma\left(\frac{1}{x^2}\right) \right]^N \frac{x^{-N^2}}{\prod_1^{N-1} p!} \det \prod_0^{r+s} (1 + px^2)|_{0 \leq r,s \leq N-1}$$

$$= \left[\frac{(ex^2)^{\frac{1}{x^2}}}{\sqrt{2\pi x^2}} \Gamma\left(\frac{1}{x^2}\right) \right]^N \prod_1^{N-1} (1 + px^2)^{N-p}$$

Pour compléter le calcul, il suffit d'utiliser la formule asymptotique de Stirling

$$\ell n \frac{(ex^2)^{\frac{1}{x^2}}}{\sqrt{2\pi x^2}} \Gamma\left(\frac{1}{x^2}\right) \sim \sum_{n \geq 1} \frac{B_{2n}}{2n(2n-1)} x^{4n-2}$$

où les nombres de Bernouilli B_n sont reliés aux polynômes de Bernouilli $B_n(u)$ par les expressions

$$B_n(u) = \sum_{0 \leq r \leq n} \binom{n}{r} B_r x^{n-r}$$
$$\frac{te^{tu}}{e^t - 1} = \sum_{n \geq 0} B_n(u) \frac{t^n}{n!}$$

de sorte que

$$\sum_{1 \leq p \leq N-1} p^k = \frac{B_{k+1}(N) - B_{k+1}}{k+1}$$

et

$$\phi(N,x) \sim N \sum_1^\infty \frac{B_{2n}}{2n(2n-1)} x^{4n-2} + \sum_{k \geq 1} (-1)^{k+1} \frac{x^{2k}}{k} \times$$
$$\left[\frac{N}{k+1} \sum_{r=1}^{k+1} N^r \binom{k+1}{r}) B_{k+1-r} - \frac{1}{k+2} \sum_{r=1}^{k+1} N^r \binom{k+1}{r} B_{k+2-r} \right]$$

Tenant compte de ce que pour $k \geq 1$, B_{k+1} s'annule si k est pair, on trouve

$$\phi_1(x) = \sum_{g \geq 1} \chi(\Gamma_{g,1}) x^{4g-2} \sim - \sum_{g \geq 1} x^{4g-2} \frac{B_{2g}}{2g}$$

D'où le très remarquable

Théorème (Harer-Zagier, Penner)

$$\chi(\Gamma_{g,1}) = -\frac{B_{2g}}{2g} = \zeta(1-2g)$$

où $\zeta(s)$ est la fonction ζ de Riemann.

En poussant l'analyse, on montre que la caractéristique virtuelle de l'espace des modules de surfaces de genre g à n points marqués (non distingués) n'est autre que le coefficient de N^n dans le développement de $\phi(N,x)$ et le calcul précédent montre que pour $n > 0$, $2g - 2 + n > 0$

$$\chi(\Gamma_{g,n}) = (-1)^{n-1} \frac{\zeta(1-2g)}{2g+n-2} \binom{2g+n-2}{n}$$

Cette formule est à prendre en un sens limite le cas échéant. Par exemple

$$g = 0 \qquad n = 3 \qquad \chi = \frac{1}{3!}$$
$$g = 0 \qquad n = 4 \qquad \chi = -\frac{1}{4!}$$
$$g = 1 \qquad n = 1 \qquad \chi = -\frac{1}{12}$$

Enfin lorsque l'expression a un sens, elle s'applique à $\Gamma_g \equiv \Gamma_{g,0}$ sans point marqué sous la forme

$$\chi(\Gamma_g) = \frac{\chi(\Gamma_{g,1})}{2-2g} = \frac{\zeta(1-2g)}{2-2g} \qquad g > 1$$

B. Formules d'intersection

A ce jour l'application la plus remarquable des méthodes d'intégration matricielle est due à Witten et Kontsevich. Soit $\mathcal{M}_{g,n}$ l'espace des modules des courbes de genre g à n points marqués et $\overline{\mathcal{M}_{g,n}}$ une compactification adéquate (correspondant semble-t-il aux dégénérences possibles des courbes projectives planes n'ayant au plus comme singularités que des points doubles à tangentes distinctes et tels que les points marqués ne viennent pas se confondre avec ces singularités). En chaque point marqué, on considère la première classe de Chern du fibre cotangent $\mathcal{L}$ et pour une suite $\{d_f \geq 0\}$ où f indexe les points distingués telle que

$$\sum_f d_f = 3g - 3 + n = \sum_i i\, k_i$$

$$k_i = \{\text{nombre de } d_f = i\}$$

on pose

$$\left\langle \tau_0^{k_0} ... \tau_i^{k_i} ... \right\rangle = \int_{\overline{\mathcal{M}}_{g,n}} c_1\left(\mathcal{L}_1\right)^{d_1} ... c_1\left(\mathcal{L}_n\right)^{d_n}$$

$\mathcal{M}_{g,n}$ ayant la structure d'un quotient de l'espace de Teichmüller correspondant par le groupe modulaire, ces "nombres d'intersection" sont des rationnels, d'ailleurs positifs. Comme le montre Witten, il sont reliés par des relations triangulaires à coefficients entiers positifs (et même égaux à l'unité le long de la diagonale) aux nombres d'intersection des classes de cohomologie stables sur $\overline{\mathcal{M}}_{g,0}$ introduites par Mumford, Morita et Miller qui s'expriment par des intégrales analogues sur $\overline{\mathcal{M}}_{g,0}$.

La série formelle

$$F(t.) = \ell \text{n}\, Z(t.) = \sum_{k_0, ..., k_i, ...} \left\langle \exp \sum_0^\infty \tau_i t_i \right\rangle$$

est uniquement déterminée par le

Théorème (Witten, Kontsevich)

(i) $Z(t.) = \exp F(t.)$ est une fonction τ au sens de Sato pour la hiérarchie des équations du système intégrable KdV (Kortweg et de Vries), ce qui revient à dire que $u = \frac{\partial^2 F(t.)}{\partial t_0^2}$ satisfait aux conditions

$$\frac{\partial}{\partial t_n} u = \frac{\partial}{\partial t_0} R_{n+1}\left(u, \frac{\partial u}{\partial t_0}, ...\right)$$

où les polynômes (différentiels) de Gelfand-Dikii

$$R_1 = u$$
$$R_2 = \frac{u^2}{2} + \frac{\partial^2 u}{12 \partial t_0^2}$$
$$\cdots\cdots\cdots\cdots$$
$$R_n = \frac{u^n}{n!} + ...$$

satisfont à la relation de récurrence

$$(2n+1)\frac{\partial}{\partial t_0} R_{n+1} = \left(\frac{1}{4}\frac{\partial^3}{\partial t_0^3} + 2u\frac{\partial}{\partial t_0} + \frac{\partial u}{\partial t_0}\right) R_n$$

(ii) $Z(t.)$ est un vecteur de plus haut poids pour une algèbre de Virasoro. Précisément, si pour m entier supérieur ou égal à -1 on définit $((-1)!! \equiv 1)$

$$2L_m = \sum_{k \geq m} \frac{(2k+1)!!}{(2(k-m)-1)!!} t_{k-m} \frac{\partial}{\partial t_k}$$

$$+ \frac{1}{2} \sum_{k+\ell=m-1} (2k+1)!!(2\ell+1)!! \frac{\partial^2}{\partial t_k \partial t_\ell}$$

$$-(2m+3)!! \frac{\partial}{\partial t_{m+1}} + \frac{t_0^2}{2} \delta_{m+1,0} + \frac{1}{8} \delta_{m,0}$$

où $\delta_{m,n}$ est le symbole de Kronecker, tels que

$$[L_m, L_n] = (m-n)L_{m+n}, \qquad m,n \geq -1$$

alors

$$L_m \, Z(t.) = 0, \qquad m \geq -1$$

Les assertions assez surprenantes du théorème sont à prendre au sens des séries formelles. L'algèbre de Lie infinie dont il est question dans la seconde partie est en fait une sous algèbre de l'algèbre de Virasoro qui ne pas fait apparaître la "charge centrale" c'est-à-dire la valeur du coefficient du cocycle de l'extension centrale de l'algèbre de Lie de Diff (S_1). Enfin les conditions (i) et (ii) sont surabondantes mais bien entendu compatibles.

Ce paragraphe va être consacré à la démonstration de ce théorème.

Donnons d'abord quelques indications, hélas un peu imprécises, sur le modèle combinatoire de $\mathcal{M}_{g,n}$ qui conduit à représenter $Z(t.)$ à l'aide d'intégrales sur des matrices hermitiennes. Il s'agit d'une décomposition cellulaire où chaque cellule est représentée par une carte (connexe) dont les sommets sont au moins trivalents, de genre g, ayant n faces, munie d'une "métrique" qui affecte une longueur positive à chaque arête: $a \longrightarrow \ell_a$ et à chaque face f un périmètre $p_f = \sum_{a \subset f} \ell_a$, le symbole $a \subset f$ désignant la relation d'incidence. On attribue à chaque cellule une dimension égale au nombre de variables ℓ_a indépendantes lorsque les p_f sont fixés, c'est-à-dire $A - F$ (où le nombre de faces F est fixé à la valeur n). Des relations

$$A = 2g - 2 + S + F, \qquad 2A \geq 3S$$

on tire

$$A - n \leq 2(3g - 3 + n)$$

où l'égalité n'a lieu que si tous les sommets sont trivalents, correspondant aux cellules de dimension maximale $(= \dim_{\mathbf{R}} \mathcal{M}_{g,n})$ celles auxquelles on se borne dans la suite.

A une face orientée de valence k bordée par les arêtes $a_1, a_2, ..., a_k$ dans cet ordre, à permutation cyclique près, on fait alors correspondre la 2-forme

$$\omega_f = \sum_{1 \le i < j \le k-1} d\left(\frac{\ell_i}{p_f}\right) \wedge d\left(\frac{\ell_j}{p_f}\right)$$

invariante dans un changement d'échelle et une permutation circulaire. Kontsevich montre alors que les nombres d'intersection cherchés s'expriment dans ce modèle sous la forme

$$\langle \tau_{d_1} ... \tau_{d_n} \rangle = \int \prod_f \omega_f^{d_f}$$

où le signe somme est à interpréter comme une intégrale sur les cellules de dimension $6g - 6 + 2n$ et une somme sur ces cellules, c'est-à-dire des cartes connexes à sommets trivalents affectées d'un poids égal à l'inverse de l'ordre de leur groupe d'automorphisme. A vrai dire, il est difficile de dire si dans l'article original on trouve une preuve ou une assertion de ce fait (il s'agit là d'une remarque qui n'engage que les auteurs, peu versés dans la topologie). On notera en particulier qu'une orientation cohérente est requise, compatible avec celle induite par la structure complexe de $\mathcal{M}_{g,n}$.

Ce point, capital, étant admis, le reste n'est affaire que de calcul. On obtiendra de deux façons la quantité

$$\int_0^\infty \prod_f \mathrm{d}p_f e^{-\lambda_f p_f} \int \frac{\left(\sum_f p_f^2 \omega_f\right)^{3g-3+n}}{(3g-3+n)!}$$

qui assigne à chaque point marqué une variable λ_f conjuguée du périmètre p_f. D'après l'identification précédente, ceci s'écrit

$$\sum_{d_1 + ... + d_n = 3g-3+n} \langle \tau_{d_1} ... \tau_{d_n} \rangle 2^{3g-3+n} \prod_1^n (2d_f - 1)!! \lambda_f^{-(2d_f+1)}$$

Par ailleurs au prix d'un Jacobien donné par

$$\frac{1}{(3g-3+n)!} \prod_f \mathrm{d}p_f \left(\sum_f p_f^2 \omega_f\right)^{3g-3+n} = 2^{5g-5+2n} \mathrm{d}\ell_1 ... \mathrm{d}\ell_A$$

quelle que soit la carte de genre g à n faces (à un signe près qu'il faut soigneusement déterminer), on peut intégrer directement sur les longueurs ℓ_a des arêtes communes aux faces de variables λ_f, $\lambda_{f'}$ ($a \subset f$, $a \subset f'$)

distinctes ou confondues, chacune produisant un facteur $\frac{1}{\lambda_f + \lambda_{f'}}$. Combinant les facteurs 2 on trouve en désignant par $C_{g,n}$ l'ensemble des cartes connexes de sommets trivalents de genre g à n faces

$$\sum_{\substack{d_1 + \cdots + d_n = \\ 3g-3+n}} \langle \tau_{d_1} \ldots \tau_{d_n} \rangle \prod_1^n \frac{(2d_f - 1)!!}{\lambda_f^{2d_f+1}} = \sum_{C = C_{g,n}} \frac{2^{-S}}{|\text{Aut } C|} \prod_a \frac{2}{\lambda_{f(a)} + \lambda_{f'(a)}}$$

où $|$ Aut C $|$ est l'ordre du groupe d'automorphisme de la carte C et le dernier produit est sur les arêtes de la carte.

Le membre de droite suggère une interprétation en terme du développement perturbatif d'une intégrale matricielle normalisée si l'on effectue une sommation sur tous les choix possibles des λ_f parmi des valeurs $\lambda_{(0)}, \ldots, \lambda_{(N-1)}$ considérées comme les valeurs propres d'une matrice hermitienne définie positive de taille $N \times N$. Plus précisément posons

$$t_r(\Lambda) = -(2r-1)!! \text{ tr } \Lambda^{-2r-1}, \qquad r \geq 0$$

avec la convention $(-1)!! = 1$. On a alors

$$F(t.(\Lambda)) = \sum_{\substack{n \geq 1 \\ d_1 + \cdots + d_n \geq 0}} \frac{1}{n!} \langle \tau_{d_1} \ldots \tau_{d_n} \rangle t_{d_1}(\Lambda) \ldots t_{d_n}(\Lambda)$$

$$= \sum_{g,n} \sum_{C = C_{g,n}} \sum_{\lambda_1 \ldots \lambda_n \in \{\lambda_{(0)}, \ldots, \lambda_{(N-1)}\}} \left(\frac{i}{2}\right)^S \prod_a \frac{2}{\lambda_{f(a)} + \lambda_{f'(a)}}$$

où l'on a noté que $(-)^n = i^S$. En effet la relation $2A = 3S$ implique $S = 2p$ et de $S - A + F = S - A + n = 2p - 3p + n \equiv 0 \mod 2$, il s'ensuit que $p \equiv n \mod 2$.

On a alors les résultats suivants (Witten, Kontsevitch)

Proposition 1

$$F(t.(\Lambda)) = \ell n \, Z_N(\Lambda)$$

$$Z_N(\Lambda) = \frac{\int dM \exp - \text{tr} \left(\frac{\Lambda M^2}{2} - i \frac{M^3}{6}\right)}{\int dM \exp - \text{tr} \frac{\Lambda M^2}{2}}$$

Dans cette intégrale matricielle normalisée dM est la mesure de Lebesgue sur les matrices hermitiennes $N \times N$ invariante par l'action adjointe du groupe unitaire et l'égalité est à entendre au sens du développement perturbatif (asymptotique) qui revient à remplacer $\exp i \, \text{tr} \frac{M^3}{6}$ par son développement en série de puissances.

Il est naturel d'attribuer à Λ^{-1} le degré 1 de sorte que si l'on introduit la mesure normalisée

$$d\mu_\Lambda(M) \;=\; \frac{\exp - \mathrm{tr}\frac{\Lambda M^2}{2}}{\int dM\,\exp - \mathrm{tr}\,\frac{\Lambda M^2}{2}}$$

$$Z_N(\Lambda) \;=\; \sum_{k\geq 0} Z_{N,k}(\Lambda)$$

$$Z_{N,k}(\Lambda) \;=\; \frac{(-)^k}{(2k)!}\int d\mu_\Lambda(M)\left(\mathrm{tr}\frac{M^3}{6}\right)^{2k}$$

où $Z_{N,k}(\Lambda)$ est de degré $3k$. Bien entendu à N fixé seules N des variables $t_r(\Lambda)$ sont algébriquement indépendantes, cependant l'expression que nous allons trouver pour $Z_N(\Lambda)$ montre que

Proposition 2

Exprimé en fonction des quantités $t_r(\Lambda)$, $Z_{N,k}(\Lambda)$ est indépendant de N pour $N \geq 3k$ (et n'est fonction que de $t_r(\Lambda)$, $r \leq 3k$). Si on note cette quantité $Z_{.,k}(t.(\Lambda))$ on obtiendra ainsi $Z(t.)$ sous la forme

$$Z(t.) = \sum_{k\geq 0} Z_{.,k}(t.)$$

qui ne fait plus référence à la taille des matrices.

Nous en arrivons à l'évaluation des intégrales matricielles. Pour ce faire introduisons la fonction $z(\lambda)$ correspondant au cas $N = 1$

$$z(\lambda) = \frac{\int_{-\infty}^{+\infty} dm\;e^{-\lambda\frac{m^2}{2}+\frac{im^3}{6}}}{\int_{-\infty}^{+\infty} dm\;e^{-\lambda\frac{m^2}{2}}}$$

qui à λ grand admet le développement asymptotique

$$z(\lambda) = \sum_{k\geq 0} c_k\lambda^{-3k} \qquad c_k = \left(-\frac{2}{9}\right)^k\frac{\Gamma\left(3k+\frac{1}{2}\right)}{\sqrt{\pi}}$$

et satisfait à une équation différentielle équivalente à l'équation d'Airy

$$\left(D^2 - \lambda^2\right)z(\lambda) = 0 \qquad D = -e^{\frac{1}{3}\lambda^3}\lambda^{1/2}\left(\frac{\partial}{\lambda\partial\lambda}\right)\lambda^{-1/2}\,e^{-\frac{1}{3}\lambda^3}$$

$$= \lambda + \frac{1}{2\lambda^2} - \frac{\partial}{\lambda\partial\lambda}$$

Revenant au cas général, nous pouvons exprimer $Z_N(\Lambda)$, au prix d'un prolongement analytique et d'une translation de la variable au numérateur, sous la forme

$$Z_N(\Lambda) \;=\; 2^{-\frac{N(N-1)}{2}}\prod_r \lambda_r^{1/2}\prod_{r<s}(\lambda_r+\lambda_s)\exp \mathrm{tr}\frac{\Lambda^3}{3} \times \tilde{Z}_N(\Lambda)$$

$$\tilde{Z}_N(\Lambda) \;=\; \int d\left(\frac{M}{(2\pi)^{1/2}}\right)\exp i\,\mathrm{tr}\left(\frac{M^3}{6} + \frac{M\Lambda^2}{2}\right)$$

où nous avons explicitement écrit $dM = \prod_i dM_{ii} \prod_{i<j} d\,\mathrm{Re}\,M_{ij}\,d\,\mathrm{Im}\,M_{ij}$.
La dernière intégrale est invariante si on remplace M par un conjugué
$M \longrightarrow UMU^{-1}$ et simultanément $\Lambda \longrightarrow U\Lambda U^{-1}$ (U unitaire) aussi pouvons
nous supposer Λ diagonal. Cependant il reste à effectuer l'intégrale sur les
"angles relatifs". Cette intégrale était obtenue dans un travail avec J.-B.
Zuber, par un argument de noyau de la chaleur. Nous avons appris depuis
que la même intégrale avait été considérée bien plus tôt par Harish-Chandra.
Quoiqu'il en soit on trouve

$$\tilde{Z}_N(\Lambda) = \int \prod_r \frac{dm_r}{(2\pi)^{1/2}} e^{i\left(\frac{m_r^2}{6} + \frac{m_r \lambda_r^2}{2}\right)} \prod_{r<s} \left(\frac{m_r - m_s}{\frac{i\lambda_r^2}{2} - \frac{i\lambda_s^2}{2}}\right)$$

où les indices r et s varient de 0 à $N-1$ et les m_r sont les valeurs propres
de la matrice M. On emploiera dans la suite la notation de Weyl pour le
déterminant

$$|f_0(\lambda), ..., f_{N-1}(\lambda)| = \det \begin{vmatrix} f_0(\lambda_0) & ... & f_{N-1}(\lambda_0) \\ \vdots & & \vdots \\ f_0(\lambda_{N-1}) & ... & f_{N-1}(\lambda_{N-1}) \end{vmatrix}$$

de sorte que le déterminant de Vandermonde s'écrit

$$|m^0, m^1, ..., m^{N-1}| = \prod_{r<s} (m_r - m_s)$$

On effectue sans peine les intégrales dans $\tilde{Z}_N(\Lambda)$ et reportant

$$Z_N(\Lambda) = \frac{|D^0 z(\lambda), ..., D^{N-1} z(\lambda)|}{|\lambda^0, ..., \lambda^{N-1}|}$$

C'est de cette expression légèrement réorganisée qu'on va partir pour obtenir
les propriétés cherchées. Pour cela on introduit une seconde fonction
(l'équation d'Airy est du second ordre)

$$\bar{z}(\lambda) = \frac{1}{\lambda} Dz = \sum_{k \geq 0} d_k \lambda^{-3k}, \qquad d_k = \frac{1 + 6k}{1 - 6k} c_k$$

et on note que

$$D^{2k} z = \lambda^{2k} z \mod \{D^{2k-1} z, ..., D^0 z\}$$
$$D^{2k+1} z = \lambda^{2k+1} \bar{z} \mod \{D^{2k} z, ..., D^0 z\}$$

Ainsi on peut écrire

$$Z_N(\Lambda) = \frac{|\lambda^0 z, \lambda^1 \bar{z}, \lambda^2 z, ...|}{|\lambda^0, \lambda^1, \lambda^2, ..., \lambda^{N-1}|}$$

où au numérateur, le dernier terme est $\lambda^{N-1}z$ si N est impair, ou $\lambda^{N-1}\bar{z}$ si N est pair. Factorisons au numérateur et au dénominateur le produit $(\lambda_0, ..., \lambda_{N-1})^{N-1}$ et développons les fonctions z et $\bar{z}$ en puissances de

$$x \equiv \lambda^{-1}$$

Nous trouvons en définitive la série

$$Z_N(\Lambda) = \sum_{n_0, n_1, ..., n_{N-1} \geq 0} c_{n_0}^{(0)} c_{n_1}^{(1)} ... c_{n_{N-1}}^{(N-1)} \frac{\left| x^{3n_0+N-1}, x^{3n_1+N-2}, ..., x^{3n_{N-1}} \right|}{\left| x^{N-1}, x^{N-2}, ..., x^{N-1} \right|}$$

où nous adoptons la convention

$$c_n^{(2p)} = c_n, \qquad c_n^{(2p+1)} = d_n$$

Nous reconnaissons dans le rapport des deux déterminants un caractère polynomial du groupe linéaire (exprimé en terme des valeurs propres d'un élément de GL_N et étendu à toute matrice en vertu de sa nature polynomiale). En général pour une matrice X (ici Λ^{-1}) diagonalisable de valeurs propres $x_0, ..., x_{N-1}$, nous posons

$$\mathrm{ch}_{\ell_{N-1}, ..., \ell_0}(X) == \frac{\left| x^{\ell_{N-1}}, ..., x^{\ell_0} \right|}{\left| x^{N-1}, ..., x^0 \right|}$$

dont l'expression polynomiale en fonction des traces des puissances de X

$$\theta_k(X) = \frac{1}{k} \mathrm{tr} \; X^k$$

s'appelle une fonction de Schur généralisée. Notons que

$$t_r(\Lambda) = -(2r+1)!! \; \theta_{2r+1}(\Lambda^{-1})$$

Définissons en outre les traces des puissances symétriques de X sous la forme

$$p_k(X) = \mathrm{tr} \bigotimes_{\text{sym}}^{k} X$$

telles que

$$\sum_{k \geq 0} u^k p_k(X) = \det(1 - uX)^{-1} = \exp \sum_{n \geq 1} u^k \theta_k(X)$$

L'égalité entre les termes extrêmes définit les polynômes de Schur que nous notons encore $p_k(\theta)$, polynômes dans des variables $\{\theta_1, \theta_2, \dots\}$

$$p_k(\theta) = \sum_{\sum_{r \geq 1} r\nu_r = k} \frac{\theta_1^{\nu_1}}{\nu_1!} \frac{\theta_2^{\nu_2}}{\nu_2!} \dots \quad , \quad p_0(\theta.) = 1$$

Ces polynômes, comme le remarque Sato, jouissent de la propriété fondamentale

$$\frac{\partial p_k(\theta.)}{\partial \theta_r} = \frac{\partial^r}{\partial \theta_1^r} p_k(\theta.) = p_{k-r}(\theta.)$$

où nous convenons que $p_k(\theta.)$ s'annule si son indice est négatif. Un calcul classique fondé sur la formule déterminantale de Cauchy permet alors d'exprimer les ch. en fonction des p sous la forme

$$\mathrm{ch}_{N-1+f_0, N-2+f_1, \dots, f_{N-1}} = \begin{vmatrix} p_{f_0} & & \\ & p_{f_1} & * \\ * & & \ddots \\ & & & p_{f_{N-1}} \end{vmatrix}$$

où la notation symbolique implique que les indices croissent (décroissent) d'une unité à chaque déplacement vers la droite (gauche).

On en conclut que $Z_N(\Lambda)$ admet un développement en termes de traces des puissances de Λ^{-1} de la forme

$$Z_N(\Lambda) = \sum_{n_0, \dots, n_{N-1} \geq 0} c_{n_0}^{(0)} c_{n_1}^{(1)} \dots c_{n_{N-1}}^{(N-1)} \begin{vmatrix} p_{3n_0} & & \\ & \ddots & * \\ * & & \\ & & p_{3n_{N-1}} \end{vmatrix} (\theta.(\Lambda^{-1}))$$

où les termes de degré k, $Z_{N,k}$, sont obtenus en restreignant la somme par la condition $n_0 + n_1 + \dots + n_{N-1} = k$. La proposition 2 triviale pour $k = 0$ ($Z_{N,0} = 1$) en résulte si on suppose $0 < 3k \leq N$. Considérons dans $Z_{N,k}$ un terme tel que δ soit le plus petit entier satisfaisant $r \geq \delta \longrightarrow n_r = 0$. On peut alors restreindre le déterminant correspondant à la taille $\delta \times \delta$, les indices des p. de la dernière colonne étant donnés par la suite $3n_0 + \delta - 1$, $3n_1 + \delta - 2, \dots, 3n_\delta$ dont la somme est $3k + \sum_0^{\delta-1} r$. Si $\delta > 3k$ cette somme de δ entiers positifs est inférieure à celle des δ premiers entiers positifs: deux d'entre eux doivent donc être égaux, le déterminant s'annule et le terme correspondant est donc nul. On en conclut que pour tout terme qui contribue à Z a une "profondeur" $\delta \leq 3k$ ce qui prouve que $N \geq 3k \Longrightarrow Z_{N,k} = Z_{3k,k} = Z_{.k}$ où à l'évidence lorsqu'on a pris les θ. comme variables, ce terme est indépendant de N ce qu'affirme la proposition.

En outre en posant $Z(\theta.) = \sum_{k \geq 0} Z_{.,k}(\theta.)$ on peut présenter cette série comme

$$Z(\theta.) = \sum_{\substack{\sum_{i \geq 0} n_i < \infty}} \prod_{i \geq 0} c_{n_i}^{(i)} \begin{vmatrix} p_{3n_0} & & * \\ & p_{3n_1} & \\ * & & \ddots \end{vmatrix} (\theta.)$$

Les déterminants, d'apparence infinie, se réduisent à des déterminants finis puisque pour presque tout i, $n_i = 0$ et que dans ces conditions la profondeur δ d'un terme donné est finie ce qui implique que le déterminant peut être restreint à la taille finie $\delta \times \delta$.

L'expression de la série asymptotique ainsi obtenue permet de comprendre qu'elle est indépendante des variables d'indice pair comme il se doit d'après la construction initiale. En effet dérivons Z par rapport à θ_{2p} ce qui revient à dériver successivement chaque ligne des déterminants. Or en dérivant la ligne d'indice r, en vertu de la propriété fondamentale des polynômes de Schur on substitue à

$$p_{3n_r-r}, \ p_{3n_r-r+1}, ..., p_{3n_r}, p_{3n_r+1}, ...$$

la suite

$$p_{3n_r-r-2p}, p_{3n_r-r+1-2p}, ..., p_{3n_r-2p}, p_{3n_r+1-2p}, ...$$

qui s'identifie à la ligne $r + 2p$ à l'échange près de n_r avec n_{r-2p}. Dans cet échange le déterminant où l'on a dérivé la ligne r est impair tandis que son coefficient est pair en vertu de $c_n^{(i)} = c_n^{(i+2p)}$ par définition. En conclusion

$$\frac{\partial}{\partial \theta_{2p}} Z(\theta.) = 0$$

On peut donc bien prendre comme variables les quantités $t_r = -(2r + 1)!! \theta_{2r+1}$. Cependant pour la simplicité des expressions qui vont suivre, nous conservons la notation θ.

Nous sommes maintenant en mesure de passer à la preuve du théorème. Puisque le passage à la limite $N \longrightarrow \infty$ ne présente aucune difficulté dans la série asymptotique quand on a fait choix des variables $\theta.(\Lambda^{-1})$, il est plus commode de revenir au cas N fini, quitte à faire tendre $N \longrightarrow \infty$ dans la suite. Observons qu'on peut récrire Z_N sous la forme d'un Wronskien. Posant

$$f_s(\theta.) = \sum_{n \geq 0} c_n^{(N-s-1)} p_{3n+s}(\theta.), \qquad 0 \leq s \leq N - 1$$

on a en effet

$$Z_N(\theta.) = \begin{vmatrix} \frac{\partial^{N-1}}{\partial \theta_1^{N-1}} f_{N-1} & \cdots & \frac{\partial}{\partial \theta_1} f_{N-1} & f_{N-1} \\ \vdots & & \vdots & \vdots \\ \frac{\partial^{N-1}}{\partial \theta_1^{N-1}} f_0 & \cdots & \frac{\partial}{\partial \theta_1} f_0 & f_0 \end{vmatrix} (\theta.)$$

où nous considérons les $\theta.$ comme indépendants (à la vérité $Z_N(\Lambda) \equiv Z_N\left(\theta.(\Lambda^{-1})\right)$. Cette expression suggère la considération d'un opérateur différentiel linéaire associé dont les solutions linéairement indépendantes sont les $f_s(\theta.)$ en particularisant la variable θ_1. On pose $d \equiv \frac{\partial}{\partial \theta_1}$ et on définit cet opérateur Δ_N par

$$\Delta_N F = \sum_{0 \le r \le N} w_r(\theta.)d^{N-r}F = Z_N^{-1} \begin{vmatrix} d^N F & \cdots & F \\ d^N f_{N-1} & \cdots & f_{N-1} \\ \vdots & & \vdots \\ d^N f_0 & \cdots & f_0 \end{vmatrix}$$

En toute rigueur les coefficients $w_r(\theta.)$ devraient aussi porter l'indice N $(w_0(\theta.) \equiv 1)$ mais comme on s'en convaincra sans peine à r fixé et N assez grand $w_r(\theta.)$ est indépendant de N. Il est commode en suivant Sato de remplacer l'opérateur différentiel Δ_N par un opérateur pseudo-différentiel équivalent

$$\Delta_N = W_N d^N$$
$$W_N = \sum_{0 \le r \le N} w_r(\theta.)d^{-r}$$

Nous n'utiliserons ici que les propriétés algébriques des opérateurs pseudo-différentiels. Ces derniers ont une longue histoire qui remonte au moins à la fin du siècle dernier. Pour un exposé on pourra consulter Gelfand-Dikii. Il nous suffit de mentionner l'extension de la règle de Leibniz. Notant $a^{(k)}$ la k-ième dérivée d'une fonction $a^{(0)} = a$, $a^{(k)} = \left[d, a^{(k-1)}\right]$ on a

$$d^{-r} a = \sum_{k \ge 0}(-)^k \binom{r+k-1}{k} a^{(k)}d^{-r-k}$$

$$a\, d^{-r} = \sum_{k \ge 0} \binom{r+k-1}{k} d^{-r-k}a^{(k)}$$

Désormais nous nous permettons d'être cavalier avec la limite $N \longrightarrow \infty$ et nous omettrons l'indice N. L'expression de Δ nous fournit les coefficients $w_r(\theta.)$ et on constate sans peine que

$$w_0(\theta.) = 1$$
$$w_1(\theta.) = Z^{-1}\left(-\frac{\partial Z}{\partial \theta_1}\right)$$

résultat bien connu. Plus généralement on peut écrire

$$w_r(\theta.) = \frac{1}{Z}\left(p_r\left(-\frac{\partial}{\partial \theta.}\right)Z\right)$$

où pour faire bref on a noté

$$\frac{\partial}{\partial \theta_.} \equiv \frac{\partial}{\partial \theta_1}, \frac{1}{2}\frac{\partial}{\partial \theta_2}, ..., \frac{1}{k}\frac{\partial}{\partial \theta_k}, ...$$

L'opérateur inverse W^{-1} (au sens formel) est donné par des formules analogues que l'on trouve en manipulant les identités de Plücker

$$W^{-1} = \sum_{r \geq 0} d^{-r} w_r^*(\theta_.)$$

$$w_r^*(\theta_.) = Z^{-1}\left(p_r\left(\frac{\partial}{\partial \theta_.}\right)Z\right)$$

Revenons à l'opérateur différentiel Δ_N d'ordre N et différentions le par rapport à θ_n pour $n \ll N$. Il existe alors un opérateur différentiel *d'ordre n* soit Q_n uniquement déterminé par la relation

$$\frac{\partial \Delta_N}{\partial \theta_n} = Q_n \Delta_N - \Delta_N d^n$$

En effet puisque le membre de gauche est d'ordre au plus $N - 1$ tandis que celui de droite semble a priori d'ordre $N + n$, on en déduit que

$$Q_n = \left(\Delta_N d^n \Delta_N^{-1}\right)_+ = \left(W_N d^n W_N^{-1}\right)_+$$

où pour un opérateur pseudo différentiel P le symbole $(P)_+$ désigne sa partie différentielle, et dans cette relation on peut passer à la limite $N \longrightarrow \infty$.

L'opérateur Q_n étant ainsi choisi, ce qui assure que la combinaison $Q_n \Delta_N - \Delta_N d^n$ est d'ordre inférieur à N montrons qu'elle s'identifie à $\frac{\partial \Delta_N}{\partial \theta_n}$. Pour ce faire il suffit donc d'établir que ces deux opérateurs ont la même action sur N fonctions linéairement indépendantes par exemple $f_0, ..., f_{N-1}$ qui engendrent le noyau de Δ_N, ou encore sur f, une combinaison linéaire à coefficients constants de ces fonctions, telle donc que $\Delta_N f = 0$. L'égalité annoncée sera établie si l'on montre que

$$\frac{\partial \Delta_N}{\partial \theta_n} \cdot f + \Delta_N \frac{\partial^n f}{\partial \theta_1^n} = 0$$

Or f, qui admet un développement en polynômes de Schur, satisfait comme ces derniers à $\frac{\partial^n f}{\partial \theta_1^n} = \frac{\partial f}{\partial \theta_n}$, et le membre de gauche n'est autre que $\frac{\partial}{\partial \theta_n}(\Delta_N f) = 0$.

En récrivant

$$Q_n = \left(W d^n W^{-1}\right)_+ = (L^n)_+$$

$$L = W d W^{-1} = d + O\left(d^{-1}\right)$$

les relations précédentes peuvent encore se récrire

$$\frac{\partial L}{\partial \theta_n} = [Q_n, L]$$

dont la compatibilité implique les conditions d'intégrabilité

$$\frac{\partial Q_m}{\partial \theta_n} - \frac{\partial Q_n}{\partial \theta_m} + [Q_m, Q_n] = 0$$

C'est là qu'on peut faire intervenir le fait que toutes les quantités introduites jusqu'ici sont indépendantes des θ. d'indice pair. Par exemple si

$$u = \frac{\partial^2}{\partial \theta_1^2} \ell n\, Z = \frac{\partial^2}{\partial t_0} \ell n\, Z$$

on trouve

$$\begin{aligned}
Q_2 &= d^2 + 2u \\
Q_3 &= d^3 + \frac{3}{2}(ud + du) = \left(Q_2^{3/2}\right)_+
\end{aligned}$$

tandis que l'indépendance de u par rapport à θ_2 entraîne

$$\frac{\partial Q_2}{\partial \theta_3} = [Q_3, Q_2]$$

soit traduit dans les variables t_0, t_1 la première des équations de la hiérarchie KdV

$$\frac{\partial u}{\partial t_1} = \frac{\partial}{\partial t_0}\left(\frac{1}{12}\frac{\partial^2 u}{\partial t_0^2} + \frac{1}{2}u^2\right)$$

Plus généralement on montre que $Q_n = \left(Q_2^{\frac{n}{2}}\right)_+$ et donc

$$\frac{\partial}{\partial \theta_{2p+1}} Q_2 = \left[\left(Q_2^{p+\frac{1}{2}}\right)_+, Q_2\right], \quad p \geq 1$$

qui est une forme concise de résumer la première partie du théorème.

On a déjà fait allusion à la fonction d'Airy et Z_N en est une version à argument matriciel. L'analogue de l'équation différentielle d'Airy va fournir la preuve de la seconde partie du théorème. En effet rappelons qu'on a écrit pour Λ diagonal

$$Z_N(\Lambda) = A(\Lambda)\tilde{Z}_N(\Lambda)$$

où

$$\tilde{Z}_N(\Lambda) = \int d\left(\frac{M}{(2\pi)^{1/2}}\right) \exp i\, \mathrm{tr}\left(\frac{M^3}{6} + \Lambda^2 \frac{M}{2}\right)$$

et $A(\Lambda)$ est donné explicitement ci-dessus. En écrivant que l'intégrale de la dérivée du poids exponentiel par rapport à élément diagonal de M s'annule, on obtient les relations

$$0 = \left\langle \lambda_k^2 + M_{kk}^2 + \sum_{\ell, \ell \neq k} M_{k\ell} M_{\ell k} \right\rangle$$

où les crochets désignent l'intégration avec le poids

$$d\left(\frac{M}{2\pi}\right) \exp i \, \mathrm{tr}\left(\frac{M^3}{6} + \frac{\Lambda^2 M}{2}\right)$$

Le terme $\langle \lambda_k^2 + M_{kk}^2 \rangle$ peut s'interpréter comme $\left\{ \lambda_k^2 + \left(\frac{1}{i\lambda_k}\frac{\partial}{\partial \lambda_k}\right)^2 \right\} \tilde{Z}_N(\Lambda)$. Pour les termes restant on exprime l'invariance de l'intégrale sous l'effet d'un changement infinitésimal de variable

$$M \longrightarrow M + i\epsilon[X(M), M]$$

où

$$X(M)_{ab} = \delta_{ak}\delta_{b\ell} M_{k\ell}$$

de Jacobien $1 + i\epsilon(M_{\ell\ell} - M_{kk})$ au premier ordre en ϵ. Ceci fournit la relation

$$0 = \left\langle M_{\ell\ell} - M_{kk} - \frac{i}{2}\left(\lambda_k^2 - \lambda_\ell^2\right) M_{k\ell} M_{\ell k} \right\rangle$$

et conduit à récrire les équations précédentes

$$0 = \left\langle \lambda_k^2 + M_{kk}^2 - 2i \sum_{\ell, \ell \neq k} \frac{M_{kk} - M_{\ell\ell}}{\lambda_k^2 - \lambda_\ell^2} \right\rangle$$

soit un ensemble d'équations du second ordre couplées

$$\left\{ \lambda_k^2 - \left(\frac{1}{\lambda_k}\frac{\partial}{\partial \lambda_k}\right)^2 - 2 \sum_{\ell, \ell \neq k} \frac{1}{\lambda_k^2 - \lambda_\ell^2} \left(\frac{1}{\lambda_k}\frac{\partial}{\partial \lambda_k} - \frac{1}{\lambda_\ell}\frac{\partial}{\partial \lambda_\ell}\right) \right\} \tilde{Z}_N(\Lambda) = 0$$

Transformant ces équations en équations pour $Z_N(\Lambda)$ à l'aide du facteur A explicite, développant en λ_k^{-1} à $\lambda_k \longrightarrow \infty$, faisant tendre $N \longrightarrow \infty$, on trouve la série

$$\sum_{m \geq -1} \frac{1}{(\lambda_k^2)^{2m+1}} L_m Z(t_\cdot) = 0$$

où les opérateurs L_m ont la forme indiquée dans l'énoncé du théorème. Bien entendu la relation précédente conduit à annuler séparément chaque coefficient $L_m Z(t.)$ achevant la démonstration (on vérifie sans peine les règles de commutation des L_m).

Il nous reste à conclure par quelques calculs effectifs de nombres d'intersection. Pour cela il est commode de développer $\log Z = F = \sum_{g \geq 0} F_g$, en contributions de genre donné. De même

$$u = \frac{\partial^2 F}{\partial t_0^2} = \sum_{g \geq 0} u_g$$

Posons

$$I_k(x, t.) = \sum_{p \geq 0} t_{k+p} \frac{x^p}{p!}$$

Appliquant la définition de F_g et le théorème de Witten et Kontsevich on trouve que $u_0(t.)$ est déterminée par l'équation de point fixe

$$u_0(t.) = I_0 (u_0, t.)$$

Par inversion de Lagrange on obtient u_0 puis par une double intégration F_0, ce qui fournit en genre zéro les nombres d'intersection

$$\langle \tau_{d_1} ... \tau_{d_n} \rangle = \frac{(\sum d_i)!}{d_1! ... d_n!} \quad , \qquad \sum_{1 \leq i \leq n} d_i = n - 3$$

qui sont dans ce cas des entiers positifs. Posant désormais

$$J_k(t.) = I_k (u_0(t.), t.)$$

on obtient en genre 1

$$F_1(t.) = \frac{1}{24} \ell n \frac{1}{1 - J_1(t.)} = \frac{1}{24} \ell n \left(\frac{\partial u_0}{\partial t_0} \right)$$

Plus généralement si $g > 1$ on trouve que F_g est donnée par une *somme finie*

$$F_g = \sum_{\sum_{2 \leq k \leq 2g-2} (k-1)\ell_k = 3g-3} \left\langle \tau_2^{\ell_2} ... \tau_{3g-2}^{\ell_{3g-2}} \right\rangle$$
$$\times \frac{1}{(1 - J_1)^{2(g-1) + \Sigma \, \ell_p}} \frac{J_2^{\ell_2}}{\ell_2!} \frac{J_3^{\ell_3}}{\ell_3!} ... \frac{J_{3g-2}^{\ell_{3g-2}}}{\ell_{3g-2}!}$$

où les coefficients sont eux mêmes des nombres d'intersection de genre g, déterminés par le théorème. Par exemple

$$F_2 = \frac{1}{5760} \left[\frac{5J_4}{(1 - J_1)^3} + 29 \frac{J_2 J_3}{(1 - J_1)^4} + 28 \frac{J_2^3}{(1 - J_1)^5} \right]$$

d'où

$$\langle \tau_4 \rangle = \frac{1}{1152} \ , \quad \langle \tau_2 \tau_3 \rangle = \frac{29}{5760} \ , \quad \langle \tau_2^3 \rangle = \frac{7}{240}$$

et ainsi de suite. De même on peut resommer les termes ayant au dénominateur la plus grande puissance de $1 - J_1$ à l'aide d'une équation de Painlevé, puis successivement ceux impliquant une puissance inférieure par une méthode perturbative.

Citons encore la propriété suivante des intégrales de Kontsevich, conjecturée initialement par Witten

Théorème (Witten, DF-I-Z). Il existe une application biunivoque φ des polynômes dans une infinité de variables

$$P \longmapsto \varphi P$$

telle que

$$\frac{\int dM \exp \operatorname{tr} \left\{ i \frac{M^3}{6} - \frac{\Lambda M^2}{2} \right\} P \left(\operatorname{tr} M, \ \operatorname{tr} M^3, \ \operatorname{tr} M^5, ... \right)}{\int dM \exp - \operatorname{tr} \frac{\Lambda M^2}{2}}$$
$$= \varphi P \left(\frac{\partial}{\partial t_0}, \frac{\partial}{\partial t_1}, ... \right) Z(t.)$$

où $t. \equiv t. \left(\Lambda^{-1} \right)$ et le membre de gauche doit être entendu au sens des séries formelles dans la limite $N \longrightarrow \infty$.

Mentionnons enfin qu'il existe une généralisation du problème d'intersection impliquant des recouvrements finis des espaces de module où l'algèbre de Virasoro apparaît comme sous-algèbre de structures plus complexes appelées (par les physiciens) algèbres-W.

Références

Pour la première partie

M. Asbacher, R. Guralnick. Some applications of the first cohomohology group, *J. Alg.* **90** (1984), 446-460.

G.V. Belyi, "On Galois Extensions of a Maximal Cyclotomic Field"*Math. USSR Izvestija* Vol.14 (1980) 247-256.

Marston Conder, "Hurwitz Groups: A Brief Survey", *Bulletin Am. Math. Soc.* (New series) **23** (1990) 359-370.

H.S.M. Coxeter, W.O.J. Moser, "Generators and Relations for Discrete Groups" (second edition) Springer (1965) Chap.8.

A. Grothendieck, "Esquisse d'un Programme"

Y. Ihara, "Braids, Galois Groups, and Some Arithmetic Functions", Proceedings of the International Congress of Mathematicians, Kyoto (1990) 99, 120.

"Galois Groups over **Q**", édité par Y. Ihara, K. Ribet et J.P. Serre, *Publ. MSRI* n°16 (1989) Springer.

Yu.I. Manin, "Reflections on arithmetical Physics " in "Conformal Invariance and String Theory" Academic Press, N.Y. (1989), pages 293-303.

D. Mumford, "Towards an Enumerative Geometry of the Moduli Space of Curves" in Arithmetics and Geometry Vol.II, Birkhaüser, Boston (1983).

J. Oesterlé, contribution au Colloque en hommage à P. Cartier, 54ème Rencontre de Strasbourg (1992).

G. Ringel, "Map Color Theorem" Springer New York (1974).

T.L. Saaty, P.C. Kainen, "The Four Colour Problem" Mc Graw Hill, New York (1977).

G.B. Shabat and V.A. Voevodsky, "Drawing Curves over Number Fields" in "A Grothendieck Festchrift" Birkhaüser.

J.P. Serre, "Groupes de Galois sur **Q**", Séminaire Bourbaki, Astérisque **161-162** (1988) 73-85;

——, "Representations Linéaires des Groupes Finis", Hermann, Paris (1967).

——, Topics in Galois Theory, Jones and Bartlett, 1992.

Dirk-Jan Smit, "Summations over Equilaterally Triangulated Surfaces and the Critical String Measure" *Comm. Math. Phys.* **143** (1992) 253-285.

R. Steinberg, Generators for simple groups, *Can. Math. J.* **14** (1962), 277-283.

V.A. Voevodsky and G. Shabat, "Equilateral Triangulations of Riemann Surfaces and Curves over Algebraic Number Fields" *Soviet Math. Dokl.* **39** (1989) 38-41.

J. Wolfart, "Mirror Invariant Triangulations of Riemann Surfaces, Triangle Groups and Grothendieck Dessins: Variations on a Thema of Belyi" Mathematisches Seminar der Universität, Frankfurt (1992).

Pour la seconde partie

D. Bessis, C. Itzykson, J.-B. Zuber, "Quantum Field Theory Techniques in Graphical Enumeration", *Adv. Appl. Math.* **1** (1980) 109-157.

P. Di Francesco, C. Itzykson, J.-B. Zuber, "Polynomial Averages in the Kontsevich Model" *Comm. Math. Phys.* **151** (1993) 193-219.

I.M. Gelfand, L.A. Dikii, "Asymptotic Behaviour of the Resolvent of Sturm-Liouville Equations and the Algebra of the Korteweg-De Vries Equations", *Russian Math. Surveys* **30** (1975) 77-113.

J. Harer, D. Zagier, "The Euler Characteristic of the Moduli Space of Curves", *Inv. Math.* **85** (1986) 457-485.

Harish-Chandra, "Differential Operators on a Semi-Simple Lie Algebra", *Am. J. Math.* **79** (1957) 87-120.

C. Itzykson, J.-B. Zuber, "The Planar Approximation II", *J. Math. Phys.* **21** (1980) 411-421.

——, "Matrix Integration and Combinatorics of Modular Groups" *Comm. Math. Phys.* **134** (1990) 197-207.

——, "Combinatorics of the Modular Group II, the Kontsevich Integrals", *Int. Journ. Mod. Phys.* **A7** (1992) 5661-5705.

M. Kontsevich, "Intersection Theory on the Moduli Space of Curves" *Funct. Anal. and Appl.* **25** (1991) 123-128.

——, "Intersection Theory on the Moduli Space of Curves and the Matrix Airy Function", *Comm. Math. Phys.* **147** (1992) 1-23.

R.C. Penner, "The Decorated Teichmüller Space of Punctured Surfaces" *Comm. Math. Phys.* **113** (1987) 299-339.

——, "Perturbative Series and the Moduli Space of Punctured Surfaces", *J. Diff. Geom.* **27** (1988) 35-53.

M. Sato, "Soliton Equations as Dynamical Systems on an Infinite Dimensional Grassmann Manifold" *RIMS Kokyuroku* **439** (1981) 30-46.

——, "The KP Hierarchy and Infinite Dimensional Grassmann Manifolds" *Proc. Symp. Pure Math.* **49** (1989) 51-66.

M. Sato, Y. Sato, "Soliton Equations as Dynamical Systems on an Infinite Dimensional Grassmann Manifold" Lecture Notes in *Num. Appl. Anal.* **5** (1982) 259-271.

E. Witten, "Two Dimensional Gravity and Intersection Theory on Moduli Space" *Surveys in Diff. Geom.* **1** (1991) 243-310.

——, "The N-Matrix Model and Gauged WZW Models " pretirage IAS, HEP-91/26.

*Service de Physique Théorique, † Centre d'Etudes de Saclay, 91191 Gif-sur-Yvette Cedex, France

---

† Laboratoire de la Direction des Sciences de la Matière du Commissariat à l'Energie Atomique

Dessins d'enfants and Shimura varieties

Paula Beazley Cohen[*]

In this article we present joint work in preparation with **C.Itzykson** and **J.Wolfart** [CoItWo]. Certain of the main ideas have been announced in [CoWo2]. Some joint work in preparation with **H.Shiga** and J.Wolfart will also be mentioned.

Let C be a smooth projective algebraic curve defined over a number field and $\beta : C \to \mathbf{P}_1(\mathbf{C})$ a Belyi function ramified over $0, 1, \infty$. We call (C, β) a Belyi pair. If $\mathfrak{H} = \mathfrak{H}^+$ and $\mathfrak{H}^-$ denote respectively the upper and lower half planes, then the connected components of $\beta^{-1}(\mathfrak{H}^+)$ and $\beta^{-1}(\mathfrak{H}^-)$ are the open cells of a triangulation of C with vertices the elements of $\beta^{-1}\{0, 1, \infty\}$. Let p, q, r be positive common multiples of the respective orders of ramification of β over $0, 1, \infty$ with $(\frac{1}{p} + \frac{1}{q} + \frac{1}{r}) < 1$. Generalising a discussion of Shabat and Voevodsky [ShVo], one can recover the above triangulation of C from a tesselation of $\mathfrak{H}$ by copies of a hyperbolic triangle T with angles $\frac{\pi}{p}, \frac{\pi}{q}, \frac{\pi}{r}$, as shown by Itzykson and Wolfart [Wo2] and explained in our joint article [CoItWo] (see also [CoWo2]). Indeed, let $\Delta = \Delta(p, q, r)$ be the Fuchsian triangle group of signature (p, q, r) consisting of the words of even length in the reflections in the sides of T. This group has presentation,

$$\Delta = < M_1, M_2, M_3 \mid M_1^p = M_2^q = M_3^r = M_1 M_2 M_3 = 1 > .$$

If p, q, r are all finite, then the group Δ is cocompact and the quotient space $\Delta \backslash \mathfrak{H}$ is $\mathbf{P}_1(\mathbf{C})$. The suitably normalised inverse of a Schwarz triangle function for T can be analytically continued to a Δ−automorphic function

$$j : \mathfrak{H} \to \mathbf{P}_1(\mathbf{C})$$

with ramification orders over $0, 1, \infty$ exactly given by p, q, r. For example, for $(p, q, r) = (3, 2, \infty)$ the group Δ is $\mathrm{PSL}_2(\mathbf{Z})$ (not cocompact) and $j = j(\tau)$ is the elliptic modular function. On choosing a local branch of $\beta^{-1} \circ j$ that one continues analytically, and on applying the monodromy theorem, one sees that j factorises as $j = \beta \circ \Phi$ where Φ determines a (possibly) ramified covering of C by $\mathfrak{H}$ depending on β and Δ. We call (C, β, Δ) a Belyi triple.

The images under Φ of the vertices of the copies of T tesselating $\mathfrak{H}$, which are the fixed points of Δ, comprise exactly the set $\beta^{-1}\{0, 1, \infty\}$. The associated dessin is given by the pre-image $\beta^{-1}[0, 1]$. If at each element of $\beta^{-1}\{1\}$ the order of ramification is assumed to be equal to 2 one can, following Grothendieck, Shabat, Voevodsky [Gr],[VoSh],[ShVo], consider the "clean dessin" G on the curve C given by $\beta^{-1}[0, 1]$. In this case we have $q = 2$. Denoting by G^* the dual dessin, the set of vertices of $G \cup G^*$ is $\beta^{-1}\{0, 1, \infty\}$.

We have,

Proposition : *Let* $(\mathcal{C}, \beta, \Delta)$ *be a Belyi triple. There exists a holomorphic map,*

$$\Phi : \mathfrak{H} \to \mathcal{C}$$

ramified at most over $\beta^{-1}\{0, 1, \infty\}$. *The covering group* H *of* Φ *is a subgroup of finite index in* Δ. *We therefore recover* $\mathcal{C}$ *as the quotient* $H\backslash\mathfrak{H}$.

Remarks : **1)** One may apply the above construction with $\Delta = \Delta(p, q, r)$ replaced by $\Delta_M = \Delta(M, M, M)$, $M = \text{l.c.m.}(p, q, r)$. The Fermat curve F_M has affine equation $x^M + y^M = 1$ and universal covering group $[\Delta_M, \Delta_M]$, $(M > 3)$. We can find in this way a common finite ramified cover of $\mathcal{C}$ and F_M. **2)** The map Φ is the universal covering map if and only if p, q, r are the precise ramification numbers of β in every point of $\beta^{-1}(0), \beta^{-1}(1), \beta^{-1}(\infty)$ respectively. **3)** In the particular case where the ramifications of β over 1 are all equal to 2, one can choose $q = 2$ and define β and H by means of the cartographic group of Grothendieck [Gr],[ShVo]. **4)** The above discussion goes through equally well for non-cocompact triangle groups. Indeed, with $p = q = r = \infty$ we recover the known result that every smooth projective curve defined over $\overline{\mathbb{Q}}$ can be represented as a compactified quotient $\overline{H\backslash\mathfrak{H}}$ with subgroup H of finite index in the principal subgroup of level 2 in $\text{PSL}_2(\mathbb{Z})$. Yet, the vertices of the triangulation obtained in the cocompact case play a very important role, as we shall now see.

Namely, using a method developed by Wolfart and myself in another context, the Proposition enables us to show,

Theorem : *Let* $(\mathcal{C}, \beta, \Delta)$ *be a Belyi triple. The associated covering group* H *determines an arithmetic group* Γ *acting on* $\mathfrak{H}^t$ *for some positive integer* t. *The quotient* $V = \Gamma\backslash\mathfrak{H}^t$ *is a Shimura variety parameterising isomorphism classes of polarised abelian varieties whose endomorphism algebrae contain a certain complex representation of a subfield* $\mathbf{K}$ *of the cyclotomic field* $\mathbb{Q}(exp(\frac{\pi i}{l}))$, $l = \text{l.c.m.}(p, q, r)$. *There exists an analytic injection*

$$\mathfrak{H} \hookrightarrow \mathfrak{H}^t$$

and a compatible group inclusion

$$H \hookrightarrow \Gamma$$

inducing by passage to the quotient a non-trivial morphism over $\overline{\mathbb{Q}}$,

$$\Psi : \mathcal{C} \to V$$

which sends the elements of $\beta^{-1}\{0, 1, \infty\}$ *onto points of complex multiplication by* $\mathbf{K}$.

For Grothendieck dessins G the morphism Ψ, therefore, sends the vertices of $G \cup G^*$ onto complex multiplication (CM) points.

The simplest example of a Shimura variety is $\mathbf{P}_1(\mathbf{C})$ which via the elliptic modular function $j = j(\tau) : \mathfrak{H} \to \mathbf{P}_1(\mathbf{C})$ parametrises the isomorphism classes of elliptic curves $\mathcal{E}_\tau$. The field contained in all the $\mathrm{End}_0(\mathcal{E}_\tau) = \mathrm{End}(\mathcal{E}_\tau) \otimes_{\mathbf{Z}} \mathbf{Q}$ is here just $\mathbf{Q}$. The CM points are those $\tau \in \mathfrak{H}$ for which $\mathrm{End}_0(\mathcal{E}_\tau)$ strictly contains $\mathbf{Q}$ and is therefore an imaginary quadratic field. In particular, at the fixed points $\tau = i, \frac{1}{2}(1 + i\sqrt{3})$ of $\mathrm{PSL}_2(\mathbf{Z})$, which are roots of unity, the corresponding elliptic curves have (special) CM.

For our purposes we have therefore to associate to a Belyi triple $(\mathcal{C}, \beta, \Delta)$ a (sub-)family of abelian varieties whose endomorphism algebrae have some complex multiplications. At the fixed points of Δ the corresponding abelian variety will have a stronger property (CM) : up to isogeny it factorises into a product of powers of simple abelian varieties B with $\mathrm{End}_0(B)$ a CM field of degree $2\dim(B)$ over $\mathbf{Q}$. A CM field is a totally imaginary quadratic extension of a totally real number field.

In order to provide some additional motivation for the Theorem, it is instructive to briefly explain how Wolfart and myself were able to apply what we call "the modular embedding" of $(\mathfrak{H}, H)$ into $(\mathfrak{H}^t, \Gamma)$ to a concrete problem in transcendence theory which preceded by some years our association of it to Belyi's work and has also featured in more recent work. Little is known about the transcendence properties of suitably normalised automorphic functions and their derivatives at algebraic points, except when these functions provide uniformisations of commutative algebraic groups defined over $\bar{\mathbf{Q}}$. There one has generalisations of the Hermite-Lindemann theorem that e^α is transcendental for $\alpha \in \bar{\mathbf{Q}}, \alpha \neq 0$. A theorem of Schneider [Schn] on the elliptic modular function $j = j(\tau)$ says that both $\alpha \in \mathfrak{H}$ and $j(\alpha)$ are algebraic if and only if the elliptic curve $\mathcal{E}_\alpha$ has CM. Extending some work of H.Shiga [Sh], in joint work with him and J. Wolfart [CoSWo] we have proved the analogous result for the automorphic functions defined over $\bar{\mathbf{Q}}$ coming from the existence of canonical models of Shimura varieties parametrising polarised abelian varieties whose endomorphism algebrae contain a given endomorphism type. In particular, we provide a less awkward proof of a result announced in [Sh] that if A is an n-dimensional abelian variety defined over $\bar{\mathbf{Q}}$ with modulus τ in the Siegel upper half plane $\mathfrak{H}_n$ of degree n, then the matrix τ has all its coefficients algebraic if and only if A has CM. In [CoSWo] these and other complex multiplication criteria come from observations about linear independence over $\bar{\mathbf{Q}}$ of periods of differential forms of the first kind on abelian varieties. Results of Shimura [Shi] and Siegel [Sie] also show that the derivatives of certain modular functions are transcendental at CM points (see also [Wo1]).

For suitably normalised functions j automorphic with respect to a Fuchsian triangle group Δ (which include the elliptic modular function) (see also [Sh]) the techniques of the proof of the Theorem combined with the above results of [CoSWo] enable one to show that $\alpha \in \mathfrak{H}$ and $j(\alpha)$ are both algebraic if and only if α corresponds, via the modular embedding, to an abelian variety with CM. For cocompact Δ (special) CM points (in a cyclotomic field; compare to $\text{PSL}_2(\mathbf{Z})$) occur at the elliptic fixed points of Δ. For these vertices α one knows that $j'(\alpha)$ is transcendental. This was shown by J. Wolfart and G. Wüstholz [WW] in 1985. They found that these $j'(\alpha)$ are ratios of a period of the first kind by a period of the second kind on abelian varieties with CM having nothing to do with the only abelian varieties immediately to hand, namely the Jacobians of curves of the form $H\backslash\mathfrak{H}$ for H a subgroup of Δ (and with H acting without fixed points). Wolfart and myself developed the modular embedding technique in 1987/8 precisely to explain this phenomenon : these abelian varieties occur at the images of the fixed points under the modular embedding of the Theorem!

Some comments about the proof of the Theorem. Several different constructions of the modular embedding determined by a subgroup H of finite index in a Fuchsian triangle group Δ of signature (p, q, r) are detailed in [CoWo1] and the marriage with the Proposition is made in [CoWo2]. To capture the essence of why a family of abelian varieties with some complex multiplications by a subfield $\mathbf{K}$ of a cyclotomic field arises, observe that the triangulation of C with vertices $\beta^{-1}\{0, 1, \infty\}$ arises from the image under Φ of the fundamental domain for H, which is just finitely many copies of a hyperbolic triangle $T^{(1)}$ with angles $\frac{\pi}{p}, \frac{\pi}{q}, \frac{\pi}{r}$. This triangle $T^{(1)}$ (minus its vertices) can be obtained as the image loci of the $x \in \mathbf{P}_1(\mathbf{C}) - \{0, 1, \infty\}$ under the ratio

$$\left(\int_0^x \omega^{(1)} \right) \Big/ \left(\int_1^\infty \omega^{(1)} \right)$$

where

$$\omega^{(1)} = \omega^{(1)}(x) = u^{-\mu_0}(u-1)^{-\mu_1}(u-x)^{-\mu_2} du$$

and $\mu_0 = \frac{1}{2}(1 - \frac{1}{p} + \frac{1}{q} + \frac{1}{r}), \mu_1 = \frac{1}{2}(1 + \frac{1}{p} + \frac{1}{q} - \frac{1}{r}), \mu_2 = \frac{1}{2}(1 - \frac{1}{p} - \frac{1}{q} - \frac{1}{r})$. Let l be the least common multiple of the denominators of μ_0, μ_1, μ_2 and let $\mu_3 = 2 - (\mu_0 + \mu_1 + \mu_2)$. The smooth projective curve $X_l(x)$ with (singular) affine model

$$w^l = u^{l\mu_0}(u-1)^{l\mu_1}(u-x)^{l\mu_2}, \qquad (*)$$

on which $\omega^{(1)}$ is a differential of the first kind, has an automorphism $(u, w) \mapsto (u, \zeta^{-1}w), \zeta = exp(\frac{2\pi i}{l})$, so that $\text{End}_o(B(x))$ contains $\mathbf{K} = \mathbf{Q}(\zeta)$ where $B(x)$ is the abelian variety of dimension $\phi(l)$ in

$$Jac(X_l(x)) = B(x) \bigoplus \sum_{d|l, d\neq l} Jac(X_d(x)).$$

(For the definition of $X_d(x)$, replace w^l by w^d in the equation (*)). We say that $B(x)$ has generalised CM by $\mathbf{K}$. The induced action of $\mathbf{K}$ on $H^0(B(x),\Omega)$ induces a decomposition of the latter into $\phi(l)$ eigenspaces where this action is multiplication by $\sigma_\nu(\mathbf{K}), \nu \in (\mathbf{Z}/l\mathbf{Z})^*, \sigma_\nu : \zeta \mapsto \zeta^\nu$. This eigenspace has dimension

$$r_\nu =: -1 + \{\nu\mu_0\} + \{\nu\mu_1\} + \{\nu\mu_2\} + \{\nu\mu_3\}$$

where $0 \leq \{a\} < 1, a \in \mathbf{Q}$, denotes the fractional part of a. The Shimura variety V of the Theorem parametrises isomorphism classes of (principally polarised) abelian varieties A of dimension $\phi(l)$ whose endomorphism algebra contains $\mathbf{K}$ represented on $H^0(A,\Omega)$ as

$$\sum_{\nu \in (\mathbf{Z}/l\mathbf{Z})^*} r_\nu \sigma_\nu.$$

The dimension of V is $\frac{1}{2}\sum_{\nu \in (\mathbf{Z}/l\mathbf{Z})^*} r_\nu r_{-\nu}$. As $r_\nu + r_{-\nu} = 2$, the variety V has dimension $t = \frac{1}{2}\mathrm{Card}R$ where $R = \{\nu \in (\mathbf{Z}/l\mathbf{Z})^* \mid r_\nu = 1\}$. This is the integer t of the Theorem. By results of Takeuchi [T], the group algebra $k[H]$ is a quaternion algebra where k is the trace field $\mathbf{Q}(\mathrm{tr}H)$ with $\mathrm{tr}H = \{\mathrm{tr}(\gamma) \mid \gamma \in H\}$. The group Γ is the norm unit group of this quaternion algebra, and one has $V = \Gamma\backslash\mathfrak{H}^t$. The component of the modular group embedding for $\nu \in R$, defined on an appropriate conjugate of Δ in $SL_2(\mathbf{R})$, is given by the Galois embedding $\Delta \to \Delta^{\sigma_\nu}$ and the compatible modular space embedding by the analytic continuation of the map $T^{(1)} \to T^{(\nu)}$ where $T^{(\nu)}$ is the triangle with angles the multiples of π :

$$\mid \{\nu\mu_0\} + \{\nu\mu_2\} - 1 \mid, \mid \{\nu\mu_1\} + \{\nu\mu_2\} - 1 \mid, \mid \{\nu\mu_2\} + \{\nu\mu_3\} - 1 \mid .$$

To understand how one obtains CM points from fixed points, suppose $(\mu_0 + \mu_2) < 1$ and consider the point $x = 0$. Then the curve $X_l(0)$ with affine model

$$w^l = u^{l(\mu_0+\mu_2)}(u-1)^{l\mu_1}$$

is an image of the Fermat curve of degree l. The abelian variety $B(0)$ splits into a product of abelian varieties of dimension $\phi(l)/2$ with CM by $\mathbf{K}$ as was detailed by N.Koblitz and D.Rohrlich [KR].

These comments are enough to give the spirit of the proof of the Theorem which is given in detail in the references stated above. The triangle groups Δ are the monodromy groups of the Gauss hypergeometric differential equations. The modular embedding technique has been extended by Wolfart and myself [CoWo3] to the hypergeometric functions of several variables which are solutions of a system of partial differential equations with

242 Paula Beazley Cohen

monodromy groups that have been studied by Picard, Terada, Deligne and Mostow. As an application, we have completely classified in [CoWo4] the algebraic Appell-Lauricella functions in several variables, so extending the famous list of H.A.Schwarz of the algebraic Gauss hypergeometric functions.

References

[CoItWo] P.B. Cohen, C. Itzykson, J. Wolfart : Fuchsian triangle groups and Grothendieck dessins : Variations on a theme of Belyi, *to appear*.

[CoSWo] P.B. Cohen, H. Shiga, J. Wolfart : Criterea for complex multiplication and transcendence properties of automorphic functions, to appear.

[CoWo1] P. Cohen, J. Wolfart : Modular Embeddings for some non-arithmetic Fuchsian groups, Acta Arithmetica 56, 1990, 93-110.

[CoWo2] P.B. Cohen, J. Wolfart : Dessins de Grothendieck et variétés de Shimura, C.R.Acad.Sci.Paris,t.315,Série I,p.1025-1028,1992.

[CoWo3] P.B. Cohen, J. Wolfart : Fonctions hypergéométriques en plusieurs variables et espaces des modules de variétés abéliennes, to appear in Annales Sci. de l'Ecole Normale Supérieure.

[CoWo4] P.B. Cohen, J. Wolfart : Algebraic Appell-Lauricella functions, Analysis 12,359-376 1992.

[Gr] A. Grothendieck : Esquisse d'un programme, 1984, unpublished.

[KR] N. Koblitz, D. Rohrlich : Simple factors in the Jacobian of a Fermat curve, Can. J. Math. 30 1183-1205 1978.

[Schn] Th. Schneider : Arithmetische Unterschungen elliptischer Integrale, Math. Ann. 113 1-13 1937

[Sh] H. Shiga : On the transcendency of the values of the modular function at algebraic points, S.M.F. Astérisque 209*∗ (1992).

[Shi] G. Shimura : Automorphic forms and the periods of abelian varieties, J. Math. Soc. Japan 31,561-592 1979.

[Sie] C.L. Siegel : Bestimmung der elliptischen Modulfunktion durch eine Transformationgleichung, Abh. Math. Sem. Hamburg 27,32-38 1964.

[ShVo] G.B. Shabat, V.A. Voevodsky : Drawing curves over number fields, pp. 199-227 in the Grothendieck Festschrift Vol III, ed. P. Cartier et al., Progress in Math. 88, Birkhäuser 1990.

[T]   K. Takeuchi : Arithmetic triangle groups, J. Math. Soc. Japan
        29, 91-106 1977.

[VoSh] V.A. Voevodsky, G.B. Shabat : Equilateral triangulations of
        Riemann surfaces, and curves over algebraic number fields, Soviet
        Math. Dokl. **39**, 1989, No.1, 38-41.

[Wo1]   J. Wolfart : Taylorenwicklungen automorpher Formen und ein
        Transzendenzproblem aus der Uniformisierungstheorie, Abh. Math.
        Sem. Univ. Hamburg 54, 25-33 1984.

[Wo2]   J. Wolfart : Mirror-invariant triangulations of Riemann surfaces,
        triangle groups and Grothendieck dessins :Variations on a thema
        of Belyi, Preprint of the Dept. Math. Joh. Wolf. Goethe-Univers.,
        Frank./Main.

[Wo3]   J. Wolfart : Eine arithmetische Eigenschaft automorpher Formen
        zu gewissen nicht-arithmetische Gruppen, Math. Ann. 262, 1-21
        1983.

[WW]   J. Wolfart, G. Wüstholz : Der Überlagerungs radius gewisser al-
        gebraischer Kurven und die Werte der Betafunktion an rationalen
        Stellen, Math. Ann. 273, 1-15 1985.

*Collège de France, 3 rue d'Ulm, 75005 Paris, France

Horizontal divisors on arithmetic surfaces
associated with Belyi uniformizations

Yasutaka Ihara*

For finite surjective morphisms $f : Y \to X$ between some *arithmetic* surfaces and *horizontal* divisors D on X, we consider questions related to connectedness of $f^{-1}(D)$. The results will then be applied to fundamental groups of related surfaces. This article owes much to Harbater's work [Hb] (his lemma cited in §1 plays a key role), and contains an appendix on a proof by T. Saito.

By an arithmetic surface, we mean any two dimensional integral scheme having structure of a proper flat $\mathfrak{D}$-scheme, where $\mathfrak{D}$ is the ring of integers of a number field k (the dimension relative to $\mathfrak{D}$ is 1). Horizontal divisors are effective Cartier divisors that are finite over $\mathfrak{D}$. Let us begin by describing some special examples. First, if $\mathbf{P}^1_{\mathbf{Z}}$ is the projective line over $\mathbf{Z}$, $f : \mathbf{P}^1_{\mathbf{Z}} \to \mathbf{P}^1_{\mathbf{Z}}$ is defined by $y \to y^N = x$ ($N \geq 1$), and D is defined by $x = 1$, then $f^{-1}(D) \simeq \mathrm{Spec}\,(\mathbf{Z}[y]/(y^N - 1))$ is connected, being the spectrum of the ring of virtual characters of a finite group ($\simeq \mathbf{Z}/N$ in this case; cf [S] 11·4). Each irreducible component of $f^{-1}(D)$ meets some other components on the special fibres $\mathbf{P}^1_{\mathbf{Z}} \otimes \mathbf{F}_p$ at $p|N$, to make $f^{-1}(D)$ connected. This remains valid if $\mathbf{Z}$ is replaced by any $\mathfrak{D}$. Secondly, if $f : Y \to X$ is *everywhere etale* and D is normal, then distinct irreducible components of $f^{-1}(D)$ cannot meet each other (cf. e.g. [G] Exp I, Cor 9·11). As these examples show, when $f^{-1}(D)$ splits into the union of several irreducible components, the connectedness of $f^{-1}(D)$ is closely related to ramifications of f at special fibres (vertical prime divisors) of Y. In a sense, each connectedness assertion for such a divisor gives a "horizontal connection" of information on such ramifications. Its delicacy is best explained in the Example in §2.

The main results proved in this note are as follows. Let $X = \mathbf{P}^1_{\mathfrak{D}}$ be the projective t-line over $\mathfrak{D}$ ($\mathfrak{D}, k$ being as above), $L/k(t)$ be a finite extension and $f : Y \to X$ be the integral closure of X in L. The subscript k will denote the effect of $\otimes k$. For example, $f_k : Y_k \to X_k$ denotes the finite branched covering of $X_k = \mathbf{P}^1_k$ corresponding to the field extension $L/k(t)$. For $a \in k^\cup(\infty)$, denote by D_a the prime divisor on X defined by $t = a$. It is the closure on X of the point of X_k defined by $t = a$ (considered as a reduced closed subscheme of X).

Theorem A (Cor 1 of Th 1; §2). *If f_k is unramified outside $t = b_1, \ldots, b_r$ with $b_1, \ldots, b_r \in \mathbf{Z}^{\cup}(\infty)$, and $a \in \mathbf{Q}$, then $f^{-1}(D_a)$ is connected.*

The case of modular curves $(b_1, b_2, b_3 = 0, 12^3, \infty)$ will be discussed in more detail (§2). $f^{-1}(D_\infty)$ is sometimes disconnected.

Theorem B (Cor 2, 3 of Th 1; §3). *Assume that f_k is unramified outside $t = 0, 1, \infty$. Then: (i) If $a = 0, 1, \infty, f^{-1}(D_a)$ is connected; (ii) if $a \in \mathbf{Q}, f^{-1}(D_a)$ is again connected; (iii) there exists $\mathfrak{O}$ and $a \in \mathfrak{O}$, such that $a, 1-a$ are both units of $\mathfrak{O}$ (so that D_a does not meet $D_0^{\cup} D_1^{\cup} D_\infty$), and that $f^{-1}(D_a)$ is connected for any f.*

As applications, we obtain, for example:

Theorem C (T. Saito) (Cor 1 of Th 2; §4).

$$\pi_1(\mathbf{P}^1_{\mathfrak{O}} - D_0^{\cup} D_1^{\cup} D_\infty) \simeq \pi_1(\text{Spec } \mathfrak{O}).$$

This is a direct application of Th B (iii). Saito's original proof is quite different (see §4 and the Appendix). Applying Th B (i) for $f \circ \varphi$ in place of f, where $\varphi : Y' \to Y$ is any finite etale connected covering, we obtain:

Theorem D (Cor 2 of Th 2; §4). *If f_k is unramified outside $t = 0, 1, \infty$ and totally ramified at one of $0, 1, \infty$, then $\pi_1(Y) \simeq \pi_1(\text{Spec } \mathfrak{O})$.*

According to Belyi [B](Th 4 and its proof), every algebraic function field of one variable L over a number field k contains an element t such that $L/k(t)$ is unramified outside $t = 0, 1, \infty$. Moreover, when L has a prime divisor of degree 1 over k, t can be so chosen that $L/k(t)$ is totally ramified at $t = \infty$. So, Theorem D implies that *every arithmetic surface over $\mathfrak{O}$ having a section over $\mathfrak{O}$ has a normal model Y such that $\pi_1(Y) \simeq \pi_1(\text{Spec } \mathfrak{O})$.*

In §1, we shall prove a criterion for connectedness of $f^{-1}(D)$ when $X = \mathbf{P}^1_{\mathfrak{O}}$ (Theorem 1). This is a direct consequence of Harbater's criterion [Hb] for an algebraic function given as power series over $\mathfrak{O}$ to be rational (a modification of Dwork's criterion). Logically, this is just a simple remark. But the author could not find a reference with an explicit statement on this connection, and so he thought it necessary that it be presented. We note here that in the *geometric* cases (geometric surfaces, etc.), the connectedness of $f^{-1}(D)$ was established under some mild conditions (such as $(D^2) > 0$) in Hironaka-Matsumura [H-M] cf. also [Ht]. There, the main point was the extendability of any formal-rational function on the completion of X along D to a global rational function on X. In our arithmetic case, one must also take care of neighborhoods of D above *archimedean* places of $\mathfrak{O}$ which is the role of archimedean radii of convergence appearing in the criterion.

In §§2-3, we restrict ourselves to the case where only $t = b_i \in \mathbf{Z}^{\cup}(\infty)$ $(1 \leq i \leq r)$ can be ramified in f_k, and obtain Cor 1 ~ 3 of Theorem 1 using suitable coordinate transformations.

In §4, we prove Theorem 2 and its corollaries as direct applications of §3.

The next problems would be (i) to find out whether Theorem 1 extends to more general arithmetic surfaces and a full arithmetic analogue of Hironaka-Matsumura criterion can be described using an appropriate Arakelov type theory, and (ii) to give further applications to fundamental groups of arithmetic surfaces. We hope to be able to discuss these problems more concretely in the near future.

The author wishes to thank Kyoji Saito and Takeshi Saito for helpful discussions.

§1. In what follows, k will denote an algebraic number field, $\mathfrak{O}$ the ring of integers of k, and Σ the set of all distinct embeddings $\sigma : k \hookrightarrow \mathbf{C}$. We denote by $K = k(t)$ the rational function field of one variable, and by L/K a finite extension which may contain constant field extensions. Let $X = \mathbf{P}_{\mathfrak{O}}^1 = \operatorname{Spec} \mathfrak{O}[t]^{\cup} \operatorname{Spec} \mathfrak{O}[t^{-1}]$, and $f : Y \to X$ be the integral closure of X in L. For each $\sigma \in \Sigma$, let $f_\sigma : Y_\sigma \to X_\sigma$ denote the base change $\otimes_{k,\sigma} \mathbf{C}$ of f. Each f_σ defines a finite branched covering $Y_\sigma(\mathbf{C}) \to X_\sigma(\mathbf{C}) = \mathbf{P}^1(\mathbf{C})$ between (not necessarily connected) compact Riemann surfaces.

Theorem 1. *Let D_0 be the prime divisor of $X = \mathbf{P}_{\mathfrak{O}}^1$ defined by the equation $t = 0$. Assume that there exists $r_\sigma > 0$ for each $\sigma \in \Sigma$ such that f_σ is unramified above $\{t \in \mathbf{C}; 0 < |t| < r_\sigma\}$ and $\Pi_\sigma r_\sigma \geq 1$. Then the $\mathfrak{O}$-scheme $f^{-1}(D_0) = Y \times_X D_0$ is connected.*

(Note that if L/K is a constant field extension, then $f^{-1}(D_0)$ is the spectrum of the corresponding ring of integers.)

This theorem is a direct consequence of the following result of Harbater ([Hb] Prop 2.1 and the preceding remarks).

Lemma (Harbater). *Let k be a number field with normalized absolute values $| \ |_v$ (so that $\Pi_v |a|_v = 1$ for all $a \in k^\times$). Suppose that $F(t) \in k[[t]]$ is algebraic over $k(t)$. Then one can choose $r_v > 0$ for each place v of k, with $r_v = 1$ for almost all v, such that $F(t)$ is v-adically convergent on the open disc of radius r_v (w.r.t. $| \ |_v$). If, moreover, one can choose r_v's such that $\Pi_v r_v \geq 1$, then $F(t)$ is rational, i.e. $F(t) \in k(t)$.*

Remark 1. For a complex archimedean place v corresponding to $\sigma, \bar\sigma \in \Sigma, r_v$ in this lemma corresponds to $r_\sigma r_{\bar\sigma} = r_\sigma^2$ in Theorem 1.

Remark 2. We shall only need the case where $F(t)$ belongs to $\mathfrak{O}[[t]]$ and is integral over $\mathfrak{O}[t]$. In this case, since we may choose $r_v = 1$ for all non-archimedean v, the assumption is $\Pi_\sigma r_\sigma \geq 1$. (It is not easy to make use of non-archimedean v with $r_v > 1$; see Remark 4 at the end of §1.) In this case, the proof in [Hb] is easy enough to be sketched. For each σ, $F_\sigma \in C[[t]]$ is not only holomorphic in the open disc of radius r_σ, but extends to a continuous function on its closure, because F_σ is integral over $C[t]$. Therefore, by the Riemann-Lebesgue lemma, one obtains $|\, a_n^\sigma \,|\, r_\sigma^n \to 0(n \to \infty)$. Therefore, $\Pi_\sigma \,|\, a_n^\sigma \,| = N(a_n) \to 0$. But since $a_n \in \mathfrak{O}$, and hence $N(a_n) \in Z$, this implies $N(a_n) = 0$ for $n \gg 0$, hence $F(t) \in \mathfrak{O}[t]$. For more details, and for comparison with classical Dwork criterion, see [Hb] §2.

Proof of Theorem 1. First, by using $L(t^{1/N})$ and $k(t^{1/N})$ for suitable N, we can reduce to the case where f_k is unramified above $t = 0$. Now choose any geometric point η of $Y_k = Y \otimes_\mathfrak{O} k$ above $t = 0$, and use the completion of L at η to embed L into $\bar{k}((t))$ ($\bar{k}$: an algebraic closure of k).

Claim 1A. $L \cap \mathfrak{O}[[t]] \subset k(t)$.

Proof. Take any $F = F(t) = \Sigma a_n t^n \in L \cap \mathfrak{O}[[t]]$, and by multiplying a suitable element $\neq 0$ of $\mathfrak{O}[t]$, we assume F to be integral over $\mathfrak{O}[t]$. Let $\sigma \in \Sigma$. Then $F_\sigma(t) = \Sigma a_n^\sigma t^n \in C[[t]]$ extends to a holomorphic function on $t \in C$, $|t| < r_\sigma$ (and hence converges there), because F_σ is integral over $C[t]$ and f_σ is unramified above $|t| < r_\sigma$. Since $\Pi_\sigma r_\sigma \geq 1$, the above lemma gives $F(t) \in k[t]$.

Claim 1B. Let E be the quotient field of $\mathfrak{O}[[t]]$ ($k(t) \subset E \subset k((t))$). Then $L \cap E = k(t)$.

Proof. Since $L \cap E$ is finite over $k(t)$, every element of $L \cap E$ is a $k(t)^\times$-multiple of some $g \in L \cap E$ which is integral over $\mathfrak{O}[t]$. Since $g \in E$ and integral over $\mathfrak{O}[[t]]$, $g \in \mathfrak{O}[[t]]$. Hence $g \in L \cap \mathfrak{O}[[t]] \subset k(t)$ by Claim 1A.

Claim 1C. L and E are linearly disjoint over $k(t)$.

Proof. Apply Claim 1B to the Galois closure of L over $k(t)$ (which does not change r_σ's).

Claim 1D. Let B be the integral closure of $\mathfrak{O}[t]$ in L. Then $B \otimes_{\mathfrak{O}[t]} \mathfrak{O}[[t]] \simeq \varprojlim(B/t^N B)$ is an integral domain.

Proof. Since $B \to L$ is injective and $\mathfrak{O}[[t]]/\mathfrak{O}[t]$ is flat, $B \otimes_{\mathfrak{O}[t]} \mathfrak{O}[[t]] \to L \otimes_{\mathfrak{O}[t]} \mathfrak{O}[[t]]$ is also injective. On the other hand, $\mathfrak{O}[[t]] \to E$ is injective and $L/\mathfrak{O}[t]$ is flat; hence $L \otimes_{\mathfrak{O}[t]} \mathfrak{O}[[t]] \to L \otimes_{\mathfrak{O}[t]} E = L \otimes_{k(t)} E$ is also injective. By Claim 1C, $L \otimes_{k(t)} E$ is a field. Therefore, $B \otimes_{\mathfrak{O}[t]} \mathfrak{O}[[t]]$ is a domain.

The last isomorphism follows from a general fact; if A is a Noetherian ring, M is a (not necessarily free) finite A-module, and I is an ideal of A, then $M \otimes \varprojlim(A/I^n) \simeq \varprojlim(M/I^n M)$ (cf [A-M] p108).

Claim 2. *If J, J' are ideals of B such that (i) $J + J' = (1)$, (ii) $J, J' \supset (t)$, (iii) $(JJ')^n \subset (t)$ for some $n \geq 1$, then either $J = (1)$ or $J' = (1)$.*

Proof. By these conditions,

$$\varprojlim(B/t^N B) \simeq \varprojlim(B/J^N) \oplus \varprojlim(B/J'^N)$$

which reduces the Claim to Claim 1D.

Completing the proof of Theorem 1. If $f^{-1}(D_0) = \mathrm{Spec}(B/tB)$ were not connected, it must be a disjoint union of two non-empty open subsets S, S'. Let J (resp. J') be the intersection of all (minimal) primes of B belonging to S (resp. S'). Then J, J' satisfies the conditions of Claim 2. Therefore, J or $J' = (1)$, a contradiction. $\qquad\qquad\qquad\square$

Remark 3. Perhaps we should show some example where $f^{-1}(D)$ is disconnected. This is the case when $L = \mathbf{Q}(t, y)$, with $y^2 - y = t$ and D is defined by $t = 0$. In fact, then $f^{-1}(D) \simeq \mathrm{Spec}(\mathbf{Z}[y]/y(y - 1)) \cong \mathrm{Spec}\,\mathbf{Z} \sqcup \mathrm{Spec}\,\mathbf{Z}$. Note that the branch point $t = -\frac{1}{4}$ is "archimedean close" to $t = 0$.

Remark 4. At non-archimedean primes $\mathfrak{p}$, the radius of convergence can be strictly smaller than the distance from the center of the nearest branch point (cf. [Hb] §3 Remark 2, [D-R]). For this reason, we could not use non-archimedean primes to loosen the assumption of Theorem 1.

Remark 5. Since the connectedness of $f^{-1}(D)$ remains unchanged if we replace Y by Y', where either $Y' \to Y$ or $Y \to Y'$ is a proper birational morphism, our assumptions that Y be normal and $Y \to X$ be finite can be loosened to some extent. At any rate the general case follows immediately from our case.

§2. Let $k, \mathfrak{O}$, L/K, $f : Y \to X$ ($X = \mathbf{P}^1_{\mathfrak{O}}$) be as at the beginning of §1. We shall give some applications of Theorem 1.

Corollary 1. *If f_k is unramified outside $t = b_1, \ldots, b_r$ where $b_1, \ldots, b_r \in \mathbf{Z}^\cup(\infty)$, and $a \in \mathbf{Q}$, then $f^{-1}(D_a)$ is connected.*

Remark 6. $f^{-1}(D_\infty)$ need not be connected, as the example below shows. (See also Remark 3 (apply for t^{-1} instead of t).)

Proof. The group $GL_2(\mathfrak{O})$ acts on X by linear fractional transformations

$$\gamma = \begin{pmatrix} p & q \\ p' & q' \end{pmatrix} : t \longrightarrow (pt + q)(p't + q')^{-1}.$$

We shall find such $\gamma \in GL_2(\mathbf{Z})$ that $\gamma(a) = 0$ and $|\gamma(b)| \geq 1$ for any $b \in \mathbf{Z}^\cup(\infty)$, $b \neq a$. This will reduce Corollary 1 to Theorem 1.

Define $p, q \in \mathbf{Z}$ by

$$pa + q = 0, \quad (p, q) = 1, \quad p > 0,$$

and then $p', q' \in \mathbf{Z}$ by

$$pq' - p'q = 1, \quad 0 \le p' < p.$$

Now let $b \in \mathbf{Z}$, $b \ne a$. Then

$$p(q' + bp') - p'(q + bp) = 1$$

and $q + bp \ne 0$ (as $b \ne a = -q/p$); hence

$$\frac{q' + bp'}{q + bp} = \frac{p'}{p} + \frac{1}{p(q + bp)}.$$

Since the first term on the right side lies in the closed interval $[0, 1 - p^{-1}]$ and the second term has absolute value $\le p^{-1}$, the left side has absolute value ≤ 1. Therefore $|\gamma(b)| \ge 1$. Note that this holds also when $b = \infty$. $\square$

Example. Let X be the modular curve of level one, the projective j-line over $\mathbf{Z}$, and $f : Y \to X$ be a modular curve with any level, e.g., $Y = X_0(N)$ ($N \ge 1$). ($f : Y \to X$ can be the integral closure of X in any finite extension of $\mathbf{Q}(j)$ unramified outside $j = 0, 12^3, \infty$). Then by Corollary 1, $f^{-1}(D_a)$, for any $a \in \mathbf{Q}$, is connected. On the other hand, $f^{-1}(D_\infty)$ is not always connected. In fact, if $Y = X_0(p)$, $f^{-1}(D_\infty)$ consists of two distinct connected components. (The two $\mathbf{Q}$-rational cusps of $X_0(p) \otimes \mathbf{Q}$ do not meet even at p, as cusps are not supersingular.) The above connectedness in the modular case does not seem to be so obvious a consequence of known results on modular curves. Observe: let $Y = X_0(p)$ and $a \in \mathbf{Q}$, $a \ne 0, 12^3$, with a (mod p) *not* supersingular. Then $f^{-1}(D_a)$ consists of at most two connected components, because its fibre above p consists only of two points P_1, P_2. Call P_1 (resp. P_2) the fibre isomorphic to $\operatorname{Spec} \mathbf{F}_p$ (resp. $\operatorname{Spec} \mathbf{F}_p[x]/x^p$). Then the connectedness of $f^{-1}(D_a)$ is equivalent to the following assertion; either

(i) there is no $\mathbf{Q}$-rational point of $f^{-1}(D_a) \otimes \mathbf{Q}$ whose reduction mod p coincides with P_1,

or else (call P the unique such $\mathbf{Q}$-rational point)

(ii) there exists a prime number $l \ne p$ such that $a \equiv 0, 12^3$ or ∞ (mod l), and that $f \otimes \mathbf{F}_l$ is ramified at the point of $Y \otimes \mathbf{F}_l$ obtained as the specialization of P.

One sees that this is a delicate horizontal property!

§3. Now we assume that $f_k; Y_k \to X_k$ is unramified outside $t = 0, 1, \infty$. A prime divisor of X defined by $t = 0, 1$, or ∞ will be called *cuspidal*.

Corollary 2. *If f_k is unramified outside $t = 0, 1, \infty$, and D is a cuspidal prime divisor of $X = \mathbf{P}^1_{\mathfrak{O}}$, then $f^{-1}(D)$ is connected.*

Proof. We may assume that D is the cusp defined by $t = 0$. But then the connectedness of $f^{-1}(D)$ is an immediate consequence of Theorem 1. □

For the closure D_a in $\mathbf{P}^1_{\mathfrak{O}}$ of other rational points $t = a \in k$ of $\mathbf{P}^1_k$, we can only prove:

Corollary 3. *If f_k is unramified outside $t = 0, 1, \infty$, and $a \in k$ ($a \neq 0, 1$), $f^{-1}(D_a)$ is connected at least in the following cases; (i) $a \in \mathbf{Q}$; (ii) $a = (1 - \zeta)^{-1}$, where ζ is a root of unity whose order is not a prime power; (ii)$'$ $a = (1 - \zeta')(\zeta - \zeta')^{-1}$, where ζ, ζ' are roots of unity such that none of the orders of $\zeta, \zeta', \zeta'\zeta^{-1}$ are prime powers.*

Remark 6. In cases (ii)(ii)$'$, a is a *special unit*, i.e., a and $1 - a$ are both units. This means that D_a does not meet any cuspidal prime divisor. An example of (ii): $a = (1 + \omega)^{-1} = -\omega$, where ω is a cubic root of unity.

By Theorem 1, $f^{-1}(D_a)$ is connected if there exists $\gamma \in GL_2(\mathfrak{O})$ (acting on $\mathbf{P}^1_{\mathfrak{O}}$ by linear fractional transformations) such that $\gamma(a) = 0$ and

$$\prod_{\sigma \in \Sigma} \mathrm{Min}(|\gamma(0)^\sigma|, |\gamma(1)^\sigma|, |\gamma(\infty)^\sigma|) \geq 1.$$

We shall show, in each of the cases (i)(ii)(ii)$'$, that such an element γ exists.

Actually, we can also show that when a is a special unit, (ii)(ii)$'$ *are the only cases* where there exists some field $k \ni a$ and some $\gamma \in GL_2(\mathfrak{O})$ satisfying these conditions. Thus, in particular, when a is (a special unit which is) non-abelian over $\mathbf{Q}$, or when (for example) $a = \frac{1}{2}(1 + \sqrt{5})$, there does not exist any such γ. We do not know whether $f^{-1}(D_a)$ is still connected in such cases.

(i) *The case $a \in \mathbf{Q}$ ($a \neq 0, 1$).* This is a special case of Corollary 1.

(ii) In this case, it is enough to take $\gamma(t) = 1 - a^{-1}t$. In fact, then $\gamma(a) = 0$, $\gamma(0) = 1$, $\gamma(1) = \zeta$, $\gamma(\infty) = \infty$.

(ii)$'$ In this case, it is enough to take

$$\gamma = \begin{pmatrix} \zeta - \zeta' & \zeta' - 1 \\ \zeta - \zeta' & \zeta(\zeta' - 1) \end{pmatrix}.$$

In fact, then $\det \gamma = (\zeta - 1)(\zeta' - 1)(\zeta - \zeta') \in \mathfrak{O}^\times$, $\gamma(a) = 0$, $\gamma(0) = \zeta^{-1}$, $\gamma(1) = \zeta'^{-1}$, $\gamma(\infty) = 1$. □

§4. In general, let Y, Z be connected locally Noetherian schemes, $f : Z \to Y$ be a morphism and $f_* : \pi_1(Z, \zeta) \to \pi_1(Y, \eta)$ be the induced homomorphism between their fundamental groups, where ζ is any geometric point of Z

and $\eta = f(\zeta)$. Then by their definitions [G], f_* is *surjective* if and only if $Z' = Z \times_Y Y'$ is *connected* for any finite etale connected covering Y'/Y of Y. We apply this to the determination of $\pi_1(Y)$ for some special o arithmetic surfaces Y, by using horizontal divisors $Z \hookrightarrow Y$ and the results of §3.

The following is a direct application.

Theorem 2. *Let k be a number field, $\mathfrak{O}$ its ring of integers, and $X = \mathbf{P}^1_{\mathfrak{O}}$ (the projective t-line over $\mathfrak{O}$). Let $L/k(t)$ be a finite extension field, which is unramified outside $t = 0, 1, \infty$, and $f : Y \to X$ be the normalization of X in L. Let $a \in k^{\cup}(\infty)$ be either $a \in \mathbf{Q}^{\cup}(\infty)$ (including $0, 1, \infty$) or of the form (ii) or (ii)' of Cor 3 of Th 1, and D_a be the prime divisor on X defined by $t = a$. Let E be any closed subscheme of Y contained in (the support of) $f^{-1}(D_0 {}^{\cup} D_1 {}^{\cup} D_\infty)$, which does not meet $f^{-1}(D_a)$ (for example, $E = \emptyset$). Then the natural homomorphism*

$$\pi_1\left(f^{-1}(D_a)^{\mathrm{red}}\right) \longrightarrow \pi_1(Y - E)$$

is surjective. In particular, (i) if $f^{-1}(D_a)^{\mathrm{red}} \xrightarrow{\sim} \mathrm{Spec}\,\mathfrak{O}$, then $\pi_1(Y - E) \xrightarrow{\sim} \pi_1(\mathrm{Spec}\,\mathfrak{O})$;
(ii) if $f^{-1}(D_a)^{\mathrm{red}}$ is a tree-like union of $\mathrm{Spec}\,\mathfrak{O}$ (see below) and $\pi_1(\mathrm{Spec}\,\mathfrak{O}) = (1)$, then $\pi_1(Y - E) = (1)$.

Here, $f^{-1}(D_a)^{\mathrm{red}}$ (the reduced part of $f^{-1}(D_a)$) is called *tree-like* if its graph (edges = irreducible components, vertices on an edge = closed points on the corresponding irreducible component) is a tree.

Proof. The divisor $F = f^{-1}(D_a)^{\mathrm{red}}$ is a closed subscheme of $Y_1 = Y - E$. If Y_1'/Y_1 is any connected finite etale covering, $Y_1' \times_{Y_1} F \simeq Y' \times_Y F$, where Y' is the integral closure of Y (and also of $\mathbf{P}^1_{\mathfrak{O}}$) in the function field of Y_1'. By Cor 3 of Th 1, $Y' \times_Y f^{-1}(D_a) = Y' \times_X D_a$ is connected; hence $Y' \times_Y F$ is also connected. Therefore, $\pi_1(F) \to \pi_1(Y_1)$ is surjective.

When $F \xrightarrow{\sim} \mathrm{Spec}\,\mathfrak{O}$, this defines a section $\mathrm{Spec}\,\mathfrak{O} \to Y_1$, and hence we have a surjection $\alpha : \pi_1(\mathrm{Spec}\,\mathfrak{O}) \to \pi_1(Y_1)$, and the structural homomorphism $\beta : \pi_1(Y_1) \to \pi_1(\mathrm{Spec}\,\mathfrak{O})$, with $\beta \circ \alpha = \mathrm{id}$. Therefore, $\pi_1(Y_1) \xrightarrow{\sim} \pi_1(\mathrm{Spec}\,\mathfrak{O})$. In case (ii), F has no non-trivial connected finite etale coverings, because each irreducible component $\simeq \mathrm{Spec}\,\mathfrak{O}$ is simply connected, and there can be no non-trivial connected "mock coverings" (graph-theoretically produced finite connected etale coverings) because F is tree-like. $\qquad\square$

Corollary 1 (T. Saito). $\pi_1(\mathbf{P}^1_{\mathfrak{O}} - D_0 {}^{\cup} D_1 {}^{\cup} D_\infty) \simeq \pi_1(\mathrm{Spec}\,\mathfrak{O})$.

This fact may well have been known, but the author could not find any reference, except that Example 3.1 in [Hb] §3 is quite close. (It gives $\pi_1(\mathrm{Spec}\,\mathbf{Z}[t, (t^N - 1)^{-1}]) = (1)$, to which the case $\mathfrak{O} = \mathbf{Z}$ reduces directly, and [Hb] contains enough tools for treating the case of general $\mathfrak{O}$.) As far as the author knows, the first proof of this was provided by T. Saito. It is a direct application of the generalized Abhyankar lemma (see Appendix). Our argument gives an alternative proof which is more archimedean in nature.

Proof. First, take some a as in Cor 3 of Th 1 (ii) or (ii)′, and choose k such that $k \ni a$. In Th 2, take $Y = X$, $E = D_0 {}^{\cup} D_1 {}^{\cup} D_\infty$. Since $D_a \cap E = \emptyset$, Th 2 (i) applies to this case, and we conclude that $\pi_1(\mathbf{P}^1_{\mathfrak{O}} - E) \simeq \pi_1(\operatorname{Spec} \mathfrak{O})$ for $\mathfrak{O}$: big enough. But then, for any $\mathfrak{O}$, $\mathbf{P}^1_{\mathfrak{O}} - E$ cannot have finite etale connected coverings other than constant ring extensions (which must be etale). Therefore, our assertion holds for any $\mathfrak{O}$. □

Corollary 2. *Let $f : Y \to X$ be as at the beginning of Th 2 (the first two sentences preserved). Suppose that one of the cusps, say $t = \infty$, is totally ramified in $f_k = f \otimes k : Y_k \to X_k$. Then $\pi_1(Y) \xrightarrow{\sim} \pi_1(\operatorname{Spec} \mathfrak{O})$, or more strongly,*

$$\pi_1(Y - f^{-1}(D_0 {}^{\cup} D_1)) \cong \pi_1(\operatorname{Spec} \mathfrak{O}).$$

Proof. In fact, in this case $f^{-1}(D_\infty)^{\mathrm{red}} \simeq \operatorname{Spec} \mathfrak{O}$.

In particular,

Corollary 3. *Let p be a prime, $a, b, c \in \mathbf{Z}$, $a+b+c = 0$, $abc \not\equiv 0 \pmod{p}$, and $L = \mathbf{Q}(t, y)$, where*

$$y^p = (-1)^c t^a (1 - t)^b$$

(a "primitive Fermat curve"). Let $f : Y \to \mathbf{P}^1_{\mathbf{Z}}$ be the normalization of $\mathbf{P}^1_{\mathbf{Z}}$ (the t-line) in L. Then for $i, j \in \{0, 1, \infty\}$, $i \neq j$,

$$\pi_1\big(Y - f^{-1}(D_i {}^{\cup} D_j)\big) = (1).$$

[Appendix] T. Saito's original proof of Cor 1 of Th 2

It proceeds as follows. Let $L/k(t)$, $f : Y \to X = \mathbf{P}^1_{\mathfrak{O}}$ be as at the beginning of Theorem 2. Suppose that $f : Y \to X$ is etale outside $D_0 {}^{\cup} D_1 {}^{\cup} D_\infty$. Let $\mathfrak{p}$ be any prime ideal of $\mathfrak{O}$, and put $X_{\mathfrak{p}} = X \otimes_{\mathfrak{O}} (\mathfrak{O}/\mathfrak{p})$. Choose any cuspidal prime divisor D_i $(i = 0, 1, \infty)$ on X, and let P be the intersection of D_i with $X_{\mathfrak{p}}$, which is a closed point on $X_{\mathfrak{p}}$. Then the only prime divisor on X passing through P, along which f can possibly be ramified, is D_i. From this follows, by the generalized Abhyankar lemma ([G] Exp. XIII §5), that the ramification indices of $f_k = f \otimes k$ above $t = i$ cannot be divisible by the residue characteristic of $\mathfrak{p}$. Since $\mathfrak{p}$ and i are arbitrary, f must be etale

254   Yasutaka Ihara

also above D_0, D_1, D_∞; hence $\pi_1(X - D_0 \cup D_1 \cup D_\infty) \simeq \pi_1(X) \simeq \pi_1(\text{Spec } \mathfrak{O})$, as desired.

Saito has also noted that the same argument holds for a somewhat more general case; $\mathbf{P}^1_{\mathfrak{O}} - \bigcup_{a \in A} D_a$ where A is a finite set of elements of $k^\cup(\infty)$ satisfying the following conditions. For each pair of $\mathfrak{p}$ and $a \in A$, put $P(a, \mathfrak{p}) = D_a \cap X_\mathfrak{p}$ (a closed point on $X_\mathfrak{p}$). Then for each pair $(a, \mathfrak{p})$, either $P(a, \mathfrak{p}) \neq P(a', \mathfrak{p})$ for all $a' \neq a$ ($a' \in A$), or there exists exactly one $a' \in A$, $a' \neq a$ with $P(a', \mathfrak{p}) = P(a, \mathfrak{p})$, *and* in this case the maximal ideal of the local ring of X at $P(a, \mathfrak{p})$ is generated by two elements defining D_a and $D_{a'}$ at $P(a, \mathfrak{p})$. (Roughly speaking, the conditions require that the only singularities of $\bigcup D_a$ be "ordinary double points".)

An example: $\mathfrak{O} = \mathbf{Z}$, $A = \{0, 1, 2, 3, \infty\}$.

References

[A-M]   Atiyah M.F. & Macdonald I.G., Introduction to commutative algebra, Addison-Wesley

[B]   Belyi, G.V., On Galois extensions of a maximal cyclotomic field, Izv. Akad. Nauk USSR **43** (1979), 267-276; Transl. Math. USSR Izv. **14** (1980), 247-256.

[D-R]   Dwork, B. and P. Robba, On natural radii of p-adic convergence, Trans. AMS **256** (1979), 199-213

[G]   Grothendieck, A., Revêtements etales et groupe fondamental (SGA 1), Lecture Notes in Math. **224**, Springer

[Hb]   Harbater, D., Galois covers of an arithmetic surface, Amer. J. Math. **110** (1988), 849-885

[Ht]   Hartshorne, R., Ample subvarieties of algebraic varieties, Lecture Notes in Math. **156**, Springer

[H-M]   Hironaka, H. & H. Matsumura, Formal functions and formal embeddings, J. Math. Soc. Japan **20** (1968), 52-67

[S]   Serre, J.-P., Linear representations of finite groups, GTM **42**, Springer

*Research Institute for Mathematical Sciences, Kyoto University

Algebraic representation of the Teichmüller spaces

Kyoji Saito*

For integers $g \geq 0$ and $n \geq 0$, consider $(X, \underline{x})$, where X is a compact Riemann surface of genus g and $(\underline{x}) = (x_1, \ldots, x_n)$ is an n-tuple of distinct ordered points of X. Recall ([Mu2]) that the moduli space $\mathfrak{M}_{g,n}$ is the set of isomorphic classes of $(X, \underline{x})$ and that the Teichmüller space $\mathfrak{T}_{g,n}$ is the set of isomorphic classes (in a suitable sense) of $(X, \alpha, \underline{x})$, where α is an isomorphism from the group $\Gamma_{g,n}$ (6.2.1) to $\pi_1(X \backslash \{\underline{x}\})$ with certain marking properties (recalled in §6).

We recall some well known facts on the spaces, which led the author to the present work. The $\mathfrak{M}_{g,n}$ has a structure of a quasi-projective algebraic variety defined over $\mathbf{Q}$ ([Mu]) and $\mathfrak{T}_{g,n}$ a smooth real manifold so that the natural projection $\mathfrak{T}_{g,n} \to \mathfrak{M}_{g,n}$ is a ramified covering map, having $\mathrm{Out}^+(\Gamma_{g,n})$ (6.2.3) as the covering transformation group. Since $\mathfrak{T}_{g,n}$ is homeomorphic to a cell, it is the universal covering space of $\mathfrak{M}_{g,n}$ in the sense of orbifold. The covering map cannot be algebraic (in any sense) when the group $\mathrm{Out}^+(\Gamma_{g,n})$ is infinite, in particular when $2 - 2g - n < 0$. Then, it is interesting to ask for a construction of an automorphic function (or a form in a suitable sense) on $\mathfrak{T}_{g,n}$ with respect to the action of $\mathrm{Out}^+(\Gamma_{g,n})$. After such an attempt ([S2]), the author was necessarily led to fix certain real algebraic coordinates of $\mathfrak{T}_{g,n}$. This motivated the present paper.

In this paper, we describe $\mathfrak{T}_{g,n}$ for $2 - 2g - n < 0$ as a component of the $\mathbf{R}$-valued point set of an affine scheme $\mathrm{Spec}(R_{g,n})$ over $\mathbf{Z}$. Here $R_{g,n}$ is introduced abstractly as *the ring of universal characters* (see (6.5) theorem). The image of $R_{g,n}$ in $\mathbf{R}$ at a point $(X, \alpha, \underline{x})$ of $\mathfrak{T}_{g,n} \subset \mathrm{Hom}(R_{g,n}, \mathbf{R})$ is an invariant of $(X, \underline{x})$ called the *ring of uniformization* ((6.3.1)).

Historically, such real algebraic description for the Teichmüller space is due to Fricke [F-K], who started to use characters of discrete subgroups in $\mathrm{SL}_2(\mathbf{R})$ in order to determine the moduli space of compact Riemann surfaces. So some authors call it the Fricke moduli. Since then there were numerous results in this direction. They lead to a semialgebraic description of the Teichmüller spaces and to its application to the Thurston compactification $\overline{\mathfrak{T}}_g$ (cf. [K], [H], [Br], [C-S], [O], [S-S], [Ko]).

The present paper follows the same idea (in particular, of [H]) with some refinements motivated by the reasons explained above. For a group Γ, we describe the variety of representations of Γ in $\mathrm{SL}_2(R)$ and its quotient variety by the adjoint action of PGL_2 as schemes over $\mathbf{Z}$ (free from the scalar ring

R). The result (including an introduction of the universal character ring in §3) is exposed in §1-4 (see §4 for relevant theorems). Then they are applied in §6 to obtain the coordinate ring $R_{g,n}$ of the Teichmüller space $\mathfrak{T}_{g,n}$. The original problem (on automorphic forms on $\mathfrak{T}_{g,n}$) still seems impenetrable. We restrict ourselves in the present paper to posing a related problem: *Compare the algebraic variety structure on $\mathfrak{M}_{g,n}$ and that on $\mathfrak{T}_{g,n}$.* See §6 for details.

The construction of the paper is as follows. The first 4 paragraphs §1-4 develops a general functorial scheme for the quotient of variety of representations of any group Γ in SL_2 and GL_2. This part summarizes [S4] and proofs are sketchy or omitted. Then §5 contains some detailed study of the case of the field of real numbers **R** coefficients. In §6, the previous results are applied to describe the Teichmüller spaces ((6.5) Theorem). We ask two questions in §6, one on the algebraic points of $\mathfrak{T}_{g,n}$ and the other on "arithmetic compactification" of $\mathfrak{T}_{g,n}$.

A hurrying reader may be suggested to start with §6 (or §4) and to return to previous paragraphs according to necessity.

§1 Universal representation of a group Γ in SL_n and in GL_n

§2 PGL_2-invariants for pairs of 2×2 matrices

§3 Universal character rings $R(\Gamma)$ and $\tilde{R}(\Gamma)$

§4 Invariant morphisms π_Γ and $\tilde{\pi}_\Gamma$

§5 Representation variety with real coefficients, Appendix

§6 Teichmüller spaces $\mathfrak{T}_{g,n}$

§1. Universal representation of a group Γ in SL_n or in GL_n

As explained in the introduction, in §1-4, we prepare a general scheme for a quotient of the representation variety of a group Γ in SL_2 and GL ([S4] for proofs and details). In this §, we prepare basic terminologies on representation variety of a group in SL_n or GL_n (see [Pr] [Ba] [L-M]).

(1.1) Let Γ be a group. The set of all homomorphisms (representations) of Γ in the other group G is denoted by $\mathrm{Hom}(\Gamma, G)$. We sometimes use Hom^{gr} to distinguish it from the set Hom^{ring} of ring homomorphisms.

For a fixed n and Γ, we let $\mathrm{Hom}(\Gamma, \mathrm{SL}_n)$ and $\mathrm{Hom}(\Gamma, \mathrm{GL}_n)$ denote the functors: $R \in \{\text{commutative ring with } 1\} \mapsto \mathrm{Hom}^{gr}(\Gamma, \mathrm{SL}_n(R))$ and $\mathrm{Hom}^{gr}(\Gamma, \mathrm{GL}_n(R))$. An elementary but key fact is that these functors are representable ((1.2) lemma). Let us explain this in the case $n = 1$. Let $\mathbf{Z} \cdot \Gamma^{ab}$ be the group ring for $\Gamma^{ab} := \Gamma/[\Gamma, \Gamma]$, where the image of $\gamma \in \Gamma$ is denoted by $d(\gamma)$ (so $d(e) = 1$). Then for any commutative ring R with 1, the map:

(1.1.1) $\quad \varphi \in \mathrm{Hom}^{ring}(\mathbf{Z} \cdot \Gamma^{ab}, R) \mapsto \rho \in \mathrm{Hom}^{gr}(\Gamma, \mathrm{GL}_1(R))$

given by $\rho(\gamma) = \varphi(d(\gamma))$ $(\gamma \in \Gamma)$ is a bijection. The trivial representation corresponds to the ring homomorphism factored by the augmentation map $\mathbf{Z} \cdot \Gamma^{ab} \to \mathbf{Z} \cdot \Gamma^{ab}/J \simeq \mathbf{Z}$, where the augmentation ideal J is given by

(1.1.2) $\quad\quad\quad\quad J := (d(\gamma) - 1, \gamma \in \Gamma).$

(1.2) The representability of the functors is the next goal of this §.

Lemma 1. *For a given Γ and $n \in \mathbf{Z}_{>0}$, there exists a pair $(\widetilde{A}_n(\Gamma), \sigma)$ of a commutative ring $\widetilde{A}_n(\Gamma)$ with 1 over $\mathbf{Z} \cdot \Gamma^{ab}$ and a representation $\sigma : \Gamma \to \mathrm{GL}_n(\widetilde{A}_n(\Gamma))$ called the universal representation for Γ, such that for any commutative ring R with 1, one has the bijection:*

(1.2.1) $\quad\quad \mathrm{Hom}^{ring}(\widetilde{A}_n(\Gamma), R) \quad \simeq \quad \mathrm{Hom}^{gr}(\Gamma, \mathrm{GL}_n(R)).$
$\quad\quad\quad\quad\quad\quad\quad \varphi \quad\quad\quad \longmapsto \quad\quad\quad \varphi \circ \sigma$

2. Put

(1.2.2) $\quad\quad\quad\quad A_n(\Gamma) := \widetilde{A}_n(\Gamma)/J\widetilde{A}_n(\Gamma)$

for the augmentation ideal J (1.1.2). Then the following diagram is commutative.

(1.2.3)
$$
\begin{array}{ccc}
\mathrm{Hom}^{ring}(A_n(\Gamma), R) & \simeq & \mathrm{Hom}^{gr}(\Gamma, \mathrm{SL}_n(R)) \\
\cap & & \cap \\
\mathrm{Hom}^{ring}(\widetilde{A}_n(\Gamma), R) & \simeq & \mathrm{Hom}^{gr}(\Gamma, \mathrm{GL}_n(R)) \\
\downarrow & & \downarrow \mathrm{det}_* \\
\mathrm{Hom}^{ring}(\mathbf{Z} \cdot (\Gamma)^{ab}, R) & \simeq & \mathrm{Hom}^{gr}(\Gamma, \mathrm{GL}_1(R)).
\end{array}
$$

Here the maps in the first column are induced by $\mathbf{Z} \cdot \Gamma^{ab} \subset \widetilde{A}_n(\Gamma) \to A_n(\Gamma)$.

3. The pair $(\widetilde{A}_n(\Gamma), \sigma)$ is unique up to an isomorphism of the ring $\widetilde{A}_n(\Gamma)$ over $\mathbf{Z} \cdot \Gamma^{ab}$ commuting with the universal representation σ.

4. If Γ is a finitely generated group, then $\widetilde{A}_n(\Gamma)$ and $A_n(\Gamma)$ are finitely generated algebras over $\mathbf{Z}$. Hence they are Noetherian.

Proofs are elementary and omitted. For later uses, we construct the $\mathbf{Z} \cdot \Gamma^{ab}$-algebra $\widetilde{A}_n(\Gamma)$ and the representation σ explicitly. For each $\gamma \in \Gamma$, consider an $n \times n$ matrix:

(1.2.4) $\quad\quad\quad\quad \sigma(\gamma) := \left(a_{ij}(\gamma) \right)_{ij=1,\dots,n},$

Kyoji Saito

whose entries $a_{ij}(\gamma)$ are regarded as indeterminates. Then $\widetilde{A}_n(\Gamma)$ is defined to be an algebra generated by the indeterminates $a_{ij}(\gamma)$ over $\mathbf{Z}$ modulo the ideal I generated by the polynomials of all entries of the matrices $\sigma(e) - I_n$ (I_n = the $n \times n$ unit matrix) and $\sigma(\gamma\delta) - \sigma(\gamma)\sigma(\delta)$. That is:

$$I := (a_{ij}(e) - \delta_{ij}, a_{ij}(\gamma\delta) - \sum_k a_{ik}(\gamma)a_{kj}(\delta) \text{ for } 1 \le i,j \le n \text{ and } \gamma, \delta \in \Gamma)$$

$$(1.2.5) \quad \widetilde{A}_n(\Gamma) := \mathbf{Z}[a_{ij}(\gamma) \quad \text{for} \quad \gamma \in \Gamma \quad \text{and} \quad 1 \le ij \le n] / I(\Gamma).$$

By definition, the correspondence

$$(1.2.6) \qquad \sigma : \gamma \in \Gamma \longrightarrow \sigma(\gamma) \in \mathrm{GL}_n(\widetilde{A}_n(\Gamma))$$

is a representation of Γ. In particular, one has $\sigma(\gamma)\sigma(\gamma^{-1}) = I_n$ and $\det \sigma(\gamma) \cdot \det \sigma(\gamma^{-1}) = 1$. Hence $\det \sigma(\gamma)$ is invertible in the ring $\widetilde{A}_n(\Gamma)$. The $\mathbf{Z} \cdot \Gamma^{ab}$ algebra structure on $\widetilde{A}_n(\Gamma)$ is defined by the correspondence:

$$(1.2.7) \qquad d(\gamma) \in \mathbf{Z} \cdot \Gamma^{ab} \longrightarrow \det \sigma(\gamma) \in \widetilde{A}_n(\Gamma).$$

(1.3) An element $X \in \mathrm{PGL}_n(R)$ acts on $\mathrm{Hom}(\Gamma, \mathrm{SL}_n(R))$ and on $\mathrm{Hom}(\Gamma, \mathrm{GL}_n(R))$ by the adjoint action: $\rho \in \mathrm{Hom}(\Gamma, \mathrm{SL}_n(R))$ (or $\mathrm{Hom}(\Gamma, \mathrm{GL}_n(\mathbf{R}))$) $\mapsto \rho \cdot Ad(X)$, where $\rho \cdot Ad(X)(\gamma) := X^{-1}\rho(\gamma)X$. The adjoint action induces the dual action (or, sometimes called the comodule structure over the Hopf algebra $A(PGL_n)$) on the coordinate rings $\widetilde{A}_n(\Gamma)$ (1.2.5) and $A_n(\Gamma)$ (1.2.2), as given below.

Let PGL_n be the group scheme and $A(\mathrm{PGL}_n)$ its coordinate ring ([SGA III], [D-G]). The $A(\mathrm{PGL}_n)$ is the subring $A_0(\mathrm{GL}_n)$ of the coordinate ring $A(\mathrm{GL}_n) := \mathbf{Z}[x_{ij} 1 \le i,j \le n]_{\det(X)}$ (for $X := (x_{ij})_{ij=1}^n$) of GL_n consisting of elements of degree 0. Then the dual adjoint action is a ring homomorphism:

$$(1.3.1) \qquad Ad : \widetilde{A}_n(\Gamma) \longrightarrow \widetilde{A}_n(\Gamma) \otimes_{\mathbf{Z}} A(\mathrm{PGL}_n)$$

$$(1.3.2) \qquad Ad : A_n(\Gamma) \longrightarrow A_n(\Gamma) \otimes_{\mathbf{Z}} A(\mathrm{PGL}_n)$$

sending (i,j)-entry of $\sigma(\gamma)$ to the same (i,j)-entry of $X^{-1}\sigma(\gamma)X$.

Theorems in §4 state that *the universal categorical quotient of the functors* $\mathrm{Hom}(\Gamma, \mathrm{GL}_2)$ *and* $\mathrm{Hom}(\Gamma, \mathrm{SL}_2)$ *by the adjoint action of* PGL_2 *exist up to the loci of abelian or reducible representations.* This fact is shown by showing the existence of the universal invariant subrings of $\widetilde{A}_2(\Gamma)$ and $A_2(\Gamma)$ over $\mathbf{Z}$ of the adjoint action up to some localizations (4.3.3).

(1.4) From now on, we switch to $n = 2$. Put $G := \{\det \sigma(\gamma), \mathrm{tr}\, \sigma(\gamma) \mid \gamma \in \Gamma\}$, where σ is the universal representation (1.2.6). Here are questions:

1. whether the elements of G generates $\left(R \otimes \tilde{A}_2(\Gamma)\right)^{\text{PGL}_2}$ for any scalar ring R? 2. what are the basic relations among the elements of G? Those questions are (partially) answered by introducing the universal character rings $\tilde{R}(\Gamma)$ and $R(\Gamma)$ in §3 and by §4 Theorem 2 (4.3.3). Let us list up some relations among the $\det \sigma(\gamma)$ and $\operatorname{tr} \sigma(\gamma)$. The first one is obvious:

$$(1.4.1) \qquad \operatorname{tr}(\sigma(e)) = 2.$$

The Cayley-Hamilton relation for the matrix $\sigma(\gamma)$ is rewritten as: $\sigma(\gamma) + \det(\sigma(\gamma)) \cdot \sigma(\gamma^{-1}) = \operatorname{tr}(\sigma(\gamma)) \cdot I_2$. Multiply $\sigma(\delta)$ from the right and take traces. So we obtain:

$$(1.4.2) \qquad \operatorname{tr}(\sigma(\gamma\delta)) + \det \sigma(\gamma) \cdot \operatorname{tr}(\sigma(\gamma^{-1}\delta)) = \operatorname{tr}(\sigma(\gamma)) \cdot \operatorname{tr}(\sigma(\delta))$$

for γ and $\delta \in \Gamma$ (cf. [F-K, formulas (2), p.338]). Many more relations were obtained by Fricke ([F-K]).

§2. PGL$_2$-invariants for pairs of 2×2 matrices

In this paragraph, we study the invariants of the diagonal adjoint action of PGL$_2$ on the space $M_2 \times M_2$ of pairs of 2×2 matrices. By the terminology in §1, it is a study of *the quotient of the representation variety for $\Gamma = F_2$* (the free group of 2 generators).

(2.1) Let $\mathbf{Z}[M_2 \times M_2]$ be the coordinate ring for the space of a pair $(A, B) \in M_2 \times M_2$ of 2×2 matrices. Let us fix the coordinates explicitly as $A = \begin{bmatrix} a & b \\ c & d \end{bmatrix}$ and $B = \begin{bmatrix} e & f \\ g & h \end{bmatrix}$ so that $\mathbf{Z}[M_2 \times M_2] = \mathbf{Z}[a, b, c, d, e, f, g, h]$. The group scheme PGL$_2$ acts on $M_2 \times M_2$ diagonally from the right by

$$(2.1.1) \qquad Ad(A, B) := (A, B) \circ Ad(X) = (X^{-1}AX, X^{-1}BX)$$

for $X \in \text{PGL}_2$. So we get the dual homomorphism among the coordinate rings

$$(2.1.2) \qquad Ad : \mathbf{Z}[M_2 \times M_2] \longrightarrow \mathbf{Z}[M_2 \times M_2] \otimes_{\mathbf{Z}} A(\text{PGL}_2)$$

sending the entries of (A, B) to the corresponding entries of $Ad(A, B)$.

(2.2) Consider the polynomial ring $\mathbf{Z}[T_1, T_2, T_3, D_1, D_2]$ of 5 variables, which will be denoted by $\mathbf{Z}[\underline{T}, \underline{D}]$. Consider the ring homomorphism:

$$(2.2.1) \qquad \iota : \mathbf{Z}[\underline{T}, \underline{D}] \longrightarrow \mathbf{Z}[M_2 \times M_2],$$

given by

$$(2.2.2) \qquad \iota(T_1) := \operatorname{tr}(A), \quad \iota(T_2) := \operatorname{tr}(B), \quad \iota(T_3) := \operatorname{tr}(AB),$$
$$\iota(D_1) := \det(A), \quad \iota(D_2) := \det(B).$$

The image of ι are invariants by the PGL$_2$-action.

Lemma A ([S4]). $\mathbf{Z}[M_2 \times M_2]$ *is a free module over* $\mathbf{Z}[\underline{T}, \underline{D}]$.

In fact, one can choose monomials $c^p f^q m_r a^\varepsilon e^\nu$ for $p, q \in \mathbf{Z}_{\geq 0}$ $r \in \mathbf{Z}$ and $\varepsilon, \nu \in \{0, 1\}$ as basis, where $m_r := b^r$ for $r \in \mathbf{Z}_{\geq 0}$ and $m_r := g^{-r}$ for $r \in \mathbf{Z}_{\leq 0}$. As a consequence of the lemma A, $\mathbf{Z}[M_2 \times M_2]$ *is faithfully flat over* $\mathbf{Z}[\underline{T}, \underline{D}]$. In particular, the map ι (2.2.1) is injective. Therefore, we shall regard $\mathbf{Z}[\underline{T}, \underline{D}]$ as a subring of $\mathbf{Z}[M_2 \times M_2]$.

(2.3) Let R be a commutative $\mathbf{Z}[\underline{T}, \underline{D}]$-algebra with 1. The PGL$_2$ action on $\mathbf{Z}[M_2 \times M_2]$ (2.1.2) extends to an action on $R \otimes_{\mathbf{Z}[T,D]} \mathbf{Z}[M_2 \times M_2]$ by letting PGL$_2$ act trivially on R. The next result is the key to the existence of the universal quotient $M_2 \times M_2 // \text{PGL}_2$.

Lemma B ([S4]). *Let R be any $\mathbf{Z}[\underline{T}, \underline{D}]$-algebra with 1 as above. Then,*

$$(2.3.1) \qquad R \simeq \left(R \otimes_{\mathbf{Z}[\underline{T},\underline{D}]} \mathbf{Z}[M_2 \times M_2] \right)^{\text{PGL}_2}.$$

Remark. It is well known that the invariants $\mathbf{Q}[M_2 \times M_2]^{\text{PGL}_2}$ with rational coefficients is generated by traces $\text{tr}(W)$ for $W \in \{$the monoid generated by the A and $B\}$ ([G-Y]). But this is not true for the integral coefficients. For instance, $\det(A)$ cannot be expressed by *an integral coefficient polynomial* in traces. In fact, one has

$$2 \det(A) = \text{tr}(A)^2 - \text{tr}(A^2).$$

This relation implies also an algebraic dependence relation between $\text{tr}(A^2)$ and $\text{tr}(A)$ for the case char.$= 2$, whereas $\det(A)$ and $\text{tr}(A)$ are universally algebraically independent (cf. [D] and [S4, (2.3) Remark 3.1]).

(2.4) In (2.4), (2.5) and (2.7), we introduce a concept of the discriminant Δ of the extension $\mathbf{Z}[\underline{T}, \underline{D}] \subset \mathbf{Z}[M_2 \times M_2]$ in three approaches.

Commutator of A and B. For a matrix $A = \begin{bmatrix} a & b \\ c & d \end{bmatrix}$, the A^* denotes the matrix $\begin{bmatrix} d & -b \\ -c & a \end{bmatrix}$ of cofactors so that $A^* A = A A^* = \det(A) I_2$ and $A + A^* = \text{tr}(A) I_2$. One has also the following: $\text{tr}(A^* B) = \text{tr}(A B^*) = \text{tr}(A) \text{tr}(B) - \text{tr}(AB)$ for two matrices A and B. It is straightforward to see the equalities:

$$(2.4.1) \quad \text{tr}(ABA^*B^*) = \text{tr}(BAB^*A^*) = \text{tr}(A^*B^*AB) = \text{tr}(B^*A^*BA)$$
$$= T_1^2 D_2 + T_2^2 D_1 + T_3^2 - T_1 T_2 T_3 - 2 D_1 D_2.$$

Definition. We shall call

$$(2.4.2) \qquad \Delta(A, B) := \text{tr}(ABA^*B^*) - 2 \det(A) \det(B)$$
$$= T_1^2 D_2 + T_2^2 D_1 + T_3^2 - T_1 T_2 T_3 - 4 D_1 D_2$$

the *discriminant* of the extension $\mathbf{Z}[M_2 \times M_2] \supset \mathbf{Z}[\underline{T}, \underline{D}]$ for the reason explained in (2.5). This polynomial in the specialized form $D_1 = D_2 = 1$ has been studied by many authors. One has also a useful formula:

$$(2.4.3) \qquad 4\Delta = (2T_3 - T_1 T_2)^2 - (T_1^2 - 4D_1)(T_2^2 - 4D_2).$$

(2.5) Jacobian ideal of the invariant morphism. Recall that $a, b, c, d,$ e, f, g and h are generators of the ring $\mathbf{Z}[M_2 \times M_2]$ and that T_1, T_2, T_3, D_1 and D_2 are generators of the invariant ring. The Jacobi matrix is given by

$$\frac{\partial(T_1, T_2, T_3, D_1, D_2)}{\partial(a, b, c, d, e, f, g, h)} = \begin{bmatrix} 1 & 0 & 0 & 1 & 0 & 0 & 0 & 0 \\ 0 & 0 & 0 & 0 & 1 & 0 & 0 & 1 \\ e & g & f & h & a & c & b & d \\ d & -c & -b & a & 0 & 0 & 0 & 0 \\ 0 & 0 & 0 & 0 & h & -g & -f & e \end{bmatrix}.$$

Let J be the ideal of $\mathbf{Z}[M_2 \times M_2]$ generated by all its of 5×5 minors. Then the next lemma justifies the name discriminant for Δ.

Lemma C ([S4]). *The intersection of the Jacobi ideal J with the invariant subring $\mathbf{Z}[\underline{T}, \underline{D}]$ is a principal ideal generated by the Δ.*

$$(2.5.1) \qquad J \cap \mathbf{Z}[\underline{T}, \underline{D}] = (\Delta).$$

The proof of the lemma C, using the next formula, is omitted.
(2.5.2)

$$\Delta(A, B) = \left(bg - cf\right)^2 - \left((a - d)g - (e - h)c\right)\left((a - d)f - (e - h)b\right).$$

(2.6) The previous description (2.5.1) of the discriminant Δ together with the well known Jacobian criterion implies that the extension $\mathbf{Z}[\underline{T}, \underline{D}]_\Delta \subset \mathbf{Z}[M_2 \times M_2]_\Delta$ is smooth. That is: the associated morphism between the schemes:

$$(2.6.1) \qquad \mathrm{Spec}(\mathbf{Z}[M_2 \times M_2]_\Delta) \longrightarrow \mathrm{Spec}(\mathbf{Z}[\underline{T}, \underline{D}]_\Delta)$$

is smooth. More strongly, we prove the next lemma.

Lemma D ([S4]). *The morphism (2.6.1) is a principal PGL_2-bundle with respect to the etale topology. In another word, the homomorphism:*
(2.6.2)

$$\mathbf{Z}[M_2 \times M_2] \otimes_{\mathbf{Z}[\underline{T}, \underline{D}]} \mathbf{Z}[M_2 \times M_2] \longrightarrow \mathbf{Z}[M_2 \times M_2] \otimes_{\mathbf{Z}} A(\mathrm{PGL}_2)$$
$$f \otimes g \longmapsto (f \otimes 1) \cdot Ad(g)$$

induces an isomorphism after localization by the discriminant Δ.

(2.7) Degeneration of the quadruple (I_2, A, B, AB). In this paragraph, we regard M_2 as a linear space of rank 4 spanned by the basis $\begin{bmatrix} 1 & 0 \\ 0 & 0 \end{bmatrix}, \begin{bmatrix} 0 & 1 \\ 0 & 0 \end{bmatrix}, \begin{bmatrix} 0 & 0 \\ 1 & 0 \end{bmatrix}, \begin{bmatrix} 0 & 0 \\ 0 & 1 \end{bmatrix}$. The adjoint action of $T = \begin{bmatrix} p & q \\ r & s \end{bmatrix} \in$ PGL_2 on M_2 induces $SL(M_2)$ action. For a pair (A, B) of matrices in $M_2 \times M_2$, we consider two R-linear maps:

(2.7.1) $p : R^4 \to M_2(R)$, ${}^t(x_0, x_1, x_2, x_3) \mapsto x_0 I_2 + x_1 A + x_2 B + x_3 AB$,

(2.7.2) $q : M_2(R) \to R^4$, $X \mapsto {}^t(\operatorname{tr} X, \operatorname{tr} AX, \operatorname{tr} BX, \operatorname{tr} ABX)$.

A direct calculation shows that *the determinant of the map p (resp. q) with respect to the basis of M_2 is equal to the discriminant $\Delta(A, B)$ (resp. $-\Delta(A, B)$)*. The composition map $q \circ p : R^4 \to R^4$ is represented by a matrix $QP := (\operatorname{tr}(U_i U_j))_{i,j=1}^4$ for $U_1 = I_2$, $U_2 = A$, $U_3 = B$ and $U_4 = AB$, whose entries are expressed by polynomials in T_1, T_2, T_3, D_1, D_2. Its inverse matrix is given explicitly as follows.

(2.7.3) $(QP)^{-1} = -\Delta(A, B)^{-1} T$

where

(2.7.4) $T := \begin{bmatrix} -\Delta(A,B) - 2D_1 D_2 & T_1 D_2 & T_2 D_1 & T_3 - T_1 T_2 \\ T_1 D_2 & -2D_2 & -T_3 & T_2 \\ T_2 D_1 & -T_3 & -2D_1 & T_1 \\ T_3 - T_1 T_2 & T_2 & T_1 & -2 \end{bmatrix}$

As a consequence of these considerations, we obtain the next assertion, which will be used in a proof of the main theorem in §4.

Assertion. *Let R be a commutative ring with 1, and A, $B \in M_2(R)$. Assume that $\Delta(A, B)$ is invertible in R. Then, for X, $Y \in M_2(R)$, one has the representation:*

(2.7.6) $X = \dfrac{1}{\Delta(A,B)}(I_2, A, B, AB)\, T \begin{bmatrix} \operatorname{tr}(X) \\ \operatorname{tr}(AX) \\ \operatorname{tr}(BX) \\ \operatorname{tr}(ABX) \end{bmatrix}.$

(2.7.7)

$\operatorname{tr}(XY) = \dfrac{1}{\Delta(A,B)}(\operatorname{tr}(Y), \operatorname{tr}(AY), \operatorname{tr}(BY), \operatorname{tr}(ABY))\, T \begin{bmatrix} \operatorname{tr}(X) \\ \operatorname{tr}(AX) \\ \operatorname{tr}(BX) \\ \operatorname{tr}(ABX) \end{bmatrix}.$

§3. The universal character rings $\widetilde{R}(\Gamma)$ and $R(\Gamma)$.

We introduce rings $R(\Gamma)$ and $\widetilde{R}(\Gamma)$, which we shall call *the universal charcter rings for representations of* Γ *in* SL_2 *and* GL_2 respectively. They are abstractly defined by generators and relation in (3.1). The goal of this paragraph is to prove formula (3.7.3), which is essentially used in the proof of theorem (4.3).

(3.1) Recall that we denote by $d(\gamma)$ the class of γ in $\mathbf{Z}\cdot\Gamma^{ab} = \mathbf{Z}\cdot\Gamma/[\Gamma,\Gamma]$. By definition, one has $d(e) = 1$ and $d(\gamma\delta) = d(\gamma)d(\delta) = d(\delta)d(\gamma)$ for $\gamma,\delta \in \Gamma$.

Definition. The *universal character ring* $\widetilde{R}(\Gamma)$ of representations of Γ in GL_2 is the ring over $\mathbf{Z}\cdot\Gamma^{ab}$ generated by the indeterminates $s(\gamma)$ for $\gamma \in \Gamma$ modulo the ideal generated by $s(e) - 2$ and $s(\gamma)s(\delta) - s(\gamma\delta) - d(\gamma)s(\gamma^{-1}\delta)$ for $\gamma,\delta \in \Gamma$.
(3.1.1)

$$\widetilde{R}(\Gamma) := \mathbf{Z}\cdot\Gamma^{ab}[s(\gamma),\gamma \in \Gamma]/\left(s(e) - 2, s(\gamma)s(\delta) - s(\gamma\delta) - d(\gamma)s(\gamma^{-1}\delta)\right).$$

The *universal character ring* $R(\Gamma)$ of representations of Γ in SL_2 is the quotient of $\widetilde{R}(\Gamma)$ by the ideal $J\widetilde{R}(\Gamma)$, where J is the augmentation ideal (1.1.2) of $\mathbf{Z}\cdot\Gamma^{ab}$.

$$(3.1.2)\quad R(\Gamma) := \widetilde{R}(\Gamma)/J\widetilde{R}(\Gamma)$$
$$= \mathbf{Z}[s(\gamma),\gamma \in \Gamma]/\left(s(e) - 2, s(\gamma)s(\delta) - s(\gamma\delta) - s(\gamma^{-1}\delta)\right).$$

Let us list up some relations among the indeterminates. By definition:

$$(3.1.3)\qquad\qquad s(e) = 2,$$
and
$$(3.1.4)\qquad\qquad s(\gamma)s(\delta) = s(\gamma\delta) + d(\gamma)s(\gamma^{-1}\delta)$$

for $\gamma,\delta \in \Gamma$. Substitute $\delta = e$ in (3.1.4) and apply (3.1.3), so:

$$(3.1.5)\qquad\qquad s(\gamma) = d(\gamma)s(\gamma^{-1})$$

for $\gamma \in \Gamma$. Take the difference of (3.1.4) and the (3.1.4) with γ and δ exchanged. Since $d(\gamma)s(\gamma^{-1}\delta) = d(\delta)s(\delta^{-1}\gamma)$ ((3.1.5)), one obtains:

$$(3.1.6)\qquad\qquad s(\gamma\delta) = s(\delta\gamma),$$

for $\gamma,\delta \in \Gamma$. In another words: $s(\gamma) = s(\delta^{-1}\gamma\delta)$. This means that the class of $s(\gamma)$ in $\widetilde{R}(\Gamma)$ depends only on the *conjugacy class* of γ in Γ.

(3.2) The following three term relation (cf [F-K] formula (5), p.366, [H] 2.d) and [Ho] (2.3)) can be imediately shown by a use of (3.1.4).
(3.2.1)
$$s(\alpha\beta\gamma) + s(\gamma\beta\alpha) = s(\alpha)s(\beta\gamma) + s(\beta)s(\gamma\alpha) + s(\gamma)s(\alpha\beta) - s(\alpha)s(\beta)s(\gamma).$$

(3.3) We show that $\tilde{R}(\Gamma)$ is finitely generated over $\mathbf{Z} \cdot \Gamma^{ab}$ and $R(\Gamma)$ is finitely generated over $\mathbf{Z}$, if Γ is a finitely generated group. Such finiteness was asserted for the ring a of traces by Fricke [F-K] and proven in [H, 2.f)], [Ho, Theo.3.1.] and [C-S].

Proposition. *Let A be a linearly ordered subset of Γ, which generates Γ. Then $\tilde{R}(\Gamma)$ (resp. $R(\Gamma)$) is generated by* $\bigcup\limits_{m\in\mathbf{N}}\{s(\alpha_1\cdots\alpha_m)\big|\alpha_i \in A, \alpha_1 < \cdots < \alpha_m\}$ *over $\mathbf{Z} \cdot \Gamma^{ab}$ (resp. $\mathbf{Z}$).*

Let us prove this, since it is easy. It is sufficent to prove the proposition only for $\tilde{R}(\Gamma)$. Let R' be the subring of $\tilde{R}(\Gamma)$ generated by $\bigcup\limits_{m\in\mathbf{N}}\{s(\alpha_1\cdots\alpha_m)\big|$ $\alpha_i \in A, \alpha_1 < \cdots < \alpha_m\}$ over $\mathbf{Z} \cdot \Gamma^{ab}$. We show $s(\gamma) \in R'$ for $\gamma \in \Gamma$ by inunction on the length $l(W)$ of a word W in letters A expressing γ. The hypothesis for $l(W) \leq 1$ is true, for $s(\alpha) = d(\alpha)s(\alpha^{-1})$. Assume $s(W) \in R'$ for words W with $l(W) < l$. Let $\gamma \in \Gamma$ be expressed by a word $\alpha_1^{\epsilon(1)} \cdot \alpha_2^{\epsilon(2)} \cdot \ldots \cdot \alpha_l^{\epsilon(l)}$ for $\alpha_j \in A, \epsilon(j) \in \{\pm 1\}$ $1 \leq j \leq l$. For any permutation $\sigma \in \mathfrak{S}_n$, using (3.1.6) and (3.2.1) repeatedly, we see that there is $\epsilon \in \{\pm 1\}$ such that

$$s\left(\alpha_1^{\epsilon(1)} \cdot \alpha_2^{\epsilon(2)} \cdot \ldots \cdot \alpha_k^{\epsilon(k)}\right) - \epsilon s\left(\alpha_{\sigma(1)}^{\epsilon(\sigma(1))} \cdot \alpha_{\sigma(2)}^{\epsilon(\sigma(2))} \cdot \ldots \cdot \alpha_{\sigma(k)}^{\epsilon(\sigma(k))}\right)$$

is expressed as a polynomial of $s(\alpha)$ for α with $l(\alpha) < l$ and belongs to R'. Hence, we may assume that γ is of the form $\alpha_1^{k(1)} \cdot \alpha_2^{k(2)} \cdot \ldots \cdot \alpha_n^{k(n)}$ for $\alpha_i \in A$ with $\alpha_1 < \ldots < \alpha_n$ and $k(1), \ldots, k(n) \in \mathbf{Z}\backslash\{0\}$ with $|k(1)|+\ldots+|k(m)| = l$. If $k(j) = 1$ for all j, then $s(\gamma) \in R'$ by definition. Assume $k(J) \neq 1$ for some J. Put $U := \alpha_1^{k(1)}\cdots\alpha_{J-1}^{k(J-1)}$ $(= 1$ if $J = 1)$, $\alpha := \alpha_J$ and $V := \alpha_{J+1}^{k(J+1)}\cdots\alpha_n^{k(n)}$ $(=1$ if $J = 1)$ so that $\gamma = U\alpha^{k(J)}V$. Then for any integer t, one has

$$s(\gamma) = s(U\alpha^{k(J)}V) = s(\alpha^{k(J)}VU) = s(\alpha^t\alpha^{k(J)-t}VU)$$
$$= -d(\alpha^t)s(\alpha^{k(J)-2t}VU) + s(\alpha^t)s(\alpha^{k(J)-t}VU).$$

Put $t = [k(J)/2]$ so that $k(J) - 2t = 0$ or 1 according to whether $k(J)$ is even or odd. Then the right hand side belongs to R' by the induction hypothesis on the word length. □

Corollary 1. *If the group* Γ *is finitely generated, then the rings* $\widetilde{R}(\Gamma)$ *and* $R(\Gamma)$ *are finitely generated over* $\mathbf{Z}$. *Hence they are Noetherian.*

 2. *Let* $F_2 = \langle \alpha, \beta \rangle$. *Then* $\widetilde{R}(F_2)$ *is isomorphic to the polynomial ring over* $\mathbf{Z} \cdot F_2^{ab} \simeq \mathbf{Z}[d(\alpha)^{\pm 1}, d(\beta)^{\pm 1}]$ *generated by* $s(\alpha), s(\beta)$ *and* $s(\alpha\beta)$.

(3.5) For a pair α and β of elements of Γ, we introduce a concept of *discriminant* $\Delta(\alpha, \beta) \in \widetilde{R}(\Gamma)$ inspired by (2.4.2).
(3.5.1)

$$\Delta(\alpha, \beta) := \big(s(\alpha\beta\alpha^{-1}\beta^{-1}) - 2 \big) d(\alpha\beta).$$

$$= d(\beta)s(\alpha)^2 + d(\alpha)s(\beta)^2 + s(\alpha\beta)^2 - s(\alpha)s(\beta)s(\alpha\beta) - 4d(\alpha)d(\beta).$$

This expression of Δ is specialized in $R(\Gamma)$ as

$$(3.5.2) \quad \Delta(\alpha, \beta) = s(\alpha)^2 + s(\beta)^2 + s(\alpha\beta)^2 - s(\alpha)s(\beta)s(\alpha\beta) - 4.$$

Assertion. *Define a* 4×4 *matrix* $QP := (s(\xi_i\xi_j))_{i,j=1}^4$ *where* $\xi_1 = e$, $\xi_2 = \alpha$, $\xi_3 = \beta$ *and* $\xi_4 = \alpha\beta$, *Then* i) $\det(QP) = -\Delta(\alpha, \beta)^2$, *and* ii) *the inverse matrix of* QP *is given by* $-\Delta(\alpha, \beta)^{-1}T$ *in* $\widetilde{R}(\Gamma)_{\Delta(\alpha,\beta)}$, *where*

(3.5.3)

$$T = \begin{bmatrix} -\Delta(\alpha, \beta) - 2d(\alpha\beta) & s(\alpha)d(\beta) & s(\beta)d(\alpha) & s(\alpha\beta) - s(\alpha)s(\beta) \\ s(\alpha)d(\beta) & -2d(\beta) & -s(\alpha\beta) & s(\beta) \\ s(\beta)d(\alpha) & -s(\alpha\beta) & -2d(\alpha) & s(\alpha) \\ s(\alpha\beta) - s(\alpha)s(\beta) & s(\beta) & s(\alpha) & -2 \end{bmatrix}$$

We shall use the matrix T in a proof of the theorem in §4.

(3.6) The next concept of a *form* is introduced for the purpose of proving the formula (3.7.3) of this paragraph, which plays a crucial role in a proof of the main theorem in (4.4).

Definition. Let M be a left $\widetilde{R}(\Gamma)$-module. A map $h : \Gamma \to M$ is called a *form* with values in M, if for any γ and $\delta \in \Gamma$, one has a relation:

$$(3.6.1) \qquad h(\gamma\delta) + d(\gamma)h(\gamma^{-1}\delta) = s(\gamma)h(\delta).$$

An example of a form is given by $h(\gamma) := \sum s(\gamma p_i)q_i$ where $p_i \in \Gamma$ and $q_i \in M$. A form h will be called *homogeneous*, if $h(e) = 0$.

(3.7) This subsection is the goal of §3.

Lemma. *Let* h *be a homogeneous form with values in* M. *Suppose* $h(\alpha) = h(\beta) = h(\alpha\beta) = 0$ *for* $\alpha, \beta \in \Gamma$. *Then for any* $\gamma \in \Gamma$ *one has*

$$(3.7.1) \qquad\qquad \Delta(\alpha\beta)h(\gamma) = 0,$$

where $\Delta(\alpha, \beta)$ is the discriminant (3.5). Therefore, the composition of h with the localization $M \to M_{\Delta(\alpha,\beta)}$ is identically zero.

Corollary. *Let h be a form with values in M. Then, for any $\alpha, \beta, \gamma \in \Gamma$,*

$$(3.7.2) \quad h(\gamma) = -\frac{1}{\Delta(\alpha, \beta)}(s(\gamma), s(\gamma\alpha), s(\gamma\beta), s(\gamma\alpha\beta))T \begin{bmatrix} h(e) \\ h(\alpha) \\ h(\beta) \\ h(\alpha\beta) \end{bmatrix}$$

in $M_{\Delta(\alpha,\beta)}$. Here T is the 4×4 matrix (3.5.3). In particular, one has

$$(3.7.3) \quad s(\gamma\delta) = -\frac{1}{\Delta(\alpha, \beta)}(s(\gamma), s(\gamma\alpha), s(\gamma\beta), s(\gamma\alpha\beta))T \begin{bmatrix} s(\delta) \\ s(\alpha\delta) \\ s(\beta\delta) \\ s(\alpha\beta\delta) \end{bmatrix},$$

in the localization $\widetilde{R}(\Gamma)_{\Delta(\alpha,\beta)}$ for any $\alpha, \beta, \gamma, \delta \in \Gamma$.

(3.8) Remark. Magnus [M] introduced the Fricke character ring as the homomorphic image of the ring $R(F_n)$ (F_n = the free group of n generators) in the ring of functions on $\mathrm{Hom}(F_n, SL_n(\mathbf{C}))$ ($= A_n(F_n) \otimes_{\mathbf{Z}} \mathbf{C}$ /radical). The work [G-M] determined the defining ideal for the Fricke character ring.

§4. The invariant morphism π_Γ and $\tilde{\pi}_\Gamma$ for Γ

In this section, we state a main result on the adjoint action of PGL_2 on the open parts, consisting of non-abelian and irreducible representations, of the representable functors $\mathrm{Hom}(\Gamma, GL_2)$ ($= \mathrm{Spec}(\widetilde{A}_2(\Gamma))$) and $\mathrm{Hom}(\Gamma, GL_2)$ ($= \mathrm{Spec}(A_2(\Gamma))$) (cf. (1.1), (1.2)). We show that the universal quotient spaces over $\mathbf{Z}$ of these open parts exist and are given by the complements of the discriminant loci of the affine schemes for the universal character rings $\widetilde{R}(\Gamma)$ and $R(\Gamma)$ introduced in §3. The notation is recalled in (4.3) and the result is formulated in a Theorem in (4.3).

(4.1)   Let Γ be a group. Let us define a $\mathbf{Z} \cdot \Gamma^{ab}$-homomorphism:

$$(4.1.1) \qquad\qquad \widetilde{\Phi} : \widetilde{R}(\Gamma) \to \widetilde{A}_2(\Gamma),$$

by

$$s(\gamma) \mapsto tr(\sigma(\gamma)) \quad \text{and} \quad d(\gamma) \mapsto det(\sigma(\gamma))$$

from $\widetilde{A}_2(\Gamma)$ (1.2) to $\widetilde{R}(\Gamma)$ (3.1). This is well defined, since the defining relations (3.1.3) and (3.1.4) for $\widetilde{R}(\Gamma)$ are satisfied by traces $tr(\sigma(\gamma))$ (1.4). In view of $R(\Gamma) := \tilde{R}(\Gamma)/J\tilde{R}(\Gamma)$ and $A_2(\Gamma) := \tilde{A}_2(\Gamma)/J\tilde{A}_2(\Gamma)$ for the augmentation ideal J (1.1.2), $\tilde{\Phi}$ induces a homomorphism:

$$(4.1.2) \qquad\qquad \Phi : R(\Gamma) \to A_2(\Gamma).$$

Let us consider the morphisms associated to (4.1.1) and (4.1.2)

$$(4.1.3) \qquad \tilde{\pi}_\Gamma : \mathrm{Hom}(\Gamma, GL_2) = \mathrm{Spec}(\tilde{A}_2(\Gamma)) \to \mathrm{Spec}(\tilde{R}(\Gamma))$$

$$\qquad\qquad\qquad \cup \qquad\qquad\qquad \cup \qquad\qquad\qquad \cup$$

$$(4.1.4) \qquad \pi_\Gamma : \mathrm{Hom}(\Gamma, GL_2) = \mathrm{Spec}(A_2(\Gamma)) \to \mathrm{Spec}(R(\Gamma)),$$

which we shall refer as the invariant morphisms for Γ. They are the main objects of study of this paragraph. The map π_Γ is the pull-back of $\tilde{\pi}_\Gamma$, since we have

$$(4.1.5) \qquad\qquad \tilde{A}_2(\Gamma) \otimes_{\tilde{R}(\Gamma)} R(\Gamma) \simeq A_2(\Gamma).$$

(4.2)   First, we introduce the discriminant loci of the invariant morphisms as the closed subscheme in the target spaces $\mathrm{Spec}(\tilde{R}(\Gamma))$ and $\mathrm{Spec}(R(\Gamma))$ given by

$$(4.2.1) \qquad\qquad \tilde{D}_\Gamma := \bigcap_{\alpha,\beta\in\Gamma} \{\Delta(\alpha,\beta) = 0\}$$

where $\{\Delta(\alpha,\beta) = 0\} := \{\mathfrak{p} \in \mathrm{Spec}(\tilde{R}(\Gamma)) | \Delta(\alpha,\beta) \in \mathfrak{p}\}$ for $\alpha, \beta \in \Gamma$ (cf. (3.5)). The inverse image $\tilde{\pi}_\Gamma^{-1}(\tilde{D}_\Gamma)$ of the discriminant loci in the representation variety is determined in the next assertion. For a proof of it, we note the fact $\tilde{\Phi}(\Delta(\alpha,\beta)) = \Delta(\sigma(\alpha),\sigma(\beta))$ due to the definition of $\tilde{\Phi}$(cf. (2.4.2) and (1.2.4)).

Assertion. Let $\mathfrak{p} \in \mathrm{Spec}(\tilde{A}_2(\Gamma))$. Then $\mathfrak{p}$ belongs to $\tilde{\pi}_\Gamma^{-1}(\tilde{D}_\Gamma)$, if and only if the image $\sigma_\mathfrak{p}(\Gamma)$ in $GL_2(K_\mathfrak{p})$ is either abelian or reducible. Here $K_\mathfrak{p}$ is the fraction field of the integral domain $\tilde{A}_2(\Gamma)/\mathfrak{p}$ and $\sigma_\mathfrak{p}$ is the representation $\sigma_\mathfrak{p} : \Gamma \to SL_2(K_\mathfrak{p})$ induced from σ by the map $\tilde{A}_2(\Gamma) \to K_\mathfrak{p}$.

(4.3)   To state theorems below, we recall the notation:

$\Gamma := a\ group,$

$\tilde{A}_2(\Gamma)$ and $A_2(\Gamma)\text{;} = the\ rings\ representing\ the\ functors\ \mathrm{Hom}(\Gamma, GL_2)$ and $\mathrm{Hom}(\Gamma, SL_2)$ $((1.2.5), (1.2.2)$ and $(1.2.3))$, where we identify Spec $(\tilde{A}_2(\Gamma))$ and $\mathrm{Spec}(A_2(\Gamma))$ with the functors respectively,

$\sigma : \Gamma \to \mathrm{Hom}(\Gamma, GL_2(\tilde{A}_2(\Gamma))) := the\ universal\ representation\ of\ \Gamma\ (1.2.6),$

$\tilde{R}_2(\Gamma)$ and $R_2(\Gamma) := the\ universal\ character\ rings\ ((3.1.1), (3.1.2)),$

$\tilde{\pi}_\Gamma$ and $\pi_\Gamma := the\ invariant\ morphisms\ \mathrm{Spec}(\tilde{A}_2(\Gamma)) \to \mathrm{Spec}(\tilde{R}(\Gamma))$ and $\mathrm{Spec}(A_2(\Gamma)) \to \mathrm{Spec}(R(\Gamma))$ $((4.1.3)$ and $(4.1.4))$,

$\Delta(\alpha,\beta) := the\ discriminant \in \tilde{R}(\Gamma)\ for\ \alpha,\beta \in \Gamma\ ((3.5)),$

$\tilde{D}_\Gamma := the\ discriminant\ loci\ \bigcap_{\alpha,\beta\in\Gamma}\{\Delta(\alpha,\beta) = 0\}\ of\ \mathrm{Spec}(\tilde{R}(\Gamma))\ (4.2.1),$

$D_\Gamma := \tilde{D}_\Gamma \cap \mathrm{Spec}(R_2(\Gamma)).$

Theorem. *Let the notation be as above. The restrictions of the invariant morphisms $\tilde{\pi}_\Gamma$ and π_Γ to the complements of the inverse images of the discriminant loci:*

(4.3.1) $\qquad \operatorname{Hom}(\Gamma, GL_2) \setminus \tilde{\pi}_\Gamma^{-1}(\tilde{D}_\Gamma) \to \operatorname{Spec}(\tilde{R}(\Gamma)) \setminus \tilde{D}_\Gamma$

(4.3.2) $\qquad \operatorname{Hom}(\Gamma, GL_2) \setminus \tilde{\pi}_\Gamma^{-1}(\tilde{D}_\Gamma) \to \operatorname{Spec}(\tilde{R}(\Gamma)) \setminus \tilde{D}_\Gamma$

are principal PGL_2-bundles (defined over $\mathbb{Z} \cdot \Gamma^{ab}$ and $\mathbb{Z}$ respectively) with respect to the etale topology. The value of $d(\gamma)$ and $s(\gamma)$ (=generators of the rings $\tilde{R}(\Gamma)$ and $R(\Gamma)$) represents the value of the determinant and the trace of $\sigma(\gamma)$ $\gamma \in \Gamma$ of the representations σ in the fibre.

Proof. Since π_Γ is a pull-back of $\tilde{\pi}_\Gamma$ (cf. (4.1)), it is sufficient to show that (4.3.1) is a principal PGL_2 bundle. Since one has $\operatorname{Spec}(\tilde{R}(\Gamma)) \setminus \tilde{D}_\Gamma = \bigcup_{\alpha,\beta \in \Gamma}(\operatorname{Spec}(\tilde{R}(\Gamma)) \setminus \{\Delta(\alpha,\beta) = 0.\})$, the proof is reduced to each affine open set, stated in the next lemma. We shall confuse $\Delta(\alpha,\beta)$ with $\Delta(\sigma(\alpha),\sigma(\beta))$ and denote it by Δ.

Lemma. *For any pair α and β of Γ, consider the localizations $\tilde{R}(\Gamma)_\Delta$ of $\tilde{R}(\Gamma)$ by $\Delta = \Delta(\alpha,\beta)$ and $\tilde{A}_2(\Gamma)_\Delta$ of $\tilde{A}_2(\Gamma)$ by $\Delta = \Delta(\sigma(\alpha),\sigma(\beta))$. Then there exists a PGL_2-equivariant isomorphism:*

(4.3.3) $\qquad \tilde{A}_2(\Gamma)_\Delta \simeq \tilde{R}(\Gamma)_\Delta \otimes_{\mathbb{Z}[\underline{T},\underline{D}]} \mathbb{Z}[M_2 \times M_2].$

such that

$$\sigma(\alpha) = A, \quad \sigma(\beta) = B, \quad tr(\sigma(\gamma)) = s(\gamma) \quad \text{and} \quad det(\sigma(\gamma)) = d(\gamma)$$

by this identification. As consequences, one has the followings.
1. $\tilde{A}_2(\Gamma)_\Delta$ *is a free module over* $\tilde{R}(\Gamma)_\Delta$.
2. $\tilde{R}(\Gamma)_\Delta$ *is the universal PGL_2-invariant subring of* $\tilde{A}_2(\Gamma)_\Delta$. *That is: $\tilde{\Phi}$ induces a $\mathbb{Z} \cdot \Gamma^{ab}$-isomorphism:*

(4.3.4) $\qquad\qquad \tilde{R}_\Delta \simeq (\tilde{R}_\Delta \otimes_{\tilde{R}(\Gamma)} \tilde{A}_2(\Gamma))^{PGL_2}$

for any commutative ring $\tilde{R}$ with 1 over $\tilde{R}(\Gamma)$.
3. *The morphism*

(4.3.5)

$$\tilde{\pi}_{\Gamma,\Delta} : \operatorname{Hom}(\Gamma, GL_2) \setminus \{\Delta = 0\} = \operatorname{Spec}(\tilde{A}_2(\Gamma)_\Delta) \to \operatorname{Spec}(\tilde{R}(\Gamma)_\Delta)$$

is a principal bundle with respect to the adjoint action of PGL_2. That is: (4.3.5) is smooth and we have an isomorphism:

$$id \otimes Ad : \tilde{A}_2(\Gamma)_\Delta \otimes_{\tilde{R}(\Gamma)_\Delta} \tilde{A}_2(\Gamma)_\Delta \simeq \tilde{A}_2(\Gamma)_\Delta \otimes_{\mathbb{Z}} A(PGL_2)$$

over $Z \cdot \Gamma^{ab}$. Looking modulo the augmentation ideal J, one has also an isomorphism:

$$id \otimes Ad : A_2(\Gamma)_\Delta \otimes_{R(\Gamma)_\Delta} A_2(\Gamma)_\Delta \simeq A_2(\Gamma)_\Delta \otimes_Z A(GL_2).$$

(4.4) An outline of a proof of the lemma is given in (4.4)–(4.5). Lemmas 1, 2, and 3 follows from the Lemma A, B and D in §2 respectively, if we show the existence of a PGL_2-equivariant isomorphism (4.3.3).

The isomorphism (4.3.3) is shown as follows. Consider

(4.4.1) $$\tilde{A}(\alpha, \beta) := \tilde{R}(\Gamma) \otimes_{Z[\underline{T}, \underline{D}]} Z[M_2 \times M_2]$$

where the homomorphism: $Z[\underline{T}, \underline{D}] \to \tilde{R}(\Gamma)$ is given by:
(4.4.2)

$$T_1 \mapsto s(\alpha), \quad T_2 \mapsto s(\beta), \quad T_3 \mapsto s(\alpha\beta), \quad D_1 \mapsto d(\alpha) \quad \text{and} \quad D_2 \mapsto d(\beta),$$

Since $Z[M_2 \times M_2]$ is faithfully flat over $Z[\underline{T}, \underline{D}]$ (cf §2 (2.2) Lemma iii)), $\tilde{A}(\alpha, \beta)$ is faithfully flat over $\tilde{R}(\Gamma)$. This implies that the natural map $\tilde{R}(\Gamma) \to \tilde{A}(\alpha, \beta)$ is injective. So we shall regard $\tilde{R}(\Gamma)$ as a subring of $\tilde{A}(\alpha, \beta)$. We remark also the fact that the elements $D_1 = det(A)$ and $D_2 = det(B)$ are automatically invertible in $\tilde{A}(\alpha, \beta)$, since they are identified with the invertible elements $d(\alpha)$ and $d(\beta)$ in $\tilde{R}(\Gamma)$. The adjoint action of PGL_2 on $Z[M_2 \times M_2]$ (2.1.2) induces an action $\tilde{A}(\alpha, \beta) \to \tilde{A}(\alpha, \beta) \otimes_Z A(PGL_2)$ by letting PGL_2 act trivially on $\tilde{R}(\Gamma)$.

Let us construct a natural PGL_2-equivariant homomorphism

(4.4.3) $$\tilde{\Psi} : \tilde{A}(\alpha, \beta) \to \tilde{A}_2(\Gamma)$$

as follows. Let $Z[M_2 \times M_2] \to \tilde{A}_2(\Gamma)$ be the homomorphism induced by the entry-wise correspondence of $A \mapsto \sigma(\alpha)$ and $B \mapsto \sigma(\beta)$. On the other hand, we already have a homomorphism $\tilde{\Phi}$ from $\tilde{R}(\Gamma)$ to $\tilde{A}_2(\Gamma)$ (4.1.1), whose image is contained in the invariant subring. The composition of the map $Z[\underline{T}, \underline{D}] \to \tilde{R}(\Gamma)$ (4.4.1) with $\tilde{\Phi}$ gives the map $T_1 \mapsto tr(\sigma(\alpha))$, $T_2 \mapsto tr(\sigma(\beta))$, $T_3 \mapsto tr(\sigma(\alpha\beta))$, $D_1 \mapsto det(\sigma(\alpha))$, and $D_2 \mapsto det(\sigma(\beta))$. So the universality of the tensor product implies the existence of the PGL_2-equivariant map $\tilde{\Psi}$ (4.4.3) with the following commutativities:

(4.4.4) $$\tilde{\Psi}(A) = \sigma(\alpha), \tilde{\Psi}(B) = \sigma(\beta)$$
$$\tilde{\Psi}(s(\gamma)) = tr(\sigma(\gamma)), \tilde{\Psi}(d(\gamma)) = det(\sigma(\gamma)) \quad \text{for} \quad \gamma \in \Gamma.$$

(4.5) To show the isomorphy of (4.3.3), we construct a homomorphism $\tilde{\varphi}$ from $\tilde{A}_2(\Gamma)$ to $\tilde{A}(\alpha, \beta)_\Delta$ such that its localization $\tilde{\varphi}_\Delta$ by Δ induces the inversion map to $\tilde{\Psi}_\Delta$. In view of the representability of the functor $Hom(\Gamma, GL_2)$

(1.2.3), giving a homomorphism $\widetilde{\varphi}$ is equivalent to giving a representation $\sigma^* : \Gamma \to GL_2(\widetilde{A}(\alpha,\beta)_\Delta)$. We now describe a construction of a representation σ^*.

Define a map $\sigma^* : \Gamma \to M_2(\widetilde{A}(\alpha,\beta)_\Delta)$ by

$$(4.5.1) \qquad \sigma^*(\gamma) := -\frac{1}{\Delta}(I_2, A, B, AB)T \begin{bmatrix} s(\gamma) \\ s(\alpha\gamma) \\ s(\beta\gamma) \\ s(\alpha\beta\gamma) \end{bmatrix},$$

where T is a 4×4 matrices with coefficients in $\widetilde{R}(\Gamma)$ such that $-\frac{1}{\Delta}T$ is the inverse matrix of $(s(\xi_i\xi_j))_{ij=1}^4$ for $\xi_1 = e$, $\xi_2 = \alpha$, $\xi_3 = \beta$ and $\xi_4 = \alpha\beta$ (cf. (3.5.3)). One can show that *the map σ^* is a representation of Γ into $GL_2(\widetilde{A}(\alpha,\beta)_\Delta)$* (see [S4] where the formula (3.7.3) is essentially used) *such that* i) $tr(\sigma^*(\gamma)) = s(\gamma)$ *and* $det(\sigma^*(\gamma)) = d(\gamma)$ *for any* $\gamma \in \Gamma$, ii) $\sigma^*(\alpha) = A$ *and* $\sigma^*(\beta) = B$. Apply the representability (1.2.3) to σ^*. So there is a homomorphism $\widetilde{\varphi} : \widetilde{A}_2(\Gamma) \to \widetilde{A}(\alpha,\beta)_\Delta$ which is characterized by $\widetilde{\varphi}(\sigma(\gamma)) = \sigma^*(\gamma)$ for all $\gamma \in \Gamma$. In the other words, $\widetilde{\varphi}$ has the properties: $\widetilde{\varphi}(\sigma(\alpha)) = A$, $\widetilde{\varphi}(\sigma(\beta)) = B$, and $\widetilde{\varphi}(tr(\sigma(\gamma)) = s(\gamma)$, $\widetilde{\varphi}(det(\sigma(\gamma)) = d(\gamma)$ for $\gamma \in \Gamma$.

Thus we obtain the commutative diagram.

$$
\begin{array}{ccc}
\mathbf{Z}[\underline{T},\underline{D}] & \subset & \mathbf{Z}[M_2 \times M_2] \\
\downarrow & \square & \downarrow \\
\widetilde{R}(\Gamma) & \subset & \widetilde{A}(\alpha,\beta) \xrightarrow{\widetilde{\Psi}} \widetilde{A}_2(\Gamma) \\
\downarrow & \square & \downarrow \quad \swarrow{\widetilde{\varphi}} \quad \downarrow \\
\widetilde{R}(\Gamma)_\Delta & \subset & \widetilde{A}(\alpha,\beta)_\Delta \xrightarrow[\widetilde{\Psi}_\Delta]{} \widetilde{A}_2(\Gamma)_\Delta
\end{array}
$$

For any $\gamma \in \Gamma$, one has:

$$\widetilde{\Psi}_\Delta(\widetilde{\varphi}(\sigma(\gamma))) = \widetilde{\Psi}_\Delta(\sigma^*(\gamma)) = \widetilde{\Psi}_\Delta(\frac{1}{\Delta}(I_2, A, B, AB)T \begin{bmatrix} s(\gamma) \\ s(\alpha\gamma) \\ s(\beta\gamma) \\ s(\alpha\beta\gamma) \end{bmatrix})$$

$$= \frac{1}{\Delta}(\sigma(e), \sigma(\alpha), \sigma(\beta), \sigma(\alpha\beta))T \begin{bmatrix} tr\,\sigma(\gamma) \\ tr\,\sigma(\alpha\gamma) \\ tr\,\sigma(\beta\gamma) \\ tr\,\sigma(\alpha\beta\gamma) \end{bmatrix} \qquad (4.4.4)$$

$$= \sigma(\gamma). \qquad (2.7.6)$$

This implies $\widetilde{\Psi}_\Delta \circ \widetilde{\varphi}_\Delta = id_{\widetilde{A}_2(\Gamma)_\Delta}$. The facts $\widetilde{\varphi} \circ \widetilde{\Psi}(A) = \widetilde{\varphi}(\sigma(\alpha)) = A$, $\widetilde{\varphi} \circ \widetilde{\Psi}(B) = \widetilde{\varphi}(\sigma(\beta)) = B$, $\widetilde{\varphi} \circ \widetilde{\Psi}(s(\gamma)) = \widetilde{\varphi}(tr(\sigma(\gamma))) = s(\gamma)$ and $\widetilde{\varphi} \circ \widetilde{\Psi}(d(\gamma))$

$= \widetilde{\varphi}(det(\sigma(\gamma))) = d(\gamma)$ (4.4.5) imply $\widetilde{\varphi}_\Delta \circ \widetilde{\Psi}_\Delta = id_{\widetilde{A}(\alpha,\beta)_\Delta}$. Thus $\widetilde{\Psi}_\Delta$ (4.3.3) is an isomorphism. $\square$

(4.6) We state an immediate consequence of the Theorem in (4.3) for a fixed scalar ring R. Let R be any commutative ring with the unit 1. To a representation $\rho : \Gamma \to SL_2(R)$, we associate a ring homomorphism $R(\Gamma) \to R$ by putting $s(\gamma) \mapsto tr(\rho(\gamma))$ for $\gamma \in \Gamma$. So we get a map

(4.6.1) $\qquad \pi_\Gamma(R) : \mathrm{Hom}^{gr}(\Gamma, SL_2(R)) \to \mathrm{Hom}^{ring}(R(\Gamma), R)$

invariant by the adjoint action of $PGL_2(R)$ on $\mathrm{Hom}(\Gamma, SL_2(R))$.

Theorem. *For a point $t \in \mathrm{Hom}^{ring}(R(\Gamma), R)$, assume that there is a pair α, β of Γ such that $t(\Delta(\alpha, \beta))$ is invertible in R and that there exist matrices $A, B \in SL_2(R)$ with $t(\alpha) = tr(A)$, $t(\beta) = tr(B)$ and $t(\alpha\beta) = tr(AB)$. Then there exists a representation $\rho : \Gamma \to SL(2, R)$ such that $t = \pi_\Gamma(R)(\rho)$. If there exist two such representations ρ and ρ', then there is a unique $C \in PGL_2(R)$ such that $\rho'(\gamma) = C^{-1}\rho(\gamma)C$ for $\gamma \in \Gamma$.*

Supplement. *Two elements t and $t' \in \mathrm{Hom}(R(\Gamma), R)$ coincide if and only if $t(s(\alpha_1 \cdots \alpha_m)) = t'(s(\alpha_1 \cdots \alpha_m))$ for all $\alpha_i \in A$ with $\alpha_1 < \cdots < \alpha_m$ and $m \in \mathbb{Z}_{>0}$, where $A \subset \Gamma$ is an ordered set generating Γ.*

One asks a question: find a criterion for $(t_1, t_2, t_3) \in R^3$ for there to exist matrices A and $B \in SL_2(R)$ such that $t_1 = tr(A)$, $t_2 = tr(B)$ and $t_3 = tr(AB)$. In §5, we shall answer the question in the case where R is a real field.

§5. Representation variety with real coefficients

We study the representation variety with the coefficients in $\mathbf{R}$ ((5.5) Theorem). For the purpose we analyse the discriminant Δ over $\mathbf{R}$ ((5.3) Lemma E). See also [K], [G], [H1] and [Ko] for the geometry of the discriminant.

(5.1) Let Γ be a group and let $\widetilde{R}(\Gamma)$ and $R(\Gamma)$ be its universal character rings (3.1). Then we have the following invariant morphisms ((4.1) and (4.6)) defined over $\mathbf{R}$.

(5.1.1) $\qquad \widetilde{\pi}_\Gamma = \widetilde{\pi}_\Gamma(\mathbf{R}) : \mathrm{Hom}(\Gamma, \mathrm{GL}_2(\mathbf{R})) \to \mathrm{Hom}(\widetilde{R}(\Gamma), \mathbf{R}),$

$\qquad\qquad\qquad\qquad \cup \qquad\qquad\qquad\qquad\qquad \cup$

(5.1.2) $\qquad \pi_\Gamma = \pi_\Gamma(\mathbf{R}) : \mathrm{Hom}(\Gamma, \mathrm{SL}_2(\mathbf{R})) \to \mathrm{Hom}(R(\Gamma), \mathbf{R}),$

associating to a representation ρ a homomorphism φ given by $\varphi(d(\gamma)) :=$ $\det \rho(\gamma)$ and $\varphi(s(\gamma)) := \operatorname{tr}\rho(\gamma)$ for $\gamma \in \Gamma$. Recall the discriminant (4.2.1):

$$\widetilde{D}_\Gamma(\mathbf{R}) := \{\varphi \in \operatorname{Hom}(\widetilde{R}(\Gamma), \mathbf{R}) | \Delta(\alpha, \beta)(\varphi)(:= \varphi(\Delta(\alpha, \beta))) = 0 \ \forall \alpha, \beta \in \Gamma\}$$

$$D_\Gamma(\mathbf{R}) := \widetilde{D}_\Gamma(\mathbf{R}) \cap \operatorname{Hom}(R(\Gamma), \mathbf{R})$$

in the target spaces. It was shown that their fibres $\tilde{\pi}_\Gamma^{-1}(\widetilde{D}_\Gamma(\mathbf{R}))$ and $\pi_\Gamma^{-1}(D_\Gamma$ $(\mathbf{R}))$ consist of representation ρ such that $\rho(\Gamma)$ is either abelian or reducible, and that the restrictions of $\tilde{\pi}_\Gamma$ and π_Γ on their complements are principal $\mathrm{PGL}_2(\mathbf{R})$-bundles. We will describe $\operatorname{Image}(\tilde{\pi}_\Gamma) \backslash \widetilde{D}_\Gamma$ and $\operatorname{Image}(\pi_\Gamma) \backslash D_\Gamma$ as open semialgebraic sets ((5.5) Lemma F).

(5.2) First, we study the case of $\Gamma = F_2$. Recall that $\mathbf{Z}[\underline{T}, \underline{D}] = \mathbf{Z}[T_1, T_2, T_3, D_1, D_2]$ is the ring of invariants of the PGL_2-action on the space of a pair (A, B) of 2×2 matrices (2.2), where $T_1 = \operatorname{tr}(A)$, $T_2 = \operatorname{tr}(B)$, $T_3 = \operatorname{tr}(AB)$, $D_1 = \det(A)$ and $D_2 = \det(B)$. Put $\mathbf{Z}[\underline{T}] := \mathbf{Z}[\underline{T}, \underline{D}]/(D_1 - 1, D_2 - 1)$. The invariant morphisms:

$$(5.2.1) \qquad \tilde{\pi} : M_2(\mathbf{R}) \times M_2(\mathbf{R}) \to V_5 := \operatorname{Hom}(\mathbf{Z}[\underline{T}, \underline{D}], \mathbf{R})$$

$$\cup \qquad\qquad\qquad \cup$$

$$(5.2.2) \qquad \pi : \mathrm{SL}_2(\mathbf{R}) \times \mathrm{SL}_2(\mathbf{R}) \to V_3 := \operatorname{Hom}(\mathbf{Z}[\underline{T}], \mathbf{R})$$

associates to a pair (A, B) of matrices its 5 invariants $\tilde{\pi}(A, B) := (\operatorname{tr}(A),$ $\operatorname{tr}(B), \operatorname{tr}(AB), \det(A), \det(B))$. Let $\Delta := D_2 T_1^2 + D_1 T_2^2 + T_3^2 - T_1 T_2 T_3 - 4 D_1 D_2$ be the discriminant (2.4.2) of the map $\tilde{\pi}$, whose zero locus is denoted by

$$\widetilde{D}_\Delta(\mathbf{R}) := \{\varphi \in V_5 : \Delta(\varphi) := \varphi(\Delta) = 0\} \quad \text{and} \quad D_\Delta(\mathbf{R}) := \widetilde{D}_\Delta(\mathbf{R}) \cap V_3.$$

Since we treat only $\mathbf{R}$ coefficients, we shall denote them simply by $\widetilde{D}_\Delta$ and D_Δ.

(5.3) In Lemma E, we determine the image loci of the maps $\tilde{\pi}$ and π. Let us introduce an open semialgebraic subset of V_5:

$$(5.3.1) \qquad \widetilde{T}_\Delta := \{\delta_1 < 0, \delta_2 < 0, \delta_3 < 0, \Delta < 0\},$$

where we set

$$(5.3.2) \quad \delta_1 = T_1^2 - 4D_1, \ \delta_2 = T_2^2 - 4D_2 \ \text{and} \ \delta_3 = T_3^2 - 4D_1 D_2.$$

Lemma E. i) $\widetilde{T}_\Delta$ (resp. $T_\Delta := V_3 \cap \widetilde{T}_\Delta$) is a connected component of $V_5 \backslash \widetilde{D}_\Delta$ (resp. $V_3 \backslash D_\Delta$), which is topologically contractible to a point.

ii)
$$V_5 \backslash \widetilde{T}_\Delta = \tilde{\pi}(M_2(\mathbf{R}) \times M_2(\mathbf{R})),$$
$$V_3 \backslash T_\Delta = \pi(\mathrm{SL}_2(\mathbf{R}) \times \mathrm{SL}_2(\mathbf{R})).$$

iii)
$$\tilde{\pi} : M_2(\mathbf{R}) \times M_2(\mathbf{R}) \backslash \tilde{\pi}^{-1}(\tilde{D}_\Delta) \to V_5 \backslash (\tilde{D}_\Delta \cup \tilde{T}_\Delta)$$
$$\pi : \mathrm{SL}_2(\mathbf{R}) \times \mathrm{SL}_2(\mathbf{R}) \backslash \pi^{-1}(D_\Delta) \to V_3 \backslash (D_\Delta \cup T_\Delta)$$

are principal $\mathrm{PGL}_2(\mathbf{R})$-bundles.

Proof. Since iii) is a consequence of ii) and (2.6) Lemma D, we prove only i) and ii).

Put $V_2 := \mathrm{Hom}(\mathbf{Z}[D_1, D_2], \mathbf{R}) \simeq \mathbf{R}^2$. By an abuse of notation, we shall use $(\underline{T}, \underline{D}) = (T_1, T_2, T_3, D_1, D_2)$, $(\underline{T}) = (T_1, T_2, T_3)$ and $(\underline{D}) = (D_1, D_2)$ not only for the coordinates of V_5, V_3 and V_2 but also for a point of V_5, V_3 and V_2 respectively. Let $\mathfrak{p} : V_5 \to V_2$ be the natural projection. For any point $\underline{D} \in V_2$, denote by $V_{\underline{D}}$ the fibre $\mathfrak{p}^{-1}(\underline{D})$ (a 3 dimensional affine space). So $V_3 = V_{(1,1)}$.

i) Let us consider a domain in V_5:

$$(5.3.3) \qquad \tilde{C} := \{\delta_1 \le 0, \delta_2 \le 0, \delta_3 \le 0\}$$

and its projection to V_2. The image $\mathfrak{p}(\tilde{C})$ is equal to the domain $V_2^+ := \{D_1 \ge 0, D_2 \ge 0\}$. The fibre $C_{\underline{D}} := \tilde{C} \cap V_{\underline{D}}$ over a point $\underline{D} = (D_1, D_2) \in V_2^+$ is the solid cube $[-2\sqrt{D_1}, 2\sqrt{D_1}] \times [-2\sqrt{D_2}, 2\sqrt{D_2}] \times [-2\sqrt{D_1 D_2}, 2\sqrt{D_1 D_2}]$ (which degenerates to an edge or a point if $D_1 D_2 = 0$). We denote by $(\varepsilon_1, \varepsilon_2, \varepsilon_3)$ for $\varepsilon_i \in \{\pm 1\}$ the 8 vertices of the cube indicating the sign of the coordinates T_1, T_2 and T_3 at the vertex. We see easily that $D_\Delta \cap C_{\underline{D}}$ (=the discriminant cut by the cube $C_{\underline{D}}$) is homeomorphic to a *tetrahedron* such that 4 vertices of the tetrahedron are the vertices of the cube $(\varepsilon_1, \varepsilon_2, \varepsilon_3)$ with $\varepsilon_1 \varepsilon_2 \varepsilon_3 = 1$, 6 edges of the tetrahedron are the diagonals of the faces of the cube connecting the 4 vertices, and 4 faces of the tetrahedron are the smooth surfaces inside the cube $\overset{\circ}{C}_{\underline{D}}$ with the boundary described above. (Let H be the Hessian matrix of Δ as a function in $(\underline{T})$. Then all of its principal 2×2 submatrices are positive definite and $\det(H) = -2\Delta$.) Therefore, $C_{\underline{D}} \backslash \tilde{D}_\Delta$ decomposes into 5 connected components: $T_{\underline{D}} := \tilde{T}_\Delta \cap V_{\underline{D}}$ = the solid tetrahedron surrounded by $\tilde{D}_\Delta \cap C_{\underline{D}}$ and 4 cones joining the vertices $(\varepsilon_1, \varepsilon_2, \varepsilon_3)$ $(\varepsilon_1 \varepsilon_2 \varepsilon_3 = -1)$ with the corresponding faces of the tetrahedron. (In fact, these 4 cones belong to the same connected component in $V_D \backslash \tilde{D}_\Delta$.) Of course $T_{\underline{D}} = \phi$ unless $\underline{D} \in V_2^+$. Thus the $\tilde{T}_\Delta$ (resp. $T_{(1,1)}$) is shown to be a connected component of $V_5 \backslash \tilde{D}_\Delta$ (resp. $V_3 \backslash D_\Delta$). The $T_{(1,1)}$ is a solid tetrahedron and is contractible. The map $(\underline{T}, \underline{D}) \in \tilde{T}_\Delta \mapsto (T_1/\sqrt{D_1}, T_2/\sqrt{D_2}, T_3/\sqrt{D_1 D_2}, D_1, D_2) \in [-2, 2]^3 \times V_2^+$ induces a homeomorphism: $\tilde{T}_\Delta \simeq T_{(1,1)} \times V_2^+$. So $\tilde{T}_\Delta$ is contractible.

ii) We need to show two facts: 1) for any $(\underline{T}, \underline{D}) \in V_5 \backslash \tilde{T}_\Delta$ there exists $(A, B) \in M_2(\mathbf{R}) \times M_2(\mathbf{R})$ such that $\pi(A, B) = (\underline{T}, \underline{D})$, and 2) there does not exist a pair of matrices $(A, B) \in M_2(\mathbf{R})^2$ whose invariant $\pi(A, B)$ belongs to $\tilde{T}_\Delta$.

Case. $\delta_1 > 0$. The equation $\lambda^2 - T_1\lambda + D_1 = 0$ has two real distinct roots, say λ and μ. Then one finds a normal form of a pair of matrices $\left(\begin{bmatrix} \lambda & 0 \\ 0 & \mu \end{bmatrix}, \begin{bmatrix} p & q \\ 1 & r \end{bmatrix} \right)$ with $\begin{bmatrix} p \\ r \end{bmatrix} = \frac{1}{\lambda - \mu} \begin{bmatrix} \mu & -1 \\ -1 & 1 \end{bmatrix} \begin{bmatrix} T_1 \\ T_3 \end{bmatrix}$ and $q = -\Delta/\delta_1$, whose invariants are the given $(\underline{T}, \underline{D})$ (cf. [S4 Appendix I.2)]).

Case. $\delta_1 = 0$ and $\Delta \neq 0$. The equation $\lambda^2 - T_1\lambda + D_1 = 0$ has a real multiple root λ. Then one finds a normal form of a pair of matrices $\left(\begin{bmatrix} \lambda & 0 \\ 0 & \lambda \end{bmatrix}, \begin{bmatrix} 0 & p \\ q & r \end{bmatrix} \right)$ with $r = T_2$, $q = T_2 - \lambda T_2$ and $p = -D_2/q$ (where $\Delta = q^2$) whose invariants are the given $(\underline{T}, \underline{D})$ (cf. [S4 Appendix I.5)]).

Similarly, the set $\{\delta_2 > 0\}$ and the set $\{\delta_2 = 0, \Delta \neq 0\}$ are in Image($\tilde{\pi}$).

Case. $\delta_1 < 0$ and $\Delta \geq 0$. Then one finds a normal form of a pair of matrices $\left(\begin{bmatrix} c & s \\ -s & c \end{bmatrix}, \begin{bmatrix} p & q \\ r & p \end{bmatrix} \right)$ where $c = T_1/2, s = \sqrt{-\delta_1}/2, q - r = \frac{2}{\sqrt{-\delta}}(T_3 - T_1 T_2/2)$ and $q + r = \frac{2}{\sqrt{-\delta}}\sqrt{\Delta}$, whose invariants are the given $(\underline{T}, \underline{D})$ (cf. Appendix 1) of §5).

Similarly, the set $\{\delta_2 < 0, \Delta \geq 0\}$ is contained in the Image($\tilde{\pi}$).

The remaining sets to consider are $\{\delta_1 = \delta_2 = \Delta = 0\}$ and $\{\delta_1 < 0, \delta_2 < 0, \Delta < 0\}$.

The first set $\{\delta_1 = \delta_2 = \Delta = 0\}$ is exactly the set of vertices of the tetrahedrons (and their degeneration for $D_1 D_2 = 0$) considered in i). One has $D_1 \geq 0$ and $D_2 \geq 0$. Clearly the pair of matrices

$$\left(\varepsilon_1 \begin{bmatrix} \sqrt{D_1} & * \\ 0 & \sqrt{D_1} \end{bmatrix}, \varepsilon_2 \begin{bmatrix} \sqrt{D_2} & * \\ 0 & \sqrt{D_2} \end{bmatrix} \right)$$

has the required invariants at the vertex $(\varepsilon_1, \varepsilon_2, \varepsilon_1\varepsilon_2)$.

Thus we have shown: $\{\delta_1 < 0, \delta_2 < 0, \Delta < 0\} \cup \text{Image}(\tilde{\pi}) = V_5$.

We shall see in the Appendix of this §5 that if $(A, B) \in M_2(\mathbf{R})^2$ with $\pi(A, B) = (\underline{T}, \underline{D})$ satisfies $\delta_1 < 0$, then automatically $\Delta(A, B) \geq 0$. This implies that $\{\delta_1 < 0, \Delta < 0\} \cap \text{Image}(\tilde{\pi}) = \phi$. Therefore we have the inclusions

$$V_5 \backslash \text{Image}(\tilde{\pi}) \subset \{\delta_1 < 0, \delta_2 < 0, \Delta < 0\} \subset \{\delta_1 < 0, \Delta < 0\} \subset V_5 \backslash \text{Image}(\tilde{\pi})$$

and hence the equality. Finally the fact $\{\delta_1 < 0, \delta_2 < 0, \Delta < 0\} = \tilde{T}_\Delta$ follows from the fact $\{\delta_1 < 0, \delta_2 < 0, \Delta < 0\} \subset \{\delta_3 < 0\}$, the proof of which is left to the reader.

These complete a proof of the lemma. □

Lemma E bis. *The domain $\tilde{T}_\Delta$ has the expressions:*
(5.3.4)
$$\tilde{T}_\Delta = \{\delta_1 < 0, \delta_2 < 0, \delta_3 < 0, \Delta < 0\}$$
$$= \{\delta_1 < 0, \delta_2 < 0, \Delta < 0\} = \{\delta_1 < 0, \Delta < 0\} = \{\delta_2 < 0, \Delta < 0\}.$$

Let us decompose the spaces V_5 and V_3 into semialgebraic sets.

(5.3.5) $\qquad V_5 = \tilde{D}_\Delta \amalg \tilde{H}_\Delta \amalg \tilde{T}_\Delta \quad$ and $\quad V_3 = D_\Delta \amalg H_\Delta \amalg T_\Delta,$

where $\tilde{T}_\Delta, T_\Delta := V_3 \cap \tilde{T}_\Delta$, $\tilde{H}_\Delta := V_5 \backslash (\tilde{D}_\Delta \amalg \tilde{T}_\Delta)$ and $H_\Delta := V_3 \backslash (D_\Delta \amalg T_\Delta) = V_3 \cap \tilde{H}_\Delta$ are open semialgebraic subsets of V_5 and V_3 respectively. As we have seen, T_Δ is relatively compact but H_Δ consists of 5 non-bounded components. In this notation, lemma E can be paraphrased as Image($\tilde{\pi}$) = $\tilde{D}_\Gamma \amalg \tilde{H}_\Gamma$ and Image(π) = $D_\Delta \amalg H_\Delta$.

(5.4) We return to the analysis of the target spaces of $\tilde{\pi}_\Gamma$ and π_Γ (5.1). For any given $\alpha, \beta \in \Gamma$, consider the homomorphism $F_2 \to \Gamma$ associating the two generators of F_2 to α and β. This induces commutative diagrams:

(5.4.1)
$$\begin{array}{ccc} \mathrm{Hom}(\Gamma, \mathrm{GL}_2(\mathbf{R})) & \longrightarrow & \mathrm{Hom}(F_2, \mathrm{GL}_2(\mathbf{R})) \\ \downarrow \tilde{\pi}_\Gamma & & \downarrow \tilde{\pi} \\ \mathrm{Hom}(\tilde{R}(\Gamma), \mathbf{R}) & \xrightarrow{\tilde{h}_{\alpha,\beta}} & \mathrm{Hom}(\mathbf{Z}[\underline{T},\underline{D}], \mathbf{R}) = V_5 \end{array}$$

(5.4.2)
$$\begin{array}{ccc} \mathrm{Hom}(\Gamma, \mathrm{SL}_2(\mathbf{R})) & \longrightarrow & \mathrm{Hom}(F_2, \mathrm{SL}_2(\mathbf{R})) \\ \downarrow \pi_\Gamma & & \downarrow \pi \\ \mathrm{Hom}(R(\Gamma), \mathbf{R}) & \xrightarrow{h_{\alpha,\beta}} & \mathrm{Hom}(\mathbf{Z}[\underline{T}], \mathbf{R}) = V_3 \end{array}$$

where morphisms $\tilde{h}_{\alpha,\beta}$ and $h_{\alpha,\beta}$ are induced from ring homomorphisms $T_1 \mapsto s(\alpha)$, $T_2 \mapsto s(\beta)$, $T_3 \mapsto s(\alpha\beta)$, $D_1 \mapsto d(\alpha)$ and $D_2 \mapsto d(\beta)$. The key fact is that (5.4.1) *and* (5.4.2) *are Cartesian diagrams*, that is: $\tilde{\pi}_\Gamma$ (resp. π_Γ) is a pull-back of $\tilde{\pi}$ (resp. π) on the complement of $\tilde{D}_\Delta \subset V_5$ (§4 Theorem (4.3.3)). This fact together with lemma E implies: *For $t \in \mathrm{Hom}(\tilde{R}(\Gamma), \mathbf{R})$,* a) *if $\tilde{h}_{\alpha,\beta}(t) \in \tilde{H}_\Delta$ for some $\alpha, \beta \in \Gamma \Rightarrow \tilde{\pi}_\Gamma^{-1}(t) \simeq \pi^{-1}(\tilde{h}_{\alpha,\beta}(t)) \neq \phi$, and* b) *if $\tilde{\pi}_\Gamma^{-1}(t) \neq \phi \Rightarrow \tilde{h}_{\alpha,\beta}(t) \in \tilde{D}_\Delta \amalg \tilde{H}_\Delta$ for any $\alpha, \beta \in \Gamma$.* Therefore, for the given t only one of the next three cases occurs.

$$\begin{array}{lll} 1) & \tilde{h}_{\alpha,\beta}(t) \in \tilde{D}_\Delta & \text{for all} \quad \alpha, \beta \in \Gamma. \\ 2) & \tilde{h}_{\alpha,\beta}(t) \in \tilde{H}_\Delta & \text{for some} \quad \alpha, \beta \in \Gamma. \\ 3) & \tilde{h}_{\alpha,\beta}(t) \in \tilde{T}_\Delta & \text{for some} \quad \alpha, \beta \in \Gamma. \end{array}$$

According to the latter two cases, we introduce open subsets of target spaces.

(5.4.3)
$$\tilde{H}_\Gamma := \{t \in \mathrm{Hom}(\tilde{R}(\Gamma), \mathbf{R}) | \exists \alpha, \beta \in \Gamma \text{such that } \tilde{h}_{\alpha,\beta}(t) \in \tilde{H}_\Delta\},$$
$$\tilde{T}_\Gamma := \{t \in \mathrm{Hom}(\tilde{R}(\Gamma), \mathbf{R}) | \exists \alpha, \beta \in \Gamma \text{such that } \tilde{h}_{\alpha,\beta}(t) \in \tilde{T}_\Delta\},$$

$$H_\Gamma := \tilde{H}_\Gamma \cap \mathrm{Hom}(R(\Gamma), \mathbf{R})$$

(5.4.4)
$$= \{t \in \mathrm{Hom}(R(\Gamma), \mathbf{R}) \mid \exists \alpha, \beta \in \Gamma \text{ such that } h_{\alpha, \beta}(t) \in H_\Delta\},$$

$$T_\Gamma := \tilde{T}_\Gamma \cap \mathrm{Hom}(R(\Gamma), \mathbf{R})$$

$$= \{t \in \mathrm{Hom}(R(\Gamma), \mathbf{R}) \mid \exists \alpha, \beta \in \Gamma \text{ such that } h_{\alpha, \beta}(t) \in T_\Delta\}.$$

By definition, one has the disjoint union decompositions:
(5.4.5)

$$\mathrm{Hom}(\tilde{R}(\Gamma), \mathbf{R}) = \tilde{D}_\Gamma(\mathbf{R}) \amalg \tilde{H}_\Gamma \amalg \tilde{T}_\Gamma \text{ and } \mathrm{Hom}(R(\Gamma), \mathbf{R}) = D_\Gamma(\mathbf{R}) \amalg H_\Gamma \amalg T_\Gamma.$$

We show that all components of the above decompositions are semialgebraic. First, $\tilde{D}_\Gamma(\mathbf{R})$ is a Zariski closed set given by the system of equations $\Delta(\alpha, \beta) = 0$ for $\alpha, \beta \in \Gamma$. So, one can find a system $\{(\alpha_i, \beta_i)\}_{i \in I}$ with $\sharp I < \infty$ such that $\tilde{D}_\Gamma(\mathbf{R}) = \underset{i \in I}{\cap} \{\Delta(\alpha_i, \beta_i) = 0\}$ by the Hilbert basis Theorem. Then for the same index set I, one can show the equalities:

(5.4.6) $$\tilde{H}_\Gamma = \underset{i \in I}{\cup} \tilde{h}_{\alpha_i \beta_i}^{-1}(\tilde{H}_\Delta), \quad \tilde{T}_\Gamma = \underset{i \in I}{\cup} \tilde{h}_{\alpha_i \beta_i}^{-1}(\tilde{T}_\Delta),$$

$$H_\Gamma = \underset{i \in I}{\cup} h_{\alpha_i \beta_i}^{-1}(H_\Delta) \quad \text{and} \quad T_\Gamma = \cup_{i \in I} h_{\alpha_i \beta_i}^{-1}(T_\Delta).$$

(5.5) Let us state the goal of this paragraph: Lemmas F,G and the Theorem.

Lemma F. *Let $\tilde{\pi}_\Gamma$ and π_Γ be the invariant morphisms (5.1). Then*

(5.5.1) $$\mathrm{Image}(\tilde{\pi}_\Gamma) \backslash \tilde{D}_\Gamma(\mathbf{R}) = \tilde{H}_\Gamma \quad \text{and} \quad \mathrm{Image}(\pi_\Gamma) \backslash D_\Gamma(\mathbf{R}) = H_\Gamma.$$

The proof is done already as we have introduced the sets $\tilde{H}_\Gamma$ and H_Γ. A meaning of the set T_Γ is given by the next lemma. Let us consider a morphism:

(5.5.2) $$u_\Gamma : \mathrm{Hom}(\Gamma, SU(2)) \longrightarrow \mathrm{Hom}(R(\Gamma), \mathbf{R})$$
$$\rho \longmapsto (s(\gamma) \longmapsto \mathrm{tr}(\rho(\gamma)).$$

The morphism u_Γ is well defined by the same calculation as §1(1.4).

Lemma G. *The image set of the morphism u_Γ is given by*

(5.5.3) $$\mathrm{Image}(u_\Gamma) \backslash D_\Gamma(\mathbf{R}) = T_\Gamma.$$

Proof. In view of (5.4.5) and (5.5.1), it is sufficient to show:
a) $\mathrm{Image}(u_\Gamma) \cup \mathrm{Image}(\pi_\Gamma) \supset (\mathrm{Hom}(R(\Gamma), \mathbf{R}) \backslash D_\Gamma(\mathbf{R}))$ and b) $\mathrm{Image}(u_\Gamma) \cap \mathrm{Image}(\pi_\Gamma) \subset D_\Gamma(\mathbf{R})$. A proof of a) (cf. [M-S, Prop.III.1.1]). For any $\varphi \in$

$\operatorname{Hom}(R(\Gamma), \mathbf{R}) \backslash D_\Gamma(\mathbf{R})$ there exists $\alpha, \beta \in \Gamma$ with $\varphi(\Delta(\alpha, \beta)) \neq 0$. Since C is algebraically closed, there are $A, B \in SL_2(\mathbf{C})$ such that $Ad(A, B) = \varphi(\underline{T}, 1, 1)$. Applying the proof of (4.5), we see that $\rho(\gamma) :=$

$$-\frac{1}{\varphi(\Delta(\alpha, \beta))} (I_2, A, B, AB) \cdot \varphi T \cdot {}^t(\varphi(s(\gamma)), \varphi(s(\alpha\gamma)), \varphi(s(\beta\gamma)), \varphi(s(\alpha\beta\gamma)))$$

is a representation of Γ into $SL_2(\mathbf{C})$ s.t. $\rho(\alpha) = A$, $\rho(\beta) = B$ and $\operatorname{tr} \rho(\gamma) = \varphi(s(\gamma))$ for $\gamma \in \Gamma$. The linear span $R := \mathbf{R}I_2 + \mathbf{R}A + \mathbf{R}B + \mathbf{R}AB$ is a simple algebra of rank 4 over $\mathbf{R}$, which (according to Wedderburn's classification) is isomorphic to either $M_2(\mathbf{R})$ or to a division algebra. Accordingly, one sees that either R is conjugate to $M_2(\mathbf{R}) \subset M_2(\mathbf{C})$ or $\ker(|\det|) \subset R\backslash\{0\}$ is conjugate to $SU(2)$. A proof of b). Let ρ be a representation which is conjugate to that in $SL_2(\mathbf{R})$ and that in $SU(2)$. We may assume that (A, B) is the normal form 1) of the Appendix and that there exists $X \in PGL_2(\mathbf{C})$ with $X^{-1}AX = A$ and $X^{-1}BX \in SU(2)$. This implies $q + r = 0$ and $A, B \in SO(2)$, which contradicts $\Delta(A, B) = \varphi(\Delta(\alpha, \beta)) \neq 0$. $\square$

Question. $\operatorname{Image}(\tilde{\pi}_\Gamma) = (\tilde{H}_\Gamma)^-$, $\operatorname{Image}(\pi_\Gamma) = \bar{H}_\Gamma$ and $\operatorname{Image}(u_\Gamma) = \bar{T}_\Gamma$? Does the set of representations $\operatorname{Hom}(\Gamma, SO(2))$ naturally map onto $\bar{H}_\Gamma \cap \bar{T}_\Gamma$? (Here we mean by $(*)^-$ the closure of $*$ in the classical topology.)

Combining the above (5.5.1) and (5.5.3) with the Theorem in (4.3), one has the following statements on fibrations over the open semialgebraic sets.

Theorem. *The following restrictions of the maps $\tilde{\pi}_\Gamma$, π_Γ and u_Γ*

$$\tilde{\pi}_\Gamma : \operatorname{Hom}(\Gamma, GL_2(\mathbf{R})) \backslash \tilde{\pi}_\Gamma^{-1}(\tilde{D}_\Gamma(\mathbf{R})) \to \tilde{H}_\Gamma \qquad (5.1.1)$$

$$\pi_\Gamma : \operatorname{Hom}(\Gamma, SL_2(\mathbf{R})) \backslash \pi_\Gamma^{-1}(D_\Gamma(\mathbf{R})) \to H_\Gamma \qquad (5.1.2)$$

and

$$u_\Gamma : \operatorname{Hom}(\Gamma, SU(2)) \backslash u_\Gamma^{-1}(D_\Gamma(\mathbf{R})) \to T_\Gamma \qquad (5.5.2)$$

are principal $PGL_2(\mathbf{R})$-bundles and a principal $U(2)$-bundle respectively. The values of $d(\gamma)$ and $s(\gamma)$ (= generators of the rings $\tilde{R}(\Gamma)$ and $R(\Gamma)$) at a point in the base space are equal to the values of $\det \rho(\gamma)$ and $\operatorname{tr} \rho(\gamma)$ for any representation ρ in its fibre.

Appendix. Normal form for $(A, B) \in M_2(\mathbf{R})^2$ with $\delta_1 < 0$.

Assertion. i) For any point $(\underline{T}, \underline{D}) \in \{\delta_1 < 0, \Delta \geq 0\} \in V_5$ the pair of matrices:

1)
$$\left(\begin{bmatrix} c & s \\ -s & c \end{bmatrix}, \begin{bmatrix} p & q \\ r & p \end{bmatrix} \right),$$

where $c = T_1/2$, $s = \sqrt{-\delta_1}/2$, $p = T_2/2$, $q - r = \frac{2}{\sqrt{-\delta}}(T_3 - T_1T_2/2)$ and $q + r = \frac{2}{\sqrt{-\delta}}\sqrt{\Delta}$ gives an element of $M_2(\mathbf{R})^2$ having $(\underline{T}, \underline{D})$ as the invariants. Particularly,

2) $$\Delta = -\frac{\delta}{4}(q + r)^2, \ \delta = \delta_1 = -4s^2 \text{ and } \delta_2 = 4qr.$$

ii) Let $(A, B) \in M_2(\mathbf{R})^2$ with $\delta := \delta_1 < 0$. Then $\Delta = \Delta(A, B)$ is automatically ≥ 0 and there exists $X \in \mathrm{PGL}_2(\mathbf{R})$ such that $X^{-1}(A, B)X$ is the normal form 1), denoted by $\Phi(A, B)$. If $\Delta > 0$, then such X is unique in $\mathrm{PGL}_2(\mathbf{R})$, and if $\Delta = 0$ then such X is unique in $\mathrm{PGL}_2(\mathbf{R})/\mathbf{R}^\times \mathrm{SO}(2)$.

Proof. i) This is a matter of calculation and we omit the proof.

ii) The condition implies that $4bc = \delta_1 - (a - d)^2 \leq 0$ and hence $bc \neq 0$. Consider matrices $U_1 := \begin{bmatrix} 0 & -2b \\ \sqrt{-\delta} & a - d \end{bmatrix}$, $U_2 := \begin{bmatrix} \sqrt{-\delta} & d - a \\ 0 & -2c \end{bmatrix}$. We see that $AU_i = U_i \begin{bmatrix} c & s \\ -s & c \end{bmatrix}$ for $c = T_1/2$ and $s = \sqrt{-\delta}/2$ and that $\det(U_1) = 2b\sqrt{-\delta}$ and $\det(U_2) = -2c\sqrt{-\delta}$. Therefore at least one of the U_i is invertible. For such U_i, put

$$B \cdot U_i = U_i \cdot \begin{bmatrix} p_i & q_i \\ r_i & x_i \end{bmatrix}.$$

Then the discriminant $\Delta = \Delta(A, B)$ is calculated in terms of this:

$$\Delta = (-\delta/4)((q_i + r_i)^2 + (p_i - x_i)^2).$$

In particular, this implies $\Delta(A, B) \geq 0$. A matrix commuting with $\begin{bmatrix} c & s \\ -s & c \end{bmatrix}$ has the form $\begin{bmatrix} u & v \\ -v & u \end{bmatrix}$ for $u, v \in \mathbf{R}$. Let us see that there exists a matrix X of the form $U_i \begin{bmatrix} u & v \\ -v & u \end{bmatrix}$ such that $(X^{-1}AX, X^{-1}BX)$ is the normal form 1) $=: \Phi(A, B)$ (the difference of two diagonal elements of $X^{-1}BX$ is equal to $(u^2 - v^2)(p_i - x_i) - 2uv(q_i + r_i)$ so that its discriminant, as a binary quadratic form in u and v, is equal to $(q_i + r_i)^2 + (p_i - x_i)^2 \geq 0$). If $\Delta = 0$, then any matrix X of the form $U_i \begin{bmatrix} u & v \\ -v & u \end{bmatrix}$ will work. If $\Delta = -\delta(q - r)^2 \neq 0$ then any matrix which commutes with the normal form belongs to $\mathbf{R}^\times I_2$. $\square$

§6. Teichmüller spaces $T_{g,n}$

Fricke [F-K] started to describe the moduli of Riemann surfaces in terms of characters of discrete subgroups of $\mathrm{PGL}_2(\mathbf{R})$. This idea was followed by many authors, which led to a semialgebraic representation of the Teichmüller space (which sometimes is called the Fricke moduli space) (see [K],[H],[C-S],[O],[S-S],[Ko] and others).

Following the same idea, we describe $T_{g,n}$ (= the Teichmüller space for Riemann surfaces X of genus g with ordered sets $\{\underline{x}\}$ of n distinct points of X) as a component of real-valued point sets of an affine scheme $\mathrm{Spec}(R_{g,n})$ (Theorem (6.5)). Here $R_{g,n}$ is an abstractly defined ring of universal characters over $\mathbf{Z}$ (6.5.1) such that the values (in $\mathbf{R}$) of $R_{g,n}$ at a point t of $T_{g,n}$ is an invariant of $(X,\underline{x})$, called the ring of uniformization $R(x,\underline{x})$ introduced in (6.3.1). The description leads to some questions asked in (6.8) and (6.9).

(6.1) First we recall definitions of the moduli space of curves and the Teichmüller space from Mumford [Mu2, Lec.II].

For non-negative integers g and n, we put

$$(6.1.1) \quad \mathfrak{M}_{g,n} := \left\{ \begin{array}{l} \text{isomorphism classes of } (X, x_1, \ldots, x_n) \\ \text{where } X \text{ is a curve/}\mathbf{C} \text{ of genus } g, \text{ and } x_1, \ldots, x_n \\ \text{are distinct ordered points of } X. \end{array} \right\}$$

The space $\mathfrak{M}_{g,n}$ is equipped with a structure of the $\mathbf{C}$-valued point set of a quasi-projective algebraic variety defined over $\mathbf{Q}$ (see [Mu1,2]). The structure is characterized by the property that for any smooth family $\pi : X \to S$ of curves of genus g with n disjoint sections, the induced map $S \to \mathfrak{M}_{g,n}$, associating to each $s \in S$ the class of the fibre $\pi^{-1}(s)$, is a morphism of the variety.

(6.2) It is well-known that $\mathfrak{M}_{g,n}$ has singular points. Taking the so-called level N structure ($N \geq 3$), we obtain a smooth space $\mathfrak{M}_{g,n}(N)$ which covers $\mathfrak{M}_{g,n}$ ramifying along the singularities. Among such smooth covering spaces, the Teichmüller space gives the most universal one. So let us introduce the Teichmüller space and the mapping class group acting on it. First, introduce an abstract group
(6.2.1)
$$\Gamma_{g,n} := \langle a_1, \ldots, a_g, b_1, \ldots, b_g, c_1, \ldots, c_n, \left| \left(\prod_{i=1}^{g} a_i b_i a_i^{-1} b_i^{-1} \right) \prod_{j=1}^{n} c_j \right)$$

which is isomorphic to the fundamental group of n-punctured oriented surface of genus g. Then the Teichmüller space is introduced by $T_{g,n} :=$
(6.2.2)
$$\left\{ \begin{array}{l} \text{set of } (X, \alpha, x_1, \cdots, x_n), \text{ where } X \text{ is a curve/}\mathbf{C} \text{ of genus } g, x_1, \cdots, x_n \\ \text{are distinct points of } X \text{ and } \alpha : \Gamma_{g,n} \simeq \pi(X \setminus \{x_1, \cdots, x_n\}) \text{ is an} \\ \text{isomorphism such that } \alpha(c_i) \text{ is freely homotopic to a small loop} \\ \text{around } x_i \text{ in the positive sense [and if } n = 0, \alpha \text{ is orientation pre-} \\ \text{serving], modulo the equivalence } (X, \alpha, \underline{x}) \sim (X', \alpha', \underline{x}') \text{ if there is} \\ \text{an isomorphism } \phi : X \simeq X' \text{ such that } \phi(x_i) = x_i' \text{ and such that} \\ \phi_* \circ \alpha \text{ differs from } \alpha' \text{ by an inner automorphism.} \end{array} \right\}.$$

This set obtains a complex manifold structure via deformation theory of compact complex manifolds [K-S] in such a way that the natural projection $T_{g,n} \to \mathfrak{M}_{g,n}$ is a ramified complex analytic covering. Set $\mathrm{Out}^+(\Gamma_{g,n}) :=$

(6.2.3)
$$\left\{ \sigma \in \mathrm{Aut}(\Gamma_{g,n}) \,\middle|\, \begin{array}{l} \sigma(c_i) \text{ is conjugate to } c_i \text{ (and if} \\ n = 0, \ \sigma \text{ is orientation preserving)} \end{array} \right\} \middle/ \text{Inner auts.}$$

(This is an abuse of notation, since the abstract group $\Gamma_{g,n}$ alone does not determine $\mathrm{Out}^+(\Gamma_{g,n})$). Then $\mathrm{Out}^+(\Gamma_{g,n})$ acts on $T_{g,n}$ properly discontinuously by letting $\sigma \cdot (X, \alpha, \underline{x}) = (X, \alpha \circ \sigma^{-1}, \underline{x})$, such that one has:

$$(6.2.4) \qquad\qquad \mathfrak{M}_{g,n} \simeq \mathrm{Out}^+(\Gamma_{g,n}) \setminus T_{g,n}.$$

From now on, let us assume that the Euler number $2 - 2g - n$ of $X \setminus \{\underline{x}\}$ is negative. Then $T_{g,n}$ is realized as a holomorphically convex bounded domain in $\mathbb{C}^{3g-3+n}$ (shown via the Bers embeddings).

(6.3) Since a (compact) curve/$\mathbb{C}$ is equivalent data to a compact Riemann surface, we switch terminologies from now on. Let $(X, \underline{x})$ be a pair of a compact Riemann surface X of genus g with n distinct ordered points $\underline{x}$ on it. The Poincaré uniformization theorem asserts that there exists a discrete subgroup Γ of $\mathrm{PSL}_2(\mathbb{R})$ acting on the complex upper half plane $\mathbf{H} := \{z \in \mathbb{C} \mid \mathrm{Im}(z) > 0\}$ fixed point freely such that $X \setminus \{x_1, \cdots, x_n\}$ is complex analytically isomorphic to the quotient $\Gamma \setminus \mathbf{H}$. In fact, then, Γ is naturally isomorphic to $\pi_1(X \setminus \{x_1, \cdots, x_n\}) \overset{\alpha}{\simeq} \Gamma_{g,n}$ in such a way that the $\bar{\rho}(c_i) \in \Gamma$ corresponding to c_i are parabolic (i.e. $|\mathrm{tr}\,\bar{\rho}(c_i)| = 2$).

For the given $(X, \underline{x})$, the group $\Gamma \subset \mathrm{PSL}_2(\mathbb{R})$ of uniformization of $X \setminus \{\underline{x}\}$ is unique up to conjugacy by the adjoint action of $\mathrm{PGL}_2(\mathbb{R})$. Therefore, associated to the $(X, \underline{x})$ we introduce the following subring of $\mathbb{R}$.

(6.3.1) $R(X, \underline{x}) :=$ subring of $\mathbb{R}$ generated by $\pm \mathrm{tr}(\gamma)$ for all $\gamma \in \Gamma$ over $\mathbb{Z}$.

This ring does not depend on the choice of the group Γ but depends only on the equivalence class of $(X, \underline{x})$, that is, the point in $\mathfrak{M}_{g,n}$ associated to $(X, \underline{x})$. There seems not to be a name yet for this ring in the literature. For brevity, let us call $R(X, x)$ *the ring of uniformization for* $(X, \underline{x})$. We shall see that the ring of uniformization is finitely generated over $\mathbb{Z}$ and its transcendental degree is bounded. In order to see this, we shall find a *universal ring* $R_{g,n}$ such that the ring of uniformization $R(X, \underline{x})$ *for any* $(X, x) \in \mathfrak{M}_{g,n}$ *is a homomorphic image of* $R_{g,n}$. In fact, this is the way in which we embed the Teichmüller space into the affine variety $\mathrm{Hom}(R_{g,n}, \mathbb{R})$.

(6.4) Before we state the result, we need to discuss spin structure on X. A spin on a Riemann surface X is, by definition, a line bundle $\mathcal{L}$ on X such that $\mathcal{L}^2 \simeq$ canonical bundle of X. The group $\text{Hom}(\pi_1(X), \mathbf{Z}/2\mathbf{Z})$ ($\simeq$ the 2-torsions of the Jacobi variety of X) acts simply and transitively on the set of spins over X. Replacing $(X, \underline{x})$ by a spin Riemann surface $(X, \underline{x}, \mathcal{L})$, we reformulate the definitions of $\mathfrak{M}_{g,n}$ and $T_{g,n}$ and obtain the moduli space $\widetilde{\mathfrak{M}}_{g,n}$ and the Teichmüller space $\tilde{T}_{g,n}$ for spin Riemann surfaces respectively. One can show that the spin group $\text{Hom}(\Gamma_g, \mathbf{Z}/2\mathbf{Z})$ acts freely on the set of components of $\widetilde{\mathfrak{M}}_{g,n}$ and $\tilde{T}_{g,n}$ such that their quotient spaces are $\mathfrak{M}_{g,n}$ and $T_{g,n}$ respectively. (Actually, this follows from the next description of $\tilde{T}_{g,n}$.)

(6.5) We now state the main theorem of the present paper. For the given $g \geq 0$ and $n \geq 0$ with $2 - 2g - n < 0$, let us introduce the ring of universal characters by

$$(6.5.1) \qquad R_{g,n} := R(\Gamma_{g,n})/(s(c_1) + 2, \ldots, s(c_n) + 2)$$

where the ring $R(\Gamma_{g,n})$ for a group $\Gamma_{g,n}$ ((6.2.1)) is introduced in (3.1.2). Since $R(\Gamma_{g,n})$ is finitely generated ((3.3) Prop.), so is $R_{g,n}$. The spin group $\text{Hom}(\Gamma_g, \mathbf{Z}/2\mathbf{Z})$ (= the subgroup of $\text{Hom}(\Gamma_{g,n}, \mathbf{Z}/2\mathbf{Z})$ through the identification $\Gamma_g \simeq \Gamma_{g,n}/\langle c_i \rangle$) and the mapping class group (6.2.3) act naturally on the ring $R_{g,n}$ by letting:

$$(6.5.2) \qquad (\chi, s(\gamma)) \in \text{Hom}(\Gamma_g, \mathbf{Z}/2\mathbf{Z}) \times R_{g,n} \mapsto \chi(\gamma)s(\gamma) \in R_{g,n}$$

$$(6.5.3) \qquad (\alpha, s(\gamma)) \in \text{Out}^+(\Gamma_{g,n}) \times R_{g,n} \mapsto s(\alpha^{-1}(\gamma)) \in R_{g,n}.$$

Theorem. *The spin Teichmüller space $\tilde{T}_{g,n}$ is real analytically isomorphic to a union of connected components (in classical topology) of the real affine variety $\text{Hom}(R_{g,n}, \mathbf{R})$, where the embedding map*

$$(6.5.4) \qquad \iota : \tilde{T}_{g,n} \subset \text{Hom}(R_{g,n}, \mathbf{R})$$

is equivariant with the actions of the spin group $\text{Hom}(\Gamma_g, \mathbf{Z}/2\mathbf{Z})$ and of the mapping class group $\text{Out}^+(\Gamma_{g,n})$. Furthermore, ι satisfies the property

$$(6.5.5) \qquad R(X, \underline{x}) = \iota(t)(R_{g,n})$$

for any point t (= the class of $(X, \alpha, \underline{x})$) in $\tilde{T}_{g,n}$. That is, the ring of uniformization $R(X, \underline{x})$ of $(X, \underline{x})$ ((6.3.1)) is equal to the homomorphic image of $R_{g,n}$ at the point in $T_{g,n}$ representing the class of $(X, \alpha, \underline{x})$ for any isomorphism $\alpha : \Gamma_{g,n} \simeq \pi_1(X \setminus \{\underline{x}\})$.

Corollary. *For any $(X, \underline{x}) \in \mathfrak{M}_{g,n}$, the ring of uniformization $R(X, \underline{x})$ is finitely generated over $\mathbf{Z}$ and*

$$(6.5.6) \qquad 0 \leq \text{tr} . \deg_{\mathbf{Q}}(\mathbf{Q} \otimes_{\mathbf{Z}} R(X, \underline{x})) \leq 6g - 6 + n.$$

An outline of a proof of the theorem.

We repeat the ideas of description of the Teichmüller spaces, started by Fricke [F-K] and developed by many authors including Keen [K], Helling [H], Okumura [O] and Sepällä-Sorvalli [S-S].

In view of the Poincaré uniformization theorem, the $T_{g,n}$ can be described as

$$\left\{ \begin{array}{l} \text{set of } (\Gamma, \bar{\rho}), \text{ where } \Gamma \text{ is a discrete subgroup of } \mathrm{PSL}_2(\mathbf{R}) \text{ and } \bar{\rho} \\ \text{is an isomorphism } \bar{\rho} : \Gamma_{g,n} \simeq \Gamma \text{ such that } |\operatorname{tr} \bar{\rho}(c_i)| = 2, \text{ modulo} \\ \text{the equivalence } (\Gamma, \bar{\rho}) \sim (\Gamma', \bar{\rho}') \text{ if there is } g \in \mathrm{PGL}_2(\mathbf{R}) \text{ such that} \\ g^{-1} \Gamma g = \Gamma' \text{ and } g^{-1} \bar{\rho} g = \bar{\rho}'. \end{array} \right\}$$

In order to adapt to the invariant theory developed in §1–5, we want to "lift" the $\bar{\rho} : \Gamma_{g,n} \to \mathrm{PSL}_2(\mathbf{R})$ to $\rho : \Gamma_{g,n} \to \mathrm{SL}_2(\mathbf{R})$. This is possible for any group Γ without 2-torsion [Cu] (cf. [A-A-S],[F],[Kr2],[Pa]). The ambiguity of the lifting $\bar{\rho}$ to ρ is obviously described by the action of $\chi \in \mathrm{Hom}(\Gamma_{g,n}, \mathbf{Z}/2\mathbf{Z})$ on ρ by letting $(\chi\rho)(\gamma) = \chi(\gamma)\rho(\gamma)$. On the other hand, it is shown by Okumura [O1,2] that the signs of $\operatorname{tr}(\rho(c_i))$ can be chosen to be simultaneously negative: $\operatorname{tr}(\rho(c_i)) = -2$ for $1 \leq i \leq n$. Choosing only such liftings, the ambiguity is reduced to $\mathrm{Hom}(\Gamma_g, \mathbf{Z}/2\mathbf{Z})$. On the other hand, the subgroup $\rho(\Gamma_{g,n}) \subset \mathrm{SL}_2(\mathbf{R})$ is non-abelian and irreducible. So in view of Assertion (4.2), we see that the restriction of the map $\pi_{\Gamma_{g,n}}$ (5.1.2) on the set $\{\rho \in \mathrm{Hom}(\Gamma_{g,n}, \mathrm{SL}_2(\mathbf{R})) |$ faithful, discrete and $\operatorname{tr}\rho(c_i) + 2 = 0$ for $1 \leq i \leq n\}$ induces a $\mathrm{PGL}_2(\mathbf{R})$-fibration over the base space of $\mathrm{Hom}(R(\Gamma_{g,n}), \mathbf{R}) \cap \{s(c_1) + 2 = \cdots = s(c_n) + 2 = 0\}$ $(= \mathrm{Hom}(R_{g,n}, \mathbf{R}))$ (Theorem (5.5)). So the $\tilde{T}_{g,n}$ (as the $\mathrm{PGL}_2(\mathbf{R})$ = quotient set) is embedded into the real affine algebraic variety $\mathrm{Hom}(R_{g,n}, \mathbf{R})$ by the map $\pi_{\Gamma_{g,n}}(\mathbf{R}) =: \iota$ ((5.1.2)). The fact that $\tilde{T}_{g,n}$ is an open subset of $\mathrm{Hom}(R_{g,n}, \mathbf{R})$ follows from the fact that $\pi^{-1}(\tilde{T}_{g,n})$ in open in the representation variety ([W] in case of $n = 0$, [M-S] in case of $n > 0$) and by an application of Luna's transversal slice theorem ([Lu],[Sl]) to the fibration π_Γ (Theorem (5.5)). Closedness follows from the Jørgensen's inequalities for discrete subgroups of $\mathrm{PSL}_2(\mathbf{R})$ in terms of traces [Jø] and the fact that $\mathrm{Hom}(A_2(\Gamma_{g,n})^{\mathrm{PSL}_2}, \mathbf{R})$ is a closed subset of $\mathrm{Hom}(R(\Gamma_{g,n}), \mathbf{R})$ (cf. (4.6) 1.).

The equality (6.5.5) is an immediate consequence of the definitions of ι and of the ring of uniformization (6.3.1). The corollary is a consequence of the fact $R_{g,n}$ is finitely generated and $\operatorname{tr} . \deg_{\mathbf{Q}}(R_{g,n} \otimes \mathbf{Q}) = 6g - 6 + n$.

$$\text{QED}$$

Question: Does the property (6.5.5) characterize the embedding ι (6.5.4) up to the actions of $\mathrm{Hom}(\Gamma_g, \mathbf{Z}/2\mathbf{Z})$ and $\mathrm{Out}^+(\Gamma_{g,n})$?

(6.6) *Remark.* Here one should be careful about the fact that the semi-algebraic structure on $\tilde{T}_{g,n}$ may not determine the ring $R_{g,n}$, but only the ring $R_{g,n}/I_{g,n}$ where

$$I_{g,n} := \bigcap_{t \in T_{g,n}} \ker(\iota(t)).$$

It seems to be an interesting question to determine the ideal $I_{g,n}$. Can it be a zero ideal or nilradical, or, at least, a minimal prime?

(6.7) *Example.* Let $g = n = 1$. Then $\tilde{T}_{1,1}$ is expressed as follows.

$$R_{1,1} \simeq \mathbf{Z}[x,y,z]/(x^2 + y^2 + z^2 - xyz).$$
$$\mathrm{Hom}(R_{1,1}, \mathbf{R}) = \{0\} \coprod H_2 \coprod H_2 \coprod H_2 \coprod H_2$$

where $x = s(a), y = s(b)$ and $z = s(ab)$ and the component H_2 is the hyperbolic plane. One has $\iota : \tilde{T}_{1,1} \simeq 4 \times H_2$. The spin group $(\mathbf{Z}/2)^2$ acts transitively on the components. The point $(x,y,z) \in \tilde{T}_{1,1}$ corresponds to the $PGL_2(\mathbf{R})$-orbit of a representation ρ of $\Gamma_{1,1}$ with $\rho(a) = \begin{bmatrix} \lambda & 0 \\ 0 & \mu \end{bmatrix}$ and $\rho(b) = \begin{bmatrix} p & q \\ 1 & r \end{bmatrix}$ where λ, μ are solutions of the equation $\lambda^2 - \chi\lambda + 1 = 0$, $\begin{bmatrix} p \\ r \end{bmatrix} = \frac{1}{\lambda-\mu}\begin{bmatrix} \mu & -1 \\ -\lambda & 1 \end{bmatrix}\begin{bmatrix} x \\ z \end{bmatrix}$, $q = -\frac{\Delta}{\delta}$ and $\Delta = -4$. The point $\{0\}$ corresponds to a representation $\rho : \Gamma_{1,1} \to SU(2)$ such that $\mathrm{tr}\,\rho(a) = \mathrm{tr}\,\rho(ab) = 0$ and $\mathrm{tr}\,\rho(aba^{-1}b^{-1}) = -2$.

(6.8) We have now two algebraic structures on the moduli spaces: the classical one on $\mathfrak{M}_{g,n}(N)$ $(N \geq 3)$ coming from the field of definition of the curves, and the new one on $T_{g,n}$ coming from the ring of uniformization of the curves. Symbolically speaking, they are transcendentally different, since the group $\mathrm{Out}^+(\Gamma_{g,n})$ is infinite. The attempt of the author, which motivated him to introduce the real algebraic coordinate system $R_{g,n}$ on $T_{g,n}$ studied in the present paper, is to construct real automorphic forms on $T_{g,n}$ with respect to the mapping class group (see [S2]). The original goal is as yet impenetrable. We restrict ourselves here to posing some related problems.

The first one is the following.

Problem. Study the point $(X, \underline{x})$ of $\mathfrak{M}_{g,n}$ (and $T_{g,n}$), which is algebraic in both the sense of $\mathfrak{M}_{g,n}$ and $T_{g,n}$. (Here the $(X, \underline{x})$ is algebraic in the sense of $T_{g,n} \underset{\mathrm{def}}{\Longleftrightarrow}$ if $\mathrm{tr}\,.\deg(R(X, \underline{x})) = 0$.) Can this point be characterized in terms of certain *symmetry* of (X, x), its Jacobian, or ..., etc.? Is there any process to list such points?

In case of $n = 0$, there is an intermediate space between $\mathfrak{M}_g$ and T_g. That is the period domain (inside Siegel upper half space) of abelian integrals. Does the algebraicity of a $(X, \underline{x})$ in the above two senses of $\mathfrak{M}_g$ and T_g imply also the algebraicy of the periods? Even stronger, any two algebraicities imply the third?

For instance, it is classical that in the case $g = 1, n = 0$, the proportion ω_1/ω_2 of basic periods of an elliptic curve E is imaginary quadratic if and only if $J(E)$ has a non trivial endomorphism, and then the absolute invariant $J(E) \in \mathfrak{M}_1$ is an algebraic integer. Concerning these questions, one may refer to the classical literature on complex multiplication as well as to [Co] [Wolf] [Shi] [Ta] and their references.

(6.9) To formulate the second problem, we introduce a completion of the universal character ring. For any normal subgroup Γ' of Γ, we define

$$I(\Gamma', \Gamma) := \text{ideal in } R(\Gamma) \text{ generated by } s(\gamma\delta) - s(\delta) \; \forall \gamma \in \Gamma' \text{ and } \forall \delta \in \Gamma'.$$

Let $\{\Gamma_i\}_{i \in \Lambda}$ be an inductive system (i.e. $\forall i, j \in \Lambda$, $\exists k \in \Lambda$ s.t. $\Gamma_k \subset \Gamma_i \cap \Gamma_j$) of normal subgroups of Γ and let $\widehat{\Gamma} := \varprojlim_{i \in \Lambda} \Gamma/\Gamma_i$. Then we define

$$R(\widehat{\Gamma}) := \varprojlim_{i \in \Lambda} R(\Gamma)/I(\Gamma_i).$$

Problem. Study the variety $\text{Spec}(R(\widehat{\Gamma}))$ and the action of $\text{Out}(\widehat{\Gamma})$ on it.

Example. Let Γ be the infinite cyclic group $\mathbf{Z}$. Then one has:

$$R(\widehat{\mathbf{Z}}) \simeq \varprojlim_n \mathbf{Z}[x]/(P_n(x)), \text{ where } P_n(x) := \prod_{0 < i \le [n/2]} (x - \cos(2i\pi/n)).$$

Therefore $\text{Hom}^c(R(\widehat{\mathbf{Z}}), \mathbf{R}) = \bigcup_{n=1}^{\infty} \{\cos(2i\pi/n) | 1 \le i \le [n/2]\}$ is a dense subset of the compact interval $[-2, 2]$. The Galois group $\text{Gal}(\widetilde{\mathbf{Q}}|\mathbf{Q}) = \widehat{\mathbf{Z}}^* = \text{Out}(\widehat{\mathbf{Z}})$ acts on this space.

References

[A]   W. Abikoff: The real analytic theory of Teichmüller space, Springer Lecture Notes Math., **820** (1980).

[A-A-S]   W. Abikoff, Appel, K. and Schupp, P.: Lifting surface groups to $SL(2, \mathbf{C})$, Lecture Notes in Math., Springer **971** (1983).

[B-L]   Hyman Bass and Alexander Lubotzky: Automorphism of groups and of schemes of finite type, Israel J. of Math., 44 no.1 (1983), 1–22.

[B]   Hyman Bass: Group of integral representation type, Pacific J. Math. **86** (1980), 15–51.

[Be]   Beardon, Alan F: The geometry of discrete groups., GTM, Springer Verlag, 1983.

[Br]   G.W. Brumfiel: The real spectrum compactification of Teichmüller space, Contem. Math., **74** (1988), 51–75.

[C-W]   Paula Cohen and Jürgen Wolfart: Modular embedding for some non-arithmetic Fuchsian groups, Preprint IHES/M/88/57, Nov. 1988.

[Cu]   Marc Culler: Lifting representations to covering groups., Adv. Math., **59** (1986), 64–70.

[C-S]   Marc Culler and Peter B. Shalen: Varieties of group representations and splittings of 3-manifolds, Annals of Math., **117** (1983), 109–146.

[D]   Stephen Donkin: Invariants of several matrices, Invent. Math. **110** (1992), 389–401.

[F]   G. Faltings: Real projective structures on Riemann surfaces, Compos. Math., **48** (1983), 223–269.

[F]   Formanec: The center of the ring of 4×4 generic matrices, J. Alg. **62** (1980), 304–319.

[F-K]   Robert Fricke and Felix Klein: Vorlesungen uber die Theorie der Automorphen Funktionen, **1** pp.365–370, Leipzig: B.G. Teubner 1897. Reprint: New York: Johnson Reprint Corporation, Academic Press, 1965.

[G]   William Goldman: Topological components of spaces of representations, Inventiones Math., **93** (1988), 557–607.

[G-M]   F. Gonzáles-Acuña and José María Montesinos-Amilibia: On the character variety of group representations in $SL(2, C)$ and $PSL(2, C)$, Math. Z. **214** (1993), 627–652.

[G-Y]   J.H. Grace and A. Young: The algebra of invariants, Cambridge Univ. Press, New York, 1903.

[H1]   Heinz Helling: Diskrete Untergruppen von $SL_2(\mathbf{R})$, Inventiones math., **17** (1972), 217–229.

[H2]   Heinz Helling: Ueber den Raum der kompakten Riemannschen flächen vom Geschlecht 2, J. reine angew. Math., **268/269** (1974), 286–293.

[Ho1]   Robert Horowitz: Characters of free groups represented in the two dimensional linear group, Comm. Pure Apll. Math. **25** (1972), 635–649.

[Ho2]   Robert Horowitz: Induced automorphisms of Fricke characters of free groups, Trans. Am. Math. Soc., **208** (1975), 41–50.

[J]   J. Jørgensen: On discrete groups of Möbius Transformations, Amer. J. Math., **98** (1976), 739–749.

[K1] Linda Keen: Intrinsic Moduli, Ann. of Math., **84** (1966), 404–420.

[K2] Linda Keen: On Fricke Moduli, Advances in the theory of Riemann surfaces, Ann. of Math. Studies **66** (1971), 205–2024.

[K3] Linda Keen: A correction to "On Fricke Moduli", Proc. Amer. Math. Soc., **40** (1973), 60–62.

[K4] Linda Keen: A rough fundamental domain for Teichmüller spaces, Bull. Amer. Math. soc., **83** (1977), 1199–1226.

[K-S] Kodaira, K. and Spencer, D.C.: On deformation of complex analytic structures, Ann. of Math., **67** (1958), 328–466.

[Ko] Yohei Komori: Semialgebraic description of Teichmüller space, Thesis, RIMS, Kyoto Univ., Jan., 1994.

[Kr1] Irvin Kra: Automorphic Forms and Kleinean Groups, W.A. Benjamin Inc., 1972.

[Kr2] Irvin Kra: On lifting of Kleinian groups to $SL(2, \mathbf{C})$ in differential geometry and complex analysis (Rauch H. E. Memorial Volume), Springer, Berlin, Heidelberg, New York, (1985), 181–193.

[Kr3] Irvin Kra: Maskit coordinate, Holom. Funct. & Moduli 2, Math. Sci. Research Inst. Publication **11**, Springer, 1992.

[L-M] Alexander Lubotzky and Andy R. Magid: Varieties of representations of finitely generated groups, Memoirs of the AMS, **58** No. 336, (1985).

[Lu] Luna: Slices etale, Bull. Soc. Math. France, Memoire, **33** (1973), 81–105.

[Ma] William Magnus: Rings of Fricke characters and automorphism groups of free groups, Math. Z. **170** (1980), 91–103.

[Mu1] David Mumford: Geometric Invariant Theory, Springer-Verlag, Berlin, Heidelberg 1965, Library of Congress Catalog Card Number 65–16690.

[Mu2] David Mumford: Curves and their Jacobians, c by University of Michigan 1975, All right reserved, ISBN 0-472-66000-4, Library of Congress Catalog Card No. 75–14899.

[N] Laussen Nyssen: Thesis, Lyon (1993)?

[O1] Yoshihide Okumura: Fricke moduli and Keen moduli for Fuchsian groups and a certain class of quasi-Fuchsian groups, Master Thesis, Shizuoka Univ., 1986.

[O2] Yoshihide Okumura: Global real analytic coordinates for Teichmüller spaces, Thesis, Kanazawa Univ., 1989.

[O3] Yoshihide Okumura: On lifting problem of Kleinean group into $SL(2, \mathbf{C})$, Summer Seminar on Function Theory, July 1992.

[O4] Yoshihide Okumura: On the global real analytic coordinates for Teichmüller spaces, J. Math. Soc. Japan, **42** (1990), 91–101.

[P-B-K] V.P. Platonov and V.V. Benyash-Krivets: Characters of representations of finitely generated groups, Proc. of Steklov Inst. of Math., (1991) 203–213.

[Pr1] Claudio Procesi: Finite dimensional representations of algebras, Israel J. of Math. **19** (1974), 169–182.

[Pr2] Claudio Procesi: Invariant theory of $n \times n$ matrices, Adv. Math. **19** (1976), 306–381.

[S1] Kyoji Saito: Moduli space for Fuchsian groups, algebraic analysis, II Academic Press (1988), 735–787.

[S2] Kyoji Saito: The limit element in the configuration algebra for a discrete group, A précis: Proc. ICM90, Kyoto (1990), 931–942. Preprint: RIMS-726, Nov. 1990.

[S3] Kyoji Saito: The Teichmüller space and a certain modular function from a view point of group representations, Alg. Geom. and relate Topics, Proc. Int. Symp., Inchoen, Republic of Korea, 1992.

[S4] Kyoji Saito: Representation variety of a finitely generated group into SL_2 or GL_2, Preprint RIMS-958, 1993.

[Sc] Paul Schmutz: Die Parametriserung den Teichmüllerschen Räumen.

[S-S1] Mika Seppälä and Tuomas Sorvali: Parametrization of Möbius groups acting in a disk, Comment. Math. Helvetici **61** (1986) 149–160.

[S-S2] Mika Seppälä and Tuomas Sorvali: Affine coordinates for Teichmüller spaces, Math. Ann. **284** (1989) 169–176.

[S-S3] Mika Seppälä and Tuomas Sorvali: Trace commutators of Möbius transformations, Math. Scand. **68** (1991), 53–58.

[S-S4] Mika Seppälä and Tuomas Sorvali: Geometry of Riemann surfaces and Teichmüller spaces, North-Holland and Mathematics Studies **169**, 1992.

[Sl] Peter Slodowy: Der Scheibensatz für Algebraische transformationsgruppen, in DMV Seminar Band 13, Algebraic Transformation Groups and Invariant Theory, edited by H. Kraft, P. Slodowy and T.A. Springer, Birkhäuse (1989), 89–113.

[T1] Kisao Takeuchi: On some discrete subgroups of $SL_2(\mathbf{R})$, J. Fac. Sci. Univ. Tokyo, Sec. I, **16** (1969), 97–100.

[T2] Kisao Takeuchi: Fuchsian groups contained in $SL_2(\mathbf{Q})$, J. Math. Soc. Japan, Vol. 23, No.1 (1971), 82–94.

[V] H. Vogt: Sur les invariants, fondamentaux des équations différentielles linéaires du second ordre, Ann. Sci. Ecole Norm. Sup. **(3)6**, Suppl. 3–72 (1889) (Thèse, Paris).

[W] Andre Weil: On discrete subgroups of Lie groups. I. Ann. Math. **72** (1960), 369–384, II. Ann. Math. **75** (1962) 578–602.

[Wf]   Jürgen Wolfart: Eine arithmetische Eigenschaft automorpher For-
         men zu gewiisen nicht-arithmetischen Gruppen, Math. Ann. 62
         (1983), 1–21.
[Wo1]   Scott Wolpert: The Fenchel-Nielsen deformation, Ann. of Math.
         115 (1982), 501–528.
[Wo2]   Scott Wolpert: On the symplectic geometry of deformation of a
         hyperbolic surface, Ann. of Math., 117 (1983), 207–234.
[Wo3]   Scott Wolpert: Geodesic length functions and the Nielsen problem,
         J. Diff. Geom. 25 (1987), 275–296.
[Z-V-C] Zieschang, Voght and Colderway: Lecture Note Math., Springer,
         122, 835 & 875.

*RIMS, Kyoto University, Kyoto 606, Japan

On the embedding of Gal($\bar{\mathbf{Q}}/\mathbf{Q}$) into $\widehat{GT}$

Yasutaka Ihara[†]

With an Appendix by
Michel Emsalem[*] and Pierre Lochak[**]

Introduction

Let $\mathbf{Q}$ be the rational number field, $\bar{\mathbf{Q}}$ be its algebraic closure in $\mathbf{C}$ (the complex number field), and $G_{\mathbf{Q}} = \mathrm{Gal}(\bar{\mathbf{Q}}/\mathbf{Q})$ be the Galois group. Let F_2 be the free group of rank two on x, y, $\hat{F}_2$ be its profinite completion, and Aut $\hat{F}_2$ be the full automorphism group of the topological group $\hat{F}_2$. In [Dr], Drinfeld defined a certain subgroup $\widehat{GT}$ of Aut $\hat{F}_2$, called the Grothendieck-Teichmüller group, and asserted that the image of the standard representation

$$G_{\mathbf{Q}} \to \mathrm{Aut}\,\hat{\pi}_1(\mathbf{P}^1 - \{0, 1, \infty\}; (0, 1)) = \mathrm{Aut}\,\hat{F}_2$$

lies in $\widehat{GT}$, with a brief indication of his method for proof. In my ICM report [Ih], I tried to make a survey also on this subject, and provided a sketch of a proof. But still not in full detail. Since then, several colleagues have asked me for more details, and so I decided to take this opportunity to publish my own verifications on the lines suggested in [Dr], of this important result of Drinfeld. For some technical points, the idea of using Puiseux series expansions to construct Galois actions on fundamental groups comes from [A-I].

§1. Preliminaries

Our aim here is first to show that one can naturally associate to each $\sigma \in G_{\mathbf{Q}}$ a certain element $f_\sigma \in \hat{F}_2$. Then we discuss some properties of f_σ ($f_\sigma \in \hat{F}_2'$, the 2,3,5-cycle relations), leaving the proof of the 5-cycle relation until §2 . For backgrounds and more details, see [Ih] §2, 3, and also the Appendix.

1.1 Let $\mathbf{Q}(t)$ be the rational function field in one variable over $\mathbf{Q}$, and $\bar{\mathbf{Q}}\{\{t\}\} = \bigcup_{N \geq 1} \bar{\mathbf{Q}}((t^{1/N}))$ be the field of formal Puiseux series over $\bar{\mathbf{Q}}$. Here, for each integer $N \geq 1$, $t^{1/N}$ is a symbolic N-th root of t, and it is such that $(t^{1/MN})^M = t^{1/N}$, all $M, N \geq 1$. Note that $\bar{\mathbf{Q}}\{\{t\}\}$ is algebraically closed.

Let M denote the maximal Galois extension of $\bar{\mathbf{Q}}(t)$ in $\bar{\mathbf{Q}}\{\{t\}\}$ which is unramified outside $t = 0, 1, \infty$. By definition, each $h \in M$ has a formal Puiseux expansion

$$(1.1.1) \qquad\qquad h = \sum_{\substack{i \in \mathbf{Z} \\ i \geq -n}} a_i t^{i/N} \qquad (a_i \in \bar{\mathbf{Q}}),$$

and this series converges for any $t = t_0 \in \mathbf{C}$ with $0 < |t_0| < 1$ for any choice of $t_0^{1/N}$. If $(0,1)_{\mathbf{C}}$ is any simply connected open set in $\mathbf{C} - \{0,1\}$ containing the open interval $(0,1)$ of $\mathbf{R}$, e.g.

$$\mathbf{C} - \{(-\infty, 0] \cup [1, \infty)\},$$

then there is a unique meromorphic function $h_{\mathbf{C}}$ on $(0,1)_{\mathbf{C}}$ whose restriction to $(0,1)$ is given by the convergent series $(1.1.1)$, where we choose $t^{1/N}$ *positive real.*

1.2 Let $\pi_1 = \pi_1(\mathbf{P}^1(\mathbf{C}) - \{0,1,\infty\}, (0,1))$ be the topological fundamental group with "base point" $(0,1)$. Since the open interval $(0,1)$ $(\subset \mathbf{R})$ is simply connected, this makes obvious sense. Then π_1 is free on x, y, where x, y are positive simple loops around $0, 1$, respectively.

$(1.2.1)$

Accordingly, we can identify π_1 with F_2. Each element $\gamma \in \pi_1$ gives rise to an automorphism $h \to \gamma(h)$ of $M/\bar{\mathbf{Q}}(t)$, where $\gamma(h)$ denotes the element of M which, as a function on $(0,1)_{\mathbf{C}}$, is obtained from $h_{\mathbf{C}}$ by analytic continuation along (any path representing) γ. This way, π_1 can be considered as a (dense) subgroup of $\mathrm{Gal}(M/\bar{\mathbf{Q}}(t))$, and by passage to the profinite completion, we obtain an isomorphism

$$(1.2.2) \qquad\qquad (\hat{F}_2 =)\hat{\pi}_1 \overset{\sim}{\to} \mathrm{Gal}(M/\bar{\mathbf{Q}}(t)).$$

1.3 Our construction of $f_\sigma \in \hat{F}_2$ $(\sigma \in G_{\mathbf{Q}})$ starts with a seemingly meaningless construction. Let $\bar{\mathbf{Q}}\{\{1-t\}\}$ denote the field isomorphic to $\bar{\mathbf{Q}}\{\{t\}\}$ but where t is replaced by $1 - t$. So each element h' of $\bar{\mathbf{Q}}\{\{1-t\}\}$ has a Puiseux expansion

$$(1.3.1) \qquad\qquad h' = \sum_{\substack{i \in \mathbf{Z} \\ i \geq -n}} b_i (1-t)^{i/N} \qquad (b_i \in \bar{\mathbf{Q}})$$

(do not try to expand it in terms of $t!$). Embed $\bar{\mathbf{Q}}(t)$ into $\bar{\mathbf{Q}}\{\{1-t\}\}$ in the obvious manner ($t = 1 - (1-t)$, etc.). Let M' denote the maximal Galois extension of $\overline{\mathbf{Q}}(t)$ in $\overline{\mathbf{Q}}\{\{1-t\}\}$ unramified outside $t = 0, 1, \infty$. Consider each $h' \in M'$ as the meromorphic function $h'_{\mathbf{C}}$ on $(0,1)_{\mathbf{C}}$ whose restriction to $(0,1)$ is given by the convergent series (1.3.1). Here, again, $(1-t)^{1/N}$ is chosen to be the positive real N-th root of $1 - t$ for each $t \in (0,1)$.

Proposition 1.3. *There is a unique isomorphism* $p : M \to M'$ *over* $\overline{\mathbf{Q}}(t)$ *such that* $h_{\mathbf{C}} = p(h)_{\mathbf{C}}$ *for any* $h \in M$.

Proof. Take any $h \in M$, and expand $h_{\mathbf{C}}$ as a Puiseux series in $1 - t$. This is possible because $h_{\mathbf{C}}$ is an algebraic function of t. The coefficients belong to $\overline{\mathbf{Q}}$ because h is algebraic over $\mathbf{Q}(t)$. $\square$

Remark. There is another obvious isomorphism $\theta : M \xrightarrow{\sim} M'$ over $\overline{\mathbf{Q}}$ which maps $\sum_i a_i t^{i/N}$ to $\sum_i a_i (1-t)^{i/N}$. (hence t to $1-t$).

1.4 *Now* $G_{\mathbf{Q}} = \mathrm{Gal}(\overline{\mathbf{Q}}/\mathbf{Q})$ *acts on* $\overline{\mathbf{Q}}\{\{t\}\}$ *and* $\overline{\mathbf{Q}}\{\{1-t\}\}$ *via the action on Puiseux coefficients. It acts trivially on their common subfield* $\mathbf{Q}(t)$. *Therefore, these actions stabilize the subfields* M *and* M' *which are Galois over* $\mathbf{Q}(t)$, *and define the* $G_{\mathbf{Q}}$-*actions on* M *and* M', *denoted by* $\sigma_M, \sigma_{M'}$ *respectively* ($\sigma \in G_{\mathbf{Q}}$). *The element* f_σ *arises because the isomorphism* $p : M \to M'$ *is not* $G_{\mathbf{Q}}$-*equivariant. Thus, we define*

$$f_\sigma = p^{-1} \circ \sigma_{M'} \circ p \circ \sigma_M^{-1} \qquad (\sigma \in G_{\mathbf{Q}}),$$

which is an automorphism of M *over* $\overline{\mathbf{Q}}(t)$ *and therefore can be considered (via (1.2.2)) as an element of* $\hat{F}_2$.

Remark. Note that θ (§1. 3) is $G_{\mathbf{Q}}$-equivariant.

If we regard M, M' as embedded in the field of meromorphic functions on $(0,1)_{\mathbf{C}}$, the definition of f_σ can be described in the following way. It is the element of $\hat{\pi}_1$ which maps each $H_1 \in M$ to the following $H_4 \in M$. Here, Puiseux expansions at 0 (resp. 1) will always mean those in $t, 1-t$, subject to the rules: $t^{1/N}, (1-t)^{1/N}$ *positive real for* $t \in (0,1)$.

(i) *Apply* σ^{-1} *to the Puiseux coefficients of* H_1 *at* $t = 0$. *Call the resulting element* H_2.

(ii) *Take the Puiseux expansion of* H_2 *at* $t = 1$ *and apply* σ *to the coefficients* $\Rightarrow H_3$.

(iii) *Take the Puiseux expansion of* H_3 *at* $t = 0$. $\Rightarrow H_4$.

1.5 As a first exercise, we prove

Proposition 1.5. f_σ *belongs to* $\hat{F}'_2$, *the commutator subgroup.*

Proof. Since the maximal abelian subextension of $M/\overline{\mathbf{Q}}(t)$ is generated by $t^{1/N}$ and $(1-t)^{1/N} = 1 - \cdots$ ($N \geq 1$), it suffices to prove that these elements are invariant under f_σ. To see this, take e.g. $H_1 = t^{1/N}$. Then (with the notations of §1.4) $H_2 = t^{1/N} = (1-(1-t))^{1/N} = 1-(1/N)(1-t)+\cdots$, because its value at $t = 1$ must be 1 according to the "positive real root" rule. So, the coefficients of $(1 - (1 - t))^{1/N}$ belong to $\mathbf{Q}$; hence $H_3 = H_2$; hence $H_4 = \cdots = H_1 = t^{1/N}$. The other case can be proved similarly. $\square$

Exercise. Use $(1-t^{1/N})^{1/N}$ to show that f_σ is non-trivial. (For the study of $(1 - (1 - t^{1/N}))^{1/N}, \cdots$, cf. $[\text{A-I}]_{1,2}$.)

1.6 The $G_\mathbf{Q}$-action σ_M on M gives a splitting $*$ of the exact sequence

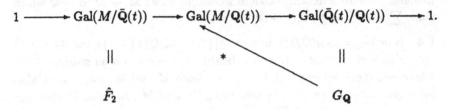

The group $\mathrm{Gal}(M/\mathbf{Q}(t))$ is thus a semi-direct product of $\hat{F}_2$ and $G_\mathbf{Q}(= *(G_\mathbf{Q}))$. Therefore, $G_\mathbf{Q}$ acts on $\hat{F}_2$ by conjugation $g \to \sigma(g) = \sigma g \sigma^{-1}$ ($\sigma \in G_\mathbf{Q}, g \in \hat{F}_2$).

Proposition 1.6. $G_\mathbf{Q}$ acts on the generators x, y of $\hat{F}_2$ as

$$\sigma(x) = x^{\chi(\sigma)}, \quad \sigma(y) = f_\sigma^{-1} y^{\chi(\sigma)} f_\sigma,$$

where $\chi : G_\mathbf{Q} \to \hat{Z}^\times$ is the cyclotomic character.

Proof. The place $M \to \bar{\mathbf{Q}} \cup (\infty)$ defined by $\sum a_i t^{i/N} \to a_0$ has $\langle x \rangle$ as its inertia group, and is $G_\mathbf{Q}$-equivariant. Therefore, $\sigma(x) = x^\lambda$ with some $\lambda \in \hat{Z}^\times$. By letting this act on $t^{1/N}$ ($N \geq 1$) we obtain $\lambda = \chi(\sigma)$. The formula for y is obtained directly from this, because $y = p^{-1} \circ \theta(x) \circ p$. (Recall that θ (§1.3) is $G_\mathbf{Q}$-equivariant (§1.4).) $\square$

1.7 We cite from [Ih] §3.1:

Theorem. *The element* $f_\sigma \in \hat{F}_2$ ($\sigma \in G_\mathbf{Q}$) *satisfies the following relations.*

(I) (2-cycle relation) $f_\sigma(x,y)f_\sigma(y,x) = 1$,
(II) (3-cycle relation) $f_\sigma(z,x)z^m f_\sigma(y,z)y^m f_\sigma(x,y)x^m = 1$
          if $xyz = 1$, $m = \frac{1}{2}(\lambda - 1)$,
(III) (5-cycle relation)

$$f_\sigma(x_{12}, x_{23})f_\sigma(x_{34}, x_{45})f_\sigma(x_{51}, x_{12})f_\sigma(x_{23}, x_{34})f_\sigma(x_{45}, x_{51}) = 1.$$

For the notation $f_\sigma(\xi, \eta)$, x_{ij}, etc., see [Ih] §2.3, 3.1. For the proofs of (I) (II), loc. cit. §3.2, and the Appendix.

The equivalence of (III) with the Drinfeld's $\hat{B}_4$-relation [Dr] is proved in the Appendix of [Ih-Ma]. (They are equivalent modulo (I) and Prop. 1.5.)

We also refer to [Dr] and [Ih] for the definition of $\widehat{GT}$. A justification of the assertion in [Dr] "$\widehat{GT}$ forms a group" is also explained in the Appendix of [Ih-Ma]. See also [LS].

So, in this note, it remains to give details for the proof of the 5-cycle relation.

§2. A geometric proof of the 5-cycle relation

This whole section is devoted to the detailed proof of the 5-cycle relation. It supplements [Dr]§4 and [Ih]§3.2. See [N] Th(A20) for a more group-theoretic proof.

2.1 Let PGL(2) be the (linear fractional) automorphism group of $\mathbf{P}^1$. It acts on $(\mathbf{P}^1)^5$ diagonally. Consider the (geometric) quotients X and X^* ($X \subset X^*$) defined by

$$(2.1.1) \quad \begin{aligned} X &= \{(x_1, \ldots, x_5) \in (\mathbf{P}^1)^5; x_i \neq x_j \text{ if } i \neq j\}/\text{PGL}(2), \\ X^* &= (\mathbf{P}^1)^5_{\text{st}}/\text{PGL}(2). \end{aligned}$$

Here, $(\mathbf{P}^1)^5_{\text{st}}$ denotes the set of all *stable* points of $(\mathbf{P}^1)^5$ w.r.t. the PGL(2)-action. Recall ([M]Ch 1) that $(x_1, \ldots, x_5) \in (\mathbf{P}^1)^5$ is stable iff there is no three distinct indices i, j, k such that $x_i = x_j = x_k$. As is well-known, and as we see below, X can be regarded as an affine open subvariety of $(\mathbf{P}^1)^2$, and X^* is a smooth compactification of X. (With the notation of [GHP]§5, $X^* = B_5 = Q_5$). And the symmetric group S_5 acts on X and X^* by substitution of coordinates.

We employ the following (non-symmetric) construction.

$$(2.1.2) \quad X = \{(\lambda, t) \in (\mathbf{P}^1)^2; \lambda, t \neq 0, 1, \infty, \ \lambda t \neq 1\}$$
$$(0, \lambda, 1, t^{-1}, \infty) \bmod \text{PGL}(2) \leftrightarrow (\lambda, t).$$

This choice of ordering of indices $(0, \lambda, 1, \ldots)$ is made because we want to have, on $X(\mathbf{R})$, $0 < \lambda < 1 < t^{-1} < \infty$ iff $0 < \lambda, t < 1$. Note that X is affine, with the coordinate ring

$$(2.1.3) \quad \begin{aligned} \Gamma(X, \mathcal{O}_X) &= \mathbf{Q}[\lambda, t, \lambda^{-1}, t^{-1}, (1-\lambda)^{-1}, (1-t)^{-1}, (1-\lambda t)^{-1}] \\ &\subset \mathbf{Q}[\lambda, t][\lambda^{-1}, t^{-1}]. \end{aligned}$$

To construct X^*, first embed X into $(\mathbf{P}^1)^2$, the product of the λ-line and the t-line, via (2.1.2). Then X^* is obtained by blowing up $(\mathbf{P}^1)^2$ at the three points $(\lambda, t) = (\infty, 0), (0, \infty), (1, 1)$. For a point (λ_0, t_0) of $(\mathbf{P}^1)^2$ outside the three points, the corresponding point of $(\mathbf{P}^1)^5_{\rm st}/{\rm PGL}(2)$ is still $(0, \lambda_0, 1, t_0^{-1}, \infty)$ mod ${\rm PGL}(2)$. The fibers of $X^* \to (\mathbf{P}^1)^2$ above $(\infty, 0), (0, \infty), (1, 1)$ will correspond to the lines $x_1 = x_3$, $x_3 = x_5$, $x_5 = x_1$ on $(\mathbf{P}^1)^5_{\rm st}/{\rm PGL}(2)$, respectively. This can be explained as follows. Take, for example, the second case. Consider a point z on the fiber above $(0, \infty)$. Such a point corresponds to a line on $(\mathbf{P}^1)^2$ passing through $(0, \infty)$. Use t^{-1} instead of t, and let $(\lambda : t^{-1}) = (a : b)$ be the equation for this line $((a : b) \in \mathbf{P}^1)$. Then z corresponds to $\lim_{\varepsilon \to 0}$ of

$$(0, a\varepsilon, 1, b\varepsilon, \infty) \equiv (0, a, \varepsilon^{-1}, b, \infty) \bmod {\rm PGL}(2),$$

which is equal to $(0, a, \infty, b, \infty)$. So, z corresponds to the ${\rm PGL}(2)$-class of $(0, a, \infty, b, \infty)$, and this induces the identification of the fiber above $(0, \infty)$ with the line $x_3 = x_5$ on X^*.

The variety X^* decomposes as

$$X^* = X^{\cup}\{\text{one pair}\}^{\cup}\{\text{two pairs}\}$$

(one pair (resp. two pairs) of coincidence of coordinates). We shall denote by l_{ij} $(i \neq j)$ the line on X^* defined by $x_i = x_j$. There are 10 such lines, each isomorphic to $\mathbf{P}^1$. Each l_{ij} meets other l_{kl} only when $\{i, j\} \cap \{k, l\} = \phi$, and in this case, at one point only. So, there are altogether 15 crossing points $l_{ij} \cap l_{kl}$ on X^* (the "two pairs").

The symmetric group S_5 acts on X^* as $(x_1, \ldots, x_5) \to (x_{s^{-1}(1)}, \ldots, x_{s^{-1}(5)})$ $(s \in S_5)$, leaving X stable. Its action on the functions and the points on X^* are related to each other by the rule $(sh)(s(x)) = h(x)$ $(s \in S_5, x \in X^*, h$: a rational function on X^*). Note that $s(l_{ij}) = l_{s(i), s(j)}$,

$$(2.1.4) \qquad \lambda = \frac{x_3 - x_5}{x_2 - x_5} \cdot \frac{x_2 - x_1}{x_3 - x_1}, \quad t = \frac{x_4 - x_5}{x_3 - x_5} \cdot \frac{x_3 - x_1}{x_4 - x_1}$$

(recall $(x_1, \ldots, x_5) \leftrightarrow (0, \lambda, 1, t^{-1}, \infty)$), and

$$(2.1.5) \qquad s\left(\frac{x_j - x_k}{x_i - x_k} \cdot \frac{x_i - x_l}{x_j - x_l}\right) = \frac{x_{s(j)} - x_{s(k)}}{x_{s(i)} - x_{s(k)}} \cdot \frac{x_{s(i)} - x_{s(l)}}{x_{s(j)} - x_{s(l)}},$$

for any $s \in S_5$ and distinct i, j, k, l.

2.2 As in [Ih]§3.1, we consider the space of all 5-triples $(x_1, \ldots, x_5)$ of distinct points on $\mathbf{R}^{\cup}(\infty)$ satisfying the condition: x_{i+1} is next to x_i in the *positive* direction on $\mathbf{R}^{\cup}(\infty)$ for all i $(1 \leq i \leq 4)$, including the case of passing through ∞. Then, $\mathrm{PGL}_2^+(\mathbf{R})$, the real projective linear group of degree two with positive determinant, acts on this space diagonally, and the quotient space $\mathcal{B}$ is naturally embedded into $X(\mathbf{C})$. The space $\mathcal{B}$ is simply connected, and the fundamental group $P_5 = \pi_1(X(\mathbf{C}), \mathcal{B})$ is generated by the elements x_{ij} $(= x_{ji})$ $(1 \le i, j \le 5, i \ne j)$ shown below (the circle is $\mathbf{R}^{\cup}(\infty)$; the inside of the circle corresponds to the lower half plane).

(2.2.1)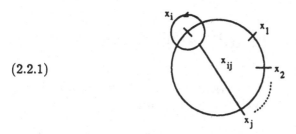

In the (λ, t)-coordinates,

$$\mathcal{B} = \{(\lambda, t) \in \mathbf{R}^2;\ 0 < \lambda, t < 1\},$$

and its closure $\mathcal{B}^*$ in $X^*(\mathbf{C})$ is given by

$$\mathcal{B}^* = \mathcal{B}^{\cup} \bigcup_{i (\mathrm{mod}\, 5)} I_{i,i+1},$$

where $I_{i,i+1}$ is the unit interval on $l_{i,i+1} \cong \mathbf{P}^1$ with respect to a suitable coordinate system.

(2.2.2)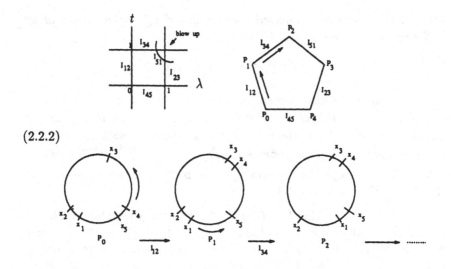

2.3 Now we consider the category Et/X of finite etale connected coverings $F : Y \to X$ of X (in the category of X-schemes). The compactification of $F : Y \to X$ is defined to be the integral closure $F^* : Y^* \to X^*$ of X^* in the function field of Y. Then Y^* is normal, and Y is a smooth open subscheme of Y^*. As X is affine, Y is also affine, whose affine ring $\Gamma(Y, \mathcal{O}_Y)$ is the integral closure of $\Gamma(X, \mathcal{O}_X)$ (the ring of (2.1.3)) in the function field of Y.

We denote by

$$\bar{\mathbf{Q}}\{\{x, y\}\} = \left(\bigcup_{N \geq 1} \bar{\mathbf{Q}}[x^{1/N}, y^{1/N}][x^{-1}, y^{-1}] \right)$$

the ring of formal Puiseux series in x, y over $\bar{\mathbf{Q}}$ (finite number of negative power terms allowed). The symbolic roots of x, y satisfy $(x^{1/NM})^M = x^{1/N}$, etc.

Put $F_{\mathbf{C}} = F \otimes_{\mathbf{Q}} \mathbf{C}$, and consider $F_{\mathbf{C}} : Y(\mathbf{C}) \to X(\mathbf{C})$ as a complex analytic mapping between complex surfaces. Then since $F_{\mathbf{C}}$ is a topological covering map and $\mathcal{B}$ is simply connected, $F_{\mathbf{C}}^{-1}(\mathcal{B})$ is a disjoint union of d $(= \deg F)$ connected components, each of which is homeomorphically mapped onto $\mathcal{B}$ by $F_{\mathbf{C}}$. We shall need the following

Proposition 2.3. *Let $F : Y \to X$ be any object of Et/X. Then there is a unique bijection*

$$\pi_0\left(F_{\mathbf{C}}^{-1}(\mathcal{B})\right) \to \mathrm{Hom}_{\Gamma(X, \mathcal{O}_X)}(\Gamma(Y, \mathcal{O}_Y), \bar{\mathbf{Q}}\{\{\lambda, t\}\})$$
$$\mathcal{D} \to \psi$$

between the set of all connected components of $F_{\mathbf{C}}^{-1}(\mathcal{B})$ and that of all $\Gamma(X, \mathcal{O}_X)$-algebra homomorphisms

$$\psi : \Gamma(Y, \mathcal{O}_Y) \to \bar{\mathbf{Q}}\{\{\lambda, t\}\},$$

characterized by the following condition (∗):
(∗) *For any $h \in \Gamma(Y, \mathcal{O}_Y)$ and $0 < \lambda_0, t_0 < 1$, the value of h at the unique point of $\mathcal{D}$ lying above $(\lambda_0, t_0) \in \mathcal{B}$ coincides with the value of the absolutely convergent power series $\sum_{i,j \geq -n} a_{ij}(\lambda_0^{1/N})^i (t_0^{1/N})^j$, where $\psi(h) = \sum_{i,j \geq -n} a_{ij} \lambda^{i/N} t^{j/N}$, and $\lambda_0^{1/N}$ (resp. $t_0^{1/N}$) denotes the positive real N-th root of λ_0 (resp. t_0).*

Proof. First, we note that $\Gamma(Y, \mathcal{O}_Y)$ is a finite $\Gamma(X, \mathcal{O}_X)$-module and hence the image of each ψ must be contained in $\bar{\mathbf{Q}}[[\lambda^{1/N}, t^{1/N}]][\lambda^{-1}, t^{-1}]$ for some N. We also note that for a given $\mathcal{D}$ (resp. ψ), there is at most one ψ (resp. $\mathcal{D}$) such that the pair $(\mathcal{D}, \psi)$ satifies (∗). The uniqueness of ψ follows directly by reducing the case to power series, using polydisc convergence criterion, and using the uniqueness theorem for complex analytic functions. The uniqueness of $\mathcal{D}$ is obvious.

Now, to prove the Proposition, it suffices to establish, for each given $\mathcal{D}$, the existence of ψ satisfying (*), because then (i) $\mathcal{D} \to \psi$ is injective and (ii) it is also clear that $\#\{\mathcal{D}\} = \deg F \geq \#\{\psi\}$. This means also that we may replace F by any object of Et/X that contains F as a subcovering.

Let $K = \mathbf{Q}(\lambda, t)$ be the function field of X, and $v_{ij}(1 \leq i, j \leq 5, i \neq j)$ be the discrete valuation of K corresponding to l_{ij} (§2.1). (Its valuation ring is the local ring $O_{X^\bullet, l_{ij}}$.) Note that v_{45}, v_{12}, v_{34} have $t, \lambda, 1 - t$, as prime elements, respectively. Let L be the function field of Y, and N be the *lcm* of ramification indices over K of all extensions of v_{45}, v_{12}, v_{34} to L. Put $L' = L(\zeta_N, t^{1/N}, \lambda^{1/N}, (1 - t)^{1/N})$, where ζ_N is a primitive N-th root of unity, and replace Y by its integral closure in L'. (So, L will be replaced by L'.) According to this construction, L contains

$$K^{(N)} = K(\zeta_N, t^{1/N}, \lambda^{1/N}, (1 - t)^{1/N}),$$

and moreover, all extensions of v_{45}, v_{12}, v_{34} to L are *unramified* in $L/K^{(N)}$. (Taking care of v_{34} is only for later purposes; not necessary for the present purpose.)

Now consider the following intermediate affine scheme X', with $X \subset X' \subset X^*(\subset$: open immersion), obtained from X^* by removing all lines l_{ij} except l_{45}, l_{12}, l_{34}. It is given as

$$X' = \operatorname{Spec} \mathbf{Q}[\lambda, t, (1 - \lambda)^{-1}, (1 - \lambda t)^{-1}].$$

Call $X^{(N)'}, Y'$ its integral closures in $K^{(N)}, L$, respectively. Then the finite morphism $Y' \to X^{(N)'}$ is etale in codimension one, and $X^{(N)'}$ is regular, being an open subscheme of $\operatorname{Spec} \mathbf{Q}[\zeta_N, \lambda^{1/N}, t^{1/N}, (1-t)^{1/N}]$. Therefore, by the purity of branch locus ([SGA 1]Exp.X §3), we conclude that $Y' \to X^{(N)'}$ is etale. Hence $Y'(\mathbf{C}) \to X^{(N)'}(\mathbf{C})$ is a topological covering map.

Now let $\mathcal{D}$ be any element of $\pi_0(F_{\mathbf{C}}^{-1}(\mathcal{B}))$, i.e., a connected component of the inverse image of $\mathcal{B}$ in $Y(\mathbf{C})$. Call $\mathcal{D}_N$ its projection on $X^{(N)}(\mathbf{C})$. We shall choose ζ_N, $t^{1/N}$, $\lambda^{1/N}$, $(1 - t)^{1/N}$ according to the rule: at each point of $\mathcal{D}_N$, *the value of ζ_N is $e^{2\pi i/N}$, and those of $t^{1/N}, \lambda^{1/N}, (1 - t)^{1/N}$ belong to the interval* $(0, 1)$ *on* $\mathbf{R}$. If $(\mathcal{D}_N)_{\mathbf{C}}$ is any simply connected open set such that

$$\mathcal{D}_N \subset (\mathcal{D}_N)_{\mathbf{C}} \subset X^{(N)'}(\mathbf{C}),$$

then the restriction of the covering map $Y'(\mathbf{C}) \to X^{(N)'}(\mathbf{C})$ to the fiber above $(\mathcal{D}_N)_{\mathbf{C}}$ splits into the disjoint union of trivial coverings of $(\mathcal{D}_N)_{\mathbf{C}}$. Call $\mathcal{D}_{\mathbf{C}}$ the unique component containing $\mathcal{D}$;

$$\mathcal{D} \subset \mathcal{D}_{\mathbf{C}} \subset Y'(\mathbf{C}).$$

Each element $h \in \Gamma(Y', O_{Y'})$ can be considered as a holomorphic function on $Y'(\mathbf{C})$; hence that on $\mathcal{D}_{\mathbf{C}}$; hence also that on $(\mathcal{D}_N)_{\mathbf{C}}$, via the projection homeomorphism. Here, we choose the following $(\mathcal{D}_N)_{\mathbf{C}}$ defined by

$$(\mathcal{D}_N)_{\mathbf{C}} : \zeta_N = e^{2\pi i/N}, \ |t^{1/N}| < 1, \ |\lambda^{1/N}| < 1, \ |arg(1-t)^{1/N}| < \pi/2N.$$

More precisely, $(\mathcal{D}_N)_{\mathbf{C}}$ consists of all points of $X^{(N)'}(\mathbf{C})$ at which the values of the *rational* (and holomorphic) functions $\zeta_N, \lambda^{1/N}, t^{1/N}, (1-t)^{1/N}$ on $X^{(N)'}$ satisfy these (in)equalities. The last condition is only giving a choice of a connected component of the open set on $X^{(N)'}(\mathbf{C})$ defined by the first three; $(\mathcal{D}_N)_{\mathbf{C}}$ is homeomorphic to the polydisc

$$\Pi : \{(\lambda^{1/N}, t^{1/N}) \in \mathbf{C}^2; \ |\lambda^{1/N}| < 1, \ |t^{1/N}| < 1\}$$

on $\mathbf{C}^2$. So, h as a holomorphic function on Π admits an absolutely convergent Taylor expansion

$$(2.3.1) \qquad \psi(h) = \sum_{i,j=0}^{\infty} a_{ij} (\lambda^{1/N})^i (t^{1/N})^j$$

valid all over Π. According to our construction, for any $(\lambda_0, t_0) \in \mathcal{B}$, the value of h at the corresponding point of $\mathcal{D}$ is given by the value of the absolutely convergent series $\sum_{i,j \geq 0} a_{ij} \lambda_0^{i/N} t_0^{j/N}$, where $\lambda_0^{1/N}, t_0^{1/N}$ are chosen to be positive real. Since $\Gamma(Y, O_Y) = \cup_n (\lambda t(1-t))^{-n} \Gamma(Y', O_{Y'})$, this proves the existence of ψ satisfying (*), except for the $\bar{\mathbf{Q}}$- rationality of coefficients.

To check the last point, consider the point P on X' with coordinates $(\lambda, t) = (0,0)$, and let Q be the unique $\bar{\mathbf{Q}}$-rational point of Y' lying above P such that the corresponding $\mathbf{C}$-rational point belongs to $\mathcal{D}_{\mathbf{C}}$. Then the maximal ideal m_Q of the local ring $O_{Y',Q}$ is generated by $\lambda^{1/N}$ and $t^{1/N}$, and (2.3.1) is nothing but the m_Q-adic expansion of an element of $\Gamma(Y', O_{Y'})$. Therefore, all coefficients a_{ij} belong to the residue field of m_Q. $\qquad \square$

2.4   We shall also need the following

Proposition 2.4. *Notations being as in Prop. 2.3, each ψ factors through $M\{\{\lambda\}\} = \cup_{N \geq 1} M((\lambda^{1/N}))$, where M is (as in §1.1) the maximal Galois extension of $\bar{\mathbf{Q}}(t)$ in $\bar{\mathbf{Q}}\{\{t\}\}$ unramified outside $t = 0, 1, \infty$.*

Proof. We keep the notations in the proof of Prop. 2.3. Let $F^* : Y^* \to X^*$ be the compactification of F (§2.3). Consider the line $l_{12} \subset X^*$ as the projective t-line $\mathbf{P}^1$, and let $l_{12}^{(N)}$ be the irreducible curve on $X^{(N)*}$ defined by the equation $\lambda^{1/N} = 0$. Let C be any irreducible component of $Y^* \times_{X^{(N)*}} l_{12}^{(N)}$, which is a prime divisor of Y^* equipped with the projection

$f : C \to l_{12}$. This $f : C \to l_{12}$ is obviously finite, and generically etale. Moreover, since $Y' \to X^{(N)'}$ (loc. cit.) is etale, and since all points of C outside $t = \infty$ belong to Y', $C \to l_{12}^{(N)}$ is etale outside $t = \infty$. But as $l_{12}^{(N)} \to l_{12}$ is etale outside $t = 0, 1, \infty$, we conclude that $f : C \to l_{12}$ *is etale outside* $t = 0, 1, \infty$. Now choose C to be the unique irreducible component of $Y^* \times_{X^{(N)_*}} l_{12}^{(N)}$ which passes through the point Q (loc. cit.). Rewrite the formula (2.3.1) as

$$\psi(h) = \sum_{i=0}^{\infty} \left(\sum_{j=0}^{\infty} a_{ij} (t^{1/N})^j \right) (\lambda^{1/N})^i = \sum_{i=0}^{\infty} a_i(t)(\lambda^{1/N})^i.$$

Then $a_0(t)$ can be obtained by restricting h to the curve C and expanding it at Q in terms of $t^{1/N}$. Therefore, $a_0(t)$ belongs to M.

To deduce $a_i(t) \in M$ for higher i, we note the following. Consider the unique derivation $D = \partial/\partial(\lambda^{1/N}) : L \to L$ of L defined by $\lambda^{1/N} \to 1, t^{1/N} \to 0$. Then it is easy to see that D maps $\Gamma(X^{(N)'}, O_{X^{(N)'}})$ into itself. We claim that D also stabilizes $\Gamma(Y', O_{Y'})$. This is because $Y' \to X^{(N)'}$ is etale and hence the sheaf of relative differentials $\Omega_{Y'/X^{(N)'}}$ vanishes. (Apply this to the vanishing of $D(\Gamma(Y', O_{Y'})) \bmod \Gamma(Y', O_{Y'})$.) Therefore, D is $(\lambda^{1/N}, t^{1/N})$-adically continuous and $\psi(Dh)$ can be obtained by termwise derivations of the series on the RHS of (2.3.1). Therefore, the question on $a_i(t)$ (for h) is reduced to that on $a_0(t)$ (for $(i!)^{-1}D^i(h)$). $\qquad \square$

Remark. As is clear from the above proof, elements $a_i(t)$ of M appearing in the expansion $\psi(h) = \sum_{i \geq -n} a_i(t)\lambda^{i/N}$ all belong to some finite extension of $\mathbf{Q}(t)$ which depends only on $F : Y \to X$.

2.5

Now let $V(\mathcal{B})$ denote the collection of 5 vertices of the pentagon $\mathcal{B}$, and C_5 be the cyclic subgroup of S_5 generated by the cycle (13524). Then C_5 acts simply transitively on $V(\mathcal{B})$, and $s(\mathcal{B}) = \mathcal{B}$ for any $s \in C_5$. For each $P \in V(\mathcal{B})$, denote by s_P the unique element of C_5 such that $P = s_P(P_0)$, where P_0 is the point $(\lambda, t) = (0, 0)$. Put $\lambda_P = s_P(\lambda), t_P = s_P(t)$. Then $\mathcal{B}$ is defined also by $0 < \lambda_P, t_P < 1$, and P is defined by $\lambda_P = t_P = 0$.

For each $F : Y \to X$ in Et/X, put

$$F^{-1}(\mathcal{B})^{top} = \pi_0(F_{\mathbf{C}}^{-1}(\mathcal{B})),$$
$$F^{-1}(\mathcal{B}, P)^{alg} = \mathrm{Hom}_{\Gamma(X, O_X)}(\Gamma(Y, O_Y), \bar{\mathbf{Q}}\{\{\lambda_P, t_P\}\}).$$

By the obvious modification of Prop. 2.3, we see that there is a canonical bijection

$$\iota_P : F^{-1}(\mathcal{B})^{top} \xrightarrow{\sim} F^{-1}(\mathcal{B}, P)^{alg}.$$

The Galois group $G_{\mathbf{Q}} = \mathrm{Gal}(\bar{\mathbf{Q}}/\mathbf{Q})$ acts on $\bar{\mathbf{Q}}\{\{x,y\}\}$ via its action on Puiseux coefficients, and hence on $F^{-1}(\mathcal{B}, P)^{alg}$ by $\psi \to \sigma_P(\psi) := \sigma \circ \psi$ $(\sigma \in G_{\mathbf{Q}})$.

Now we define the groupoid

$$(2.5.1) \qquad \{\hat{\pi}_1(X(\mathbf{C}); \mathcal{B}_P, \mathcal{B}_{P'})\}_{P,P' \in V(\mathcal{B})},$$

with $G_{\mathbf{Q}}$- and C_5-actions. As a groupoid, it is the profinite completion of the topological fundamental groupoid

$$\{\pi_1(X(\mathbf{C}); \mathcal{B}_P, \mathcal{B}_{P'})\}_{P,P' \in V(\mathcal{B})},$$

where $\mathcal{B}_P$ is a simply connected open set in $\mathcal{B}$ whose closure in $\mathcal{B}^*$ contains P, chosen so that $s(\mathcal{B}_P) = \mathcal{B}_{s(P)}$ $(s \in C_5, P \in V(\mathcal{B}))$. (Although we may choose $\mathcal{B}_P = \mathcal{B}$, all P, it is convenient here to distinguish each $\pi_1(X(\mathbf{C}); \mathcal{B}_P, \mathcal{B}_{P'})$ from the group $\pi_1(X(\mathbf{C}); \mathcal{B})$.) The group C_5 acts on this groupoid (2.5.1) via its natural action on paths on $X(\mathbf{C})$. For any $\gamma \in \pi_1(X(\mathbf{C}); \mathcal{B}_P, \mathcal{B}_{P'})$ and $F : Y \to X$ in Et/X, γ induces a fiber bijection $\gamma_F : \pi_0(F_{\mathbf{C}}^{-1}(\mathcal{B}_P)) \xrightarrow{\sim} \pi_0(F_{\mathbf{C}}^{-1}(\mathcal{B}_{P'}))$ which is defined by tracing above γ on $Y(\mathbf{C})$. The system $\{\gamma_F\}_{F \in Et/X}$ is compatible with the morphisms in Et/X. As Grothendieck pointed out, the profinite set $\hat{\pi}_1(X(\mathbf{C}); \mathcal{B}_P, \mathcal{B}_{P'})$ can be identified with the collection of all such compatible systems of fiber bijections.

Now identify $\pi_0(F_{\mathbf{C}}^{-1}(\mathcal{B}_P))$ with $\pi_0(F_{\mathbf{C}}^{-1}(\mathcal{B}))$ (via $\mathcal{B}_P \hookrightarrow \mathcal{B}$) and also with $F^{-1}(\mathcal{B}, P)^{alg}$ (via ι_P). Call σ_P the action of $\sigma \in G_{\mathbf{Q}}$ on $\pi_0(F_{\mathbf{C}}^{-1}(\mathcal{B}_P))$ induced from its action on $F^{-1}(\mathcal{B}, P)^{alg}$. Define the $G_{\mathbf{Q}}$-action on $\hat{\pi}_1(X(C); \mathcal{B}_p, \mathcal{B}'_p)$ by $\{\gamma_F\}_F \to \{\sigma_{P'} \circ \gamma_F \circ \sigma_P^{-1}\}_F$ $(\sigma \in G_{\mathbf{Q}})$. Then this $G_{\mathbf{Q}}$-action is obviously compatible with the groupoid compositions. Moreover,

Proposition 2.5. *The actions of an element of C_5 and of $G_{\mathbf{Q}}$ on the groupoid (2.5.1) commute with each other.*

Proof. The basic reason for this is the $\mathbf{Q}$-rationality of the S_5-action on X, which is hidden beneath the following formal argument.

First, note the following. Let $\gamma \in \hat{\pi}_1(X(\mathbf{C}); \mathcal{B}_P, \mathcal{B}_{P'}), s \in C_5$, and $F : Y \to X$ in Et/X. Then the two fiber bijections γ_F and $s(\gamma)_{s \circ F}$ are related to each other via the commutative diagram

$$
\begin{array}{ccc}
\pi_0(F_{\mathbf{C}}^{-1}(\mathcal{B}_P)) & \xrightarrow{\gamma_F} & \pi_0(F_{\mathbf{C}}^{-1}(\mathcal{B}_{P'})) \\
\| & & \| \\
\pi_0((s \circ F)_{\mathbf{C}}^{-1}(\mathcal{B}_{s P})) & \xrightarrow{s(\gamma)_{s \circ F}} & \pi_0((s \circ F)_{\mathbf{C}}^{-1}(\mathcal{B}_{s P'})).
\end{array}
$$

If ψ, ψ' are elements of $F^{-1}(\mathcal{B}, P)^{alg}, (s \circ F)^{-1}(\mathcal{B}, s(P))^{alg}$, respectively, which correspond to the mutually identified elements of $\pi_0(F_{\mathbf{C}}^{-1}(\mathcal{B}_P))$ and $\pi_0((s \circ F)_{\mathbf{C}}^{-1}(\mathcal{B}_s P))$, then for any $h \in \Gamma(Y, \mathcal{O}_Y), \psi'(h)$ is obtained from $\psi(h)$ (a Puiseux series in λ_P, t_P) simply by substituting λ_P, t_P by $\lambda_{s(P)}, t_{s(P)}$, respectively. Therefore, the actions of $G_{\mathbf{Q}}$ on $\pi_0(F_{\mathbf{C}}^{-1}(\mathcal{B}_P))$ and $\pi_0((s \circ F)_{\mathbf{C}}^{-1}(\mathcal{B}_s P))$ are compatible with the above identification of these two sets. The rest is obvious. □

2.6   Now we shall name the elements of $V(\mathcal{B})$ as follows; P_0 is the point $(\lambda, t) = (0, 0)$, and $P_j = s^j(P_0)$ for $j \in \mathbb{Z}/5$, where $s = (13524)$. We write $\lambda_j, t_j, \iota_j, \mathcal{B}_j$ for $\lambda_P, t_P, \iota_P, \mathcal{B}_P$ with $P = P_j$ ($j \in \mathbb{Z}/5$). So, $\lambda_j = s^j(\lambda), t_j = s^j(t)$. In particular,

$$(2.6.1) \qquad \begin{cases} \lambda_0 = \lambda, \quad t_0 = t; \quad \lambda_1 = (1-t)(1-\lambda t)^{-1}, \quad t_1 = \lambda; \\ 1 - t = \lambda_1(1-t_1)(1-\lambda_1 t_1)^{-1}. \end{cases}$$

Now define $q \in \hat{\pi}_1(X(\mathbf{C}); \mathcal{B}_0, \mathcal{B}_1)$ to be the element corresponding to (the image of) the homotopy class in $\pi_1(X(\mathbf{C}); \mathcal{B}_0, \mathcal{B}_1)$ of (any)path from $\mathcal{B}_0$ to $\mathcal{B}_1$ inside $\mathcal{B}$. In other words, if we identify $\hat{\pi}_1(X(\mathbf{C}); \mathcal{B}_0, \mathcal{B}_1)$ with $\hat{\pi}_1(X(\mathbf{C}); \mathcal{B})$ via $\mathcal{B}_0, \mathcal{B}_1 \hookrightarrow \mathcal{B}$, then q is the element of $\hat{\pi}_1(X(\mathbf{C}); \mathcal{B}_0, \mathcal{B}_1)$ corresponding to the identify elements of $\hat{\pi}_1(X(\mathbf{C}); \mathcal{B})$. The following is the key lemma for the proof of the 5-cycle relation.

Lemma 2.6.

$$q^{-1} \circ \sigma(q) = f_\sigma(x_{45}, x_{34}) \quad (\sigma \in G_{\mathbf{Q}}).$$

Here, $f_\sigma \in \hat{F}_2'$ is as in §1, $x_{ij} \in \pi_1(X(\mathbf{C}); \mathcal{B})$ (§2.2) are now considered as elements of $\pi_1(X(\mathbf{C}); \mathcal{B}_0)$ (via $\mathcal{B}_0 \subset \mathcal{B}$), and $f_\sigma(x_{45}, x_{34})$ is the element of the profinite group $\hat{\pi}_1(X(\mathbf{C}); \mathcal{B}_0)$ obtained by (profinite) substitution of x, y by x_{45}, x_{34}.

Proof.   It suffices to prove that

$$(2.6.2) \qquad (q^{-1} \sigma q \sigma^{-1}) \psi = f_\sigma(x_{45}, x_{34}) \psi$$

holds for all $\psi \in F^{-1}(\mathcal{B}, P_0)^{alg}$, for all "sufficiently large" objects $F : Y \to X$ of Et/X. Moreover, as for the test functions h, we may restrict to elements of $\Gamma(Y', \mathcal{O}_{Y'})$ (with the notations as in the proof of Prop 2.3). This is because $\Gamma(Y, \mathcal{O}_Y)$ is obtained from $\Gamma(Y', \mathcal{O}_{Y'})$) by allowing denominators that are powers of $\lambda t(1-t) (\in \Gamma(X, \mathcal{O}_X))$ and hence ψ is determined by its restriction to $\Gamma(Y', \mathcal{O}_{Y'})$.

Thus, take any $h \in \Gamma(Y', \mathcal{O}'_Y)$ and put

$$(2.6.3) \qquad \psi(h) = \sum_{i,j=0}^{\infty} a_{ij} \lambda^{i/N} t^{j/N}$$

$$(2.6.3') \qquad = \sum_{i=0}^{\infty} a_i(t) \lambda^{i/N}$$

with $a_i(t) \in M \cap \bar{\mathbf{Q}}[[t^{1/N}]]$ (Prop 2.4). ($a_i(t)$ is actually a power series of $t^{1/N}$, but we write it this way because we want to consider it as a holomorphic function of t on $(0,1)_{\mathbf{C}}$ (§1.1).)

Recall (§2.3) that the power series on the RHS of (2.6.3) represents a holomorphic function of $(\lambda^{1/N}, t^{1/N})$ on the unit polydisc Π, which coincides with the restriction of h on $\mathcal{D}_{\mathbf{C}}$ considered as a function on $(\mathcal{D}_N)_{\mathbf{C}} \simeq \Pi$. Regard $\psi(h)$ as a function of $\lambda^{1/N}$ and t, with $t \in (0,1)_{\mathbf{C}}$; then as a function of $\lambda^{1/N}$ and $1 - t$ ($t \in (0,1)_{\mathbf{C}}$); and finally, as a function of $\lambda^{1/N}$ and $(1-t)^{1/N}$. Then, since $h \in \Gamma(Y', \mathcal{O}_{Y'})$ it follows that $\psi(h)$ is a univalent holomorphic function of $(\lambda^{1/N}, (1-t)^{1/N})$ on $|\lambda^{1/N}| < \varepsilon, |(1-t)^{1/N}| < \varepsilon$, where $\varepsilon = 2^{-1/N}$. (The restriction by ε is to avoid the line $\lambda t = 1$.) As for $a_i(t)$'s, since these are restrictions of some elements of $\Gamma(Y', \mathcal{O}_{Y'})$ to $\lambda = 0$, the above property of $\psi(h)$ implies that $a_i(t)$, as a function $(1-t)^{1/N}$, is holomorphic on $|(1-t)^{1/N}| < \varepsilon$. In other words, $(pa_i)(1-t)$ can be expanded as a power series of $(1-t)^{1/N}$ convergent on $|(1-t)^{1/N}| < \varepsilon$. Here, $p: M \xrightarrow{\sim} M'$ is as in §1.3.

Now we are ready to show that

$$(2.6.4) \qquad (q\psi)(h) = \sum_{i=0}^{\infty} (pa_i)(\lambda_1(1-t_1)(1-\lambda_1 t_1)^{-1}) t_1^{i/N}.$$

Here, the coefficient of $t_1^{i/N}$ is a power series in $\lambda_1^{1/N}$ and t_1, obtained by substituting $(1-t)^{1/N}$ by

$$(2.6.5) \quad (\lambda_1(1-t_1)(1-\lambda_1 t_1)^{-1})^{1/N} = \lambda_1^{1/N}(1 + (1/N)(\lambda_1 - 1)t_1 + \cdots)$$

in the Puiseux series

$$(pa_i)(1-t) = \sum_{j=0}^{\infty} b_{ij}(1-t)^{j/N}.$$

To see (2.6.4), first note that $(q\psi)(h)$ is nothing but the Puiseux expansion of the holomorphic function $\psi(h)$ in terms of (λ_1, t_1). Recall that

$$\lambda = t_1, 1 - t = \lambda_1(1-t_1)(1-\lambda_1 t_1)^{-1} \qquad \text{(cf. (2.6.1))}$$

Therefore,

$$\psi(h) = \sum_{i=0}^{\infty} a_i(t)\lambda^{i/N} = \sum_{i=0}^{\infty} (pa_i)(1-t)\lambda^{i/N}$$

$$(2.6.6) \qquad = \sum_{i,j} b_{ij}(1-t)^{j/N}\lambda^{i/N},$$

and the last series is convergent on $|(1-t)^{1/N}| < \varepsilon, |\lambda^{1/N}| < \varepsilon$. Since the transformation $(\lambda^{1/N}, (1-t)^{1/N}) \to (\lambda_1^{1/N}, t_1^{1/N})$ is biholomorphic near $(0,0)$, rewriting (2.6.6) into (2.6.4) gives a convergent power series expression for $(q\psi)(h)$ in terms of $(\lambda_1^{1/N}, t_1^{1/N})$. This proves (2.6.4).

Now we proceed, using (2.6.4). Since the Puiseux series (2.6.5) has all coefficients in $\mathbf{Q}$, we obtain

$$((\sigma q\sigma^{-1})\psi)(h) = \sum_{i=0}^{\infty} (\sigma p\sigma^{-1})(a_i)(\lambda_1(1-t_1)(1-\lambda_1 t_1)^{-1})t_1^{i/N}.$$

Therefore,

$$(q^{-1}\sigma(q)\psi)(h) = \sum_{i=0}^{\infty} f_\sigma(x,y)a_i(t)\lambda^{i/N} \quad (\sigma \in G_\mathbf{Q}).$$

So, it suffices to prove that

(2.6.7) $$(f(x_{45}, x_{34})\psi)(h) = \sum_{i=0}^{\infty}(f(x,y)a_i(t))\lambda^{i/N}$$

holds for any $f \in \hat{F}_2 = \hat{\pi}_1$, where (as in §1.2) $\pi_1 = \pi_1(\mathbf{P}^1(\mathbf{C}) - \{0,1,\infty\}; (0,1))$.

By continuity, we only need to prove (2.6.7) when $f \in \pi_1$. (Here, recall §2.4. Remark.) It is also clear that we are reduced to the case $f = x$ and $f = y$. Therefore, it suffices to prove

(2.6.8) $$(x_{45}\psi)(h) = \sum_{i=0}^{\infty} x(a_i(t))\lambda^{i/N},$$

(2.6.8') $$(x_{34}\psi)(h) = \sum_{i=0}^{\infty} y(a_i(t))\lambda^{i/N}.$$

Now recall the definition of x_{ij} and recall that (λ, t) parametrizes $(0, \lambda, 1, t^{-1}, \infty)$ mod $PGL(2)$. It is clear that in (λ, t)-coordinates x_{45} (resp. $x_{34} = x_{43}$) is represented by a path $(\lambda_\theta, t_\theta)_{0 \le \theta \le 1}$, where $\lambda_\theta = \varepsilon$ is constant $(0 < \varepsilon < 1)$, and t_θ is a path representing x (resp. y) (sufficiently small loops around 0 resp. 1).

As for x_{45}, since the path $(\lambda_\theta, t_\theta)$ is contained in the domain of convergence of (2.6.3), we obtain

$$(x_{45}\psi)(h) = \sum_{i,j=0}^{\infty} a_{ij}\zeta_N^j \lambda^{i/N} t^{j/N} = \sum_{i=0}^{\infty} x(a_i(t))\lambda^{i/N},$$

where $\zeta_N = e^{2\pi i/N}$. As for x_{34}, using (2.6.6) we obtain

$$(x_{34}\psi)(h) = q^{-1}(\sum_{i,j} b_{ij}\zeta_N^j(1-t)^{j/N}\lambda^{i/N});$$

$$= \sum_i p^{-1}(\sum_j b_{ij}\zeta_N^j(1-t)^{j/N}) \cdot \lambda^{i/N}$$

$$= \sum_i y(a_i(t))\lambda^{i/N}.$$

This settles both (2.6.8)(2.6.8′).   □

2.7 (*Proof of the 5-cycle relation*) Take any $\sigma \in G_Q$. Then

$$\sigma(q) = qf_\sigma(x_{45}, x_{34}) \qquad \text{(lemma 2.6)}$$

By the commutativity of the actions of σ and $s = (13524)$ on $\{\hat{\pi}_1(X(\mathbf{C});$ $\mathcal{B}_P, \mathcal{B}_{P'})\}_{P,P' \in V(\mathcal{B})}$, and by the obvious formula

$$s(x_{j,j+1}) = qx_{j+2,j+3}q^{-1} \qquad (j \in \mathbf{Z}/5),$$

we obtain

$$\sigma(s(q)) = s(q)qf_\sigma(x_{12}, x_{51})q^{-1},$$
$$\sigma(s^2(q)) = s^2(q)s(q)qf_\sigma(x_{34}, x_{23})q^{-1}s(q)^{-1},$$
$$\sigma(s^3(q)) = s^3(q)s^2(q)s(q)qf_\sigma(x_{51}, x_{45})q^{-1}s(q)^{-1}s^2(q)^{-1},$$
$$\sigma(s^4(q)) = s^4(q)s^3(q)s^2(q)s(q)qf_\sigma(x_{23}, x_{12})q^{-1}s(q)^{-1}s^2(q)^{-1}s^3(q)^{-1}.$$

But since $s^4(q)s^3(q)s^2(q)s(q)q = 1$, we obtain

$$f_\sigma(x_{23}, x_{12})f_\sigma(x_{51}, x_{45})f_\sigma(x_{34}, x_{23})f_\sigma(x_{12}, x_{51})f_\sigma(x_{45}, x_{34}) = 1.$$

But since $f(x,y) = f(y,x)^{-1}$ (§1.7), we obtain the desired formula

$$f_\sigma(x_{12}, x_{23})f_\sigma(x_{34}, x_{45})f_\sigma(x_{51}, x_{12})f_\sigma(x_{23}, x_{34})f_\sigma(x_{45}, x_{51}) = 1.$$

                                                                        □

References

[A-I]$_{1,2}$ Anderson, G and Y. Ihara; Pro-l branched coverings of $\mathbf{P}^1$ and higher circular l-units, Ann of Math. **128** (1988), 271-293; Part 2, Int'l J. Math.1 (1990), 119-148.

[B]   Belyi, G. V; On Galois extensions of a maximal cyclotomic field, Math USSR Izv. **14** (1980), pp 247-256.

[De]   Deligne, P; Le groupe fondamental de la droite projective moins
          trois points, in Galois groups over $\mathbf{Q}$, MSRI Publ 16, pp 79-297,
          Springer 1989.

[Dr]   Drinfeld, V. G; On quasi-triangular quasi-Hopf algebras and some
          group closely associated with $Gal(\overline{\mathbf{Q}}/\mathbf{Q})$. Algebra and Analysis 2
          (1990), 4; Leningrad Math. J. 2 (1991), 4 pp 829-860.

[GHP]   Gerritzen, L, F. Herrlich and M. Van der Put; Stable n-pointed
          trees of projective lines, Indag math. 50 (1988), pp 131-163.

[Gr]   Grothendieck, A; Esquisse d'un programme. Mimeographed note
          (1984).

[Ih]   Ihara, Y; Braids, Galois groups, and some arithmetic functions,
          Proc Intern'l Congress of Math., Kyoto 1990, Vol 1 pp 99-120,
          Springer.

[Ih-Ma]   Ihara, Y and M. Matsumoto; On Galois actions on braid groups,
          submitted to Contemp. Math.; RIMS-961, 1994.

[L-S]   Lochak, P and L. Schneps; The Grothendieck-Teichmüller group
          and automorphisms of braid groups; in this Volume.

[M]   Mumford, D; Geometric invariant theory, Ergebnisse der Math.
          Vol 34, Acad. Press.

[N]   Nakamura, H; Galois rigidity of pure sphere braid groups and
          profinite calculus; to appear in J. Math. Sci., the Univ. of Tokyo,
          Vol 1.

[SGA 1]   Grothendieck, A; Revêtements étales et groupe fondamental; SLN
          224, Springer.

†Research Institute for Mathematical Sciences, Kyoto University, 606-01
Kyoto, Japan

*Département de Mathématiques, Université de Lille, Villeneuve-sur-Ascq,
France

**DMI, Ecole Normale Supérieure, 45 rue d'Ulm, 75005 Paris, France

Appendix

The action of the absolute Galois group
on the moduli space of spheres with four marked points

Michel Emsalem and Pierre Lochak

A large part of this volume is devoted to the theory of "dessins d'enfants" which is concerned with the action of (finite quotients of) the absolute Galois group on (finite quotients of) $\hat{F}_2$, the profinite completion of the free group on two generators. This last group appears as the algebraic fundamental group of $\mathbf{P}^1\overline{\mathbf{Q}}\backslash\{0,1,\infty\}$, which is the moduli space of spheres with 4 ordered marked points. As explained in paragraph 2 of Grothendieck's *Esquisse d'un programme*, this is but the first nontrivial instance of the action of the absolute Galois group on the fundamental group of a moduli space $\mathcal{M}_{g,n}$ (genus g, n marked points), namely here $\mathcal{M}_{0,4}$. We refer to Grothendieck's grandiose sketch for a – very – broad perspective. We only note that in genus 0, these fundamental groups are quotients of the Artin braid groups, and that the action of $\mathrm{Gal}(\overline{\mathbf{Q}}/\mathbf{Q})$ lifts to an action on the profinite completions of the braid groups. This has been interpreted in a strikingly different way by Drinfeld in [Dr]; we refer to the article by Lochak and Schneps in this volume, and also to [Ih-Ma], for more details.

Our aim in this appendix is to provide the necessary setting for a concrete study of the action of the Galois group on $\pi_1(\mathbf{P}^1\overline{\mathbf{Q}}\backslash\{0,1,\infty\})$. The main properties are summarized in Theorem 1 below. Although Theorem 1 is no more than Ihara's Proposition 1.6 and I and II of Theorem 1.7 from the main body of this article, we have put a particular emphasis on proving the results from scratch, recalling the definitions and constructions of the basic objects (such as tangential base points for example), and in particular, seeking to lay the groundwork for generalizations to the study of the higher dimensional moduli spaces, such as is done by Ihara in the main body of this paper for the next simplest case, namely the two-dimensional moduli space of spheres with 5 marked points. In studying the one dimensional situation, we follow Ihara's sketch in [Ih] (sections 2,3). The concrete realisation and use of tangential base points by means of convergent Puiseux series was introduced by Anderson and Ihara in [A-I] and we note that it allows for implementation on a computer and actual computations of dessins (see the paper by J.-M.Couveignes and L.Grandboulan in this volume).

The Galois action on $\pi_1(\mathbf{P}^1\overline{\mathbf{Q}}\backslash\{0,1,\infty\})$. For a field k, we write $X_4(k) := \mathbf{P}^1 k \backslash \{0,1,\infty\}$; similarly $X_n(k)$ will denote the moduli space over k of spheres with n ordered marked points. Let $\mathbb{\Gamma} := \mathrm{Gal}(\overline{\mathbf{Q}}/\mathbf{Q})$, as in Grothendieck's *Esquisse*.

Recall a basic property of the algebraic fundamental group (cf. [SGA 1]): given a smooth absolutely irreducible variety X over a field k of characteristic 0, we have the exact sequence of algebraic fundamental groups:

$$(1) \qquad 1 \to \pi_1(X \times_k \overline{k}) \to \pi_1(X) \to \mathrm{Gal}(\overline{k}/k) \to 1,$$

where $\overline{k}$ denotes an algebraic closure of k. When $k = \mathbf{Q}$, $\pi_1(X \times_k \overline{k})$ coincides with the profinite completion of $\pi_1^{\mathrm{top}}(X(\mathbf{C}))$, the topological fundamental group of the analytic variety $X(\mathbf{C})$, where we choose an embedding $\overline{\mathbf{Q}} \hookrightarrow \mathbf{C}$ (we write π_1 for the algebraic fundamental group, π_1^{top} for the topological fundamental group – when it exists). So we get in particular:

$$(2) \qquad 1 \to \hat{F}_2 \to \pi_1(\mathbf{P}^1\mathbf{Q} \setminus \{0, 1, \infty\}) \to \mathrm{I\!\Gamma} \to 1,$$

because $\pi_1^{\mathrm{top}}(X_4(\mathbf{C})) \simeq F_2$, the free group on two generators. More explicitly, $\pi_1^{\mathrm{top}}(X_4(\mathbf{C}))$ is generated by x and y where x (resp. y) is a closed loop around 0 (resp. 1); this is made more precise below. So here $F_2 = \langle x, y \rangle$ and we write the elements of $\hat{F}_2$, the profinite completion of F_2, as "prowords" $f(x, y)$. This notation allows in particular to make sense of substitutions $(x, y) \mapsto (\xi, \eta)$, where ξ and η are elements of any profinite group. As does any exact sequence, (2) defines an outer action of $\mathrm{I\!\Gamma}$ on $\hat{F}_2$, that is a map $\mathrm{I\!\Gamma} \to \mathrm{Out}(\hat{F}_2)$. But (2) is actually split, and by choosing a splitting (see below), we can define a map $\mathrm{I\!\Gamma} \to \mathrm{Aut}(\hat{F}_2)$; if $\sigma \in \mathrm{I\!\Gamma}$, we write $\phi_\sigma \in \mathrm{Aut}(\hat{F}_2)$ for the corresponding automorphism or simply $\sigma \cdot \gamma := \phi_\sigma(\gamma)$ for the action of σ on $\gamma \in \hat{F}_2$. We may now state

Theorem 1: *There exists a faithful action of* $\mathrm{I\!\Gamma} = \mathrm{Gal}(\overline{\mathbf{Q}}/\mathbf{Q})$ *on* $\hat{F}_2$, *that is, an injective map*

$$\sigma \in \mathrm{I\!\Gamma} \to \phi_\sigma \in \mathrm{Aut}(\hat{F}_2).$$

Writing $F_2 = \langle x, y \rangle$, *we have:*

$$(3) \qquad \phi_\sigma(x) = x^{\chi(\sigma)}, \quad \phi_\sigma(y) = f_\sigma^{-1}(x, y) y^{\chi(\sigma)} f_\sigma(x, y);$$

here $\chi : \mathrm{I\!\Gamma} \to \hat{\mathbf{Z}}^\times$ *denotes the cyclotomic character and* $f_\sigma \in \hat{F}_2$. *Actually* $f_\sigma \in \hat{F}_2'$, *the derived group of* $\hat{F}_2$, *so that we have an injective map:*

$$(4) \qquad \sigma \in \mathrm{I\!\Gamma} \to (\chi(\sigma), f_\sigma) \in \hat{\mathbf{Z}}^\times \times \hat{F}_2'.$$

Moreover, f_σ *satisfies the following two equations:*

$$(I) \quad f_\sigma(x, y) f_\sigma(y, x) = 1$$
$$(II) \quad f_\sigma(z, x) z^m f_\sigma(y, z) y^m f_\sigma(x, y) x^m = 1$$

where $z = (xy)^{-1}$ *and* $m = m(\sigma) = \frac{1}{2}(\chi(\sigma) - 1)$.

The rest of this appendix is essentially devoted to introducing the basic notions behind Ihara's proof of this theorem. Let $\tilde{X}_4$ be the (unique up to isomorphism) universal covering of X_4; to the family of finite coverings of X_4, which appear as quotients of $\tilde{X}_4$, there corresponds a sequence of extensions of function fields

$$(5) \qquad \mathbf{Q}(T) \hookrightarrow \overline{\mathbf{Q}}(T) \hookrightarrow \Omega$$

where Ω is the maximal extension of $\overline{\mathbf{Q}}(T)$ unramified outside $\{0,1,\infty\}$; actually (2) is nothing but the sequence of Galois groups corresponding to this sequence of extensions. We let $Ext(\overline{\mathbf{Q}}(T))$ denote the set – actually the category – of the finite extensions of $\overline{\mathbf{Q}}(T)$ unramified outside $0, 1, \infty$; they correspond to the finite connected unramified algebraic coverings of X_4.

Tangential base points (following Anderson-Ihara [A-I]). We shall first interpret the *base points* of the fundamental groups (which have been purposefully, if hastily, omitted from (1)), in terms of function fields. Let $t_0 \in \mathbf{Q} \setminus \{0,1\} = X_4(\mathbf{Q})$ be a rational point, to be taken as base point (in the general case, fix a rational point of $X_n(\mathbf{Q})$). Although we shall ultimately need that $t_0 \in \mathbf{Q}$, we first work geometrically, i.e. over $\mathbf{C}$. Let $Y(\mathbf{C})$ be a finite connected unramified covering of X_4, F its function field, p the projection $Y \to X_4$. Identifying $\mathbf{Q}(X_4)$ with $\mathbf{Q}(t)$ (and thus $\mathbf{C}(X_4)$ with $\mathbf{C}(t)$), to every point $y_0 \in p^{-1}(t_0)$, the fibre over t_0, we now associate an embedding $\phi : F = \mathbf{C}(Y) \hookrightarrow \mathbf{C}((t-t_0))$, the formal series in $t - t_0$. Indeed, any $f \in F$ can be viewed locally (in the analytic sense) as a function of $(t - t_0)$, through the local isomorphism p (between a neighbourhood of y_0 and one of t_0); it is uniquely determined by its Laurent expansion at any such point $y_0 \in Y$ and fixing y_0 determines such an embedding, where the image of F is actually contained in the ring of *convergent* Laurent series, of which $\mathbf{C}((t-t_0))$ is the completion (with respect to the obvious valuation). Also, if $f \in \overline{\mathbf{Q}}(Y)$, $\phi(f) \in \overline{\mathbf{Q}}((t-t_0))$.

The – discrete – group $\pi_1^{top}(X_4;t_0) \simeq F_2$ defines an action on this set of embeddings $\{\phi_{y_0}\}, y_0 \in p^{-1}(t_0)$, via the *monodromy* at t_0, as follows. Let $\gamma \in \pi_1^{top}(X_4;t_0)$ be a loop based at t_0, $f \in F$ a rational function, $\phi : F \hookrightarrow \mathbf{C}((t-t_0))$ an embedding (equivalently a choice of $y_0 \in p^{-1}(t_0)$). Then define $\gamma \cdot \phi$ by

$$\gamma.\phi(f) = \phi(\gamma^{-1}.f) \in \mathbf{C}((t-t_0)),$$

where $\gamma^{-1}f$ is the analytic continuation of f (viewed locally as a function on $\mathbf{P}^1\mathbf{C}$ as explained above) along γ. Since we denote $\gamma' \circ \gamma$ the compostition of γ' with γ (γ' *after* γ) we need to have γ^{-1} in the above definition so that the action be covariant.

Notice now that the above contructions are algebraic (defined over $\overline{\mathbf{Q}}$) and yield a coherent system when varying the covering Y. We may thus pass to the projective limit on the coverings and their fundamental groups, and to the inductive limit on the algebraic function fields. As a result we get:

Proposition 2: *(i) To a coherent family of places of Ω "over" t_0 is associated an embedding $\phi : \Omega \hookrightarrow \overline{\mathbf{Q}}((t - t_0))$.*

(ii) The image of a finite extension $E \in Ext(\overline{\mathbf{Q}}(T)) \subset \Omega$ is contained in the field of convergent Laurent series.

(iii) Any $\gamma \in \pi_1(X_4 \times_{\mathbf{Q}} \mathbf{C}; t_0)$, the profinite algebraic fundamental group, determines an action $\phi \to \gamma \cdot \phi$ on the set of these embeddings.

In the above, t_0 was an arbitrary rational point of X_4. To remove this arbitrariness, it is tempting to take $\{0, 1, \infty\}$ as base points. Although they are at infinity (not in X_4), Deligne described in [De] a method of using them at least as starting points: let us recall the description of his *tangential base points*. Fix $t_0 = 0$ for definiteness; the trouble is that if we look at a neighbourhood of 0 in $X_4^* = \mathbf{P}^1$ and intersect it with X_4, we find a punctured disk, which is not simply connected, so that we can't use such a region as a "base point" for a fundamental group (a similar phenomenon occurs in any dimension). So consider instead the point 0 *together with* the direction $0 \to 1$; denote this as $\overrightarrow{01}$, and similarly $\overrightarrow{0\infty}$ for the opposite direction. Altogether, we find a set $\mathcal{B}$ of six such tangential base points, namely

$$\mathcal{B} = \{\overrightarrow{01}, \overrightarrow{0\infty}, \overrightarrow{10}, \overrightarrow{1\infty}, \overrightarrow{\infty0}, \overrightarrow{\infty1}\}.$$

Any $\vec{u} = \overrightarrow{AB} \in \mathcal{B}$ defines a sector $S_{\vec{u}}$, obtained from a small disk around A by removing the semiaxis *opposite* the direction $\overrightarrow{AB}$. Picking $\overrightarrow{01}$ again as an example, it defines $S_{\overrightarrow{01}}$, obtained from a small disk around 0 by removing $(0, \infty)$.

Puiseux series. Now let $p : Y \to X_4$ again be a finite connected covering, $F = \overline{\mathbf{Q}}(Y) \in Ext(\overline{\mathbf{Q}}(T))$ the corresponding function field. There is a unique compactification Y^* of Y as a covering of $\mathbf{P}^1$ ramified (at most) at $\{0, 1, \infty\}$. The geometric situation is as follows; take $\vec{u} = \overrightarrow{01} \in \mathcal{B}$ for ease of notation, and also to make the connection with the "dessins d'enfants" clearer. Pick $y_0 \in p^{-1}(0)$; if the ramification index of Y^* at y_0 is e, there are e edges of type $\overrightarrow{01}$ originating from y_0. We define the fibre $\mathcal{F}_{\overrightarrow{01}}(Y^*)$ of Y (or Y^*) above $\overrightarrow{01}$ as the set of pairs formed by a point $y_0 \in p^{-1}(0)$ and an edge of type $\overrightarrow{01}$ originating at y_0. The set $\mathcal{F}_{\vec{u}}(Y^*)$ is defined analogously for any $\vec{u} \in \mathcal{B}$.

Note that the cardinal of $\mathcal{E}_{\vec{u}}(Y^*)$ is equal to the degree of the covering Y. Let us now explain (again assuming $\vec{u} = \overrightarrow{01}$ for convenience) how the choice of an element in $\mathcal{F}_{\vec{u}}(Y^*)$ determines an embedding of $F = \overline{\mathbf{Q}}(Y)$ in the field of Puiseux series, namely here $\overline{\mathbf{Q}}\{\{T\}\}$, the field of series with fractional exponents.

First we can find a local (in the analytic sense) coordinate z on Y^* near y_0 such that the covering $Y^*(\mathbf{C}) \to \mathbf{P}^1\mathbf{C}$ is locally isomorphic to the cyclic covering $z \to t = z^e$ of the unit disk. The choice of an edge of type $\overrightarrow{01}$ at y_0 is equivalent to the choice of a uniformizing parameter z such that its restriction to that given edge is real positive. Any $f \in F = \overline{\mathbf{Q}}(Y)$ can be expanded near y_0 as a convergent Laurent series in z,

$$f(z) = \sum_{n \geq -N} a_n z^n, \text{ for some } N \geq 0.$$

Define $\tilde{f} = \sum_n a_n T^{n/e}$ to be the corresponding formal Puiseux series. The map $f \to \tilde{f}$ defines an embedding $F \hookrightarrow \overline{\mathbf{Q}}\{\{T\}\}$, extending the natural embedding $\overline{\mathbf{Q}}(T) = \overline{\mathbf{Q}}(\mathbf{P}^1) \hookrightarrow \overline{\mathbf{Q}}\{\{T\}\}$. More precisely, any $\tilde{f} = \sum_n a_n T^{n/e}$ associated to $f \in \overline{\mathbf{Q}}(Y)$ is convergent and thus defines a germ of analytic function in the sector $S_{\overrightarrow{01}}$ (we use the word "analytic" to refer to functions into $\mathbf{P}^1$, i.e. meromorphic functions if one prefers). We denote $\mathcal{P}_{\overrightarrow{01}}(\overline{\mathbf{Q}})$ the field of such germs of analytic functions on $S_{\overrightarrow{01}}$ or, equivalently, convergent Puiseux series with coefficients in $\overline{\mathbf{Q}}$.

To summarize the above, we find that any edge of type $\overrightarrow{01}$, that is any element of $\mathcal{F}_{\overrightarrow{01}}(Y)$, defines an embedding:

$$F = \overline{\mathbf{Q}}(Y) \overset{\phi}{\hookrightarrow} \mathcal{P}_{\overrightarrow{01}}(\overline{\mathbf{Q}}) \subset \overline{\mathbf{Q}}\{\{T\}\}$$

which extends the natural embedding $\overline{\mathbf{Q}}(T) = \overline{\mathbf{Q}}(\mathbf{P}^1) \subset \overline{\mathbf{Q}}\{\{T\}\}$.

Conversely, any such embedding first defines a place in F above T, i.e. a point $y_0 \in \overline{Y}$ in the fibre over 0, together with a uniformizing parameter z, namely the pullback of $T^{1/e}$, in the completion of F at this place. The uniformizing parameter in turn determines an edge of the dessin originating at y_0, that is an element of $\mathcal{F}_{\overrightarrow{01}}(Y)$, by requiring that it be real positive on it.

This can be done for any of the six tangential base points $\vec{u} \in \mathcal{B}$; $\mathcal{P}_{\vec{u}}(\overline{\mathbf{Q}}) = \mathcal{P}_{\vec{u}}$ will denote the field of convergent Puiseux series (with coefficients in $\overline{\mathbf{Q}}$) in the sector $S_{\vec{u}}$, considered as an extension of $\overline{\mathbf{Q}}(\mathbf{P}^1)$. One should however beware of the fact that this refers to six *different* embeddings of $\overline{\mathbf{Q}}(\mathbf{P}^1)$ into fields of Puiseux series. Let us illustrate this important caveat in the case of $\vec{u} = \overrightarrow{0\infty}$. In that case we expand functions in the sector $S_{\overrightarrow{0\infty}}$ around the *negative* real axis. We now look for z such that $z^e = -t$, with z real positive for t real negative. The choice of an element of $\mathcal{F}_{\overrightarrow{0\infty}}(Y)$, the fibre over $\overrightarrow{0\infty}$, determines an embedding

$$F \hookrightarrow \mathcal{P}_{\overrightarrow{0\infty}}(\overline{\mathbf{Q}}) \subset \overline{\mathbf{Q}}\{\{T\}\}$$

extending the injective morphism $\overline{\mathbf{Q}}(T) \to \overline{\mathbf{Q}}\{\{T\}\}$ such that T is mapped to $-T$. The same phenomenon occurs at $\overrightarrow{1\infty}$, $\overrightarrow{10}$, $\overrightarrow{\infty 0}$ and $\overrightarrow{\infty 1}$, corresponding to series in fractional powers of $(T-1)$, $(1-T)$, $-T^{-1}$ and T^{-1} respectively.

We have now proved the first assertion in the following proposition.

Proposition 3: *(i) For any $\vec{u} = \overrightarrow{AB} \in \mathcal{B}$ and any finite connected unramified covering Y of X_4 (i.e. a covering of $\mathbf{P}^1$ unramified outside $0, 1, \infty$), there is a one to one correspondence between $\mathcal{F}_{\vec{u}}(Y)$, the fibre of the compactified covering $\overline{Y}$ over $\vec{u}$, and the set $\mathcal{E}_{\vec{u}}(\overline{\mathbf{Q}}(Y))$ of embeddings of the function field $\overline{\mathbf{Q}}(Y)$ in $\mathcal{P}_{\vec{u}}$, above $\overline{\mathbf{Q}}(T)$.*

(ii) $\mathcal{F}_{\vec{u}}$ defines a covariant functor from the category of finite connected unramified coverings of X_4 to the category Ens of sets.

(iii) $\mathcal{E}_{\vec{u}}$ defines a contravariant functor from the category $Ext(\overline{\mathbf{Q}}(T))$ of the finite extensions of $\overline{\mathbf{Q}}(T)$ unramified outside $0, 1, \infty$ to the category Ens of sets.

The abstract formulation of (ii) and (iii) turns out to be useful in order to phrase the analogue of (iii) in Proposition 2 (see Proposition 4 below). Proving these two assertions essentially amounts to proving that we can compose morphisms. We sketch it for (iii). For two extensions F, F', a morphism $\alpha : F \to F'$ is a morphism of fields fixing $\overline{\mathbf{Q}}(T)$, corresponding to a map between the associated coverings. Given α and $\phi \in \mathcal{E}_{\vec{u}}(F')$, we define $\mathcal{E}_{\vec{u}}(\alpha)(\phi) = \phi \circ \alpha \in \mathcal{E}_{\vec{u}}(F)$. With this in mind, the proof is reduced to some rather trivial verifications.

We note that the corresponding constructions behave nicely with respect to the inverse (resp. direct) system of coverings (resp. field extensions). This is actually included in (ii) and (iii). In particular, in much the same way as in Proposition 2 (i), we obtain a set of embeddings:

$$\mathcal{E}_{\vec{u}} : \Omega \hookrightarrow Puis_{\vec{u}}.$$

Moreover, one has the following property of surjectivity at any finite level: given $F_0 \in Ext(\overline{\mathbf{Q}}(T))$, and $\phi_0 \in \mathcal{E}_{\vec{u}}(F_0)$, there exists a coherent family $(F, \phi \in \mathcal{E}_{\vec{u}}(F))$ of function fields and embeddings, representing an element of $\mathcal{E}_{\vec{u}}(\Omega)$, and containing (F_0, ϕ_0).

The fundamental groupoid. Now, instead of the fundamental group, we want to consider the *fundamental groupoid* $\pi_1(X_4; \mathcal{B})$ based at the set $\mathcal{B}$ of the six tangential base point. A path $\gamma \in \pi_1^{\text{top}}(X_4; \vec{u}, \vec{v})$ from $\vec{u} = \overrightarrow{AB}$ to $\vec{v} = \overrightarrow{CD}$ is a path in X_4 from A to C (the endpoints being excluded), starting tangentially to $\overrightarrow{AB}$ and arriving tangentially to $\overrightarrow{CD}$ (in the "direction" $\overrightarrow{DC}$). The path γ is allowed to go through "points" of $\mathcal{B}$ but it has to "arrive" and "leave" along the same direction. We leave it to the reader to describe the corresponding notion of homotopy, which allows to define the topological groupoid $\pi_1^{\text{top}}(X_4; \mathcal{B})$ as the groupoid of these paths modulo homotopy. As an example, the path γ_0 obtained from going from 0 to 1 and back along the interval $(0,1)$ is well-defined and null homotopic (equivalent to the unit of the *group* $\pi_1^{\text{top}}(X_4; \overrightarrow{01}, \overrightarrow{01})$). Similarly, the path γ_1 obtained by going from 0 to 1, then from 1 to ∞ and back to 0 along straight lines with curls at the corners to switch direction is also well defined and null homotopic. γ_0 and γ_1 will play an important role in what follows, ultimately giving equations (I) and (II) in Theorem 1.

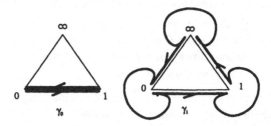

Figure 1

As for the ordinary fundamental group, the above can be expressed in terms of coverings and, passing to the projective limit yields the algebraic – profinite – groupoid $\pi_1(X_4; \mathcal{B})$. For $\vec{u} \in \mathcal{B}$, we write $\pi_1(X_4; \vec{u}) = \pi_1(X_4; \vec{u}, \vec{u})$ for the *group* of (homotopy classes of) paths from $\vec{u}$ to itself.

The next step (and the last one before describing the action of Γ!) consists in viewing any $\gamma \in \pi_1(X_4; \vec{u}, \vec{v})$ as a map from $\mathcal{E}_{\vec{u}}$ to $\mathcal{E}_{\vec{v}}$, actually as a natural transformation between these two functors. Again we describe the geometric finite situation and sketch what is needed to pass to the algebraic and profinite case. So let Y be a finite covering again, $f \in F = \overline{\mathbb{Q}}(Y)$, $\phi \in \mathcal{E}_{\vec{u}}(F)$ an embedding $F \hookrightarrow \mathcal{P}_{\vec{u}}$, and $\gamma \in \pi_1^{\text{top}}(X_4; \vec{u}, \vec{v})$ a geometric path. As in the non ramified case, $\gamma \cdot \phi \in \mathcal{E}_{\vec{v}}$ is defined by analytic continuation. Namely, for $f \in F$, $\phi(f)$ defines a germ of analytic function on $S_{\vec{u}}$, which can be continued along γ; as a result, we get a germ on $S_{\vec{v}}$, whose expansion is by definition $\gamma \cdot \phi(f) \in \mathcal{P}_{\vec{v}}$. Before looking briefly at morphisms, we state this formally:

Proposition 4: *For any $\gamma \in \pi_1(X_4; \vec{u}, \vec{v})$, the assignment $\phi \in \mathcal{E}_{\vec{u}} \to \gamma \cdot \phi \in \mathcal{E}_{\vec{v}}$ defines a natural transformation between $\mathcal{E}_{\vec{u}}$ and $\mathcal{E}_{\vec{v}}$.*

We have defined the action at the discrete level above. As usual one should check that it is compatible with the directed system of coverings, which is a consequence of the naturality of the transformation, that is, of the compatibility with the morphisms. This in turn amounts to the following. Let Y and Y' be two finite coverings, with compactifications $\overline{Y}$ and $\overline{Y}'$; let $a : Y \to Y'$ be a covering map, $\alpha : F' \to F$ the corresponding morphism of the function fields over $\overline{\mathbb{Q}}(T)$. Given ϕ and γ as above, we have defined $\mathcal{E}_{\vec{u}}(\alpha)(\phi)$ by $\mathcal{E}_{\vec{u}}(\alpha)(\phi) = \phi \circ \alpha \in \mathcal{E}_{\vec{u}}(F')$, and we want to check that for any ϕ, α, γ,

$$(6) \qquad \mathcal{E}_{\vec{u}}(\alpha)(\gamma \cdot \phi) = \gamma \cdot \big(\mathcal{E}_{\vec{u}}(\alpha)(\phi)\big).$$

This boils down in effect to a simple path lifting property, granted that we only need to prove this at a finite discrete level. Pick an oriented edge $\vec{U}$ over $\vec{u}$ in Y (think in terms of drawings and assume that $\vec{u} = \overrightarrow{01}$). $\vec{U}$ uniquely determines a lift Γ of γ in $\overline{Y}$; (6) asserts that $a(\Gamma)$ is a lift of γ to $\overline{Y}'$, which is obvious since a is a covering map that is, preserves the natural projections of the coverings onto $\mathbf{P}^1$.

There is a rich dictionary between the various ways of describing coverings, of which we have already used quite a few. In the simplest case of unramified finite connected coverings $p : Y \to X$ and ordinary, discrete, fundamental group, the monodromy action on the fibre over the base point x_0 of the fundamental group $\pi_1^{\text{top}}(X; x_0)$ sets up a category equivalence between the connected coverings of X and the finite sets endowed with a transitive action of the fundamental group. To a covering is associated the finite fibre over the base point with the monodromy action and, conversely, any finite set with a transitive π_1 action gives rise to such a connected covering. Starting from a covering, yet another way to describe this situation is to consider the subgroup $H \subset \pi_1$ of finite index having trivial action on the fibre at x_0 and to view this fibre as the homogeneous space π_1/H.

Using the above constructions, essentially the same description can be given, using the fibre at a *ramified* point. Specifically, the category of the connected unramified coverings of X_4 is equivalent to that of sets together with a transitive action of the group $\pi_1^{\text{top}}(X_4; \overrightarrow{01}) \simeq F_2$. Given a covering Y, the action of $\pi_1(X_4; \overrightarrow{01})$ is of course that on $\mathcal{E}_{\overrightarrow{01}}(\overline{\mathbb{Q}}(Y))$ which we have described above. Here we used $\overrightarrow{01}$ only to underline the connection with dessins d'enfants; we may naturally use any $\vec{u} \in \mathcal{B}$ and also, slightly less trivially, replace the monodromy at $\vec{u}$ by the "transport" from $\vec{u}$ to $\vec{v}$ via elements of $\pi_1(X_4; \vec{u}, \vec{v})$ which induce a map $\mathcal{E}_{\vec{u}} \to \mathcal{E}_{\vec{v}}$, as shown above.

The point of all this resides in the fact that, just as in the case of the fundamental group (cf. [SGA 1]), we have the following statement, which appears as a converse to Proposition 4 and could actually serve as a definition for the algebraic fundamental groupoid:

Proposition 5: *Any natural transformation from $\mathcal{E}_{\vec{u}}$ to $\mathcal{E}_{\vec{v}}$ is induced by an element of $\pi_1(X_4; \vec{u}, \vec{v})$.*

Explicit action of the Galois group on the fundamental groupoid.
We may now turn to the action of the Galois group Γ and to the proof of Theorem 1. First we make the identification $\pi_1(X_4; \overrightarrow{01}) \simeq \hat{F}_2$ explicit by choosing a generating set $\{x, y\}$ (those which occur in the statement of Theorem 1) as in Figure 2; x (resp. y) winds around 0 (resp. 1).

Figure 2

We then turn back to the exact sequence (2) which may now be rewritten more accurately as

$$(7) \qquad 1 \to \pi_1(X_4(\overline{\mathbf{Q}}); \overrightarrow{01}) \to \pi_1(X_4(\mathbf{Q}); \overrightarrow{01}) \to \Gamma \to 1.$$

This is equivalent, as we already noticed, to the sequence of Galois groups:

$$(8) \qquad 1 \to \mathrm{Gal}(\Omega/\overline{\mathbf{Q}}(T)) \to \mathrm{Gal}(\Omega/\mathbf{Q}(T)) \to \Gamma \to 1.$$

Again, (7) (or (8)) determines an outer action of Γ on $\hat{F}_2$, i.e. a map $\Gamma \to \mathrm{Out}(\hat{F}_2)$; that this action is faithful, i.e. that the map is injective, is a direct consequence of Belyi's theorem for which we refer to [B] (see also the article by L. Schneps in this volume). Here we want to take advantage of the fact that (7) and (8) are actually split to define an explicit map $\Gamma \to \mathrm{Aut}(\hat{F}_2)$, as in the statement of Theorem 1. Of course it will then still be injective.

If $F \in Ext(\overline{\mathbf{Q}}(T))$ is a finite extension of $\overline{\mathbf{Q}}(T)$ unramified outside $0, 1, \infty$, if $\vec{u} \in \mathcal{B}$, $\phi \in \mathcal{E}_{\vec{u}}(F)$ and $\sigma \in \Gamma$, we define $\sigma \cdot \phi : F \to \mathcal{P}_{\vec{u}}$ by "action on the coefficients". Namely, fixing $\vec{u} = \vec{01}$ for definiteness again, for $f \in F$, $\phi(f) = \sum_n a_n t^{n/e}$ with $a_n \in \overline{\mathbf{Q}}$ for all n and $\sigma \cdot \phi(f) := \sum_n \sigma(a_n) t^{n/e}$.

Note that $\sigma \cdot \phi$ is *not* an element of $\mathcal{E}_{\vec{u}}(F)$ because it is an embedding $F \hookrightarrow \mathcal{P}_{\vec{u}}$ over $\mathbf{Q}(T)$, *not* over $\overline{\mathbf{Q}}(T)$. For $\gamma \in \pi_1^{\mathrm{top}}(X_4; \vec{u}, \vec{v})$, we now define $\sigma \cdot \gamma := \sigma \circ \gamma \circ \sigma^{-1}$, that is,

$$\sigma \cdot \gamma(\phi) := \sigma \cdot \left(\gamma \cdot (\sigma^{-1} \cdot \phi) \right). \tag{9}$$

The explicit computational recipe, which will be applied several times below, goes as follows. Pick $\phi \in \mathcal{E}_{\vec{u}}(F)$, $f \in F$, so that $\phi(f) \in \mathcal{P}_{\vec{u}}(\overline{\mathbf{Q}})$; apply σ^{-1} on the coefficients to get $\sigma^{-1} \cdot \phi(f) \in \mathcal{P}_{\vec{u}}(\overline{\mathbf{Q}})$; analytically continue $\sigma^{-1} \cdot \phi(f)$, viewed as a germ of analytic function on $S_{\vec{u}}$, along γ and expand the result at $\vec{v}$ to find $\gamma \cdot (\sigma^{-1} \cdot \phi)(f) \in \mathcal{P}_{\vec{v}}$. Lastly, apply σ to the coefficients to obtain $(\sigma \cdot \gamma)(\phi)(f)$.

Now, notice that if $\phi \in \mathcal{E}_{\vec{u}}$, $(\sigma \cdot \gamma)(\phi)$ *is* indeed an element of $\mathcal{E}_{\vec{v}}$ because the action of γ and σ commute on the elements of $\overline{\mathbf{Q}}(T)$ (with the embeddings $\overline{\mathbf{Q}}(T) \hookrightarrow \mathcal{P}_{\vec{u}}$ for $\vec{u} \in \mathcal{B}$), so that $\sigma \cdot \gamma(\phi) : F \hookrightarrow \mathcal{P}_{\vec{v}}(\overline{\mathbf{Q}})$ is an embedding over $\overline{\mathbf{Q}}(T)$. The usual argument using the coherence of the various actions allows to extend the above to the profinite setting. Any $\sigma \cdot \gamma \in \pi_1(X_4; \vec{u}, \vec{v})$ thus defines a map $\mathcal{E}_{\vec{u}} \to \mathcal{E}_{\vec{v}}$ and it is easy to check that it is in fact a natural transformation between $\mathcal{E}_{\vec{u}}$ and $\mathcal{E}_{\vec{v}}$. Applying Proposition 5 we have proved the following:

Proposition 6: $\sigma \cdot \gamma \in \pi_1(X_4; \vec{u}, \vec{v})$ for all $\gamma \in \pi_1(X_4; \vec{u}, \vec{v})$, $\sigma \in \Gamma$.

It is clear that the following associativity and commutativity relations between the arithmetic and geometric actions hold true:

$$(\sigma \tau) \cdot \gamma = \sigma \cdot (\tau \cdot \gamma), \quad \sigma \cdot (\gamma' \circ \gamma) = (\sigma \cdot \gamma') \circ (\sigma \cdot \gamma), \tag{10}$$

for all $\sigma, \tau \in \Gamma$, $\gamma, \gamma' \in \pi_1(X_4; \mathcal{B})$, whenever the composition of paths is possible. In particular, we have thus defined an action of the Galois group on the – profinite – fundamental groupoid, that is, a morphism

$$\Gamma \to \mathrm{Aut}\left(\pi_1(X_4; \mathcal{B}) \right).$$

We briefly indicate how this is related to the a priori outer action determined by the exact sequences (7) and (8). Actually, if we fix a coherent family $(\mathcal{E}_{\vec{u}}(F))$, $F \subset \Omega$, of embeddings $F \hookrightarrow \mathcal{P}_{\vec{u}}(\overline{\mathbf{Q}})$ over $\overline{\mathbf{Q}}(T)$, we get $\phi : \Omega \hookrightarrow Puis_{\vec{u}}(\overline{\mathbf{Q}})$. Any such ϕ determines a section of (8), $s_\phi : \Gamma \to \mathrm{Gal}(\Omega/\mathbf{Q}(T))$ via the defining relation

$$\phi\left(s_\phi(\sigma)(f) \right) := \sigma \cdot \phi(f),$$

for any $f \in F \subset \Omega$, which means, put in a more down-to-earth fashion, that given a "prescription" to expand a function as a Puiseux series on $S_{\vec{u}}$ or, equivalently, given a coherent system of "oriented edges" over $\vec{u}$, there is a natural action of Γ defined by the action on the coefficients of the Puiseux series. The action of Γ on $\pi_1(X_4; \mathcal{B})$ we have constructed above is obtained from the outer action defined by (8) when using this family of sections.

We now turn to the explicit computations which will complete the proof of Theorem 1. Besides x and y (see Figure 2), we shall make use of the elements $p, q, r \in \pi_1(X_4; \mathcal{B})$, defined as follows (see Figure 3):

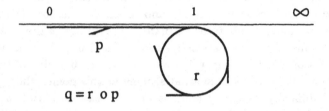

Figure 3

$-\ p \in \pi_1(X_4; \overrightarrow{01}, \overrightarrow{10})$ is the path form 0 to 1 along the interval (0,1);
$-\ r \in \pi_1(X_4; \overrightarrow{10}, \overrightarrow{1\infty})$ is obtained by rotating a half turn counterclockwise around 1;
$-\ q = r \circ p \in \pi_1(X_4; \overrightarrow{01}, \overrightarrow{1\infty})$.

Let us first compute $\sigma \cdot x$ and $\sigma \cdot y$ for $\sigma \in \Gamma$. Let $F \in Ext(\overline{\mathbf{Q}}(T))$, $\phi \in \mathcal{E}_{\overrightarrow{01}}(F)$, $f \in F$, so that $\phi(f)(t) = \sum_n a_n t^{n/e} \in \mathcal{P}_{\overrightarrow{01}}$; $x \cdot \phi(f)$ is obtained by continuing this series around 0:

(11) $\qquad x \cdot \phi(f)(t) = \sum_n a_n \zeta^n t^{n/e}$ where $\zeta = \exp(\dfrac{2i\pi}{e})$.

We now compute $\sigma \cdot x$, following the recipe detailed after formula (9):

$$\phi(f) = \sum_n a_n t^{n/e} \xrightarrow{\sigma^{-1}} \sum_n \sigma^{-1}(a_n) t^{n/e} \xrightarrow{x} \sum_n \sigma^{-1}(a_n)\zeta^n t^{n/e} \xrightarrow{\sigma}$$

(12) $\qquad \xrightarrow{\sigma} \sum_n \sigma(\sigma^{-1}(a_n)\zeta^n) t^{n/e} = \sum_n a_n \zeta^{n\chi(\sigma)} t^{n/e} = (\sigma \cdot x) \cdot \phi(f)$

Comparing (11) and (12), we find that $\sigma \cdot x = x^{\chi(\sigma)}$, as stated in Theorem 1.

We shall now compute $\sigma \cdot y$ somewhat formally, and then justify the computation. Let $\theta \in \mathrm{Aut}(X_4)$ defined by $\theta(t) = 1 - t$, and notice that $y = p^{-1} \theta(x) p \in \pi_1(X_4; \overrightarrow{01})$ (we drop the o from the notation). Define

$$f_\sigma(x, y) := p^{-1} \, \sigma \cdot p$$

and compute (we write $\sigma \cdot \gamma$ or $\sigma(\gamma)$ indifferently, hopefully making the formulae as readable as possible):

$$\sigma \cdot y = \sigma \cdot (p^{-1}\theta(x)p) = \sigma(p^{-1})p \, p^{-1}\theta(\sigma \cdot x)p \, p^{-1}\sigma(p)$$
$$= f_\sigma^{-1} \, p^{-1}\theta(x^{\chi(\sigma)})p \, f_\sigma = f_\sigma^{-1} \, p^{-1}(\theta(x))^{\chi(\sigma)}p \, f_\sigma = f_\sigma^{-1} \, y^{\chi(\sigma)} f_\sigma,$$

as in the statement of Theorem 1.

To complete the proof of this theorem, it only remains to justify the above computation, to show that $f_\sigma \in \hat{F}_2'$, and to derive equations (I) and (II). Justifying the above computation actually means studying the effect of the automorphisms of X_4 on the whole situation, and in particular justifying the intervertion $\sigma \circ \theta = \theta \circ \sigma$ which was used in the first line of the above computation. We shall be somewhat sketchy at this point, which presents no special difficulty. The key point is *that the automorphisms of X_4 are **Q**-rational*. Indeed, $\mathrm{Aut}(X_4) \simeq \mathfrak{S}_3$ is generated by $\theta : t \to 1 - t$ and $\omega : t \to (1-t)^{-1}$, corresponding to the transposition (01) and the three cycle $(0\,1\,\infty)$ respectively. Recall that more generally, for any $n \geq 5$, $\mathrm{Aut}(X_n) \simeq \mathfrak{S}_n$ and these automorphisms are **Q**-rational (that $\mathfrak{S}_n \subset \mathrm{Aut}(X_n)$ is obvious; that there are no other automorphisms is not so obvious); one can also work over other fields. For $n = 4$, the action of $\mathfrak{S}_4$ factors through $\mathfrak{S}_3$.

We shall briefly describe the action of $\mathrm{Aut}(X_4)$ on $\pi_1(X_4; \mathcal{B})$ and the commutation with the Galois action, due to the **Q**-rationality of the elements of $\mathrm{Aut}(X_4)$. In view of what was recalled above, the extension to the higher dimensional case is essentially obvious.

First any $\lambda \in \mathrm{Aut}(X_4)$ acts on $\mathcal{B}$ by looking at the derivatives of λ at the points $(0, 1, \infty)$; for example we have that $\omega(\overrightarrow{01}) = \overrightarrow{1\infty}$. Actually, $\mathrm{Aut}(X_4)$ acts simply transitively on $\mathcal{B}$ and any λ is completely determined by – say – $\lambda(\overrightarrow{01})$. Any automorphism λ also determines a map (still denoted λ):

$$Puis_{\vec{u}} \xrightarrow{\lambda} Puis_{\lambda(\vec{u})}$$

for any $\vec{u}$ in $\mathcal{B}$, simply by changing variable in the Puiseux series. Hence a map:

$$\mathcal{E}_{\vec{u}} \xrightarrow{\lambda} \mathcal{E}_{\lambda(\vec{u})}.$$

As for the action of the automorphisms on the fundamental groupoid, one first let $\lambda \in \mathrm{Aut}(X_4)$ act on $\pi_1^{\mathrm{top}}(X_4; \mathcal{B})$ by simply taking the image of a geometric path by λ; this defines a morphism:

$$\gamma \in \pi_1^{\mathrm{top}}(X_4; \vec{u}, \vec{v}) \to \lambda(\gamma) \in \pi_1^{\mathrm{top}}(X_4; \lambda(\vec{u}), \lambda(\vec{v})).$$

Now if $E \in Ext(\overline{\mathbf{Q}}(T))$ and $\phi \in \mathcal{E}_{\vec{u}}(E)$, $\gamma \cdot \phi$ stabilizes $\mathcal{E}_{\vec{v}}(E) \subset \mathcal{P}_{\vec{v}}$ and $\lambda(\gamma)$ maps $\mathcal{E}_{\lambda(\vec{u})}(\lambda(E))$ to $\mathcal{E}_{\lambda(\vec{v})}(\lambda(E))$. Since $\lambda(E) \in Ext(\overline{\mathbf{Q}}(T))$ because λ is Q-rational, we may pass to the direct limit and define

$$\pi_1(X_4; \vec{u}, \vec{v}) \xrightarrow{\lambda} \pi_1(X_4; \lambda(\vec{u}), \lambda(\vec{v})),$$

on the algebraic, profinite groupoid. Note that in the above, we more than once took $\vec{u} = \overrightarrow{01}$ for "definiteness" in the proof of some statement. Now, with the above in mind, we could also appeal to the fact that for any $\vec{u} \in \mathcal{B}$ there is a unique $\lambda \in \mathrm{Aut}(X_4)$ such that $\lambda(\overrightarrow{01}) = \vec{u}$, in order to restrict ourselves to the case $\vec{u} = \overrightarrow{01}$.

We have the following easy proposition:

Proposition 7: *For any* $\lambda \in \mathrm{Aut}(X_4)$, $\gamma \in \pi_1(X_4; \vec{u}, \vec{v})$, $\phi \in \mathcal{E}_{\vec{u}}$, *we have:*

(13) $$\lambda(\gamma \cdot \phi) = \lambda(\gamma) \cdot \lambda(\phi).$$

The two sides actually define two ways of analytically continuing the *same* function, when $\gamma \in \pi_1^{\mathrm{top}}(X_4; \vec{u}, \vec{v})$. We leave the details to the reader.

We now add in the action of the Galois group. We work with the following objects: $\sigma \in \Gamma$, $\lambda \in \mathrm{Aut}(X_4)$, $E \in Ext(\overline{\mathbf{Q}}(T))$ and $\phi \in \mathcal{E}_{\vec{u}}(E)$ an embedding $E \hookrightarrow \mathcal{P}_{\vec{u}}$ continuing the embedding $\overline{\mathbf{Q}}(T) \hookrightarrow \mathcal{P}_{\vec{u}}$ defined by $\vec{u} \in \mathcal{B}$; the first – crucial – relation is :

(14) $$\lambda(\sigma \cdot \phi) = \sigma \cdot \lambda(\phi).$$

Recall that $\sigma \cdot \phi$ is *not* an element of $\mathcal{E}_{\vec{u}}(E)$, as it does not preserve the embedding $\overline{\mathbf{Q}}(T) \hookrightarrow \mathcal{P}_{\vec{u}}$. It is defined by action on the coefficients; assuming $\vec{u} = \overrightarrow{01}$, then for $f \in E$:

$$\phi(f) = \sum_n a_n t^{n/e} \longrightarrow \sigma \cdot \phi(f) = \sum_n \sigma(a_n) t^{n/e};$$

so $\sigma \cdot \phi$ is actually an element of $\mathcal{E}_{\vec{u}}(^\sigma E)$ where $^\sigma E$ is the conjugate of the field E under σ.

In (14), $\lambda(\phi)$ is obtained by making the change of variable: $t = \lambda^{-1}(s)$, and expanding at $\lambda(\overrightarrow{01})$. As λ is defined over Q, the commutation relation (14) is clear.

We now come to the last relation we need to justify the above computation of $\sigma \cdot y$ and undertake the proof of relations (I) and (II) in Theorem 1:

Proposition 8: *For all* $\sigma \in \Gamma$, $\gamma \in \pi_1(X_4; \mathcal{B})$, $\lambda \in \mathrm{Aut}(X_4)$,

(15) $$\lambda(\sigma \cdot \gamma) = \sigma \cdot \lambda(\gamma).$$

This is a direct consequence of the defining relation (9), and relations (13) and (14); we need only work at a finite level and test both sides on $\phi \in \mathcal{E}_{\vec{u}}$:

$$\lambda(\sigma \cdot \gamma) \cdot \lambda(\phi) = \lambda((\sigma \cdot \gamma) \cdot \phi) = \lambda(\sigma \gamma \sigma^{-1}(\phi)) = \sigma \cdot \lambda(\gamma \sigma^{-1}(\phi))$$
$$= \sigma \cdot \lambda(\gamma) \lambda(\sigma^{-1} \phi) = \sigma \cdot \lambda(\gamma) \sigma^{-1} \cdot \lambda(\phi) = (\sigma \cdot \lambda(\gamma)) \cdot \lambda(\phi),$$

which proves relation (15).

Before deriving equations (I) and (II) in Theorem 1, we first prove that $f_\sigma(x, y) \in \hat{F}'_2$, the derived group of $\hat{F}_2$. In order to see this, it is enough to show that $f_\sigma \in \pi_1(X_4; \overrightarrow{01}) \simeq \mathrm{Gal}(\Omega/\overline{\mathbb{Q}}(T))$ fixes the maximal *abelian* extension of $\overline{\mathbb{Q}}(T)$ contained in Ω; this is generated by cyclic coverings over 0 and 1, that is, by $T^{1/e}$ and $(1 - T)^{1/e}$ for all integers e. Now, applying f_σ to $t^{1/e}$ for a given embedding in $\mathcal{E}_{\overrightarrow{01}}$, we have the chain

$$t^{1/e} \xrightarrow{\sigma^{-1}} t^{1/e} \xrightarrow{p} \left(1 - (1 - t)\right)^{1/e} = \sum_n a_n (1 - t)^{n/e} \quad (a_n \in \mathbb{Q})$$
$$\xrightarrow{\sigma} \left(1 - (1 - t)\right)^{1/e} \xrightarrow{p^{-1}} t^{1/e} = f_\sigma(t^{1/e}).$$

The proof that f_σ fixes cyclic coverings ramified at 1 is analogous.

Equations (I) and (II) come from the following two geometric relations, which take place in $\pi_1^{\mathrm{top}}(X_4; \overrightarrow{01})$:

$$(16) \qquad\qquad \theta(p)\, p = 1; \quad \omega^2(q)\, \omega(q)\, q = 1.$$

Recall that p and q are as on Figure 2, and that θ and ω are elements of $\mathrm{Aut}(X_4)$ of order 2 and 3 respectively ($\theta(t) = 1 - t$, $\omega(t) = (1 - t)^{-1}$). Relations (16) simply express that the paths γ_0 and γ_1 of Figure 1 are trivial in $\pi_1^{\mathrm{top}}(X_4; \overrightarrow{01})$.

The proof of equation (I) is reduced to the following computation (again we use $\sigma(\gamma)$ or $\sigma \cdot \gamma$ according to which is more readable):

$$1 = \sigma \cdot \big(\theta(p)\, p\big) = \theta(\sigma \cdot p)\, \sigma \cdot p = \theta(p f_\sigma(x, y))\, p f_\sigma(x, y) =$$
$$\theta(p)\, \theta(f_\sigma(x, y))\, p f_\sigma(x, y) = p^{-1}\, \theta(f_\sigma(x, y))\, p\, f_\sigma(x, y) =$$
$$f_\sigma\big(p^{-1}\theta(x)p, p^{-1}\theta(y)p\big)\, f_\sigma(x, y),$$

which proves equation (I) because of the geometric relations:

$$(17) \qquad\qquad p^{-1}\, \theta(x)\, p = y, \qquad p^{-1}\, \theta(y)\, p = x.$$

The proof of (II) is completely analogous, only more involved; since $q = r\, p$, we have $\sigma(q) = \sigma(r)\, \sigma(p) = \sigma(r)\, p f_\sigma$. We first prove:

$$(18) \qquad\qquad \sigma \cdot r = r\, \theta(x)^{1/2(\chi(\sigma) - 1)}.$$

This is obtained by an elementary computation, just as for $\sigma \cdot x$; the only difficulty is that one should not get mixed up with the determinations of the logarithm... We only sketch it; start from $f(t) \in \mathcal{P}_{\overrightarrow{10}}$,

$$f(t) = \sum_n a_n (1-t)^{n/e} = \sum_n a_n \exp\left(\frac{n}{e}\log(1-t)\right)$$

with the principal determination of "log" ($t \notin (1\infty)$). Continuing this along r, we find that $1-t$ goes from $\overrightarrow{01}$ to $\overrightarrow{0\infty}$, going around 0 from *above* (t goes around 1 from below). So we get, slightly abusing notation:

$$r(f) = \sum_n a_n \exp\left(\frac{n}{e}\log(t-1) + i\pi\right) = \sum_n a_n \exp\left(\frac{n}{2}\frac{2i\pi}{e}\right)\exp\left(\frac{n}{e}\log(t-1)\right),$$

with again the principal determination of "log". Now, following around $\sigma \cdot r = \sigma\, r\, \sigma^{-1}$, we find:

$$\sigma \cdot r(f) = \sum_n a_n \left(\exp\frac{i\pi}{e}\right)^{n\chi(\sigma)} \exp\left(\frac{n}{e}\log(t-1)\right).$$

Applying r^{-1} and setting $\zeta = \exp\frac{2i\pi}{e}$ we arrive at:

$$r^{-1}\sigma \cdot r(f) = \sum_n a_n \zeta^{\frac{n}{2}(\chi(\sigma)-1)}(1-t)^{n/e},$$

and since $\theta(t) = 1-t$, we find that (18) actually holds true. It immediately implies:

(19) $\qquad \sigma \cdot q = q\, y^m\, f_\sigma$, where $m = m(\sigma) = \frac{1}{2}(\chi(\sigma)-1)$.

We now have a situation with order 3 symmetry. Set $s := \omega(q)\,q$; then the analogue of relations (17) reads:

$$q^{-1}\omega(x)q = y, \quad q^{-1}\omega(y)q = z, \quad q^{-1}\omega(z)q = x, \quad \text{with } xyz = 1;$$

(20) $\qquad s^{-1}\omega^2(x)s = z, \quad s^{-1}\omega^2(y)s = x, \quad s^{-1}\omega^2(z)s = y.$

Starting from the second of relations (16) we find that:

(21) $\quad 1 = \omega^2(\sigma \cdot q)\omega(\sigma \cdot q)\sigma \cdot q = \omega^2(q\,y^m\,f_\sigma)\omega(q\,y^m\,f_\sigma)q\,y^m\,f_\sigma.$

Moreover, using the first set of relations (20), we find that:
$$\omega(q\,y^m\,f_\sigma) = \omega(q)q\,q^{-1}\omega(y^m)q\,q^{-1}f_\sigma(\omega(x),\omega(y))q = \omega(q)q\,z^m\,f_\sigma(y,z).$$
Similarly the second set of relations (20) implies:
$$\omega^2(q)\omega^2(y^m)\,f_\sigma(\omega^2(x),\omega^2(y))\,\omega(q)q = x^m\,f_\sigma(z,x).$$
Substituting into (21) gives equation (II) in Theorem 1, thus completing its proof.

The Grothendieck-Teichmüller group and automorphisms of braid groups

Pierre Lochak* and Leila Schneps**

Abstract

We show that the groups GT_ℓ and $\widehat{GT}$ defined by Drinfel'd are respectively the automorphism groups of the tower of the "pro-ℓ completions" $B_n^{(\ell)}$ of the Artin braid groups and of their profinite completions $\hat{B}_n$, equipped with certain natural inclusion and strand-doubling homomorphisms.

*

§1. Introduction

In [D], Drinfel'd introduced the groups GT_ℓ and $\widehat{GT}$ via deformations of quasi-Hopf algebras, a structure which he had defined in a previous paper. He showed, using tensor-categorical methods, that for each $n \geq 3$, $\widehat{GT}$ can be viewed as a subgroup of the automorphism group of $\hat{B}_n$, the profinite completion of the Artin braid group B_n (see §2 and the appendix for the precise definitions of braid groups). One of the main goals of this article, is to characterize this subgroup completely for $n \geq 1$ in both the pro-ℓ and the profinite cases. It is worth recalling that $\text{Aut}(B_n)$, the automorphism group of the *discrete* group B_n, is easily described: by a result of Dyer and Grossman (cf. [DG]), we know that $\text{Aut}(B_n)$ is generated by the inner automorphisms of B_n and a single other one given by the mirror reflection, which sends each generator of B_n to its inverse. In contrast to this, the automorphism groups of the pro-ℓ and profinite completions $B_n^{(\ell)}$ and $\hat{B}_n$ of B_n are large and complex groups; this emphasizes the fact that working with such completions is an essential part of the theory.

One of the most interesting features of the group $\widehat{GT}$ is that, using a result of Belyi (cf. [Be]), Drinfel'd showed that there exists an injection of the absolute Galois group $\text{Gal}(\overline{\mathbb{Q}}/\mathbb{Q})$ into $\widehat{GT}$. Characterizing the image of $\text{Gal}(\overline{\mathbb{Q}}/\mathbb{Q})$ in $\widehat{GT}$ is still a major open problem. This connection of $\widehat{GT}$ with Galois theory is not pursued in [D]; it is a surprising fact that one can even define a group containing the absolute Galois group with no reference whatsoever to the Galois theory of number fields.

A different way of perceiving the elements of $\text{Gal}(\overline{\mathbb{Q}}/\mathbb{Q})$ as elements of $\widehat{GT}$

was described by Ihara (cf. [I1]), using moduli spaces of Riemann surfaces with ordered marked points. It is known that $\mathrm{Gal}(\overline{\mathbf{Q}}/\mathbf{Q})$ acts on certain fundamental groups of these moduli spaces. By considering the two simplest cases, those of the sphere with 4 and 5 marked points respectively, Ihara shows that this classical action corresponds exactly to that of $\widehat{GT}$ restricted to certain subgroups of $\hat{B}_3$ and $\hat{B}_4$ which are isomorphic to these fundamental groups. This construction should probably be thought of as a starting point for an impressive construction sketched by A. Grothendieck in his "Esquisse d'un programme" ([G], 1984, unpublished). He describes there an action of the absolute Galois group on what he calls the tower of Teichmüller groupoids, that is, a "tower" of algebraic fundamental groupoids (using certain geometrically significant base points) of the moduli spaces for all genera and any number of marked points (and presumably over any ground field).

Let us now extract a few points and suggestions which served as motivations for the present work, and particularly as inspiration for future directions.

i) $\widehat{GT}$ (resp. GT_ℓ) acts on the profinite (resp. "pro-ℓ", see below) completions of the braid groups B_n, and these groups can be joined by the extensions of the natural inclusion homomorphisms $i_n : B_n \hookrightarrow B_{n+1}$ to form a "tower" of groups whose structure is respected by the action of $\widehat{GT}$ (resp. GT_ℓ). We may ask the following question (inspired by a question of Drinfel'd relative, not to the profinite completions but to the k-pro-unipotent completions of the B_n for a field k): is the group of tuples $(\phi_n)_{n\geq 1}$, where each $\phi_n \in \mathrm{Aut}(\hat{B}_n)$ and the ϕ_n respect the inclusion homomorphisms i_n, equal to $\widehat{GT}$? The goal of this article is to give a partial answer to this question; we show that $\widehat{GT}$ (resp. $\widehat{GT_\ell}$) are the groups of tuples $(\phi_n)_{n\geq 1}$ where the ϕ_n respect not only the inclusion homomorphisms but certain natural "string-doubling" ones as well.

ii) The B_n's are closely related to the Teichmüller modular groups in genus 0. Indeed $M(0,n)$, the modular group corresponding to spheres with n marked points, is a quotient of B_n and contains B_{n-2} and $B_{n-1}/(\text{center})$ as subgroups (cf. §2 and the appendix, as well as [Bi]). This gives the main link between the braid groups considered in this article and the geometry of moduli spaces as reflected by the modular groups which are their fundamental groups.

iii) Grothendieck considers what he calls the full Teichmüller tower, i.e. the tower consisting of the Teichmüller groupoids mentioned earlier (rather than the Teichmüller groups) in all genera. One should consider the towers of braid groups given below as very primitive versions of this tower, in genus zero and without the geometric significance of the choice of base points.

iv) A very important suggestion of Grothendieck is that the tower of Teichmüller groupoids (fundamental groupoids of the moduli spaces) should be entirely reconstructible from its first two levels ("les deux premiers étages"), levels being numbered according to their modular dimension (the complex dimension of the moduli space). This is in some sense true for the braid towers considered below, the first and second levels being embodied in B_3 and B_4 respectively.

v) Grothendieck suggests that although the absolute Galois group acts on the whole Teichmüller tower, by virtue of the preceding remark this action should be completely reflected in the restriction of this action to the first two levels. This remark corresponds precisely to what happens in the more limited situation of the actions of GT_ℓ and $\widehat{GT}$ on the braid towers considered below (see the main theorem).

From this general and speculative viewpoint (which does not claim to do more than render the faintest shadow of Grothendieck's vision), it would seem that from the profinite situation one should be able to derive a new description of the elements of $\mathrm{Gal}(\overline{\mathbf{Q}}/\mathbf{Q})$. So far, only the information coming from the action of this group on the fundamental group of the first level, namely the spheres with four marked points (whose moduli space is the much-studied $\mathbf{P}^1\mathbf{C} - \{0, 1, \infty\}$), has been studied in any detail, and just enough is known about the second level (the spheres with five marked points) to be able to derive the important new property of elements of $\mathrm{Gal}(\overline{\mathbf{Q}}/\mathbf{Q})$ which it entails (the so-called pentagon equation, corresponding to (III) below). It is still an open question whether considering higher levels, in genus zero or in any genus, should or should not provide any new constraints on the elements of $\mathrm{Gal}(\overline{\mathbf{Q}}/\mathbf{Q})$ (see [I1] for further remarks on this subject). One of the tasks which lie ahead of us may well consist in an exploration of the deep interplay between the action of the absolute Galois group and the geometry of the various moduli spaces, which Grothendieck appears to have uncovered.

Precise definitions and various necessary results on the different kinds of braid groups can be found in §2 and especially in the appendix included at the end of this article. Let us now turn to the definition of the groups GT_ℓ and $\widehat{GT}$, as given in [D]. In reading this definition, it should be kept in mind that the form (λ, f) of the elements, and the relations (I), (II) and (III) are all combinatorical properties of the elements of $\mathrm{Gal}(\overline{\mathbf{Q}}/\mathbf{Q})$ itself.

In general (i.e. with the exception of $\widehat{GT}$), we write $\hat{G}$ for the profinite completion of a group G, and G^ℓ for its pro-ℓ completion. Let Z_ℓ denote the ℓ-adic integers, $\hat{Z}$ the profinite completion of Z, F_n the free group on n generators for $n \geq 1$, and F_n^ℓ and $\hat{F}_n$ its pro-ℓ and profinite completions respectively. If x, y are the generators of $\hat{F}_2$, we write an element of $\hat{F}_2$ as a "profinite word" $f(x, y)$, although it is not generally a word in x and y. This notation allows us to give a meaning to the element $f(\bar{x}, \bar{y})$ where $\bar{x}$ and $\bar{y}$ are arbitrary elements of a profinite group.

For a group G, let $[G, G]$ denote its derived subgroup (in the topological sense if G is a topological group). Let $\underline{GT}_{\ell,0}$ (resp. $\widehat{\underline{GT}}_0$) be the set of couples $(\lambda, f) \in Z_\ell^* \times [F_2^\ell, F_2^\ell]$ (resp. $(\lambda, f) \in \hat{Z}^* \times [\hat{F}_2, \hat{F}_2]$) satisfying the two following relations:

(I) $f(x, y)f(y, x) = 1$

(II) $f(z, x)z^m f(y, z)y^m f(x, y)x^m = 1$,

where $m = \frac{1}{2}(\lambda - 1)$ and $z = (xy)^{-1}$, this set being equipped with the multiplication law defined by:

$$\big(\lambda_1, f_1(x, y)\big)\big(\lambda_2, f_2(x, y)\big) = \big(\lambda_1\lambda_2, f_2\big(f_1(x, y)x^{\lambda_1}f_1(x, y)^{-1}, y^{\lambda_1}\big)f_1(x, y)\big),$$

which makes it into a semigroup. Define $\widehat{GT}_0$ (resp. $\widehat{GT}_{\ell,0}$) to be the group of invertible elements of this semigroup. Define GT_ℓ (resp. $\widehat{GT}$) to be the subgroup of $GT_{\ell,0}$ (resp. $\widehat{GT}_0$) of couples (λ, f) satisfying the following relation, which takes place in the pro-ℓ (resp. profinite) completion of the group K_4 (the precise definition of which, together with its generators x_{ij} for $1 \leq i < j \leq 4$, is given in §2):

(III) $f(x_{12}, x_{23}x_{24})f(x_{13}x_{23}, x_{34}) = f(x_{23}, x_{34})f(x_{12}x_{13}, x_{24}x_{34})f(x_{12}, x_{23})$.

Acknowledgments: We would like to extend our warmest thanks to Pascal Degiovanni and Jean-Marc Couveignes for the hours of (extremely animated, occasionally hilarious) discussion on many aspects of the questions considered in this article, and the reams of e-mail to which these discussions gave rise. We are also particularly grateful to Professor Ihara for the patience and generosity with which he answered our questions and explained his work to us. It is a real pleasure to thank Professor Nakamura for his careful reading of the manuscript, for pointing out some errors and for making the preprint [N], containing some results which were especially helpful (cf. §4), available to us. It goes without saying that the results presented here were principally inspired by ideas of Grothendieck.

§2. The braid towers

We now introduce the groups we shall work with, starting with the discrete versions, and recall some useful facts about them; a general reference for this is [Bi], but the main reference is the appendix to this article in which detailed proofs of all the necessary technical results on braid groups (and somewhat more) are given.

For $n \geq 2$, let B_n be the Artin braid group on n strings, generated by elements $\sigma_1, \ldots, \sigma_{n-1}$ such that

$$\sigma_i\sigma_{i+1}\sigma_i = \sigma_{i+1}\sigma_i\sigma_{i+1} \text{ for } 1 \leq i \leq n-2 \quad \text{and} \quad \sigma_i\sigma_j = \sigma_j\sigma_i \text{ for } |i-j| \geq 2. \tag{1}$$

It is useful to consider the following more symmetric presentation of B_n (which is restated and proved as proposition A1 of the appendix).

Proposition 1: *The following is a presentation for the Artin braid group B_n: as generators we take σ_{ij} for $1 \leq i,j \leq n$ with the relations*

$$\sigma_{ii} = 1, \quad \sigma_{ij} = \sigma_{ji} \text{ and } \sigma_{jk}\sigma_{ij} = \sigma_{ik}\sigma_{kj}(= \sigma_{ij}\sigma_{ki}) \text{ for } i < j < k;$$

$$\sigma_{ij}\sigma_{kl} = \sigma_{kl}\sigma_{ij} \text{ for } i < j < k < l \text{ or } k < i < j < l.$$

The usual generators are given by $\sigma_i = \sigma_{i,i+1}$ for $1 \leq i \leq n-1$. An important consequence of the above relations is that σ_{ij} and σ_{kl} are conjugate for all i,j,k,l ($i \neq j, k \neq l$). Indeed, if $k = i$, σ_{ij} and σ_{il} are conjugate because $\sigma_{jl}\sigma_{ij} = \sigma_{il}\sigma_{jl}$. If $k \neq i$, σ_{ij} is conjugate to σ_{ik}, which in turn is conjugate to σ_{kl}, so that σ_{ij} and σ_{kl} are again conjugate. In particular, every σ_{ij} ($i \neq j$) is conjugate to $\sigma_{12} = \sigma_1$, that is

$$\sigma_{ij} = a_{ij}^{-1}\sigma_1 a_{ij} \text{ for } 1 < i,j < n, i \neq j, \tag{2}$$

for some elements a_{ij} which can be easily computed. This property will be used below (see sublemma 13 in §6).

We now pass to the definition and some properties of the *pure* braid groups. There exists a canonical surjection $\rho_n : B_n \to S_n$, where S_n is the group of permutations on n letters, obtained by quotienting B_n by the relations $\sigma_i^2 = 1$. The kernel $K_n = \text{Ker}\,\rho_n$, the pure plane braid group on n strings, can be described as follows: it is generated by the elements x_{ij} for $1 \leq i < j \leq n$ defined by

$$x_{ij} = \sigma_{j-1} \cdots \sigma_{i+1}\sigma_i^2\sigma_{i+1}^{-1} \cdots \sigma_{j-1}^{-1}.$$

It is useful to set $x_{ji} = x_{ij}$ and $x_{ii} = 1$; then $x_{ij} = \sigma_{ij}^2$ for the σ_{ij} as in proposition 1. The x_{ij} satisfy the following relations, where we write (a, b) for the commutator $aba^{-1}b^{-1}$:

- $x_{ij}x_{ik}x_{jk}$ commutes with x_{ij}, x_{ik} and x_{jk} for all $i < j < k$;

- $(x_{ij}, x_{kl}) = (x_{il}, x_{jk}) = 1$ for $i < j < k < l$; $\qquad\qquad$ (1')

- $(x_{ik}, x_{ij}^{-1}x_{jl}x_{ij}) = 1$ for $i < j < k < l$.

As an immediate consequence of the conjugation relations (2) above, we get:

$$x_{ij} = a_{ij}^{-1}x_{12}a_{ij} \quad \text{for} \quad 1 < i < j < n. \qquad\qquad (2')$$

Let us also define $y_i \in K_n$ and $\omega_i \in K_n$ by:

$$y_1 = 1, \quad y_i = \sigma_{i-1}\cdots\sigma_2\sigma_1^2\sigma_2\cdots\sigma_{i-1} = x_{1i}x_{2i}\cdots x_{i-1,i} \text{ for } 2 \le i \le n$$

and

$$\omega_i = y_1 y_2 \cdots y_i \, , \text{ for } 1 \le i \le n.$$

Then the y_i's commute, i.e. $(y_i, y_j) = 1$ for all i, j, and the center of B_n and of K_n is an infinite cyclic group generated by ω_n. From now on we use the notation $Z(G)$ for the center of a group G. However we generally write just G/Z for the quotient of G modulo its center.

The sphere braid group H_n (the Hurwitz braid group) is the quotient of B_n by the "sphere relation" $y_n = 1$ (sometimes called the Hurwitz relation), and the modular group $M(0, n)$ (0 because it is the modular group for genus 0 surfaces, i.e. spheres) is the quotient H_n/Z; in other words:

$$M(0, n) = H_n/\langle \tilde{\omega}_n = 1 \rangle = B_n/\langle y_n = \omega_n = 1 \rangle,$$

where $\tilde{\omega}_n$ is the image of ω_n in H_n. The pure sphere braid groups P_n and modular groups $K(0, n)$ are defined to be the kernels of the natural maps $H_n \to S_n$ and $M(0, n) \to S_n$ induced by $\rho_n : B_n \to S_n$. In particular, the subgroup U_n of K_n generated by the n elements $x_{1i}x_{2i}\cdots x_{n,i}$, for $1 \le i \le n$ is a normal subgroup and P_n is exactly K_n modulo this subgroup (cf. lemma A2 of the appendix), while $K(0, n)$ is P_n/Z. The n equations

$$x_{1i}x_{2i}\cdots x_{n,i} = 1, \quad 1 \le i \le n \qquad\qquad (3)$$

are known as the "Hurwitz relations based at i".

We use the notation $\tilde{\alpha}$ for the image in P_n (resp. its completions P_n^ℓ or $\hat{P}_n$) of any element $\alpha \in K_n$ (resp. K_n^ℓ or $\hat{K}_n$) and $\bar{\alpha}$ for the image of α in $K(0, n)$ (resp. $K(0, n)^\ell$ or $\widehat{K(0, n)}$). The following proposition, giving several relations, inclusions and homomorphisms between the various types of braid groups, is proved as proposition A4 of the appendix, but we state it here for the convenience of the reader.

Proposition 2: *Let K'_n be the subgroup of K_n generated by the x_{ij} for $1 \le i < j \le n$ with $(i,j) \ne (1,2)$. Then*

(i) $K_n = K'_n \times \langle \omega_n \rangle$.

(ii) There are two natural inclusions i_1 and i_2 of K_{n-1} into K_n. Both send x_{ij} to x_{ij} for $1 \le i < j \le n$, $(i,j) \ne (1,2)$. But i_1 is then defined by setting $i_1(x_{12}) = x_{12}$, whereas i_2 is defined by setting $i_2(\omega_{n-1}) = \omega_n$.

(iii) $P_n \simeq K'_{n-1} \times \mathbb{Z}/2\mathbb{Z}$ and $K(0, n+1) \simeq P_{n+1}/Z \simeq K'_n$.

(iv) $P_{n+1} \simeq K_{n+1}/U_{n+1} \simeq K_n/\langle \omega_n^2 \rangle$.

(v) The subgroup of P_{n+1} generated by the $\tilde{x}_{ij}$ with $1 \le i < j \le n-1$ and $(i,j) \ne (1,2)$ and the central element $\tilde{\omega}_{n+1}$ of P_{n+1} is isomorphic to P_n. Indeed, $P_{n+1} = F_n \rtimes P_n$ where P_n is this subgroup and F_n is the free group of rank $n-1$ generated by $\tilde{x}_{1,n+1}, \dots, \tilde{x}_{n,n+1}$ (whose product equals 1).

(vi) We have the inclusions $K(0, n+1) \simeq K_n/Z \subset B_n/Z \subset M(0, n+1)$.

Let us consider some low-dimensional examples which will be needed later on.

- B_3 is generated by σ_1, σ_2, with the relation $\sigma_1 \sigma_2 \sigma_1 = \sigma_2 \sigma_1 \sigma_2$; its infinite cyclic center is generated by $\omega_3 = x_{12} x_{13} x_{23} = (\sigma_1 \sigma_2)^3$.

- K_3 is the direct product of a free group on two generators, generated say by $x = x_{12}$ and $y = x_{23}$, and the infinite cyclic center generated by $x_{12} x_{13} x_{23}$. These x and y (together with $z = (xy)^{-1}$) appear in the defining relations (I) and (II) of $\widehat{GT}$ given above. The group $K(0,4) \simeq K_3/Z(K_3)$ is thus a free group on two generators.

We will need the following two relations, valid in the group $M(0,5)$:

$$\tilde{x}_{45} = \tilde{x}_{12} \tilde{x}_{13} \tilde{x}_{23} , \quad \tilde{x}_{15} = \tilde{x}_{23} \tilde{x}_{24} \tilde{x}_{34}. \tag{4}$$

Proof: In $M(0,5)$, we have $\bar{y}_5 = \bar{\omega}_5 = 1$, so

$$\bar{y}_2 \bar{y}_3 \bar{y}_4 = \tilde{x}_{12} \tilde{x}_{13} \tilde{x}_{23} \tilde{x}_{14} \tilde{x}_{24} \tilde{x}_{34} = 1;$$

moreover, we have $\tilde{x}_{14} \tilde{x}_{24} \tilde{x}_{34} \tilde{x}_{45} = 1$ in $M(0,5)$. This last identity is the Hurwitz relation based at 4 (see equation (3)). Combining it with the center relation written as above proves the first of the relations (4); the second follows by shifting the indices (which *is* a legitimate operation in this situation, as the reader can easily convince her(him)self). $\diamond$

To finish with the preliminaries, we still have to define a – nonstandard – subgroup of $\hat{B}_n$ which will be used below. For $n > 1$, let A_n denote the abstract group isomorphic to the subgroup of B_n generated by $\sigma_1^2, \sigma_2, \cdots, \sigma_{n-1}$; a presentation of A_n is given by the following sets of relations:

- the relations (1) between the generators $\sigma_2, \cdots, \sigma_{n-1}$;
- the relations $(\sigma_1^2, \sigma_i) = 1$ for $2 < i \leq n - 1$;
- the commutation relation $(\sigma_1^2, \sigma_2 \sigma_1^2 \sigma_2) = 1$, which is just $(y_2, y_3) = 1$.

We set by convention $B_1 = K_1 = A_1 = \{1\}$. Now, the pro-ℓ (resp. profinite) completions of the groups K_n, P_n and $K(0,n)$ are isomorphic to the free pro-ℓ (resp. profinite) groups on $n(n-1)/2$ generators quotiented by the relations $(1')$. Moreover (see for example [M] §2), the presentation of the profinite completions of the B_n and the A_n are identical to those of the discrete groups (so in particular all groups in Proposition 2 can be replaced by their profinite completions). This does not hold for the pro-ℓ completions of the B_n. We therefore set $B_n^{(\ell)}$ to be the free group on $n - 1$ generators quotiented by the relations (1). This is not as arbitrary a procedure as it seems. Indeed, the group $B_n^{(\ell)}$ can be obtained as a modified pro-ℓ completion of B_n as follows: consider the elements of B_n as automorphisms of K_n (by the restriction to K_n of their action as inner automorphisms); these automorphisms extend to automorphisms of K_n^ℓ. The group $B_n^{(\ell)}$ occurs as the quotient of the semi-direct product $K_n^\ell \rtimes B_n$ defined by this action by the subgroup of elements of the form (x, x^{-1}), $x \in K_n \subset K_n^\ell$. Let us define $H_n^{(\ell)}$ and $M(0,n)^{(\ell)}$ to be the quotients of $B_n^{(\ell)}$ by the same relations as in the discrete case. Proposition 2 is valid when the groups K_n, P_n, B_n and $M(0,n)$ are replaced by K_n^ℓ, P_n^ℓ, $B_n^{(\ell)}$ and $M(0,n)^{(\ell)}$.

The map ρ_n can be extended to a map $\rho_n : B_n^{(\ell)} \to S_n$ (resp. $\rho_n : \hat{B}_n \to S_n$) and we still have $K_n^\ell = \mathrm{Ker}\,\rho_n$ (resp. $\hat{K}_n = \mathrm{Ker}\,\rho_n$). From now on we simply write ρ for the canonical epimorphisms from $B_n^{(\ell)}$ or $\hat{B}_n$ to S_n for any n. A final useful remark on $[B_n, B_n]$, the derived subgroup of the Artin braid group (and its pro-ℓ and profinite completions): the quotient of B_n (resp. $B_n^{(\ell)}$, $\hat{B}_n$) by its derived subgroup is isomorphic to the free abelian group on $n - 1$ generators (resp. free pro-ℓ, free profinite).

Let us turn to the definition of the braid group tower $\mathcal{T}$.

A *tower of groups* is given by a family of groups $\{G_n\}_{n \in \mathcal{N}}$ for some index set $\mathcal{N}$, and for each pair $(i, j) \in \mathcal{N}$, of a (possibly empty) family $\mathcal{F}_{i,j}$ of homomorphisms $G_i \to G_j$. We ask that the sets $\mathcal{F}_{i,j}$ be saturated with respect to the composition of maps, i.e. we assume that $\mathcal{F}_{i,j} \supseteq \mathcal{F}_{k,j} \circ \mathcal{F}_{i,k}$ for all triples of indices i, j, k.

The *automorphism group* of a tower of groups is defined by

$$\{(\phi_n)_{n\in\mathcal{N}} \mid \phi_n \in \mathrm{Aut}(G_n) \text{ for } n \in \mathcal{N} \text{ and } f \circ \phi_i = \phi_j \circ f$$

$$\text{for all } i, j \in \mathcal{N}, \text{ and } f \in \mathcal{F}_{i,j}\}.$$

In what follows, we will frequently make use of the following maps:

- $i_n : \hat{B}_n \to \hat{B}_{n+1}$, the natural inclusion map, given by $i_n(\sigma_i) = \sigma_i$ for $1 \leq i \leq n-1$;

- $f_n : \hat{A}_n \to \hat{B}_{n+1}$ the restriction of i_n to $\hat{A}_n$ and

- $g_n : \hat{A}_n \to \hat{B}_{n+1}$, such that $g_n(\sigma_1^2) = \sigma_2\sigma_1^2\sigma_2$ and $g_n(\sigma_i) = \sigma_{i+1}$ for $2 \leq i \leq n-1$.

Remark: *The maps i_n, f_n and g_n are group homomorphisms for $n \geq 1$.*

Proof: The maps i_n and f_n are clearly group homomorphisms. In order to prove that g_n is also a group homomorphism, it suffices to show that it respects the relations defining the group $\hat{A}_n$. Since these relations are the same as those defining A_n as remarked above, it suffices to show that g_n is a homomorphism from A_n into B_{n+1}, which is a straightforward calculation. Actually, using the set of defining relations for A_n given above, one sees that only the last one, namely $(y_2, y_3) = 1$ could be a problem, but g_n maps it onto the – true – relation $(y_3, y_4) = 1$. $\qquad\qquad\diamond$

It is more enlightening to visualize the map g_n in terms of braids: the image of any braid in A_n is the same braid but with the first string doubled, in B_{n+1} (and in fact in A_{n+1}). Doubling the first string does not respect multiplication of arbitrary braids in B_n, and therefore does not induce a group homomorphism of B_n into B_{n+1}. In order for this map to be a group homomorphism, it is necessary to restrict it to those braids whose first strand, though it may wander about the braid, must return to its place at the end, i.e. the set of braids which is the pre-image under ρ_n in B_n of the subgroup of permutations in S_n which fix 1. But this is exactly the group A_n.

Let $\mathcal{T}$ be the tower of braid groups defined as follows: let the index set be the positive integers: as groups we take the $B_n^{(\ell)}$ for $n \geq 1$, and we equip this family with natural inclusions $i_n : B_n^{(\ell)} \to B_{n+1}^{(\ell)}$ given by $i_n(\sigma_i) = \sigma_i$ for $1 \leq i \leq n-1$. Saturation of these maps under composition means that in fact $\mathcal{T}$ is equipped with all the natural inclusions $i_{n,m} : B_n^{(\ell)} \to B_m^{(\ell)}$

for $n < m$, so $i_n = i_{n,n+1}$. In order to avoid a cumbersome notation, we often write $B_n^{(\ell)} \subset B_m^{(\ell)}$ instead of $i_{n,m}(B_n^{(\ell)}) \subset B_m^{(\ell)}$. We also add in the map $g_3 : \hat{A}_3 \to \hat{B}_4$ – together with its composition with the inclusion maps. For $N \geq 1$, let $\mathcal{T}_N$ denote the truncated tower given by the finite family of groups $B_n^{(\ell)}$ for $1 \leq n \leq N$, the inclusions $i_{n,m}$ for $1 \leq n < m \leq N$, and the map g_3 if $N \geq 4$.

The automorphism group of $\mathcal{T}$ is given by

$$\mathrm{Aut}(\mathcal{T}) = \{(\phi_n)_{n\geq 1} \mid \phi_n \in \mathrm{Aut}(B_n^{(\ell)}), \ i_n \phi_n = \phi_{n+1} i_n \text{ and } \phi_4 g_3 = g_3 \phi_3\};$$

for $N \geq 1$, the automorphism group $\mathrm{Aut}(\mathcal{T}_N)$ of $\mathcal{T}_N$ is given in the exact same way, with n restricted by $1 \leq n \leq N - 1$ and the compatibility condition for g_3 not included if $N \leq 3$. Note that the conditions $i_{n,m}\phi_n = \phi_n i_{n,m}$ are automatically satisfied for all $n < m$ if they are satisfied by the $i_n = i_{n,n+1}$. The reason for the appearance of the map g_3 will be clear from the proof of the corollary of proposition 6 (see §4 in fine). We do not know whether the result still holds true without this extra compatibility condition imposed on the ϕ_n.

Let now $\hat{\mathcal{T}}$ (resp. $\hat{\mathcal{T}}_N$) be the tower of groups defined as follows: as groups we take the $\hat{B}_n$ and the $\hat{A}_n$ for $n \geq 1$ (resp. $1 \leq n \leq N$), and we equip this family with all the maps i_n, f_n, g_n (and all maps obtained by composing them), for $n \geq 1$ (resp. $1 \leq n \leq N - 1$).

We shall now prove that $\phi \in \mathrm{Aut}(\mathcal{T})$ (resp. $\mathrm{Aut}(\hat{\mathcal{T}})$) preserves the permutations of the $B_n^{(\ell)}$ (resp. $\hat{B}_n$) up to a possible twist by σ_1. From now on, for any element g of a group G, we write $\mathrm{Inn}(g)$ for the inner automorphism of G given by conjugation by g: $\mathrm{Inn}(g)(h) := g^{-1}hg$.

Lemma 3: *Suppose $\phi = (\phi_n)_{n\geq 1} \in \mathrm{Aut}(\mathcal{T})$ (resp. $\mathrm{Aut}(\hat{\mathcal{T}})$) fixes the permutations of $B_3^{(\ell)}$ (resp. $\hat{B}_3$), i.e. $\rho \circ \phi_3 = \phi_3$. Then ϕ fixes the permutations globally, i.e. $\rho \circ \phi_n = \phi_n$ for all n. If ϕ_3 does not fix the permutations of $B_3^{(\ell)}$ (resp. $\hat{B}_3$), then $\tilde{\phi}_3 = \mathrm{Inn}(\sigma_1) \circ \phi_3$ does, so $\tilde{\phi} := ((\mathrm{Inn}(\sigma_1)\phi_n))_{n\geq 1}$ fixes the permutations globally.*

Proof: The proof is identical for the pro-ℓ and the profinite completions; indeed it only uses the fact that the automorphisms of the respective towers respect the inclusions i_n. We use the notation of the profinite case. Let us first suppose the following assertion true.

(*) Let ψ be an automorphism of $\hat{B}_n$ (for some $n \geq 1$); then it induces an inner automorphism of the permutation group S_n, i.e. there exists $\alpha \in S_n$ such that:
$$\rho \circ \psi(\sigma) = \alpha^{-1}\rho(\sigma)\alpha \quad \text{for all} \quad \sigma \in \hat{B}_n.$$

Write $\rho\phi_n = \mathrm{Inn}(\alpha_n)\rho$ for $n \geq 3$; since the center of S_n is trivial for $n \geq 3$, the $\alpha_n \in S_n$ are uniquely defined. Assume that ϕ_3 fixes the permutations, i.e. that $\alpha_3 = 1$. By a simple inductive argument, we show that then $\alpha_n = 1$ for all n. Assume it is true for $n-1$ ($n \geq 4$). Then α_n must lie in the centralizer of S_{n-1} in S_n, where S_{n-1} denotes the subgroup of the permutations in S_n fixing n. But as is easy to check, this centralizer is the identity when $n > 4$. So ϕ fixes the permutations globally, i.e. ϕ_n fixes the permutations of $\hat{B}_n$ for $n \geq 1$.

Note that for $n = 3$, the centralizer of S_2 in S_3 is isomorphic to $S_2 \simeq \mathbf{Z}/2\mathbf{Z}$. This shows that if ϕ_3 does not fix the permutations of S_3, then $\mathrm{Inn}(\sigma_1) \circ \phi_3$ does, which gives the last assertion in the lemma.

We now prove assertion (*), which is the generalization to the profinite case of the following result ([DG], cor. 12): for every $\psi \in \mathrm{Aut}(B_n)$, there exists an inner automorphism β of B_n such that $\rho \circ \psi \circ \beta = \rho$. We know that K_n is characteristic in B_n (see [DG], thm. 11), and if $\psi \in \mathrm{Aut}(\hat{B}_n)$ is such that $\psi(\hat{K}_n) = \hat{H}$, then setting $H := B_n \cap \hat{H}$, we see that $\hat{H}$ is the profinite completion of H (since $\hat{H}$ is of finite index in $\hat{B}_n$), and so $B_n/H = \hat{B}_n/\hat{H} = S_n$. So by corollary 12 of [DG], $H = K_n$, so $\hat{H} = \hat{K}_n$, which means that $\hat{K}_n$ is characteristic in $\hat{B}_n$. Thus if $\psi \in \mathrm{Aut}(\hat{B}_n)$, $\rho \circ \psi$ induces an automorphism of S_n, so it is inner (the exceptional cases for $n = 4$ or 6 do not occur).   ◇

Corollary: *All the elements of* $\mathrm{Aut}(\hat{T})$ *actually fix the permutations of* $\hat{B}_n$.

Proof: Let $(\phi_n)_{n\geq 1} \in \mathrm{Aut}(\hat{T})$. By Lemma 3, either $\rho \circ \phi_n = \rho$ or $\rho \circ \phi_n = \mathrm{Inn}(12)\rho$ on $\hat{B}_n$, for all $n \geq 1$. If the second possibility were verified by ϕ then the following diagram would commute:

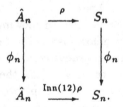

But ϕ_n preserves $\hat{A}_n$ and the image of $\hat{A}_n$ in S_n under ρ is the set of permutations fixing 1, whereas the image under $\mathrm{Inn}(12)\rho$ is the set of permutations fixing 2, so this diagram cannot commute. Thus $\rho \circ \phi_n = \rho$ for all $n \geq 1$. ◇

Let $\mathrm{Inn}(T)$ (resp. $\mathrm{Inn}(T_N)$) denote the subgroup of interior automorphisms of T (resp. T_N) which act on each $B_n^{(\ell)}$ via conjugation by a fixed $\mathbf{Z}_\ell$-power of σ_1, so that $\mathrm{Inn}(T) \simeq \mathrm{Inn}(T_N) \simeq \mathbf{Z}/2\mathbf{Z} \times \mathbf{Z}_\ell$ for all N. Set

$\mathrm{Out}(T) := \mathrm{Aut}(T)/\mathrm{Inn}(T)$ and $\mathrm{Out}(T_N) := \mathrm{Aut}(T_N)/\mathrm{Inn}(T_N)$. Lemma 3 shows that if $\Phi \in \mathrm{Out}(T)$, then there is an element $\phi = (\phi_n)_{n\geq 1} \in \mathrm{Aut}(T)$ in the class Φ, such that each ϕ_n fixes the permutations of $B_n^{(\ell)}$. The same is obviously true when T is replaced by T_N.

We can now state the main theorem of this article:

Main Theorem:

(i) $\mathrm{Out}(T_3) \simeq GT_{\ell,0}$ and $\mathrm{Aut}(\hat{T}_3) \simeq \widehat{GT}_0$;

(ii) $GT_\ell \simeq \mathrm{Out}(T_4)$ and $\widehat{GT} \simeq \mathrm{Aut}(\hat{T}_4)$;

(iii) $GT_\ell \simeq \mathrm{Out}(T) \simeq \mathrm{Out}(T_N)$ and $\widehat{GT} \simeq \mathrm{Aut}(\hat{T}) \simeq \mathrm{Aut}(\hat{T}_N)$ for $N > 4$.

Remark: In reference [I3] Ihara proves a result in the context of graded Lie algebras which, upon tensoring with $\mathbf{Q}_\ell$ (over $\mathbf{Q}$), is analogous to an "infinitesimal version" of the above theorem in the pro-ℓ case, namely a stability property of a certain Lie tower after the first two levels.

Before proving the theorem in §§3 to 6, let us give the basic underlying correspondence between elements of $\widehat{GT}$ and automorphisms of profinite completions of braid groups, due to Drinfel'd (and which also works for GT_ℓ and $B_n^{(\ell)}$). For $n \geq 1$, there is a natural map from GT_ℓ (resp. $\widehat{GT}$) into $\mathrm{Aut}(B_n^{(\ell)})$ (resp. $\mathrm{Aut}(\hat{B}_n)$) given as follows: if $(\lambda, f) \in GT_\ell$ (resp. $\widehat{GT}$) and $\phi \in \mathrm{Aut}(B_n^{(\ell)})$ (resp. $\mathrm{Aut}(\hat{B}_n)$) is its image, then

$$\phi(\sigma_1) = \sigma_1^\lambda \text{ and } \phi(\sigma_i) = f(\sigma_i^2, y_i)\sigma_i^\lambda f(y_i, \sigma_i^2) \text{ for } 2 \leq i \leq n-1. \quad (5)$$

Notice that although σ_1 has been singled out for convenience, it is in fact no exception to the rule: indeed, since f is in the *derived* group $[\hat{F}_2, \hat{F}_2]$, it satisfies $f(y_1, \sigma_1^2) = f(1, \sigma_1^2) = 1$.

The proof of the theorem consists in showing that the ϕ associated to couples (λ, f) are actually automorphisms of the $B_n^{(\ell)}$ (resp. the $\hat{B}_n$); (Drinfel'd shows this for the "k-pro-unipotent completions" of the B_n but his proof uses his tensor-categorical construction of the group GT; thus we prefer to reprove it in our cases by purely group-theoretic methods), that they respect the natural inclusion maps i_n (resp. the maps i_n, f_n and g_n), and that all outer automorphisms of T resp. $\hat{T}$ come from such couples (λ, f).

§3. $GT_{\ell,0}$ and $\widehat{GT}_0$ as automorphism groups of T_3 and $\hat{T}_3$

In this section we prove (i) of the main theorem.

Proposition 4: $GT_{\ell,0} \simeq \mathrm{Out}(T_3)$.

Proof: We begin by defining a homomorphism of $GT_{\ell,0}$ into $\mathrm{Aut}(T_3)$. Let $(\lambda, f) \in GT_{\ell,0}$; we associate to it the map ϕ_3 defined by (5) on the generators of $B_3^{(\ell)}$, namely as $\phi_3(\sigma_1) = \sigma_1^\lambda$ and $\phi_3(\sigma_2) = f(\sigma_2^2, \sigma_1^2)\sigma_2^\lambda f(\sigma_1^2, \sigma_2^2)$. Let us show first that ϕ_3 can be extended multiplicatively to an automorphism of $B_3^{(\ell)}$. Set $\omega_3 = (\sigma_1\sigma_2)^3$; this element generates the center of $B_3^{(\ell)}$. Set $x = \sigma_1^2$, $y = \sigma_2^2$, $z' = \sigma_1\sigma_2^2\sigma_1^{-1}$ and $z = z'\omega_3^{-1}$: we have $xyz = 1$ in $B_3^{(\ell)}$. As mentioned above, K_3^ℓ is isomorphic to $\langle\omega_3\rangle \times \langle x, y\rangle$, the group $\langle x, y\rangle$ being isomorphic to F_2^ℓ, so that an element $f \in F_2^\ell$ is entirely determined if we know $f(x, y) \in B_3^{(\ell)}$.

Note now that $\sigma_1 y = z'\sigma_1$ and $y\sigma_2 = \sigma_2 z'$, which gives the two following conjugation relations:

(a) $\sigma_1 f(y, x) = f(z, x)\sigma_1$, and

(b) $f(x, y)\sigma_2 = \sigma_2 f(z, y)$.

We may (and have) replaced z' by z inside f because, more generally, if α, β and $\gamma \in B_3^{(\ell)}$ and γ commute with α and β, then $f(\gamma\alpha, \beta) = f(\alpha, \gamma\beta) = f(\alpha, \beta)$. This in turn comes from f being in the derived group $[F_2^\ell, F_2^\ell]$.

Lemma 5: ϕ_3 *can be extended multiplicatively to an automorphism of* $B_3^{(\ell)}$.

Proof: It suffices to show that ϕ respects the unique relation $\sigma_1\sigma_2\sigma_1 = \sigma_2\sigma_1\sigma_2$ of $B_3^{(\ell)}$. We calculate:

$$\phi(\sigma_1)\phi(\sigma_2)\phi(\sigma_1) = \sigma_1^\lambda f(y, x)\sigma_2^\lambda f(x, y)\sigma_1^\lambda = \sigma_1^\lambda f(y, x)\sigma_2 y^m f(x, y)x^m\sigma_1$$

$$= \sigma_1^\lambda f(y, x)\sigma_2 f(z, y)z^{-m}f(x, z)\sigma_1 \text{ (since } \phi \text{ satisfies relation (II))}$$

$$= \sigma_1^\lambda f(y, x)f(x, y)\sigma_2 z^{-m}\sigma_1 f(x, y) \text{ (by (a) and (b) above)}$$

$$= \sigma_1^\lambda \sigma_2 z^{-m}\sigma_1 f(x, y) = \sigma_1^\lambda \sigma_2 \sigma_1 \sigma_2^{-2m}\omega_3^m f(x, y)$$

$$= x^m(\sigma_1\sigma_2\sigma_1)y^{-m}\omega_3^m f(x, y) = (\sigma_1\sigma_2\sigma_1)\omega_3^m f(x, y)$$

$$\text{(since } (\sigma_1\sigma_2\sigma_1)^{-1}x^k(\sigma_1\sigma_2\sigma_1) = y^k \text{ for all } k)$$

$$= (\sigma_2\sigma_1\sigma_2)\omega_3^m f(x, y) = \sigma_2\sigma_1\sigma_2^{-2m}\omega_3^m\sigma_2^\lambda f(x, y) = \sigma_2 z^{-m}\sigma_1\sigma_2^\lambda f(x, y)$$

$$= f(y, x)f(x, y)\sigma_2 z^{-m}\sigma_1\sigma_2^\lambda f(x, y) = f(y, x)\sigma_2 f(z, y)z^{-m}\sigma_1\sigma_2^\lambda f(x, y)$$

$$= f(y, x)\sigma_2 f(z, y)z^{-m}f(x, z)f(z, x)\sigma_1\sigma_2^\lambda f(x, y)$$

$$= f(y, x)\sigma_2 y^m f(x, y)x^m\sigma_1 f(y, x)\sigma_2^\lambda f(x, y) \text{ (by (a) and (II))}$$

$$= f(y, x)\sigma_2^\lambda f(x, y)\sigma_1^\lambda f(y, x)\sigma_2^\lambda f(x, y) = \phi(\sigma_2)\phi(\sigma_1)\phi(\sigma_2). \qquad \diamond$$

Set $\phi_1 = \mathrm{id}$ and define ϕ_2 on $B_2^{(\ell)}$ by $\phi_2(\sigma_1) = \sigma_1^\lambda$. Clearly $i_n\phi_n = \phi_{n+1}i_n$ for $n = 1, 2$, so $(\phi_n)_{1 \leq n \leq 3} \in \mathrm{Aut}(T_3)$.

We confirm using the multiplication law of $GT_{\ell,0}$ that the map $GT_{\ell,0}$ into $\mathrm{Aut}(T_3)$ defined in this way is actually a group homomorphism which we denote by $\bar{\eta}$. It induces a homomorphism of $GT_{\ell,0}$ into $\text{Out}(\mathcal{T}_3)$ which we denote by η.

Let us show that η is injective. Let $(\lambda, f) \in GT_{\ell,0}$, be such that $\eta(\lambda, f) = 1$ and set $\phi = (\phi_n)_{1 \le n \le 3} := \bar{\eta}(\lambda, f)$. Then there exists $\delta \in \mathbf{Z}_\ell$ such that $\phi_3 = \text{Inn}(\sigma_1^\delta)$. We then have $\phi_3(\sigma_1) = \sigma_1$ so $\lambda = 1$. Moreover we have

$$\phi_3(\sigma_2) = f(y, x)\sigma_2 f(x, y) = \sigma_1^{-\delta}\sigma_2^\lambda \sigma_1^\delta,$$

so we have an equality of the form

$$f(x, y) = C\sigma_1^\delta$$

where $C \in B_3^{(\ell)}$ commutes with σ_2.

Let us show that such a C has the form $\omega_3^\alpha \sigma_2^\gamma$ for $\alpha, \gamma \in \mathbf{Z}_\ell$. We use the well-known fact that the centralizer of y in the free pro-ℓ group $\langle x, y \rangle$ is the cyclic group $\langle y \rangle$. This shows that the centralizer of the element σ_2^2 in the group K_3^ℓ is generated by ω_3 and σ_2^2 since $K_3 \simeq \langle \sigma_1^2, \sigma_2^2 \rangle \times \langle \omega_3 \rangle$ and $\langle \sigma_1^2, \sigma_2^2 \rangle$ is a free group. Now consider an element $C \in B_3^{(\ell)}$ which centralizes σ_2. Since $\rho(C)$ must centralize $\rho(\sigma_2)$ in S_3, $\rho(C)$ must be either trivial or equal to $\rho(\sigma_2)$ (which is equal to the permutation (23)). If $\rho(C)$ is trivial then $C \in K_3^\ell$, and if C centralizes σ_2 then it centralizes σ_2^2, so it is in $\langle \sigma_2^2, \omega_3 \rangle$. If $\rho(C) = \rho(\sigma_2)$, then C can be written $\sigma_2 C'$ where $C' \in K_3^\ell$, and C' must centralize $\sigma_2 \ldots$ therefore it is again in $\langle \sigma_2^2, \omega_3 \rangle$. This gives the result.

Now, $f(x, y) \in K_3^\ell$, so the expression $f(x, y) = \omega_3^\alpha \sigma_2^\gamma \sigma_1^\delta$ must be in K_3^ℓ as well, which implies that γ and δ are congruent to 0 mod 2. But then $\sigma_2^\gamma \sigma_1^\delta = y^{\gamma/2} x^{\delta/2} \in \langle x, y \rangle = F_2^\ell$, so $\alpha = 0$ since $\omega_3 \notin \langle x, y \rangle$. But then, since f is supposed to belong to $[F_2^\ell, F_2^\ell]$, we must have $\gamma = \delta = 0$ and thus $f = 1$, which gives the injectivity of η.

Let us show that η is surjective. Let $\Phi \in \text{Out}(\mathcal{T}_3)$ and let $(\phi_n)_{1 \le n \le 3} \in \text{Aut}(\mathcal{T}_3)$ be a representative of Φ which fixes the permutations (as can always be chosen by lemma 3). Firstly, λ is determined by ϕ_2, and $\lambda \in \mathbf{Z}_\ell^*$ since ϕ_2 is invertible. Next, because ϕ_3 fixes the permutations, there exist $\alpha \in \mathbf{Z}_\ell$ and $g \in F_2^\ell$ such that $\phi_3(\sigma_1 \sigma_2 \sigma_1) = \sigma_1 \sigma_2 \sigma_1 \omega_3^\alpha g(x, y)$. Applying ϕ_3 to the relation $(\sigma_1 \sigma_2 \sigma_1)^{-1} \sigma_1 (\sigma_1 \sigma_2 \sigma_1) = \sigma_2$ in $B_3^{(\ell)}$, we obtain $\phi_3(\sigma_2) = g^{-1}(x, y)\sigma_2^\lambda g(x, y)$. Let γ, δ be the unique elements of $\mathbf{Z}_\ell$ such that $y^\gamma g(x, y)x^\delta \in [F_2^\ell, F_2^\ell]$ (where F_2^ℓ is identified with $\langle x, y \rangle$), and set $f(x, y) = y^\gamma g(x, y)x^\delta$. We thus associate a couple $(\lambda, f) \in \mathbf{Z}_\ell^* \times [F_2^\ell, F_2^\ell]$ to ϕ_3. (The automorphism of $B_3^{(\ell)}$ associated to this couple is actually $\text{Inn}(x^\delta)\phi_3$, which is in the same class as ϕ_3 modulo $\text{Inn}(\mathcal{T}_3)$). Let us show that this couple is in $GT_{\ell,0}$, i.e. that it satisfies relations (I) and (II).

Let $T := \mathrm{Inn}(\sigma_1\sigma_2\sigma_1)$ and $U := \mathrm{Inn}((\sigma_1\sigma_2)^2) \in \mathrm{Inn}(B_3^{(\ell)})$. Note that $T^2 = U^3 = 1$. Let us calculate $\mathrm{Inn}(f(y,x)^{-1})T\phi_3 T^{-1} \in \mathrm{Aut}(B_3^{(\ell)})$; under this automorphism we have

$$\sigma_1 \mapsto \sigma_1^\lambda \quad \text{and} \quad \sigma_2 \mapsto f(y,x)\sigma_2^\lambda f(y,x)^{-1}.$$

As $f(y,x) \in [F_2^\ell, F_2^\ell]$, this automorphism is equal to $\tilde\eta(\lambda, f(y,x)^{-1})$ and is thus in particular in $\mathrm{Aut}(\mathcal{T}_3)$.

Let us also calculate $\mathrm{Inn}(f(x,z)) U\phi_3 U^{-1}$ on σ_1 and σ_2; we obtain:

$$\sigma_1 \mapsto \sigma_1^\lambda \quad \text{and} \quad \sigma_2 \mapsto f(x,z)^{-1}z^m f(z,y)^{-1}\sigma_2^\lambda f(z,y)z^{-m} f(x,z);$$

thus $\mathrm{Inn}(f(x,z)) U\phi_3 U^{-1}$ is equal to $\tilde\eta(\lambda, y^{-m}f(z,y)z^{-m}f(x,z)x^{-m})$ modulo $\mathrm{Inn}(\mathcal{T}_3)$, where $z = (xy)^{-1} \in F_2^\ell$.

If we consider that $\mathrm{Aut}(\mathcal{T}_3) \subset \mathrm{Aut}(B_3^{(\ell)})$, then in fact $\mathrm{Out}(\mathcal{T}_3) \subset \mathrm{Out}(B_3^{(\ell)})$ since $\mathrm{Aut}(\mathcal{T}_3) \cap \mathrm{Inn}(B_3^{(\ell)}) = \mathrm{Inn}(\mathcal{T}_3)$. By the injectivity of $\eta : GT_{\ell,0} \to \mathrm{Out}(\mathcal{T}_3)$, we see that if (λ, f) and $(\mu, g) \in GT_{\ell,0}$ and the images of $\eta(\lambda, f)$ and of $\eta(\mu, g)$ are equal in $\mathrm{Out}(B_3^{(\ell)})$, then $\eta(\lambda, f) = \eta(\mu, g)$ in $\mathrm{Out}(\mathcal{T}_3^\ell)$. Thus we have $f(x,y) = f(y,x)^{-1}$ and $f(x,y) = y^{-m}f(z,y)z^{-m}f(x,z)x^{-m}$, so (λ, f) satisfies relations (I) and (II). By construction, $(\phi_n)_{1 \le n \le 3}$ is in the class $\eta(\lambda, f)$ in $\mathrm{Out}(\mathcal{T}_3)$, so $\eta : GT_{\ell,0} \to \mathrm{Out}(\mathcal{T}_3)$ is a bijection. This concludes the proof of proposition 4. $\qquad \diamond$

Corollary: $\mathrm{Aut}(\widehat{\mathcal{T}_3}) \simeq \widehat{GT}_0$.

Proof: Most of the calculations in the proof of proposition 4 work in the profinite case with no changes whatsoever. For the injectivity, we need to verify that every automorphism of $\hat B_3$ respecting the inclusions i_n as in the pro-ℓ situation also respects the other maps f_i and g_i. Thus let $\phi := (\phi_n)_{1 \le n \le 3}$ be a triple of automorphisms respecting the i_n for $n = 1, 2$. Define ϕ_1' to be the identity, ϕ_2' by $\phi_2'(\sigma_1^2) = \sigma_1^{2\lambda}$ and ϕ_3' to be the restriction of ϕ_3 to $\hat A_3 \subset \hat B_3$ (ϕ_3' is easily seen to be an automorphism of $\hat A_3$). Then $\phi := (\phi_i, \phi_i')_{1 \le i \le 3}$ respects the maps i_1 and i_2 as in proposition 4, f_1 and f_2 by construction and g_1 by triviality. The relation $g_2\phi_2' = \phi_3 g_2$ is a consequence of the fact that $\phi_3(y_3) = y_3^\lambda$ which is proved as follows. Recall that $\omega_3 = y_2 y_3$; we know that $\phi_3(\omega_3) = \omega_3^\lambda$ since ϕ_3 must send ω_3 to a power of itself and this power must be exactly λ; (looking at the induced action of ϕ_3 modulo the derived subgroup of $\hat B_3$, i.e. in the free profinite abelian group on 2 generators, we see that $\sigma_1 \to \sigma_1^\lambda$ and $\sigma_2 \to \sigma_2^\lambda$), we also know that y_2 and y_3 commute and that $\phi_3(y_2) = y_2^\lambda$ since $y_2 = \sigma_1^2$, which suffices to show that $\phi_3(y_3) = y_3^\lambda$. (The same argument actually shows that if ϕ_n is any automorphism of $\hat B_n$ such that $\phi_n(\hat B_m) = \hat B_m$ for all

$m < n$, then $\phi_n(y_i) = y_i^\lambda$ for $1 \leq i \leq n$.) This shows that $(\phi_n, \phi'_n)_{1 \leq n \leq 3} \in$ Aut$(\hat{T}_3)$ and thus that the map given by equation (5) does determine an injective map $\widehat{GT}_0 \rightarrow$ Aut$(\hat{T}_3)$. The surjectivity argument is easier than in the pro-ℓ case. There is no need to take a representative of an outer automorphism. Moreover the element δ introduced in order to ensure that $f(x, y)$ be in the derived group is necessarily 0 since as shown in the pro-ℓ case, $(\text{Inn}(x^\delta)\phi_n)_{1 \leq n \leq 3}$ defines an element of Aut$(\hat{T}_3)$, however no non-zero power of x can preserve the subgroup $\hat{A}_3 \subset \hat{B}_3$. The rest of the proof of surjectivity goes through as in the pro-ℓ case.   ◊

§4. $\widehat{GT}$ and GT_ℓ as automorphism groups of the braid towers $\hat{T}_4$ and T_4

In this section we prove statement (ii) of the main theorem, working first in the profinite setting and stating the straightforward adaptation to the pro-ℓ case as a corollary. As the proof is somewhat involved and depends on an unexpected introduction of the mapping class group $M(0,5)$, we give a brief description of its logical structure here. The usefulness of $M(0,5)$ (and its profinite completion) lies in the existence of a torsion element of order 5 which gives rise to the pentagon equation in a natural way. The Artin braid group B_4 is not contained in $M(0,5)$, but the quotient by the center B_4/Z is a subgroup of $M(0,5)$.

In lemma 7 (and lemma 8) we show the following result. We take a certain generating set of elements for $\widehat{M(0,5)}$, namely $\bar{\sigma}_1$, $\bar{\sigma}_2$, $\bar{\sigma}_3$, $\bar{\sigma}_4$ and $\bar{\sigma}_{15}$ (see the beginning of the proof of proposition 6). We let couples $(\lambda, f) \in \widehat{GT}_0$ act on this generating set in a particular way and show that this action extends to an automorphism of $\widehat{M(0,5)}$ if and only if the couple actually lies in $\widehat{GT}$. We obtain an (injective) map $\tilde{\iota}: \widehat{GT} \rightarrow$ Aut$(\widehat{M(0,5)})$.

In lemma 9 it is shown that the automorphisms of $\widehat{M(0,5)}$ coming from elements of $\widehat{GT}$ restrict to automorphisms of the subgroup $\hat{B}_4/Z$ and lift uniquely to automorphisms of $\hat{B}_4$ which respect the arrows of the tower $\hat{T}_4$; we thus obtain an (injective) map $\iota: \widehat{GT} \rightarrow$ Aut$(\hat{T}_4)$.

In order to show the surjectivity of this map we proceed as follows. Suppose $\Phi = (\phi_n)_{1 \leq n \leq 4} \in$ Aut$(\hat{T}_4)$; then, because Φ must respect the homomorphisms i_n, f_n and g_n for $1 \leq n \leq 3$, we have

$$\phi_4(\sigma_1) = \sigma_1^\lambda, \quad \phi_4(\sigma_2) = f(x_{23}, x_{12})\sigma_2^\lambda f(x_{12}, x_{23}),$$

$$\text{and } \phi_4(\sigma_3) = f(\sigma_3^2, y_3)\sigma_3^\lambda f(y_3, \sigma_3^2).$$

Now, any automorphism ϕ_4 of $\hat{B}_4$ which acts on σ_1, σ_2 and σ_3 in this way induces an automorphism of $\hat{B}_4/Z$; moreover if (λ, f) is the couple in $\widehat{GT}_0$ associated to ϕ_3 and a map ϕ corresponding to (λ, f) is defined on the generating set of $M(0,5)$ as in lemma 7, then ϕ restricted to the subgroup $\hat{B}_4/Z$ gives an automorphism of this subgroup since it is precisely the one induced by ϕ_4. We show that any ϕ defined on the generating set as in lemma 7 which induces an automorphism of the subgroup $\hat{B}_4/Z$ automatically extends to an automorphism of all of $\widehat{M(0,5)}$; by lemma 7 the couple (λ, f) must then lie in $\widehat{GT}$, so $\Phi = \iota(\lambda, f)$. This shows that ι is a bijection.

Proposition 6: $\widehat{GT} \simeq \mathrm{Aut}(\hat{T}_4)$.

Proof: Let $M(0,5)$ be the mapping class group in genus 0 with 5 marked points; this group is generated by elements σ_1, σ_2, σ_3 and σ_4 – please note! in order to avoid immensely long lines over all the formulae we do *not* put bars over the elements of $M(0,5)$ as we should, throughout the whole of this section, *except* in lemma 9 in which it is necessary to distinguish between elements of $\hat{B}_4$ and their images in $\hat{B}_4/Z \subset \widehat{M(0,5)}$. We introduce $\sigma_{15} := \sigma_4\sigma_3\sigma_2\sigma_1\sigma_2^{-1}\sigma_3^{-1}\sigma_4^{-1}$ as in proposition 1 (so $\sigma_{15}^2 = x_{15}$).

Set $V := \mathrm{Inn}((\sigma_4\sigma_3\sigma_2\sigma_1)^{-3}) \in \mathrm{Inn}(\widehat{M(0,5)})$, so $V^5 = 1$. The map V acts as follows on the σ_i's, σ_{15}, and the x_{ij}'s:

$$V(\sigma_1) = \sigma_3, \ V(\sigma_2) = \sigma_4, \ V(\sigma_3) = \sigma_{15}, \ V(\sigma_4) = \sigma_1, \ V(\sigma_{15}) = \sigma_2, \quad (6)$$

$$V(x_{ij}) = x_{i+2,j+2}, \qquad \text{for } i,j \in \mathbf{Z}/5\mathbf{Z}.$$

These properties could be expressed more concisely as $V(\sigma_{i,i+1}) = \sigma_{i+2,i+3}$ for all $i \in \mathbf{Z}/5\mathbf{Z}$.

We shall make use of a more symmetric form of relation (III), which holds in $\widehat{M(0,5)}$. Namely, considering the x_{ij}'s as elements of $\widehat{M(0,5)}$, relation (III) implies the following:

(III') $f(x_{34}, x_{45})f(x_{51}, x_{12})f(x_{23}, x_{34})f(x_{45}, x_{51})f(x_{12}, x_{23}) = 1$

This form of relation (III) was given by Ihara in [I1]: as he points out, to transform (III) into (III'), one uses relation (I), the relations (4) of §2 and the remarks just preceding lemma 5. We shall write this relation as $f_5 f_4 f_3 f_2 f_1 = 1$. Note that

$$f_1 = f(x_{12}, x_{23}), \quad f_{i+1} = V^{-1}(f_i) \qquad \text{for } i \in \mathbf{Z}/5\mathbf{Z}.$$

Lemma 7: *Let* $(\lambda, f) \in \widehat{GT}_0$ *and associate to it a map* ϕ *sending* σ_1, σ_2, σ_3, σ_4 *and* σ_{15} *into* $\widehat{M(0,5)}$ *as follows:*

$$\phi(\sigma_1) = \sigma_1^\lambda, \quad \phi(\sigma_2) = f(x_{23}, x_{12})\sigma_2^\lambda f(x_{12}, x_{23}),$$

$$\phi(\sigma_3) = f(x_{34}, x_{45})\sigma_3^\lambda f(x_{45}, x_{34}), \quad \phi(\sigma_4) = \sigma_4^\lambda,$$

$$\phi(\sigma_{15}) = f(x_{23}, x_{12})f(x_{51}, x_{45})\sigma_{15}^\lambda f(x_{45}, x_{51})f(x_{12}, x_{23}).$$

Then ϕ can be extended multiplicatively to an automorphism of $\widehat{M(0,5)}$ if and only if (λ, f) lies in $\widehat{GT}$; this map defines an injective group homomorphism which we denote by $\tilde{\imath} : \widehat{GT} \to \mathrm{Aut}(\widehat{M(0,5)})$.

Remark: This action of $\widehat{GT}$ on $\widehat{M(0,5)}$ is determined by considering $\widehat{M(0,5)}$ as a quotient of $\hat{B}_5$ and looking at the images of the right-hand sides of equation (5) in the quotient. The idea of passing from $\hat{B}_5$ to $\widehat{M(0,5)}$ in this way is also employed by Nakamura in the appendix of [N].

Proof: Suppose ϕ is associated to a couple $(\lambda, f) \in \widehat{GT}_0$. We must study when ϕ respects the relations defining $\widehat{M(0,5)}$, namely those of $\hat{B}_5$ and the sphere and center relations $y_5 = \omega_5 = 1$ (cf. §2). Remark first that the subgroup of $\widehat{M(0,5)}$ generated by σ_1 and σ_2 being isomorphic to $\hat{B}_3$, we know by proposition 4 that ϕ induces an automorphism of this group and thus that ϕ respects the relation $\sigma_1\sigma_2\sigma_1 = \sigma_2\sigma_1\sigma_2$. For the other relations we use the following lemma:

Lemma 8: *Let ϕ act on σ_1, σ_2, σ_3, σ_4 and σ_{15} as in lemma 7. Then*

(i) the two maps ϕV and $V \mathrm{Inn}(f_1^{-1})\phi$ take the same values when applied to the elements σ_1, σ_2, σ_4 and σ_{15}, and

(ii) the two maps ϕV^3 and $V^3 \mathrm{Inn}(f_1^{-1}f_2^{-1}f_3^{-1})\phi$ take the same values on σ_1 and σ_2 if (λ, f) lies in $\widehat{GT}$.

Proof: (i) It suffices to calculate $\mathrm{Inn}(f_1)V^{-1}\phi V$ on the given elements using (6). The calculations are all trivial, so we do it for σ_1 only: $\phi V(\sigma_1) = \phi(\sigma_3) = f_5\sigma_3^\lambda f_5^{-1} = V(f_1\sigma_1^\lambda f_1^{-1}) = V \mathrm{Inn}(f_1^{-1})\phi(\sigma_1)$.

(ii) Assume that f satisfies relation (III'), i.e. $f_5 f_4 f_3 f_2 f_1 = 1$, and let us do the calculation. By (III'), $f_1^{-1} f_2^{-1} f_3^{-1} = f_5 f_4$. Thus we have: $\phi V^3(\sigma_1) = \phi(\sigma_2) = f_1^{-1}\sigma_2^\lambda f_1$ and $V^3\mathrm{Inn}(f_5 f_4)\phi(\sigma_1) = V^3(f_4^{-1}f_5^{-1}\sigma_1^\lambda f_5 f_4)$ $= f_1^{-1}f_2^{-1}\sigma_2^\lambda f_2 f_1 = f_1^{-1}\sigma_2^\lambda f_1$ since $f_2 = f(x_{45}, x_{51})$ commutes with σ_2. Moreover, we have $\phi V^3(\sigma_2) = \phi(\sigma_3) = f_5\sigma_3^\lambda f_5^{-1}$ and $V^3\mathrm{Inn}(f_5 f_4)\phi(\sigma_2) = V^3(f_4^{-1}f_5^{-1}f_1^{-1}\sigma_2^\lambda f_1 f_5 f_4) = V^3(f_3 f_2\sigma_2^\lambda f_2^{-1}f_3^{-1}) = f_5 f_4\sigma_3^\lambda f_4^{-1}f_5^{-1} = f_5\sigma_3^\lambda f_5^{-1}$ since $f_4 = f(x_{51}, x_{12})$ commutes with σ_3.   ◇

We use lemma 8 to determine when ϕ extends to an automorphism of $\widehat{M(0,5)}$, i.e. respects the defining relations (besides $\sigma_1\sigma_2\sigma_1 = \sigma_2\sigma_1\sigma_2$, which is respected by the assumption that (λ, f) lies in $\widehat{GT}_0$):

(a) $\sigma_3\sigma_4\sigma_3 = \sigma_4\sigma_3\sigma_4,$

(b) the sphere relation $y_5 = \sigma_4\sigma_3\sigma_2\sigma_1^2\sigma_2\sigma_3\sigma_4 = 1$,

(c) the center relation $\omega_5 = (\sigma_1\sigma_2\sigma_3\sigma_4)^5 = 1$,

(d) the remaining relation $\sigma_2\sigma_3\sigma_2 = \sigma_3\sigma_2\sigma_3$.

It turns out that (a), (b) and (c) are all respected when ϕ is associated to any couple $(\lambda, f) \in \widehat{GT}_0$, for we only need to use lemma 8 (i) (and indeed, the equality of the two given maps on the elements σ_1 and σ_2). Relation (a) is proved by the following argument using the equality of the maps ϕV and $V \operatorname{Inn}(f_1^{-1})\phi$ applied to σ_1 and σ_2:

$$\phi(\sigma_3)\phi(\sigma_4)\phi(\sigma_3) = (\phi V)(\sigma_1)(\phi V)(\sigma_2)(\phi V)(\sigma_1)$$

$$= (V \operatorname{Inn}(f_1^{-1})\phi)(\sigma_1)(V \operatorname{Inn}(f_1^{-1})\phi)(\sigma_2)(V \operatorname{Inn}(f_1^{-1})\phi)(\sigma_1)$$

$$= (V \operatorname{Inn}(f_1^{-1}))(\phi(\sigma_1)\phi(\sigma_2)\phi(\sigma_1))$$

$$= (V \operatorname{Inn}(f_1^{-1}))(\phi(\sigma_2)\phi(\sigma_1)\phi(\sigma_2)) = \phi(\sigma_4)\phi(\sigma_3)\phi(\sigma_4).$$

Let us show that lemma 8 (i) implies that ϕ respects relation (b), $\sigma_4\sigma_3\sigma_2\sigma_1^2\sigma_2\sigma_3\sigma_4 = 1$, which we rewrite as $(\sigma_2\sigma_1^2\sigma_2)^{-1} = \sigma_3\sigma_4^2\sigma_3$. Here we need to use the fact that ϕ is an automorphism when restricted to $\langle\sigma_1, \sigma_2\rangle$ (since the couple (λ, f) is assumed to lie in $\widehat{GT}_0$), and that $\phi(\sigma_2\sigma_1^2\sigma_2) = (\sigma_2\sigma_1^2\sigma_2)^\lambda$, which is proved as in the proof of the corollary to proposition 4 since $\sigma_2\sigma_1^2\sigma_2 = y_3$. We have:

$$\phi(\sigma_3)\phi(\sigma_4)^2\phi(\sigma_3) = (\phi V)(\sigma_1)(\phi V)(\sigma_2)^2(\phi V)(\sigma_1) = V \operatorname{Inn}(f_1^{-1})\phi(\sigma_1\sigma_2^2\sigma_1)$$

$$= V\left(f(x_{12}, x_{23})(\sigma_1\sigma_2^2\sigma_1)^\lambda f(x_{23}, x_{12})\right) = f(x_{34}, x_{45})(\sigma_3\sigma_4^2\sigma_3)^\lambda f(x_{45}, x_{34})$$

$$= f(x_{34}, x_{45})(\sigma_2\sigma_1^2\sigma_2)^{-\lambda}f(x_{45}, x_{34}) = (\sigma_2\sigma_1^2\sigma_2)^{-\lambda} = \phi(\sigma_2\sigma_1^2\sigma_2)^{-1}.$$

Similarly, lemma 8 (i) suffices to show that ϕ respects relation (c). We work with the equivalent relation $\sigma_1^2\sigma_2\sigma_1^2\sigma_2\sigma_4^{-2} = 1$ (the left-hand side is equal to $(\sigma_1\sigma_2\sigma_3\sigma_4)^5$ in $M(0,5)$). As the three elements σ_1^2, $\sigma_2\sigma_1^2\sigma_2$ and σ_4^2 commute and ϕ takes each one to itself to the power λ, it is immediate that ϕ respects this relation.

Let us now show that ϕ extends to an automorphism of $\widehat{M(0,5)}$ if and only if (λ, f) lies in $\widehat{GT}$. First suppose that $(\lambda, f) \in \widehat{GT}$. We only need to show that ϕ respects (d), which we do using lemma 8 (ii), which tells us that when (λ, f) lies in $\widehat{GT}$, the maps ϕ and $\operatorname{Inn}(f_3f_2f_1)V^{-3}\phi V^3$ take the same values on σ_1 and σ_2. Thus we calculate

$$\phi(\sigma_2)\phi(\sigma_3)\phi(\sigma_2) = (\phi V^3)(\sigma_1)(\phi V^3)(\sigma_2)(\phi V^3)(\sigma_1) =$$

$$(V^3\mathrm{Inn}(f_3f_2f_1)^{-1}\phi)(\sigma_1)(V^3\mathrm{Inn}(f_3f_2f_1)^{-1}\phi)(\sigma_2)(V^3\mathrm{Inn}(f_3f_2f_1)^{-1}\phi)(\sigma_1)$$

$$= (V^3\mathrm{Inn}(f_3f_2f_1)^{-1})\Big(\phi(\sigma_1)\phi(\sigma_2)\phi(\sigma_1)\Big)$$

$$= (V^3\mathrm{Inn}(f_3f_2f_1)^{-1})\Big(\phi(\sigma_2)\phi(\sigma_1)\phi(\sigma_2)\Big) = \phi(\sigma_3)\phi(\sigma_2)\phi(\sigma_3),$$

which shows that (d) is respected by ϕ, so ϕ extends to an automorphism of $\widehat{M(0,5)}$ if it is associated to a couple $(\lambda, f) \in \widehat{GT}$.

Now let us show that if a couple $(\lambda, f) \in \widehat{GT}_0$, acting as in the statement of lemma 7, extends to an automorphism of $\widehat{M(0,5)}$ then it must lie in $\widehat{GT}$. This is most simply done by adapting an argument of Nakamura (see the appendix of [N]). By lemma 8 (i), since σ_1, σ_2, σ_4 and σ_{15} generate all of $\widehat{M(0,5)}$, if we assume that ϕ is an automorphism then we must have the equality $\phi = \mathrm{Inn}(f_1)V^{-1}\phi V$ on all of $\widehat{M(0,5)}$. Replacing ϕ by $\mathrm{Inn}(f_1)V^{-1}\phi V$ in the right-hand side gives $\phi = \mathrm{Inn}(f_2f_1)V^{-2}\phi V$; reiterating three more times gives $\phi =$

$$\mathrm{Inn}(f_3f_2f_1)V^{-3}\phi V^3 = \mathrm{Inn}(f_4f_3f_2f_1)V^{-4}\phi V^4 = \mathrm{Inn}(f_5f_4f_3f_2f_1)V^{-5}\phi V^5.$$

Since $V_5 = 1$, we have $\phi = \mathrm{Inn}(f_5f_4f_3f_2f_1)\phi$, so the element $f_5f_4f_3f_2f_1$ must be in the center of $\widehat{M(0,5)}$, which is trivial. But this element is exactly the left-hand side of relation (III').

We thus associate to every $(\lambda, f) \in \widehat{GT}$ an automorphism of $\widehat{M(0,5)}$; by the multiplication law, we see that this map of $\widehat{GT}$ into $\mathrm{Aut}(\widehat{M(0,5)})$, which we denote by $\tilde{\iota}$, is a group homomorphism. Restriction to the subgroup $\langle \sigma_1, \sigma_2 \rangle$ of $\widehat{M(0,5)}$, gives the injective map $\tilde{\eta} : \widehat{GT} \to \mathrm{Aut}(\hat{T}_3) \subset \mathrm{Aut}(\hat{B}_3)$ of proposition 4, which shows that $\tilde{\iota} : \widehat{GT} \to \mathrm{Aut}(\widehat{M(0,5)})$ is also injective. This concludes the proof of lemma 7.   ◊

As before, for any group G, we write (by a slight abuse of notation) G/Z for G modulo its center. We have already remarked that the subgroup of $\widehat{M(0,5)}$ generated by σ_1 and σ_2 is isomorphic to $\hat{B}_3$, and similarly, the subgroup generated by σ_1, σ_2 and σ_3 inside $\widehat{M(0,5)}$ is isomorphic, not to $\hat{B}_4$, but to $\hat{B}_4/Z$ (since the relation $(\sigma_1\sigma_2\sigma_3)^4 = 1$ holds in $\widehat{M(0,5)}$ and this element generates the center of $\hat{B}_4$). It is immediate that if ϕ is the automorphism of $\widehat{M(0,5)}$ associated to a couple $(\lambda, f) \in \widehat{GT}$ as in lemma 7, then ϕ induces an automorphism of $\hat{B}_4$ since ϕ preserves this subgroup of $\widehat{M(0,5)}$, thanks to the first of relations (4) in §2. The map $\tilde{\iota} : \widehat{GT} \to \mathrm{Aut}(\widehat{M(0,5)})$ of lemma 7 thus induces an injective map $\tilde{\iota} : \widehat{GT} \to \mathrm{Aut}(\hat{B}_4/Z)$.

Lemma 9: *The map $\tilde{\iota}$ induces an injective map $\iota : \widehat{GT} \to \mathrm{Aut}(\hat{T}_4)$.*

Proof: Let $(\lambda, f) \in \widehat{GT}$ and let $\phi := \tilde{\iota}(\lambda, f) \in \mathrm{Aut}(\hat{B}_4/Z)$. In the proof of this lemma, let σ_i and x_{ij} denote elements of $\hat{B}_4$, not to be confused with the elements of $\hat{B}_4/Z \subset \widehat{M(0,5)}$ (which ought to have been denoted $\bar{\sigma}_i$ and $\bar{x}_{ij}$ but were not for aesthetic reasons). The $\bar{\sigma}_i$ satisfy the relation $(\bar{\sigma}_1\bar{\sigma}_2\bar{\sigma}_3)^4 = 1$, and this is of course not true of the σ_i; indeed $\omega_4 = (\sigma_1\sigma_2\sigma_3)^4$ generates the infinite cyclic center of $\hat{B}_4$. Let us associate to ϕ an element $\tilde{\phi}$ of $\mathrm{Aut}(\hat{T}_4)$. Define $\tilde{\phi}$ on the generators σ_1, σ_2 and σ_3 of $\hat{B}_4$ by

$$\tilde{\phi}(\sigma_1) = \sigma_1^\lambda, \quad \tilde{\phi}(\sigma_2) = f(x_{23}, x_{12})\sigma_2^\lambda f(x_{12}, x_{23}), \quad \tilde{\phi}(\sigma_3)$$

$$= f(x_{34}, x_{45})\sigma_3^\lambda f(x_{45}, x_{34});$$

here, we write $x_{45} := x_{12}x_{13}x_{23}$ in order to consider x_{45} as an element of $\hat{B}_4$, but it is to be considered purely as a formal notation. We first show that $\tilde{\phi}$ induces an automorphism of $\hat{B}_4$. By proposition 4, it induces one on the subgroup of $\hat{B}_4$ generated by $\langle\sigma_1, \sigma_2\rangle$ since this subgroup is isomorphic to $\hat{B}_3$. Thus $\tilde{\phi}$ respects the relation $\sigma_1\sigma_2\sigma_1 = \sigma_2\sigma_1\sigma_2$ which holds in $\hat{B}_4$. Since ϕ is an automorphism of $\hat{B}_4/Z$, we know that $\tilde{\phi}$ respects the relation $\sigma_2\sigma_3\sigma_2 = \sigma_3\sigma_2\sigma_3$ modulo the center of $\hat{B}_4$, which means that there exists $\mu \in \mathbf{Z}^*$ such that

$$\tilde{\phi}(\sigma_2)\tilde{\phi}(\sigma_3)\tilde{\phi}(\sigma_2) = \tilde{\phi}(\sigma_3)\tilde{\phi}(\sigma_2)\tilde{\phi}(\sigma_3)\omega_4^\mu.$$

Define a map $\tilde{\phi}'$ on σ_1, σ_2 and σ_3 as follows: $\tilde{\phi}'(\sigma_1) = \tilde{\phi}(\sigma_1)$, $\tilde{\phi}'(\sigma_2) = \tilde{\phi}(\sigma_2)$, and $\tilde{\phi}'(\sigma_3) = \tilde{\phi}(\sigma_3)\omega_4^{-\mu}$. Then it is immediate that $\tilde{\phi}'$ respects the two relations defining $\hat{B}_4$, so it is an automorphism of $\hat{B}_4$. This automorphism clearly sends the derived subgroup of $\hat{B}_4$ into itself (since it induces an automorphism on the quotient, i.e. on the free profinite abelian group on three generators, namely sending each generator to itself to the power λ), and modulo this subgroup, σ_1, σ_2 and σ_3 become equal, which shows that μ must be equal to 0 (since the center of $\hat{B}_4$ does not intersect the commutator subgroup). Thus $\tilde{\phi}' = \tilde{\phi}$, and we have lifted $\phi \in \mathrm{Aut}(\hat{B}_4/Z)$ to $\tilde{\phi} \in \mathrm{Aut}(\hat{B}_4)$. This lifting map is clearly injective, and a simple calculation confirms that $\tilde{\phi}$ is in $\mathrm{Aut}(\hat{T}_4)$, i.e. that it respects all the homomorphisms i_n, f_n and g_n for $1 \le n \le 3$. So we have an injective map $\iota : \widehat{GT} \to \mathrm{Aut}(\hat{T}_4)$. $\qquad\diamond$

In order to finish the proof of proposition 6, it remains only to prove the surjectivity of ι. Take an element $\Phi = (\phi_n)_{1 \le n \le 4}$ in $\mathrm{Aut}(\hat{T}_4)$. We know by proposition 4 that the couple (λ, f) determined by ϕ_3 is in $\widehat{GT}_0$. By the relation $\phi_4 i_3 = i_3 \phi_3$ we see that

$$\phi_4(\sigma_1) = \sigma_1^\lambda, \quad \text{and} \quad \phi_4(\sigma_2) = f(x_{23}, x_{12})\sigma_2^\lambda f(x_{12}, x_{23}).$$

By the commutation $\phi_4 g_3 = g_3 \phi_3$ we compute that

$$\phi_4(\sigma_3) = \phi_4 g_3(\sigma_2) = g_3 \phi_3(\sigma_2) = g_3 \big(f(x_{23}, x_{12}) \sigma_2^\lambda f(x_{12}, x_{23}) \big)$$

$$= f(x_{34}, x_{45}) \sigma_3^\lambda f(x_{45}, x_{34}).$$

Let ϕ be the map associated to $(\lambda, f) \in \widehat{GT}_0$ given in the statement of lemma 7, defined on the generating set of $\widehat{M(0,5)}$ considered there. Consider ϕ restricted to the subgroup $\hat{B}_4/Z \subset \widehat{M(0,5)}$; it agrees with the automorphism of $\hat{B}_4/Z$ induced by ϕ_4, so it is an automorphism of this subgroup. We will show that this implies that ϕ actually gives an automorphism of all of $\widehat{M(0,5)}$, which in turn shows by lemma 7 that the couple (λ, f) must lie in $\widehat{GT}$. We must verify that ϕ respects the defining relations of $\widehat{M(0,5)}$. Since ϕ is an automorphism of $\hat{B}_4/Z$, it respects $\sigma_1 \sigma_2 \sigma_1 = \sigma_2 \sigma_1 \sigma_2$, $\sigma_2 \sigma_3 \sigma_2 = \sigma_3 \sigma_2 \sigma_3$ and $(\sigma_1 \sigma_2 \sigma_3)^4 = 1$ (this last relation is well-known to be equivalent to the center relation $(\sigma_1 \sigma_2 \sigma_3 \sigma_4)^5 = 1$, cf. [Bi]). The relation $\sigma_3 \sigma_4 \sigma_3 = \sigma_4 \sigma_3 \sigma_4$ and the Hurwitz relation $\sigma_4 \sigma_3 \sigma_2 \sigma_1^2 \sigma_2 \sigma_3 \sigma_4$ are respected simply because ϕ satisfies the conditions of lemma 8 (i), as in the proof of lemma 7; these two relations are respected by ϕ's associated to any couple in $\widehat{GT}_0$. Therefore ϕ is an automorphism of $\widehat{M(0,5)}$ and thus by lemma 7 (λ, f) lies in $\widehat{GT}$, so $\Phi = \iota(\lambda, f)$ and ι is a bijection. This concludes the proof of proposition 6.   ◇

Corollary: $\mathrm{Out}(\mathcal{T}_4) \simeq GT_\ell$.

Proof: Much as in the last section, this is really a corollary of the proof of proposition 6, which carries over to the pro-ℓ case. Specifically, note that the statement and proof of lemma 7 are entirely valid when $\widehat{M(0,5)}$ is replaced by $M(0,5)^{(\ell)}$. This shows that if $(\lambda, f) \in GT_\ell$ then it induces an automorphism of $B_4^{(\ell)}/Z$ which is easily seen to lift to one of $B_4^{(\ell)}$ as in the profinite case. Conversely, if ϕ_4 is an automorphism of $B_4^{(\ell)}$ which acts on σ_1, σ_2 and σ_3 as in equation (5), it induces one of $B_4^{(\ell)}/Z \subset M(0,5)^{(\ell)}$ which extends uniquely to one of $M(0,5)^{(\ell)}$ as in the proof of lemma 9, so the couple (λ, f) must belong to GT_ℓ. If ϕ_4 is associated to $(\lambda, f) \in GT_\ell$, it is immediate that defining ϕ_n for $1 \leq n \leq 3$ as the restrictions of ϕ_4 to $B_n^{(\ell)} \subset B_4^{(\ell)}$ we obtain a map $GT_\ell \to \mathrm{Aut}(\mathcal{T}_4)$ which is injective because its kernel must be contained in the kernel of the injective map $\tilde{\eta}$ defined in §3. To prove the surjectivity of this map, we again simply copy the proof of the surjectivity of the map ι given above. This is where we use the fact that the strand-doubling map g_3 is part of the tower $\mathcal{T}_4$. Needless to say, one would prefer to dispense with this extra compatibility condition.   ◇

§5. $\widehat{GT}$ is the full automorphism group of $\mathrm{Aut}(\hat{T})$

We now prove the profinite part of (iii) of the main theorem. In proposition 10 we define an injective homomorphism of $\widehat{GT}$ into $\mathrm{Aut}(\hat{T})$, which is shown to be a bijection in proposition 11.

Proposition 10: *Let* $(\lambda, f) \in \widehat{GT}$ *and associate to it for all* $n \geq 1$ *a map* $\phi_n = \phi_{n,(\lambda,f)}$ *which sends the generators of* $\hat{B}_n$ *into* $\hat{B}_n$ *as in equation (5), namely:*

$$\phi_n(\sigma_1) = \sigma_1^\lambda \text{ and } \phi_n(\sigma_i) = f(\sigma_i^2, y_i)\sigma_i^\lambda f(y_i, \sigma_i^2) \text{ for } 2 \leq i \leq n-1.$$

Then ϕ_n *can be extended by multiplicativity to an automorphism of* $\hat{B}_n$ *for all* n *which preserves the subgroup* $\hat{A}_n \subset \hat{B}_n$. *Set* $\phi'_n = \phi_n|_{\hat{A}_n}$; *then* $(\lambda, f) \mapsto (\phi_n, \phi'_n)_{n \geq 1}$ *defines an injective map* $\theta : \widehat{GT} \to \mathrm{Aut}(\hat{T})$.

Proof: Fix $(\lambda, f) \in \widehat{GT}$. Let us show that for all $n \geq 1$, ϕ_n determines an element of $\mathrm{Aut}(\hat{B}_n)$. We know by proposition 6 that ϕ_4 is an automorphism of $\hat{B}_4$. To see that ϕ_n can be extended multiplicatively to an automorphism of all of $\hat{B}_n$, we must show that ϕ_n respects relations (1) of the definition of B_n given in §2. For the commutation relations $(\sigma_i, \sigma_j) = 1$ when $|i - j| \geq 2$, it suffices to recall that the elements $y_i \in \hat{B}_n$ commute, which shows that $\phi_n(\sigma_i)\phi_n(\sigma_j) = \phi_n(\sigma_j)\phi_n(\sigma_i)$ since the factors of $\phi_n(\sigma_j)$ only contain y_j and σ_j and thus commute with those of $\phi_n(\sigma_i)$ which themselves only contain y_i and σ_i.

For the braid relations $\sigma_i\sigma_{i+1}\sigma_i = \sigma_{i+1}\sigma_i\sigma_{i+1}$, we first note that since ϕ_n restricted to $\hat{B}_4 \subset \hat{B}_n$ is an automorphism, it respects these relations for $i = 1$ and $i = 2$. For $2 < i \leq n-2$, let C_i be the subgroup of $\hat{B}_n$ generated by σ_i, σ_{i+1}, y_i and y_{i+1}. Then there is a canonical isomorphism $\psi_i : C_i \to C_2$ taking σ_i to σ_2, σ_{i+1} to σ_3, y_i to y_2 and y_{i+1} to y_3. This isomorphism can be visualized in terms of braids by noting that the braids in C_i are exactly those in C_2 where the first strand is replaced by $i - 1$ parallel strands – the situation is similar to the proof that the maps g_n are homomorphisms, cf. §2. Now, consider the map $h_i := \psi_i^{-1}\phi_n\psi_i$ on the subgroup C_i for $2 < i \leq n-2$. Since ϕ_n is an automorphism of $C_2 = \psi_i(C_i)$ because $C_2 \subset \hat{B}_4$, the map h_i is an automorphism of C_i. But

$$h_i(\sigma_i) = \psi_i^{-1}\phi_n\psi_i(\sigma_i) = \psi_i^{-1}\phi_n(\sigma_2) = \psi_i^{-1}\left(f(\sigma_2^2, y_2)\sigma_2^\lambda f(y_2, \sigma_2^2)\right) =$$

$$f(\sigma_i^2, y_i)\sigma_i^\lambda f(y_i, \sigma_i^2)$$

and

$$h_i(\sigma_{i+1}) = \psi_i^{-1}\phi_n(\sigma_3) = f(\sigma_{i+1}^2, y_{i+1})\sigma_{i+1}^\lambda f(y_{i+1}, \sigma_{i+1}^2),$$

so h_i agrees with ϕ_n on σ_i and σ_{i+1}, so ϕ_n respects the braid relations $\sigma_i\sigma_{i+1}\sigma_i = \sigma_{i+1}\sigma_i\sigma_{i+1}$ for $2 < i < n-1$ since h_i does. This shows that the ϕ_n are in $\mathrm{Aut}(\hat{B}_n)$ for $n \geq 1$. The automorphisms ϕ_n defined in this way preserve the subgroups $\hat{A}_n \subset \hat{B}_n$, so we define $\phi'_n = \phi_n|_{\hat{A}_n}$. Then it is easily checked by a simple calculation that $(\phi_n, \phi'_n)_{n\geq 1} \in \mathrm{Aut}(\hat{T})$ since the ϕ_n and the ϕ'_n respect all the arrows i_n, f_n and g_n. Thus we have a map $\theta : \widehat{GT} \to \mathrm{Aut}(\hat{T})$. Moreover, this map is injective because its kernel is contained in the kernel of the injective homomorphism $\tilde{\eta} : \widehat{GT}_0 \to \mathrm{Aut}(T_3)$ (restricted to $\widehat{GT}$) defined in §3.   ◇

Proposition 11: *The map $\theta : \widehat{GT} \to \mathrm{Aut}(\hat{T})$ is bijective.*

Proof: We prove surjectivity of θ. Let $(\phi_n, \phi'_n)_{n\geq 1} \in \mathrm{Aut}(\hat{T})$. Let $(\lambda, f) \in \widehat{GT}$ be the couple associated to $(\phi_n, \phi'_n)_{1\leq n\leq 4} \in \mathrm{Aut}(\hat{T}_4)$ by proposition 6. Let $(\psi_n, \psi'_n)_{n\geq 1} := \theta(\lambda, f)$, so $\phi_4 = \psi_4$. We must show that $\phi_n = \psi_n$ for $n > 4$, or equivalently, that ϕ_n must be as defined on $\hat{B}_n$ as in equation (5). We proceed by induction; it is true for $1 \leq n \leq 4$; suppose it true for a given n. We must determine how ϕ_{n+1} acts on σ_n. We thus calculate using the compatibility of the ϕ_n with the maps g_n that:

$$\phi_{n+1}(\sigma_n) = \phi_{n+1}g_n(\sigma_{n-1}) = g_n\phi'_n(\sigma_{n-1}),$$

$$= g_n\left(f(y_{n-1},\sigma^2_{n-1})^{-1}\sigma^\lambda_{n-1}f(y_{n-1},\sigma^2_{n-1})\right) = f(y_n,\sigma^2_n)^{-1}\sigma^\lambda_n f(y_n,\sigma^2_n),$$

since for $2 \leq i \leq n-1$, we have $g_n(y_i) = y_{i+1}$. This shows that for all $n \geq 1$, ϕ_n acts on $\hat{B}_n$ as in (5), so $\phi_n = \psi_n$ for all n, so $(\phi_n, \phi'_n)_{n\geq 1} = \theta(\lambda, f)$. Thus the map $\theta : \widehat{GT} \to \mathrm{Aut}(\hat{T})$ is bijective, which concludes the proof of the profinite part of the Main Theorem (it is obvious that we also have $\widehat{GT} \simeq \mathrm{Aut}(\hat{T}_N)$ for all $N > 4$).   ◇

§6. GT_ℓ is the full automorphism group of T

In this paragraph we finish the proof of the pro-ℓ statement (iii) of the Main Theorem. This involves proving analogues of propositions 10 and 11 in the pro-ℓ case. The statement and proof of proposition 10 are very easily adapted to the pro-ℓ case. This is not the case for proposition 11, since the strand-doubling maps g_n are not included in the tower T. One could of course include them, however we found that thanks to an injectivity theorem proved by Ihara in the pro-ℓ case (see the proof of lemma 16) which is not known to hold in the profinite case, they are actually not needed to obtain the result. Before establishing the Main Theorem in the pro-ℓ case, we prove a necessary auxiliary result on the general behavior of the automorphism groups of the $B_n^{(\ell)}$ in Proposition 12.

For all $N \geq 1$, let us denote by $\mathrm{Aut}_1(T_N)$ (resp. $\mathrm{Out}_1(T_N)$) the subgroup of elements (resp. classes of elements mod $\mathrm{Inn}(T_N)$) $(\phi_n)_{1 \leq n \leq N}$ such that $\phi_n(\sigma_1) = \sigma_1$. Note that this condition implies that $\phi_n(\omega_n) = \omega_n$ for $1 \leq n \leq N$; indeed, ϕ_n must map ω_n to a $\mathbf{Z}_\ell^*$-power μ of itself since the center of $B_n^{(\ell)}$ is cyclic, and this power is easily seen to be equal to 1 by considering the identity $\phi_n(\omega_n) = \omega_n^\mu$ modulo the derived subgroup of $B_n^{(\ell)}$.

Proposition 12: *The map* $\mathrm{Out}_1(T_N) \to \mathrm{Out}_1(T_{N-1})$ *induced by restricting automorphisms of* $B_N^{(\ell)}$ *to the subgroup* $B_{N-1}^{(\ell)} \subset B_N^{(\ell)}$ *is injective for* $N \geq 4$.

Proof: The proof uses sublemmas 13, 14 and 15 to reduce the statement to the similar statement for the pure braid groups given in lemma 16, which is an adaptation of an analogous result for the pure Hurwitz braid groups proven by Ihara in [I2].

Sublemma 13: *Let* $\phi \in \mathrm{Aut}(T_N)$. *Then for* $1 \leq i < j \leq N$ *there exists* $\alpha_{ij} \in K_N^\ell$ *such that* $\phi(x_{ij}) = \alpha_{ij}^{-1} x_{ij}^\lambda \alpha_{ij}$, *and for* $1 \leq i \leq N-1$, *there exists* $\alpha_i \in K_N^\ell$ *such that* $\phi(\sigma_i) = \alpha_i^{-1} \sigma_i^\lambda \alpha_i$.

Proof: If $\phi \in \mathrm{Aut}(T_N)$, then $\phi(\sigma_1) = \sigma_1^\lambda$ and $\phi(x_{12}) = x_{12}^\lambda$ for some $\lambda \in \hat{\mathbf{Z}}_\ell^*$ since ϕ must preserve $B_2^{(\ell)}$. Using equation (2') in §2 we find that

$$\phi(x_{ij}) = \phi(a_{ij})^{-1} x_{12}^\lambda \phi(a_{ij}) = \phi(a_{ij})^{-1} a_{ij} x_{ij}^\lambda a_{ij}^{-1} \phi(a_{ij}) = \alpha_{ij}^{-1} x_{ij}^\lambda \alpha_{ij}$$

which is the statement of the lemma, with $\alpha_{ij} = a_{ij}^{-1} \phi(a_{ij})$. For the σ_i we proceed more explicitly. Set $\pi = \sigma_1 \cdots \sigma_{n-1}$ (this is denoted π_2 in the appendix); it is easy to check that $\sigma_i = \pi^{i-1} \sigma_1 \pi^{-(i-1)}$ for $1 \leq i \leq n-1$. Thus:

$$\phi(\sigma_i) = \phi(\pi)^{i-1} \sigma_1^\lambda \phi(\pi)^{-(i-1)} = \phi(\pi)^{i-1} \pi^{-(i-1)} \sigma_i^\lambda \pi^{i-1} \phi(\pi)^{-(i-1)},$$

i.e. $\phi(\sigma_i) = \alpha_i^{-1} \sigma_i^\lambda \alpha_i$ where $\alpha_i := \pi^{i-1} \phi(\pi)^{-(i-1)}$.   ◇

Sublemma 14: *The map* $\mathrm{Aut}(B_N^{(\ell)}/Z) \to \mathrm{Aut}(K_N^\ell/Z)$ *induced by restricting automorphisms to the subgroup* $K_N^\ell/Z \subset B_N^{(\ell)}/Z$ *is injective.*

Proof: Let $K \triangleleft B$ be groups and suppose that the centralizer $\mathrm{Centr}_B(K)$ of K in B is trivial. Consider B as a normal subgroup of $\mathrm{Aut}(B)$ via inner automorphisms. Let us show that the restriction map $\mathrm{Aut}(B) \to \mathrm{Aut}(K)$ is injective. It suffices to show that that $\mathrm{Centr}_{\mathrm{Aut}(B)}(K)$ is trivial since any $\phi \in \mathrm{Aut}(B)$ whose restriction to K is the identity is in $\mathrm{Centr}_{\mathrm{Aut}(B)}(K)$. So let $x \in \mathrm{Centr}_{\mathrm{Aut}(B)}(K)$ and $b \in B$ be considered as an inner automorphism .

348   Pierre Lochak and Leila Schneps

Consider the automorphism $b^{-1}xbx^{-1}$ of B. It is inner since B is normal in $\mathrm{Aut}(B)$. But it is in $\mathrm{Centr}_B(K)$ since it acts trivially on K. So it is trivial and x commutes with b. Since this is true for all $b \in B$, $x \in \mathrm{Centr}_{\mathrm{Aut}(B)}(B)$. But this centralizer is also trivial since if $\phi \in \mathrm{Aut}(B)$, then $\phi\mathrm{Inn}(b)\phi^{-1} = \mathrm{Inn}(\phi(b))$, so if this expression is equal to $\mathrm{Inn}(b)$ for all $b \in B$, then ϕ is the identity (the center of B being trivial). So $x = 1$, which shows that $\mathrm{Aut}(B) \to \mathrm{Aut}(K)$ is injective.

The sublemma is proved by applying this result to $K_N^\ell/Z \lhd B_N^{(\ell)}/Z \lhd \mathrm{Aut}(B_N^{(\ell)}/Z)$ since the centralizer of K_N^ℓ/Z in $B_N^{(\ell)}/Z$ is trivial. ◇

Let $\mathcal{K}_N$ be the tower consisting of the pure braid groups K_n^ℓ for $1 \le n \le N$, with the restrictions to these groups of the inclusions $i_{n,m}$ for $1 \le n < m \le N$. For each N, let $\mathrm{Aut}_1^r(\mathcal{K}_N)$ denote the image of $\mathrm{Aut}_1(\mathcal{T}_N)$ under the natural restriction map $\mathrm{Aut}(\mathcal{T}_N) \to \mathrm{Aut}(\mathcal{K}_N)$. Denote by $\mathrm{Inn}(\mathcal{K}_N)$ the image of $\mathrm{Inn}(\mathcal{T}_N)$ under this map (although its elements are not really inner automorphisms of $\mathcal{K}_N$). Let $\mathrm{Out}(\mathcal{K}_N) = \mathrm{Aut}(\mathcal{K}_N)/\mathrm{Inn}(\mathcal{K}_N)$.

Sublemma 15: *The restriction map* $\mathrm{Out}_1(\mathcal{T}_N) \to \mathrm{Out}_1^r(\mathcal{K}_N)$ *is an isomorphism for $N \ge 1$.*

Proof: By definition, this map is surjective. We must show it is injective. Consider the diagram

$$\mathrm{Aut}_1(\mathcal{T}_N) \to \mathrm{Aut}_1^r(\mathcal{K}_N)$$

$$\downarrow \qquad\qquad \downarrow$$

$$\mathrm{Aut}(B_N^{(\ell)}/Z) \to \mathrm{Aut}(K_N^\ell/Z).$$

The lower map is injective by sublemma 14. The vertical maps are given by the fact that any automorphism of $B_N^{(\ell)}$ which fixes the center naturally induces an automorphism of $B_N^{(\ell)}/Z$. These maps are easily seen to be injective (see [DG], thm. 20); as usual, one considers what happens modulo the derived subgroup. So the upper map is an isomorphism. By definition, it remains an isomorphism when Aut is replaced by Out. ◇

Lemma 16: *The natural restriction map* $\mathrm{Out}_1^r(\mathcal{K}_N) \to \mathrm{Out}_1^r(\mathcal{K}_{N-1})$ *is injective for $N \ge 4$.*

Proof: Let P_n denote the pure Hurwitz braid group contained in H_n. By proposition 2 (iv) (pro-ℓ), P_{n+1}^ℓ is a semi-direct product $F_n^\ell \rtimes P_n^\ell$ for $n \ge 3$, where F_n^ℓ is the pro-free group of rank $n-1$ generated by $\tilde{x}_{1,n+1}, \ldots, \tilde{x}_{n,n+1}$ with the Hurwitz relation $\tilde{x}_{1,n+1} \cdots \tilde{x}_{n,n+1} = 1$. For $n \ge 1$, let $(K_n')^\ell \subset K_n^\ell$

be as in proposition 2. Let the groups $\mathrm{Out}^*(P_n^\ell)$ be as defined in [I2], i.e.
$\mathrm{Out}^*(P_n^\ell) =$

$$\{\phi \in \mathrm{Out}(P_n^\ell) \mid \exists \alpha_{ij} \in P_n^\ell \text{ such that } \phi(\tilde{x}_{ij}) = \alpha_{ij}^{-1} \tilde{x}_{ij} \alpha_{ij} \ \forall \ 1 \le i < j \le n\}.$$

For $n \ge 1$, let

$$h_n : \mathrm{Out}_1^r(\mathcal{K}_n) \to \mathrm{Out}((K_n')^\ell, *)$$

be the maps obtained by restricting the elements of $\mathrm{Out}_1^r(\mathcal{K}_n)$, considered as
outer automorphisms of $K_n^\ell = (K_n')^\ell \times Z(K_n^\ell)$, to the subgroup $(K_n')^\ell$ (they
are actually automorphisms of K_n', as can easily be seen by recalling that
they conjugate each x_{ij}) Let $\mathrm{Out}_1^r((K_n')^\ell) = \mathrm{Im}(h_n)$. The h_n are injective
since elements of $\mathrm{Out}_1^r(\mathcal{K}_n)$, considered as automorphisms of K_n^ℓ, fix the
center, so they are isomorphisms

$$h_n : \mathrm{Out}_1^r(\mathcal{K}_n) \overset{\sim}{\to} \mathrm{Out}_1^r((K_n')^\ell).$$

Fix $N \ge 1$. The following diagram commutes:

$$\mathrm{Out}^*(P_{N+1}^\ell) \to \mathrm{Out}^*(P_N^\ell)$$

$$\downarrow \qquad\qquad \downarrow$$

$$\mathrm{Out}_1^r((K_N')^\ell) \overset{h}{\longrightarrow} \mathrm{Out}_1^r((K_{N-1}')^\ell),$$

where the upper map corresponds to the quotient by F_N^ℓ, which induces
an automorphism of P_N^ℓ since F_N^ℓ is normal and so preserved by all $\phi \in$
$\mathrm{Out}^*(P_{N+1}^\ell)$. The lower map h is simply the restriction map $((K_{N-1}')^\ell$ is
considered to be included in $(K_N')^\ell$ via $i_{N-1,N}$). The vertical maps come
from the fact that elements of $\mathrm{Out}^*(P_N^\ell)$ fix the center of P_N^ℓ, so these
maps induce automorphisms on P_N^ℓ/Z, which is isomorphic to $(K_{N-1}')^\ell$ by
proposition 2 (iii).

Ihara showed that the upper map is injective; this theorem is the main
result of [I2]. The right-hand vertical map is also injective. For if ϕ is in
the kernel, then ϕ fixes $(K_{N-1}')^\ell$, but ϕ fixes the center of P_N^ℓ, so since
$P_N^\ell = (K_{N-1}')^\ell \times Z(P_N^\ell)$ by proposition 2 (iii), ϕ is the identity. This shows
that the lower map is injective. So the map

$$g_{N-1}^{-1} \circ h \circ g_N : \mathrm{Out}_1^r(\mathcal{K}_N) \to \mathrm{Out}_1^r(\mathcal{K}_{N-1})$$

is injective; it is the map in the statement of the lemma. $\diamond$

To conclude, we consider the commutative diagram

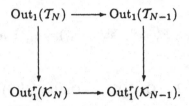

$$\begin{array}{ccc} \mathrm{Out}_1(\mathcal{T}_N) & \longrightarrow & \mathrm{Out}_1(\mathcal{T}_{N-1}) \\ \downarrow & & \downarrow \\ \mathrm{Out}_1^r(\mathcal{K}_N) & \longrightarrow & \mathrm{Out}_1^r(\mathcal{K}_{N-1}). \end{array}$$

By sublemma 14, the vertical arrows are isomorphisms and the lower arrow is injective by lemma 16, so the upper arrow is injective. This finally concludes the proof of proposition 12.   ◇

We can now easily establish pro-ℓ analogues of propositions 6, 10 and 11.

Proposition 17: *(i) There is an injective map* $\theta : GT_\ell \to \mathrm{Out}(\mathcal{T})$ *given by associating to a couple* (λ, f) *an automorphism* ϕ_n *of* $B_n^{(\ell)}$ *defined as in equation (5);*

(ii) the map θ *is a bijection.*

Proof: (i) Fix $(\lambda, f) \in GT_\ell$ and define ϕ_n on the generators of $B_n^{(\ell)}$ as in equation (5). The proof of proposition 10 can be copied directly to prove this result once the use of proposition 6 is replaced by its corollary.

(ii) We finally prove the surjectivity of the map $\theta : GT_\ell \to \mathrm{Out}(\mathcal{T})$. Let $\Phi \in \mathrm{Out}(\mathcal{T})$ and let $\phi = (\phi_n)_{n \geq 1} \in \mathrm{Aut}(\mathcal{T})$ be a representative of Φ which fixes the permutations of the $B_n^{(\ell)}$ (it exists by lemma 3). Then by proposition 4, there exists a unique couple $(\lambda, f) \in GT_{\ell,0}$ such that $\phi_3 = \tilde{\eta}(\lambda, f)$. Let $\psi = (\psi_n)_{n \geq 1} = \tilde{\theta}(\lambda, f)$. Then $\psi \in \mathrm{Aut}(\mathcal{T})$ and $\psi_3 = \phi_3$. Set $\chi = \phi\psi^{-1}$. Then writing $\chi = (\chi_n)_{n \geq 1}$, we see that χ_3 is the identity on $B_3^{(\ell)}$, so by proposition 12, χ_n is the identity on $B_n^{(\ell)}$ for $n \geq 1$, which means that $\phi = \psi = \tilde{\theta}(\lambda, f)$, so θ is surjective. Thus we obtain $\mathrm{Out}(\mathcal{T}) \simeq GT_\ell \simeq \mathrm{Out}(\mathcal{T}_N)$ for all $N \geq 4$, which concludes the proof of proposition 17, and of the Main Theorem.   ◇

Appendix

The goal of this appendix is to assemble a number of results on Artin and sphere braid groups and modular groups, which although quite elementary to prove, do not seem to be anywhere in the usual literature on braid groups ([A1], [A2], [Bi], [DG], [Ma], etc.). The basic result is proposition A3, which gives rise to a number of different isomorphisms, homomorphisms, inclusions and relations between the different braid groups described in proposition A4 (which occurs as proposition 2 of §2). We have chosen to make this appendix entirely self-contained at the price of repeating several of the definitions and restating some results already given in §§2 and 4 of the main article.

Let us begin again with notation and definitions. For $n > 2$, we write B_n for the Artin braid group on n strings, K_n for the pure Artin braid group on n strings, H_n for the Hurwitz (or sphere) braid group on n strings, P_n for the pure Hurwitz braid group, $M(0, n)$ for the modular group of the sphere with n marked points and $K(0, n)$ for the pure modular group. We will recall a presentation by generators and relations for each of these groups as well as a good many other extremely well-known facts, but we do not recall their proofs and/or geometric interpretations since these are to be found everywhere in the literature ([Bi] for example).

– B_n is generated by elements $\sigma_1, \ldots, \sigma_{n-1}$ satisfying the following relations:

$$\sigma_i\sigma_{i+1}\sigma_i = \sigma_{i+1}\sigma_i\sigma_{i+1} \text{ for } 1 \le i \le n-2 \text{ and } \sigma_i\sigma_j = \sigma_j\sigma_i \text{ for } |i-j| \ge 2. \tag{A1}$$

It is well-known that the center of B_n is an infinite cyclic group generated by $\omega_n := (\sigma_1 \cdots \sigma_{n-1})^n$.

– H_n is the quotient of B_n by normal closure of the element

$$\alpha_n := \sigma_{n-1}\sigma_{n-2} \cdots \sigma_2\sigma_1^2\sigma_2 \cdots \sigma_{n-2}\sigma_{n-1}.$$

It is well-known that the center of H_n is of order 2, generated by the image $\tilde{\omega}_n$ of $(\sigma_1 \cdots \sigma_{n-1})^n$ in H_n.

– $M(0, n)$ is the quotient of H_n by its center.

There is a natural surjective homomorphism $\rho : B_n \to S_n$, where S_n is the permutation group on n letters, giving by quotienting by the relations $\sigma_i^2 = 1$. This ρ induces homomorphisms $H_n \to S_n$ and $M(0, n) \to S_n$. The groups K_n, P_n and $K(0, n)$ are defined to be the kernels of these maps in B_n, H_n and $M(0, n)$ respectively.

Let us give a more symmetric presentation of the group B_n which we have frequently found useful.

Proposition A1: *The following is a presentation for the Artin braid group B_n: as generators we take σ_{ij} for $i, j \in \mathbb{Z}/n\mathbb{Z}$, with the relations*

$$\sigma_{ii} = 1, \quad \sigma_{ij} = \sigma_{ji}, \quad \sigma_{jk}\sigma_{ij} = \sigma_{ik}\sigma_{kj}(= \sigma_{ij}\sigma_{ki}) \quad and \quad \sigma_{ij}\sigma_{kl} = \sigma_{kl}\sigma_{ij} \quad (A2)$$

for i, j, k, l distinct in cyclic order in $\mathbb{Z}/n\mathbb{Z}$.

Proof: Let us imagine the strings of a braid to be hanging from a ring and attached at equally spaced points. The elements σ_{ij} should then be thought of as intertwining the i-th and j-th strings once. We use $1, \ldots, n$ as representatives of the elements of $\mathbb{Z}/n\mathbb{Z}$. Set $\sigma_{ij} = \sigma_{j-1} \cdots \sigma_{i+1}\sigma_i\sigma_{i+1}^{-1} \cdots \sigma_{j-1}^{-1}$ for $i, j \in \mathbb{Z}/n\mathbb{Z}$ (so $\sigma_{i,i+1} = \sigma_i$ for $1 \leq i \leq n - 1$). The symbol σ_n which appears in this expression when $i > j$ is not defined in B_n, so we define it by setting $\sigma_n := \sigma_{n1} = \sigma_{1n}$, the last quantity being well-defined. Thus, $\sigma_{ii} = 1$ holds by convention and $\sigma_{1n} = \sigma_{n1}$ is true by definition. The defining relations (A1) for B_n appear as particular cases of the relations (A2): the braid relations $\sigma_i\sigma_{i+1}\sigma_i = \sigma_{i+1}\sigma_i\sigma_{i+1}$ correspond to the "triangular" relations for $i, j = i + 1, k = i + 2$ and the commutation relations to the "four-points" relations with $i, j = k + 1, k, l = k + 1$. To prove that all the relations of the proposition hold in B_n, we first prove the relation

$$\sigma_{1,n}\sigma_{n-1,n}\sigma_{1,n}^{-1} = \sigma_{1,n-1} \quad (A3)$$

(which is just the case $i = 1$, $j = n - 1$, $k = n$ of the triangular relation) in B_n. For this we use the following principle of "index shifting". Set $\pi_1 = \sigma_{n-1} \cdots \sigma_1$ and $\pi_2 = \sigma_1 \cdots \sigma_{n-1}$. Then we have the relations

(i) $\pi_1^{-1}\sigma_{i,i+1}\pi_1 = \sigma_{i+1,i+2}$ for $i \in \mathbb{Z}/n\mathbb{Z}$ and

(ii) $\pi_2\sigma_{i,i+1}\pi_2^{-1} = \sigma_{i+1,i+2}$ for $1 \leq i \leq n - 2$.

We can now prove relation (A3) from the known relation $\sigma_{23}\sigma_{12}\sigma_{23}^{-1} = \sigma_{13}$ in B_n by conjugating both sides by π_1 to obtain π_1^2 (with the positive power on the left). The other relations are proved by induction on n, starting from the fact that the relations (A2) are valid for $n = 3$ (by inspection) and in the classical cases enumerated above which correspond to Artin's relations (A1), and checking that they remain valid if one adds a further $n + 1$-th string. This is rather obvious if one keeps in mind the geometric meaning of σ_{ij} but the computations are admittedly a bit messy and we shall not reproduce them here.   ◇

The σ_{ij} arise naturally, in particular because their squares $x_{ij} := \sigma_{ij}^2$ generate the pure braid group K_n. Since $x_{ii} = 1$ and $x_{ij} = x_{ji}$, K_n is actually generated by the x_{ij} for $1 \le i < j \le n$, with the following classical defining relations ([A1], [Bi]), where we write (a,b) for the commutator $aba^{-1}b^{-1}$:

- $x_{ij}x_{ik}x_{jk}$ commutes with x_{ij}, x_{ik} and x_{jk} for all $i < j < k$;

- $(x_{ij}, x_{kl}) = (x_{il}, x_{jk}) = 1$ for $i < j < k < l$;

- $(x_{ik}, x_{ij}^{-1}x_{jl}x_{ij}) = 1$ for $i < j < k < l$.

P_n is the image of K_n under the map $B_n \to H_n$. More precisely, we have

Lemma A2: *P_n is the quotient of K_n by the normal subgroup U_n generated by the elements α_i for $1 \le i \le n$ given by*

$$\alpha_i = x_{1i}x_{2i} \cdots x_{n-1,i}x_{n,i}. \tag{A4}$$

Proof: We use the presentation given in proposition A1 and the remark that any $n-1$ of the elements $\sigma_{12}, \sigma_{23}, \ldots, \sigma_{n-1,n}, \sigma_{n1}$ give a generating system for B_n. The lemma is then a consequence of the following identities: $(\alpha_i, \sigma_{j,j+1}) = 1$ for $j \not\equiv i$ or $i-1$ in $\mathbf{Z}/n\mathbf{Z}$, and $\sigma_{i-1,i}^{-1}\alpha_i\sigma_{i-1,i} = \alpha_{i-1}$; in particular these identities show that U_n is normal in B_n (and thus in K_n), and indeed is exactly the normal closure of the group generated by the element α_n. It suffices to prove these identities for a single value of $i \in \mathbf{Z}/n\mathbf{Z}$ since they are then all obtained by shifting indices as in the proof of proposition A1. Let us take $i = n$, $\alpha_n = x_{1n} \cdots x_{n-1,n} = \sigma_{n-1} \cdots \sigma_1^2 \cdots \sigma_{n-1}$. The fact that α_n commutes with σ_i for $1 \le i \le n-2$ is immediate to anyone having the courage to draw the the braids. We prove it explicitly by noting that $\alpha_n = \pi_2\pi_1$,

$$\alpha_n\sigma_{j,j+1}\alpha_n^{-1} = \pi_2\pi_1\sigma_{j,j+1}\pi_1^{-1}\pi_2^{-1} = \pi_2\sigma_{j+1,j+2}\pi_2^{-1} = \sigma_{j,j+1}$$

for $1 \le j \le n-2$.

For the final identity, note that $\sigma_{n-1}^{-1}\alpha_n\sigma_{n-1} = \sigma_{n-2} \cdots \sigma_1^2 \cdots \sigma_{n-2}\sigma_{n-1}^2$, which is exactly equal to $\alpha_{n-1} = x_{1,n-2} \cdots x_{n-3,n-2}x_{n-1,n-2}$. ◇

From now on, let us write $\tilde{\alpha}$ for the image in H_n of any element $\alpha \in B_n$, and $\bar{\alpha}$ for the image of α in $M(0,n)$. Thus P_n is generated by the $\tilde{x}_{ij}$. For $1 \le j \le n$, we have the sphere relations "based at i" in P_n, namely:

$$\tilde{x}_{1i} \cdots \tilde{x}_{n,i} = 1, \tag{A5}$$

which, for $1 \le i \le n-1$, can also be written

$$\tilde{x}_{i,n} = (\tilde{x}_{1i}\tilde{x}_{2i} \cdots \tilde{x}_{i,n-1})^{-1}. \tag{A5'}$$

These relations show that in fact, $\{\tilde{x}_{ij} \mid 1 \leq i < j \leq n-1\}$ is a generating set for P_n.

The center of P_n is given by the element of order 2

$$\tilde{\omega}_n := (\tilde{x}_{12})(\tilde{x}_{13}\tilde{x}_{23}) \cdots (\tilde{x}_{1n}\tilde{x}_{2n} \cdots \tilde{x}_{n-1,n}),$$

or equivalently,

$$\tilde{\omega}_n = (\tilde{x}_{12})(\tilde{x}_{13}\tilde{x}_{23}) \cdots (\tilde{x}_{1,n-1}\tilde{x}_{2,n-1} \cdots \tilde{x}_{n-2,n-1}) \qquad (A6)$$

since $\tilde{x}_{1,n}\tilde{x}_{2,n} \cdots \tilde{x}_{n-1,n} = 1$ by the sphere relation based at n. In particular, we have

$$\tilde{x}_{12} = \tilde{\omega}_n (\tilde{x}_{13} \cdots \tilde{x}_{n-2,n-1})^{-1}. \qquad (A7)$$

Let us introduce the following important sets:

$$E_n := \{x_{ij} \mid 1 \leq i < j \leq n, \ (i,j) \neq (1,2)\},$$

$$\tilde{E}_n := \{\tilde{x}_{ij} \mid 1 \leq i < j \leq n, \ (i,j) \neq (1,2)\}, \qquad (A8)$$

$$\bar{E}_n := \{\bar{x}_{ij} \mid 1 \leq i < j \leq n, \ (i,j) \neq (1,2)\},$$

where these sets are considered as subsets of K_m, P_m and $K(0,m)$ respectively, for any $m \geq n$.

By equations (A5') and (A7), $\tilde{E}_{n-1} \cup \{\tilde{\omega}_n\}$ is a generating set for P_n, and thus $\bar{E}_{n-1}$ is a generating set for $K(0,n)$. If we denote by $\langle \tilde{E}_n \rangle$ the subgroup of P_n generated by $\tilde{E}_n$, then $P_n \simeq \langle \tilde{E}_n \rangle \times \mathbb{Z}/2\mathbb{Z}$, and this implies that

$$K(0, n+1) \simeq \langle \tilde{E}_n \rangle \simeq \langle \bar{E}_n \rangle. \qquad (A9)$$

In fact we have the stronger result

Proposition A3: *Let K'_{n-1} denote the subgroup of K_{n-1} generated by E_{n-1}. Then*

$$P_n \simeq K'_{n-1} \times \mathbb{Z}/2\mathbb{Z}.$$

Proof: The main idea of the rather topological proof given here is that on the sphere, any pure braid on n-strings can be deformed into a unique braid with the following property: if the third to $n-1$-st strings are removed, the first two strings intertwine either exactly once (via $\tilde{x}_{12}$) or not at all. This implies that $P_3 = \mathbb{Z}/2\mathbb{Z}$ and in general reflects the fact that P_n is generated by a certain subgroup of braids in which the first and second strings do not intertwine at all, namely $\langle \tilde{E}_{n-1} \rangle$, and the central element $\tilde{\omega}_n$ of P_n, which intertwines them once and which is of order 2. Thus there is a bijection between the braids in P_n whose first two strings do not intertwine and the plane braids in K_n whose first two strings do not intertwine, which precisely means that $\langle E_{n-1} \rangle \simeq \langle \tilde{E}_{n-1} \rangle$.

In order to prove the result, it is best to use an other well-known interpretation of P_n, namely as the fundamental group of the following "configuration space"

$$F_{0,n} = \{Z = (z_1, \cdots, z_n), z_i \in \mathbf{P}^1 \mathbf{C}, z_i \neq z_j \text{ for all } (i,j) \in \{1, \cdots, n\}, i \neq j\}.$$

A braid is then represented as a path $Z(t), t \in (0,1)$, or rather as the class $[Z(t)] \in \pi_1(F_{0,n}) \simeq P_n$. To any path $Z(t)$ we associate a path ζ_t in $\mathrm{PGL}_2(\mathbf{C})$ as follows: ζ_t is the unique element of $\mathrm{PGL}_2(\mathbf{C})$ which maps the points $\{0, 1, \infty\}$ to $\{z_1(t), z_2(t), z_3(t)\}$. We assume below that $n \geq 3$ ($n = 1$ or 2 are trivial cases) and $\mathrm{PGL}_2(\mathbf{C})$ acts on $\mathbf{P}^1 \mathbf{C}$ via Möbius transformations as usual. To a path α_t in $\mathrm{PGL}_2(\mathbf{C})$ and a path $Z(t)$ in $F_{0,n}$ we associate the path $\alpha_t \cdot Z(t)$ in $F_{0,n}$ defined by the componentwise action of α_t, i.e. by $(\alpha_t \cdot Z(t))_i = \alpha_t z_i(t)$. It is easy to check that the corresponding braid $[\alpha_t \cdot Z(t)]$ actually depends only on the class $[\alpha_t] \in \pi_1(\mathrm{PGL}_2(\mathbf{C}))$; indeed if α_t can be homotoped to the trivial path by a homotopy $\alpha_t(s)$, $\alpha_t(s) \cdot Z(t)$ provides a homotopy between $\alpha_t \cdot Z(t)$ and $Z(t)$.

Recall now that $\pi_1(\mathrm{PGL}_2(\mathbf{C})) = \mathbf{Z}/2\mathbf{Z}$ and that a representative of the only non-trivial element can be taken to be

$$\gamma_t = \begin{pmatrix} e^{i\pi t} & 0 \\ 0 & e^{-i\pi t} \end{pmatrix}, \quad t \in (0,1).$$

Let now $[Z(t)]$ be an arbitrary braid represented by $Z(t)$ and consider $\zeta_t^{-1} \cdot Z(t)$; by the very definition of ζ_t, this path fixes $\{0, 1, \infty\}$ so in particular it can be considered as a path on the plane, representing a braid with $n-1$ strings, out of which two are kept fixed. Two cases may arise according to whether $[\zeta_t] = 1 \in \pi_1(\mathrm{PGL}_2(\mathbf{C}))$ or not. In the first case, $[\zeta_t^{-1} \cdot Z(t)] = [Z(t)] \in P_n$ and we have an explicit correspondence of this class of braids with K'_{n-1}. If on the other hand $[\zeta_t] \neq 1 \in \pi_1(\mathrm{PGL}_2(\mathbf{C}))$, we have that $[\zeta_t] = [\zeta_t^{-1}] = [\gamma_t]$.

Let $\Omega(t)$ be a path corresponding to the center $\tilde{\omega}_n$ of P_n, and ω_t the corresponding path in $\mathrm{PGL}_2(\mathbf{C})$. A possible choice of $\Omega(t)$ can be described as follows (see [Bi]; of course all choices are equivalent). We set $\Omega(0) = (0, 1, 2, \cdots, n-2, \infty)$ (these are points on $\mathbf{P}^1 \mathbf{C}$) and $\Omega_i(t) = \exp(2i\pi t)\Omega_i(0)$ (so that 0 and ∞ are actually fixed). From this, we obviously get $\omega_t = \gamma_t$ and a fortiori $[\omega_t] = [\gamma_t] \neq 1 \in \pi_1(\mathrm{PGL}_2(\mathbf{C}))$. Thus returning to the case when $[\zeta_t]$ associated to $Z(t)$ is non-trivial, we see that the path $\Omega_t \circ Z(t)$, obtained by composing $Z(t)$ with Ω_t is associated with the trivial element of $\pi_1(\mathrm{PGL}_2(\mathbb{C}))$ and we are back to the first case. But now $[\Omega_t \circ Z(t)] = \tilde{\omega}_n[Z(t)] \in P_n$, which finishes the proof and in fact provides an explicit description of the stated isomorphism. $\diamond$

Corollary: $\langle E_n \rangle \simeq \langle \tilde{E}_n \rangle \simeq \langle \bar{E}_n \rangle$ for $n > 2$.

Proposition A3 gives rise to a number of natural relations between the braid groups which we summarize as follows.

Proposition A4: *We have:*

(i) $K_n = K'_n \times \langle \omega_n \rangle$.

(ii) There are two natural inclusions i_1 and i_2 of K_{n-1} into K_n. Both send x_{ij} to x_{ij} for $1 \leq i < j \leq n$, $(i,j) \neq (1,2)$. But i_1 is then defined by setting $i_1(x_{12}) = x_{12}$, whereas i_2 is defined by setting $i_2(\omega_{n-1}) = \omega_n$.

(iii) $P_n \simeq K'_{n-1} \times \mathbb{Z}/2\mathbb{Z}$ and $K(0, n+1) \simeq P_{n+1}/Z \simeq K'_n$.

(iv) $P_{n+1} \simeq K_{n+1}/U_{n+1} \simeq K_n/\langle \omega_n^2 \rangle$.

(v) The subgroup $\tilde{P}_n$ of P_{n+1} generated by $\tilde{E}_{n-1} \cup \{\tilde{\omega}_{n+1}\}$ is isomorphic to P_n.

(vi) $P_{n+1} = F_n \rtimes \tilde{P}_n$, where F_n is the free group of rank $n-1$ generated by $x_{1,n+1}, \ldots, x_{n,n+1}$ (whose product equals 1).

(vii) We have the inclusions $K(0, n+1) \simeq K_n/\langle \omega_n \rangle \subset B_n/\langle \omega_n \rangle \subset M(0, n+1)$.

Proof: (i) Since $x_{12} = \omega_n(x_{13}x_{23} \cdots x_{n-1,n})^{-1}$, we see that K'_n and ω_n generate K_n. To see that their intersection is trivial, it suffices to consider their images modulo the derived subgroup of K_n, and notice that the quotient is just the free abelian group on the images of the x_{ij}. Finally, it is obvious that K'_n is normal in K_n since it is normalized by ω_n. This shows that $K_n = K'_n \rtimes \langle \omega_n \rangle$, so $K_n = K'_n \times \langle \omega_n \rangle$ since ω_n is central.

(ii) That $i_1(K_{n-1}) \hookrightarrow K_n$ is injective is obvious. As for $i_2(K_{n-1})$, it sends K'_{n-1} injectively into K'_n and the cyclic group $\langle \omega_{n-1} \rangle$ isomorphically onto $\langle \omega_n \rangle$, so it sends $K_{n-1} = K'_{n-1} \times \langle \omega_{n-1} \rangle$ injectively into $K_n = K'_n \times \langle \omega_n \rangle$.

(iii) The first isomorphism is just proposition A3 and the others immediate consequences of it and the definitions.

(iv) The first isomorphism is by definition. For the second, we have $K_n/\langle \omega_n^2 \rangle \simeq K'_n \times \mathbb{Z}/2\mathbb{Z} \simeq P_{n+1}$ by (i) and (iii).

(v) Consider the map $K_{n+1} \to K_{n+1}/U_{n+1} \simeq P_{n+1}$. This map induces an injection on K'_{n-1} because by proposition A3, the image of K'_{n-1} in P_{n+1} is $\langle \tilde{E}_{n-1} \rangle$ which is isomorphic to $\langle E_{n-1} \rangle$ which is isomorphic to K'_{n-1}, where we consider K'_{n-1} to be included in K_n as in (ii). The subgroup $\langle \tilde{E}_{n-1} \rangle \times \langle \tilde{\omega}_{n+1} \rangle$ of P_{n+1} is thus isomorphic to $K'_{n-1} \times \mathbb{Z}/2\mathbb{Z}$, so to P_n by proposition A3.

(vi) Let V_n denote the image of U_n under the map $K_n \to K_n/\langle \omega_n^2 \rangle \simeq P_{n+1}$ (by (iv)). This image is generated in P_{n+1} by the images $\tilde{\alpha}_i$ in P_{n+1} of the $\alpha_i = x_{1i} \cdots x_{ni} \in K_n$, so they are the products $\tilde{x}_{1i} \cdots \tilde{x}_{ni}$. But in P_{n+1}, we have the relations $x_{1i} \cdots x_{ni}x_{n+1,i} = 1$, so $x_{n+1,i}^{-1} = \tilde{\alpha}_i$, so in P_{n+1}, it is generated by the $x_{n+1,i}$ for $1 \leq i \leq n$; it is precisely $F_n \subset P_{n+1}$. Since U_n is normal in K_n, F_n is normal in P_{n+1}.

Under the isomorphism $P_{n+1} \simeq K'_n \times \mathbb{Z}/2\mathbb{Z}$ of proposition A3, F_n thus corresponds to V_n. We saw in (v) that moreover $\tilde{P}_n \subset P_{n+1}$ corresponds to the subgroup $K'_{n-1} \times \mathbb{Z}/2\mathbb{Z}$ included in $K'_n \times \mathbb{Z}/2\mathbb{Z}$ in the obvious way. The intersection $K'_{n-1} \cap V_n$ is trivial in $K'_n \times \mathbb{Z}/2\mathbb{Z}$ for the following reason: we know that $(K'_n \times \mathbb{Z}/2\mathbb{Z})/V_n \simeq P_n$ and that the image of $K'_{n-1} \times \mathbb{Z}/2\mathbb{Z}$ is $\tilde{P}_{n-1} \subset P_n$ which is isomorphic to P_{n-1} and therefore to $K'_{n-1} \times \mathbb{Z}/2\mathbb{Z}$ (by (iii) and (v)). So this quotient map is injective on $K'_{n-1} \times \mathbb{Z}/2\mathbb{Z}$ and V_n is its kernel, so these groups cannot intersect. This shows that $F_n \cap \tilde{P}_n = \{1\}$ in P_{n+1}. Since F_n and $\tilde{P}_n$ generate all of P_{n+1}, this means that $P_{n+1} \simeq F_n \rtimes \tilde{P}_n$ as stated.

(vii) The isomorphism comes from $K(0, n+1) \simeq K'_n = K_n/\langle \omega_n \rangle$ (by (i) and (iii)). The first inclusion is obvious and the second is because the relation $(\bar{\sigma}_1 \cdots \bar{\sigma}_n)^{n+1} = 1$ and the Hurwitz relation $\bar{\alpha}_n = 1$ in $M(0, n+1)$ imply that the relation $(\bar{\sigma}_1 \cdots \bar{\sigma}_{n-1})^n$ is also valid in $M(0, n+1)$ (cf. [Bi]) which means that the subgroup generated by $\bar{\sigma}_1, \ldots, \bar{\sigma}_{n-1}$ in $M(0, n+1)$ is isomorphic to $B_n/\langle \omega_n \rangle$. ◇

358   Pierre Lochak and Leila Schneps

References

[A1] E. Artin, Theory of braids, Annals of Math. **48** (1947), 101-126.

[A2] E. Artin, Braids and permutations, Annals of Math. **48** (1947), 643-649.

[Be] G.V.Belyi, On Galois extensions of a maximal cyclotomic field, Math. USSR Izv. **14** (1980), 247-256.

[Bi] J.Birman, Braids, links, and mapping class groups, Ann. of Math. Studies, vol. 82, Princeton Univ. Press, 1975.

[D] V.G.Drinfel'd, On quasitriangular quasi-Hopf algebras and a group closely connected with Gal($\overline{\mathbb{Q}}/\mathbb{Q}$), Leningrad Math. J. **2** (1991), 829-860.

[DG] J.L.Dyer and E.K.Grossman, The automorphism groups of the braid groups, American Journal of Mathematics **103** (1981), 1151-1169.

[G] A. Grothendieck, *Esquisse d'un Programme*, 1984, unpublished.

[I1] Y.Ihara, Braids, Galois groups, and some arithmetic functions, Proceedings of the International Congress of Mathematicians, Kyoto (1990), 99-120.

[I2] ———, Automorphisms of pure sphere braid groups and Galois representations, The Grothendieck Festschrift, Vol. 2, Progress in Mathematics **87** (1990), Birkhäuser, 353-373.

[I3] ———, On the stable derivation algebra associated with some braid groups, Israel J. of Math. **80** (1992), 135-153.

[M] B.H. Matzat, Zöpfe und Galoissche Gruppen, J. Crelle **420** (1991), 99-159.

[N] H.Nakamura, Galois rigidity of pure sphere braid groups and profinite calculus, to appear in J. Fac. Sc. Univ. of Tokyo.

[S] L.Schneps, Groupe de Grothendieck-Teichmüller et automorphismes de groupes de tresses, Note aux *Comptes Rendus de l'Académie de Sciences*, Paris.

*URA 762 du CNRS, Département de Mathématiques et Informatique
Ecole Normale Supérieure, 45 rue d'Ulm, 75230 Paris Cedex

**UA 741 du CNRS, Laboratoire de Mathématiques
Faculté des Sciences de Besançon, 16 route de Gray, 25030 Besançon Cedex

Moore and Seiberg equations, topological field theories and Galois theory

Pascal Degiovanni*

Abstract

This note is an attempt to summarize relations, partially conjectural, between Moore and Seiberg's equations, topological (projective) field theories in three dimensions and the second paragraph of Grothendieck's *Esquisse d'un Programme*. The first section outlines the current situation, and the second gives a summary of a review paper on the subject by the author, which could not be included in this volume because of length and scheduling problems.

*

First of all, we recall the construction of projective topological field theories in three dimensions from solutions to Moore and Seiberg's equations. We discuss the possible relation between this result and the reconstruction conjecture of the Teichmüller tower from its first two levels. Finally, we suggest an explicit translation of the natural action of $\mathrm{Gal}(\overline{\mathbf{Q}}/\mathbf{Q})$ into an action on a wide class of three-dimensional topological field theories arising from rational conformal field theories in two dimensions.

Our aim is to point out some relationships between recent developments in Topological Field Theories, the classification program of Rational Conformal Field Theories and deep ideas expressed by A. Grothendieck in the *Esquisse d'un Programme* [Groth].

First of all, we would like to stress that our present knowledge does not claim to be a definitive and complete mathematical theory since most of this wonderful story remains to be discovered. We would like to point out why, in our opinion, there is a deep connection between the world of Rational Conformal Field Theory and that of Grothendieck. In the end, the best advice we can give to the reader is to read the wonderful text by Grothendieck [Groth] and make up his own mind. Indeed, the present text is nothing more than a detailed introduction to these matters. We refer the reader to [Degio3] for a more detailed account of the subject. However, the contents of [Degio3] are described in section §2.

§1. Conformal Field Theory

Conformal field theory was originally studied for a systematic description of isotropic universality classes in two dimensions [BPZ]. A few years after their discovery, it became apparent that these theories were a prototype for the so-called geometrical quantum field theories [Segal],[A]. A special class of them, called Rational Conformal Field Theories (RCFT), attracted special attention during the late eighties. It turned out that RCFTs provided very interesting representations of various modular groups. This was discovered firstly in genus one [Cardy], and then in genus zero [Vafa3]. The important discovery of Verlinde [Ver] drew attention to this structure. Moore and Seiberg then produced an important synthesis of this subject [MS1],[MS2],[MS3]. In this work, they showed the importance of a few matrices associated with each rational conformal field theory. These matrices must satisfy polynomial equations, called Moore and Seiberg's equations. It must be mentioned that these matrices can be computed as monodromy matrices of some holomorphic multivalued functions on moduli space: see [Cardy] for the genus one case and [DF1],[DF2] for the some examples in genus zero. In passing, we remark that the Moore and Seiberg matrices represent endomorphisms of spaces associated with the following values of (g, n):

$$(0,3) \ (0,4) \ (1,1)$$

and that Moore and Seiberg's equations involve endomorphisms of spaces associated with

$$(0,5) \ (1,2).$$

Other authors [FrM],[RS],[FRS] also independently discovered the same structure but in a completely different context.

Topological Field Theories in three dimensions. At the same time, Witten discovered from the point of view of Chern-Simons theory, a deep connection between Moore and Seiberg's data associated with any RCFT and three-dimensional topological theories [Witt1]. More precisely, Chern-Simons theory associated with a compact, connected, Lie group G can be "solved" * using Moore and Seiberg's data associated with the Wess-Zumino-Witten model based on G. This mapping has been made more precise by many authors, for example [Fr],[FrK],[Dijk]. It also became clear that Moore and Seiberg's equations could be obtained from the requirement of topological invariance [Witt1],[MS4]. In fact, this result can be partially proved: one has to impose a few hypotheses and to consider only *non projective* topological field theories. In this case, only solutions to Moore and Seiberg's equations with $c \equiv 0 \pmod 8$ can be recovered (see [Degio2, Ch. 5]). On the other hand, it was expected that one could reconstruct a 3D TFT

---

* That is to say, any partition function, or any correlation function of any observable can be explicitly computed.

from any solution to Moore and Seiberg's equations. For example, topological invariants were defined by Kontsevitch in the case of undecorated closed manifolds [Kont] and also by Crane using Heegaard decompositions. The latter technique was used also by Kohno [Kohno] with some explicit solutions to Moore and Seiberg's equations coming from the WZW model based on $SU(2)$. It was shown in [Degio1] how to reconstruct a projective topological field theory from any solution to Moore and Seiberg's equations.

In a slightly different context, Reshetikhin and Turaev [RT] defined Topological Field Theories (TFT) using Kirby's calculus and quantum groups. The quantum group is an example of a modular Hopf algebra, the representation theory of which provides us with a solution to Moore and Seiberg's equations. Other works were also based on the same point of view: [Kup],[L].

Grothendieck's "Teichmüller tower". Besides this already wide-spread work, Grothendieck developed between 1981 and 1985 an extremely ambitious research program summarized in [Groth]. One of the main proposals of this program was to develop a new understanding of the absolute Galois group of the field $\mathbf{Q}$, $\mathrm{Gal}(\overline{\mathbf{Q}}/\mathbf{Q})$ by interpreting it as a group of transformations of an appropriate combinatorial object. The third paragraph of [Groth] explains how this group acts on the set of all "dessins d'enfants" which are considered at length in this volume. This is a first combinatorial approach to this description of the Galois group.

On the other hand, the second paragraph suggests that one should consider an important construction, called the Teichmüller tower. It is formed by the system of all moduli spaces * $\mathcal{M}_{g,n}$ of Riemann surfaces of any genus and with any number of punctures, together with a few fundamental operations such as the "gluing of surfaces", the "forgetting of marked points" and so on... As explained by Grothendieck, all this structure is reflected in suitable families of fundamental *groupoids* (with respect to suitable families of base points).

Two fundamental conjectures appear in [Groth,§2]:

- The reconstruction conjecture: the whole structure of the tower can be reconstructed from the two first levels (the levels are indexed by $3g-3+n$, which is the complex dimension of the corresponding moduli space). The first level provides a "system of generators" and the second one, a "system

---

* In algebraic geometry, the relevant concept is that of algebraic multiplicities, which are not schemes but algebraic stacks. This is due to the existence of Riemann surfaces with non-generic automorphism group.

of relations". These levels correspond to the following values of (g, n):

$$\begin{cases} \text{Generators}: & (0,4)\ (1,1) \\ \text{Relations}: & (0,5)\ (1,2). \end{cases}$$

• The Galois action conjecture: The structure of the tower is rigid enough for $\mathrm{Gal}(\overline{\mathbf{Q}}/\mathbf{Q})$ to act on its profinite completion, preserving all relations between the corresponding profinite groupoids. Grothendieck then suggested that one could parametrize each element of the Galois group by one or several elements of the profinite completion of the free group with two generators * subjected to certain relations. It is extremely important to find necessary and sufficient conditions for such elements to arise from the action of the absolute Galois group.

To our knowledge, these results remain conjectural, although some evidence for their validity exists.

Relationships between Conformal and Topological Field Theories and the Esquisse d'un programme. Finally, reading the *Esquisse* made it clear that there is a deep relationship between Grothendieck's unpublished work and Rational Conformal Field theory. In fact, this relationship is far from being established with all the rigour and precision suitable for this subject. The central object considered by Grothendieck – i.e. the Teichmüller tower – has not, up to now, been constructed. ** Hence, none of its properties have been proved. Our purpose will be to explain or suggest how this story should go. A great deal of work will be necessary before this "philosophy" can turn into a clean mathematical theory.

• For us, the starting point was noticing that Grothendieck's values for (g, n) in his reconstruction conjecture for the tower were exactly the values relevant in Moore and Seiberg's work. From this emerged the idea that solutions to Moore and Seiberg's equations define projective representations of the Teichmüller tower. We can refine the conjecture: Grothendick claims that the Teichmüller tower can be constructed using various systems of base points. In particular, he mentions the "small box of Legos" in which all base points arise from Thurston pants-type surfaces, i.e. surfaces of topological

---

* This is nothing but the algebraic fundamental group of $P_1(\mathbf{C}) \setminus \{0, 1, \infty\}$ with respect to some base point, which is the moduli space for Riemann surfaces of genus zero with four ordered points on it.

** It is likely that various versions of this tower exist, depending on the framework – algebraic geometry, differential geometry, topology, combinatorics ... – considered...

type $(0,3)$. We conjecture that Moore and Seiberg's matrices represent generators of the tower between such base points.

My opinion is that Moore and Seiberg's work needs to be settled on a firmer basis. A possible way of accomplishing this would be to define the Teichmüller tower, then study its projective representations, and produce Moore and Seiberg's data from such representations. The so called completeness theorem [MS3, App. B] of Moore and Seiberg should then be the expression, in representation theory, of the reconstruction conjecture of Grothendieck [Groth, §2].

Finally, starting from an axiomatic definition of a conformal field theory à la Segal, and an *intrinsic* definition – still to be found – of what a chiral algebra is, one should be able, first to define RCFTs, then to *prove* that any RCFT should provide a projective representation of the Teichmüller tower. All these steps being completed, Moore and Seiberg's work could be considered as rigorously based.

• In the *Esquisse d'un programme*, Grothendieck explained that elements of the absolute Galois group $\mathrm{Gal}(\overline{\mathbf{Q}}/\mathbf{Q})$ act as outer automorphisms of the tower itself. We were led to conjecture the existence of an action of $\mathrm{Gal}(\overline{\mathbf{Q}}/\mathbf{Q})$ on solutions to Moore and Seiberg's equations, or equivalently, on three dimensional topological field theories. More precisely, since Moore and Seiberg matrices can be computed from monodromies of "conformal blocks" along paths in Poincaré's half plane or in $P_1(\mathbf{C}) \setminus \{0, 1, \infty\}$ between $\mathbf{Q}$-rational points, one should expect an action of $\mathrm{Gal}(\overline{\mathbf{Q}}/\mathbf{Q})$ on these data. Our claim is that, under certain hypotheses, this action is nothing but the action of Galois coefficients of Moore and Seiberg matrices.

For example, in genus one, the so called "conformal blocks" are nothing but the characters of the chiral algebras. These are holomorphic functions on Poincaré's half plane $\mathfrak{H}$. Their Puiseux expansion in terms of $q = \exp(2\pi i \tau)$ (where $\tau \in \mathfrak{H}$) is of the form:

$$\chi_j(q) = q^{h_j - c/24} \sum_{n \geq 0} a_j(n) q^n \tag{1}$$

where each $a_j(n)$ is an integer since it is the dimension of a finite-dimensional vector space. The h_j's and c are rational numbers arising from the underlying RCFT. Let us define $s.\chi_j$ to be the analytic continuation of χ_j along the path $t \in]0,1[\mapsto t \in \mathfrak{H}$ that interpolates between Deligne's tangential base points $\overrightarrow{01}$ and $\overrightarrow{10}$ (here, we have mapped $\mathfrak{H}$ onto the open unit disk via $\tau \mapsto q(\tau)$). Then

$$(s.\chi_j) = \sum_k S_j{}^k \chi_k. \tag{2}$$

Let us consider $\sigma \in \mathrm{Gal}(\overline{\mathbf{Q}}/\mathbf{Q})$, and compute its action, after the fashion of Y. Ihara [Ihara]:

$$(\sigma^{-1} \cdot \chi_j)(q) = \chi_j(q)$$

$$s \cdot (\sigma^{-1} \cdot \chi_j)(q) = \sum_k S_j{}^k \chi_k(q)$$

$$\sigma \cdot (s \cdot [\sigma^{-1} \cdot \chi_j])(q) = \sum_k \sigma(S_j{}^k) \chi_k(q)$$

In this computation, the rationality of the $a_j(n)$ is used. In the end, we find that the action of σ, computed following Y. Ihara's prescription, transforms S into $\sigma(S)$. Similar reasoning on conformal blocks for the Riemann sphere with four marked points led us to our conjecture. We refer the reader to [Degio3] for more details.

Of course, what remains to be done is to explore the consequences of this program for the study of three-dimensional geometry.

§2. Further investigations

This section is devoted to outlining the contents of the review article [Degio3].

In the first section, we recall the axiomatic formulation of topological field theory in the spirit of Atiyah [A], Segal [Segal],[Segal2] and [Dijk]. Our presentation is a refined version of [Degio2, Ch. 1] and [Degio1] suitable for dealing with other ground fields than C. In a second section, we describe Moore and Seiberg's equations. We have tried to present this subject in a more intrinsic way than in the original paper [MS3]. Nevertheless, our presentation is far from being satisfactory... Still, we have tried to describe an axiomatization of the basic object handled by Moore and Seiberg, namely a certain 2-complex the vertices of which are trivalent graph with circularized vertices.

Then, we review the construction of a three-dimensional topological projective field theory [Degio1] from solutions to Moore and Seiberg's equations. We put the emphasis on representations of the modular groups which arise from this topological field theory. The proof of topological invariance using Kirby's calculus is also recalled.

The last section is devoted to the action of $\mathrm{Gal}(\overline{\mathbf{Q}}/\mathbf{Q})$ on a certain class of topological field theories. We inform the reader that it requires some familiarity with Conformal Field Theory... As explained above, we *suggest* that the translation to 3D TFTs of the action of $\mathrm{Gal}(\overline{\mathbf{Q}}/\mathbf{Q})$ discovered by Grothendieck [Groth] is nothing more than the number theory action on the matrix elements of the operators in the 3D TFT. Our reasoning is based on the computation of Moore and Seiberg's matrices from conformal blocks in RCFTs. The example of the S matrix described in the previous section contains the basic idea. As we explained before, in the case of conformal blocks for the Riemann sphere with four marked points, we have to rely on some hypotheses:

• Conformal blocks for the four-punctured sphere must be algebraic functions of the anharmonic quotient of the four points.

• Conformal blocks on the four-punctured sphere must have a Puiseux expansion near zero of a specific form: these blocks depend on the anharmonic quotient of the four points and the Puiseux expansion is assumed to have *rational* coefficients. We show that this hypothesis is satisfied by minimal models with respect to the Virasoro algebra or any non-twisted Kac-Moody algebra associated with a finite dimensional simple Lie algebra over C.

Let us mention that since no definition of a chiral algebra is available, we still do not know any good definition of RCFTs and therefore, we are not able to justify these hypotheses in a general framework!

Finally, we recall that such a Galois action has been considered in a slightly different context by Drinfel'd [Drin]. In his work, Drinfel'd described this Galois action by a pair $(\lambda, f) \in \widehat{Z}^* \times \widehat{F_2}$ satisfying particular conditions. * Equivalent results were also obtained by Y. Ihara in [Ihara]. These approaches follow Grothendieck's insight of describing elements of the absolute Galois group by outer automorphisms of the Teichmüller tower. Since a precise definition of the Teichmüller tower is still lacking, our strategy will be to rely on what is conjectured to be its representation theory – that is TFTs in 3D – and to try to translate this Galois action on the tower onto its representations. The surprise is that our final result is not expressed in terms of a pair $(\lambda, f) \in \widehat{Z}^* \times \widehat{F_2}$. We find instead the number theory action on matrix elements of operators representing elements of the various modular groups. An important question is to understand the implications of this phenomenon. In our opinion, a (good) definition of the Teichmüller tower is necessary in order to firstly formulate Grothendieck's questions in a precise way, and secondly to understand the connection between the various approaches.

Acknowledgements

Of course, all ideas explained in this short text and in the article have their origin in Grothendieck's text. I would like to thank L. Schneps for organizing this extremely stimulating conference and inviting me. I have also benefited from many useful conversations with her, P. Lochak, and

---

* See equations (4.3), (4.4) and (4.10) of [Drin].

J.-M. Couveignes.

My understanding of the Galois action on topological field theories constructed from solutions to Moore and Seiberg's equations arose from the lectures of Y. Ihara and J. Oesterlé. I would like to thank them for our discussions in Luminy. I also benefited from useful hints from G. Moore and N. Reshetikhin on the properties of Puiseux expansion of conformal blocks on the sphere with four marked points.

In December 1993, I had the opportunity to present these matters at the Université de Paris 7. I would like to thank the organizers and the audience of the seminar on Quantum Groups and Geometrical Theories for their advice and remarks.

References

[A]   M. Atiyah, Topological field theories, Publ. Math. IHES **68** (1989), 175–186.

[BPZ]   A.A. Belavin, A.B. Polyakov, and A.B. Zamolodchikov, Infinite conformal symmetry in 2d field theory, Nucl. Phys. **B. 241** (1984), 333–380.

[Cardy] J. Cardy, Operator content of 2d conformal field theories, Nucl. Phys. **B. 270** (1986), 186–204.

[Degio1] P. Degiovanni, Moore and Seiberg's equations and 3D topological field theories, Comm. Math. Phys. **145** (1992), 459–505.

[Degio2] ———, Théories des champs en dimensions deux et trois, Ph.D. thesis, Université de Paris 6, 1992.

[Degio3] ———, Equations de Moore et Seiberg, Théories Topologiques et Théorie de Galois, Preprint (ENS-Lyon) ENSLAPP-L-458/94, 1994.

[Dijk]   R. Dijkgraaf, A geometric approach to [2D] conformal field theory, Master's thesis, Utrecht University., 1989.

[DF1]   V.S. Dotsenko and V.A. Fateev, Four point correlation functions and the operator algebra in the two-dimensional conformal invariant theories with the central $c < 1$, Nucl. Phys. **B 251** (1985), 691.

[DF2]   ———, Conformal algebra and multipoint correlation functions in two-dimensional statistical models, Nucl. Phys. **B 240** (1984), 312.

[Drin]   V.G. Drinfeld, On quasi triangular algebras and a group connected with Gal($\bar{Q}/Q$), Leningrad Math. Journal **2** (1991), 829–861.

[FRS]   K. Fredenhagen, K.H. Rehren, and B. Schroer, Superselection sectors with braid group statistics and exchange algebras, Comm. Math. Phys. **125** (1989), 201–226.

[Fr]   J. Fröhlich, $2D$ conformal field theory and $3D$ topology, Int. Jour. of Modern Physics **20** (1989), 5321–5393.

[FrK]   J. Fröhlich and C. King, Chern-Simons theory and the Jones polynomial, Comm. Math. Phys. **126** (1989), 167–199.

[FrM]   J. Fröhlich and P.A. Marchetti, Quantum field theories of vortices and anyons, Comm. Math. Phys. **121** (1989), 177–223.

[Groth] A. Grothendieck, Esquisse d'un programme, Rapport Scientifique, 1984.

[Ihara] Y. Ihara, Braids, Galois groups and some arithmetic functions, International Congress of Mathematics (The mathematical society of Japan, ed.), Springer Verlag, 1992, pp. 99–120.

[Kohno] T. Kohno, Invariants of 3-manifolds based on Conformal Field Theory and Heegard splittings, Quantum groups (P.P. Kulish, ed.), Lecture Notes in Mathematics, vol. 1510, Springer-Verlag, 1990, pp. 341–349.

[Kont] M. Kontsevitch, Rational conformal field theories and invariants of 3-dimensional manifolds, Preprint CPT-88/P.2189, 1988.

[Kup]   G. Kuperberg, Involutory [Hopf algebras and 3-manifolds invariants, Preprint Berkeley 1990.

[L]   W.B.R. Lickorish, Invariants of three-manifolds from the combinatorics of the Jones polynomial, Cambridge University Preprint, 1991.

[MS1]   G. Moore and N. Seiberg, Polynomial equations for rational conformal field theories, Phys. Lett. B. **212** (1988), 451–460.

[MS2]   ——, Naturality in conformal field theory, Nucl. Phys. B. **313** (1989), 16.

[MS3]   ——, Classical and quantum conformal field theory, Comm. Math. Phys. **123** (1989), 177–255.

[MSR4] ——, Lectures on rational conformal field theories, 1990.

[RS]   K.H. Rehren and B. Schroer, Einstein causality and Artin braids,

Nucl. Phys. **B 312** (1988), 715.

[RT]   N.Y. Reshetikhin and V.G. Turaev, Invariants of 3-manifolds via link polynomials and quantum groups, Invent. Math. **103** (1991), 547–597.

[Segal] G. Segal, The definition of conformal field theory, Oxford University Preprint., 1987.

[Segal2] ——, Two dimensional conformal field theories and modular functors, IXth International Congress on Mathematical Physics (B. Simon, A. Truman, and I.M. Davies, eds.), Adam Hilger, 1989, pp. 22–37.

[Vafa3] C. Vafa, Towards classification of conformal field theories, Phys. Lett. B. **206** (1988), 421–426.

[Ver]   E. Verlinde, Fusion rules and modular transformations in 2D CFT's, Nucl. Phys. **B 300 (FS 22)** (1988), 360–376.

[Witt1] E. Witten, Quantum field theory and the Jones polynomial, Comm. Math. Phys. **121** (1989), 351–399.

*Laboratoire de Physique Théorique, URA 14-36 du CNRS (associée à l'E.N.S. de Lyon et au L.A.P.P. (IN2P3-CNRS) d'Annecy-le-Vieux)

ENS Lyon, 46 Allée d'Italie, 69007 Lyon, France

E-mail: degio at enslapp.ens-lyon.fr